Lecture Notes in Computer Science 8657

Commenced Publication in 1973
Founding and Former Series Editors:
Gerhard Goos, Juris Hartmanis, and Jan van Leeuwen

T0212625

Gethin Norman   William Sanders (Eds.)

Quantitative Evaluation of Systems

11th International Conference, QEST 2014
Florence, Italy, September 8-10, 2014
Proceedings

 Springer

Volume Editors

Gethin Norman
University of Glasgow
Glasgow, G12 8RZ, UK
E-mail: gethin.norman@glasgow.ac.uk

William Sanders
University of Illinois
Urbana, IL 61801-2918, USA
E-mail: whs@illinois.edu

ISSN 0302-9743   e-ISSN 1611-3349
ISBN 978-3-319-10695-3   e-ISBN 978-3-319-10696-0
DOI 10.1007/978-3-319-10696-0
Springer Cham Heidelberg New York Dordrecht London

Library of Congress Control Number: 2014946584

LNCS Sublibrary: SL 1 – Theoretical Computer Science and General Issues

Typesetting: Camera-ready by author, data conversion by Scientific Publishing Services, Chennai, India

Printed on acid-free paper

Springer is part of Springer Science+Business Media (www.springer.com)

Preface

Welcome to the proceedings of QEST 2014, the 11th International Conference on Quantitative Evaluation of Systems. QEST is a leading forum on quantitative evaluation and verification of computer systems and networks. QEST was first held in Enschede, The Netherlands (2004), followed by meetings in Turin, Italy (2005), Riverside, USA (2006), Edinburgh, UK (2007), St. Malo, France (2008), Budapest, Hungary (2009), Williamsburg, USA (2010), Aachen, Germany (2011), London, UK (2012), and, most recently, Buenos Aries, Argentina (2013).

This year's QEST was held in Florence, Italy, and co-located with the 12th International Conference on Formal Modeling and Analysis of Timed Systems (FORMATS 2014), the 33rd International Conference on Computer Safety, Reliability and Security (SAFECOMP 2014), the 10th European Workshop on Performance Engineering (EPEW 2014), and the 19th International Workshop on Formal Methods for Industrial Critical Systems (FMICS 2014). Together these conferences and workshops formed FLORENCE 2014, a one-week scientific event in the areas of formal and quantitative analysis of systems, performance engineering, computer safety, and industrial critical applications.

As one of the premier fora for research on quantitative system evaluation and verification of computer systems and networks, QEST covers topics including classic measures involving performance, reliability, safety, correctness, and security. QEST welcomes measurement-based as well as analytic studies, and is also interested in case studies highlighting the role of quantitative evaluation in the design of systems. Tools supporting the practical application of research results in all of the above areas are of special interest, and tool papers are highly sought as well. In short, QEST aims to encourage all aspects of work centered around creating a sound methodological basis for assessing and designing systems using quantitative means.

The Program Committee (PC) consisted of 35 experts and we received a total of 61 submissions. Each submission was reviewed by at least four reviewers, either PC members or external reviewers. In the end, 24 full papers and five tool demonstration papers were selected for the conference program. The program was greatly enriched with the QEST keynote talk of Tamer Basar (University of Illinois) and the joint keynote talk with SAFECOMP of Samuel Kounev (University of Würzburg). We believe the outcome was a high-quality conference program of interest to QEST attendees and other researchers in the field.

We would like to thank all the authors who submitted papers, as without them there simply would not be a conference. In addition, we would like to thank the PC members and the additional reviewers for their hard work and for sharing their valued expertise with the rest of the community as well as EasyChair for supporting the electronic submission and reviewing process. Also thanks to the tools chair David Parker, the publicity and publications chair Marco Paolieri and local organization chair Laura Carnevali for their dedication and excellent work. Furthermore, we gratefully acknowledge the financial support of ACM-e s.r.l. Finally, we would like to thank Joost-Pieter Katoen, chair of the QEST Steering Committee, for his guidance throughout the past year.

We hope that you find the conference proceedings rewarding and consider submitting papers to QEST 2015 in Madrid, Spain.

September 2014

Gethin Norman
William Sanders
Enrico Vicario

Organization

General Chair

Enrico Vicario   University of Florence, Italy

Program Committee Co-chairs

Gethin Norman   University of Glasgow, UK
William Sanders   University of Illinois at Urbana-Champaign,
                                 USA

Steering Committee

Nathalie Bertrand   Inria Rennes Bretagne Atlantique, France
Luca Bortolussi   Università degli Studi di Trieste, Italy
Peter Buchholz   TU Dortmund, Germany
Pedro R. D'Argenio   Universidad Nacional de Córdoba, Argentina
Holger Hermanns   Saarland University, Germany
Jane Hillston   University of Edinburgh, UK
Joost-Pieter Katoen   RWTH Aachen University, Germany
Peter Kemper   College of William and Mary, USA
William Knottenbelt   Imperial College London, UK
Gethin Norman   University of Glasgow, UK
Miklós Telek   Budapest University of Technology and
                                 Economics, Hungary

Program Committee

Alessandro Abate   University of Oxford, UK
Christel Baier   TU Dresden, Germany
Nathalie Bertrand   Inria Rennes Bretagne Atlantique, France
Robin Bloomfield   City University London, UK
Luca Bortolussi   Università degli Studi di Trieste, Italy
Jeremy Bradley   Imperial College London, UK
Peter Buchholz   TU Dortmund, Germany
Javier Campos   University of Zaragoza, Spain
Gianfranco Ciardo   University of California at Riverside, USA
Susanna Donatelli   Università di Torino, Italy
Leana Golubchik   University of Southern California, USA
Boudewijn Haverkort   University of Twente, The Netherlands
Holger Hermanns   Saarland University, Germany

Jane Hillston	University of Edinburgh, UK
András Horváth	Università di Torino, Italy
Kaustubh Joshi	AT&T Research
Mohamed Kaâniche	LAAS-CNRS, Toulouse, France
Jan Kriege	TU Dortmund, Germany
Boris Köpf	IMDEA Software Institute, Spain
Annabelle McIver	Macquarie University, Australia
Sayan Mitra	University of Illinois at Urbana-Champaign, USA
Gethin Norman	University of Glasgow, UK
Catuscia Palamidessi	Inria, CNRS and LIX, Ecole Polytechnique, France
David Parker	University of Birmingham, UK
Eric Rozier	University of Miami, USA
Gerardo Rubino	Inria Rennes Bretagne Atlantique, France
William Sanders	University of Illinois at Urbana-Champaign, USA
Evgenia Smirni	College of William and Mary, USA
Jeremy Sproston	Università di Torino, Italy
Miklós Telek	Budapest University of Technology and Economics, Hungary
Mirco Tribastone	University of Southampton, UK
Benny Van Houdt	University of Antwerp, The Netherlands
Verena Wolf	Saarland University, Germany
Lijun Zhang	Institute of Software, Chinese Academy of Sciences, China

Additional Reviewers

Abbas, Houssam	Clark, Allan
Akshay, S.	de Boer, Pieter-Tjerk
Amparore, Elvio Gilberto	Dehnert, Christian
Andreychenko, Alexander	Delahaye, Benoit
Angius, Alessio	Deng, Yuxin
Atencia, Ivan	Dreossi, Tommaso
Barbot, Benoit	Duggirala, Sridhar
Bartocci, Ezio	Fan, Chuchu
Basset, Nicolas	Fijalkow, Nathanaël
Bobbio, Andrea	Forejt, Vojtech
Bonet, Pere	Galpin, Vashti
Brihaye, Thomas	Garetto, Michele
Brodo, Linda	Gast, Nicolas
Caravagna, Giulio	Genest, Blaise
Ceska, Milan	Gerwinn, Sebastian

Ghosh, Ritwika
Gilmore, Stephen
Guenther, Marcel C.
Haddad, Serge
Haesaert, Sofie
Hahn, Ernst Moritz
Hartmanns, Arnd
Hatefi, Hassan
Horvath, Illes
Iacobelli, Giulio
Jones, Gareth
Koziolek, Anne
Kraehmann, Daniel
Krcal, Jan
Krüger, Thilo
Kyriakopoulos, Charalampos
Leon, Pedro
Loreti, Michele
Mardare, Radu
Massink, Mieke
Mikeev, Linar
Milios, Dimitrios
Murawski, Andrzej

Paolieri, Marco
Paolini, Michela
Peyronnet, Sylvain
Picaronny, Claudine
Pulungan, Reza
Rabehaja, Tahiry
Reijsbergen, Daniel
Remke, Anne
Rodríguez, Ricardo J.
Sanguinetti, Guido
Scheinhardt, Werner
Sedwards, Sean
Shi, Cong
Song, Lei
Spieler, David
Troina, Angelo
Tschaikowski, Max
Turrini, Andrea
Valero, Valentin
Van Glabbeek, Rob
Vandin, Andrea
Wang, Qinsi
Yan, Feng

Table of Contents

Models and Tools

Simulation

Queueing, Debugging and Tools

Process Algebra and Equivalences

Automata and Markov Process Theory

Applications, Theory and Tools

Probabilistic Model Checking

Quantitative Evaluation of Service Dependability in Shared Execution Environments

Samuel Kounev

Chair of Software Engineering
Department of Computer Science
University of Würzburg
Am Hubland, 97074 Würzburg, Germany
skounev@acm.org

Recent reports indicate that ICT is currently responsible for 8-10% of EU's electricity consumption and up to 4% of its carbon emissions [2,23]. By 2020, only in Western Europe, data centers will consume around 100 billion kilowatt hours each year [15] (the same as the total electricity consumption of the Netherlands), making energy a major factor in IT costs. However, according to [3], due to the growing number of underutilized servers, only 6 - 12% of the energy consumption in data centers nowadays is spent for performing computations.

Industry's answer to this challenge is cloud computing, promising both reductions in IT costs and improvements in energy efficiency. Cloud computing is a novel paradigm for providing data center resources as on demand services in a pay-as-you-go manner. It promises significant cost savings by making it possible to consolidate workloads and share infrastructure resources among multiple applications resulting in higher cost- and energy-efficiency [9]. Despite the hype around it, it is well established that if this new computing model ends up being widely adopted, it will transform a large part of the IT industry [8,17].

However, the inability of today's cloud technologies to provide dependability guarantees is a major showstopper for the widespread adoption of the cloud paradigm, especially for mission-critical applications [8,9,16,1]. The term *dependability* is understood as a combination of service *availability* and *reliability*, commonly considered as the two major components of dependability [21], in the presence of variable workloads (e.g., load spikes), security attacks, and operational failures. Given that an overloaded system appears as unavailable to its users, and that failures typically occur during overload conditions, a prerequisite for providing dependable services is to ensure that the system has sufficient *capacity* to handle its dynamic workload [22]. According to [17,16], concerns of organizations about service availability is a major obstacle to the adoption of cloud computing.

Today's cloud computing platforms generally follow a trigger-based approach when it comes to enforcing application-level service-level agreements (SLAs), e.g., concerning availability or responsiveness. Triggers can be defined that fire in a reactive manner when an observed metric reaches a certain threshold (e.g., high server utilization or long service response times) and execute certain predefined reconfiguration actions until given stopping criteria are fulfilled (e.g., response times drop).

G. Norman and W. Sanders (Eds.): QEST 2014, LNCS 8657, pp. 1–4, 2014.

Triggers are typically used to implement *elastic* resource provisioning mechanisms. The term *elasticity* is understood as the degree to which a system is able to adapt to workload changes by provisioning and deprovisioning resources in an autonomic manner, such that at each point in time the available resources match the current demand as closely as possible [6,26]. Better elasticity leads to higher availability and responsiveness, as well as to higher resource- and cost-efficiency.

However, application-level metrics, such as availability and responsiveness, normally exhibit a highly non-linear behavior on system load, and they typically depend on the behavior of multiple virtual machines (VMs) across several application tiers. Thus, for example, if a workload change is observed, the platform cannot know in advance *how much*, and at what level of granularity, additional resources in the various application tiers will be required (e.g., vCores, VMs, physical machines, network bandwidth), and *where and how* the newly started VMs should be deployed and configured to ensure dependability without sacrificing efficiency. Moreover, the platform cannot know *how fast* new resources should be allocated and *for how long* they should be reserved. Hence, it is hard to determine general thresholds of when triggers should be fired, given that the appropriate triggering points typically depend on the architecture of the hosted services and their workload profiles, which can change frequently during operation.

Furthermore, in case of contention at the physical resource layer, the availability and responsiveness of an individual application may be significantly influenced by applications running in other co-located virtual machines (VMs) sharing the physical infrastructure [7]. Thus, to be effective, triggers must also take into account the interactions between applications and workloads at the physical resource layer. The complexity of such interactions and the inability to predict how changes in application workload profiles propagate through the layers of the system architecture down to the physical resource layer render conventional trigger-based approaches unable to reliably enforce SLAs in an efficient and proactive fashion (i.e., allocating only as much resources as are actually needed and reconfiguring proactively before SLA violations have occurred).

As a result of the above described challenges, today's shared execution environments based on first generation cloud technologies rely on "best-effort" mechanisms and do not provide dependability guarantees. Nevertheless, although no guarantees are given, the provided level of dependability is a major distinguishing factor between different service offerings. To make such offerings comparable, novel metrics and techniques are needed allowing to measure and quantify the dependability of shared execution environments, e.g., cloud computing platforms or general virtualized service infrastructures.

In this keynote talk, we first discuss the inherent challenges of providing service dependability in shared execution environments in the presence of highly variable workloads, load spikes, and security attacks. We then present novel metrics and techniques for measuring and quantifying service dependability specifically taking into account the dynamics of modern service infrastructures. We consider both environments where virtualization is used as a basis for enabling resource sharing, e.g., as in Infrastructure-as-a-Service (IaaS) offerings, as well as multi-tenant Software-as-a-Service (SaaS) applications, where the whole hardware and software stack (including the application layer) is shared among different customers (i.e., tenants). We focus on evaluating three dependability aspects: i) the ability of the system to provision resources in an elastic manner, i.e., *system elasticity* [6,5,26,25,24,4], ii) the ability of the system to isolate different applications and customers sharing the physical infrastructure in terms of the performance they observe, i.e., *performance isolation* [12,13,11,14,10], and iii) the ability of the system to deal with attacks exploiting novel attack surfaces such as virtual machine monitors, i.e., *intrusion detection and prevention* [18,19,20]. We discuss the challenges in measuring and quantifying the mentioned three dependability properties presenting existing approaches to tackle them. Finally, we discuss open issues and emerging directions for future work in the area of dependability benchmarking.

References

1. Durkee, D.: Why Cloud Computing Will Never Be Free. ACM Queue 8(4), 20:20–20:29 (2010)
2. European Commission - IP/13/231 - 18/03/2013. Digital Agenda: global tech sector measures its carbon footprint (March 2013),
   http://europa.eu/rapid/press-release_IP-13-231_en.htm
3. Glanz, J.: Power, Pollution and the Internet. New York Times (September 2012)
4. Herbst, N.R., Huber, N., Kounev, S., Amrehn, E.: Self-Adaptive Workload Classification and Forecasting for Proactive Resource Provisioning. In: 4th ACM/SPEC International Conference on Performance Engineering (ICPE 2013), pp. 187–198. ACM (April 2013)
5. Herbst, N.R., Huber, N., Kounev, S., Amrehn, E.: Self-Adaptive Workload Classification and Forecasting for Proactive Resource Provisioning. Concurrency and Computation - Practice and Experience, Special Issue with extended versions of the best papers from ICPE 2013 (2014)
6. Herbst, N.R., Kounev, S., Reussner, R.: Elasticity in Cloud Computing: What it is, and What it is Not. In: 10th International Conference on Autonomic Computing (ICAC 2013). USENIX (June 2013)
7. Huber, N., von Quast, M., Hauck, M., Kounev, S.: Evaluating and Modeling Virtualization Performance Overhead for Cloud Environments. In: 1st International Conference on Cloud Computing and Services Science (CLOSER 2011), pp. 563–573. SciTePress (May 2011)
8. Jennings, B., Stadler, R.: Resource Management in Clouds: Survey and Research Challenges. Journal of Network and Systems Management (March 2014)
9. Kounev, S., Reinecke, P., Brosig, F., Bradley, J.T., Joshi, K., Babka, V., Stefanek, A., Gilmore, S.: Providing dependability and resilience in the cloud: Challenges and opportunities. In: Wolter, K., Avritzer, A., Vieira, M., van Moorsel, A. (eds.) Resilience Assessment and Evaluation of Computing Systems, XVIII. Springer, Heidelberg (2012) ISBN: 978-3-642-29031-2
10. Krebs, R., Momm, C., Kounev, S.: Architectural Concerns in Multi-Tenant SaaS Applications. In: 2nd International Conference on Cloud Computing and Services Science (CLOSER 2012). SciTePress (April 2012)

11. Krebs, R., Momm, C., Kounev, S.: Metrics and Techniques for Quantifying Performance Isolation in Cloud Environments. In: 8th ACM SIGSOFT International Conference on the Quality of Software Architectures (QoSA 2012). ACM Press (June 2012)

12. Krebs, R., Momm, C., Kounev, S.: Metrics and Techniques for Quantifying Performance Isolation in Cloud Environments. Elsevier Science of Computer Programming Journal (SciCo) (2013) (in print)

13. Krebs, R., Spinner, S., Ahmed, N., Kounev, S.: Resource Usage Control In Multi-Tenant Applications. In: 14th IEEE/ACM International Symposium on Cluster, Cloud and Grid Computing (CCGrid 2014). IEEE/ACM (May 2014)

14. Krebs, R., Wert, A., Kounev, S.: Multi-tenancy Performance Benchmark for Web Application Platforms. In: Daniel, F., Dolog, P., Li, Q. (eds.) ICWE 2013. LNCS, vol. 7977, pp. 424–438. Springer, Heidelberg (2013)

15. Kroes, N.: Greening the digital world: major companies to measure the environmental footprint of ICT (2012), http://ec.europa.eu/commission_2010-2014/kroes/en/blog/ict-footprint

16. Lohr, S.: Amazon's Trouble Raises Cloud Computing Doubts. The New York Times, April 22 (2011)

17. Armbrust, M., et al.: A View of Cloud Computing. Communications of the ACM 53(4), 50–58 (2010)

18. Milenkoski, A., Kounev, S.: Towards Benchmarking Intrusion Detection Systems for Virtualized Cloud Environments. In: 7th International Conference for Internet Technology and Secured Transactions (ICITST 2012). IEEE (December 2012)

19. Milenkoski, A., Kounev, S., Avritzer, A., Antunes, N., Vieira, M.: On Benchmarking Intrusion Detection Systems in Virtualized Environments. Technical Report SPEC-RG-2013-002 v.1.0, SPEC Research Group - IDS Benchmarking Working Group, Standard Performance Evaluation Corporation (SPEC), Gainesville, VA, USA (June 2013)

20. Milenkoski, A., Payne, B.D., Antunes, N., Vieira, M., Kounev, S.: HInjector: Injecting Hypercall Attacks for Evaluating VMI-based Intrusion Detection Systems (poster paper). In: The 2013 Annual Computer Security Applications Conference (ACSAC 2013). Applied Computer Security Associates, ACSA (December 2013)

21. Muppala, J., Fricks, R., Trivedi, K.S.: Techniques for System Dependability Evaluation. In: Computational Probability, vol. 24. Kluwer Academic Publishers (2000)

22. Nou, R., Kounev, S., Julia, F., Torres, J.: Autonomic QoS Control in Enterprise Grid Environments using Online Simulation. Journal of Systems and Software 82(3), 486–502 (2009)

23. Seneviratne, K.: ICTs Increase Carbon Footprint Which They Can Reduce. IDN (InDepthNews) (December 2013)

24. von Kistowski, J.G., Herbst, N.R., Kounev, S.: LIMBO: A Tool for Modeling Variable Load Intensities. In: 5th ACM/SPEC International Conference on Performance Engineering (ICPE 2014), pp. 225–226. ACM (March 2014)

25. von Kistowski, J.G., Herbst, N.R., Kounev, S.: Modeling Variations in Load Intensity over Time. In: 3rd International Workshop on Large-Scale Testing (LT 2014). ACM (March 2014)

26. Weber, A., Herbst, N.R., Groenda, H., Kounev, S.: Towards a Resource Elasticity Benchmark for Cloud Environments. In: 2nd International Workshop on Hot Topics in Cloud Service Scalability (HotTopiCS 2014). ACM (March 2014)

Multi-agent Networked Systems
with Adversarial Elements

Tamer Başar

Dept. Electrical and Computer Engineering & Coordinated Science Laboratory
University of Illinois at Urbana-Champaign
Urbana, IL , 61801 USA
basar1@illinois.edu

In nutshell, *multi-agent networked systems* involve the modeling framework of multiple heterogeneous agents (or decision makers, or players) connected in various ways, distributed over a network (or interacting networks) and interacting with limited information (on line and off line) under possibly conflicting objectives. We can actually view the agents as nodes in a graph or multiple graphs, which could be time varying (some edges in the network appearing or disappearing over time [11]), and the nodes themselves could be mobile [8]. In such settings, agents actually interact in a three-tiered architecture, with each tier corresponding to a different layer [21] , namely: Layer 1, where the agents operate and decisions are made; Layer 2, which is the information level where data, models, and actionable information reside and are exchanged; and Layer 3, which consists of the physical communication network that is used for Layers 1 and 2, and contains software and hardware entities, as well as sensors and actuators with which the teams interface with the dynamic physical environment. The underlying network for Layer 1 can be viewed as a *collaboration network*, where edges of the corresponding graph capture the collaboration among corresponding nodes (agents); the network for Layer 2 can be viewed as a *communication/information network*, where edges of the corresponding graph constitute communication links (uni- or bi-directional) among corresponding nodes (agents); and Layer 3 can be viewed as a *physical network*, where edges constitute the physical links.

The recent emergence of such multi-agent networked systems has brought about several non-traditional and non-standard requirements on strategic decision-making, thus challenging the governing assumptions of traditional Markov decision processes and game theory. Some of these requirements stem from factors such as: (i) limitations on memory [20], (ii) limitations on computation and communication capabilities [20], (iii) heterogeneity of decision makers (machines versus humans), (iv) heterogeneity and sporadic failure of channels that connect the information sources (sensors) to decision units (strategic agents) [13], [15], (v) limitations on the frequency of exchanges between different decision units and the actions taken by the agents [12], (vi) operation being conducted in an uncertain and hostile environment where disturbances are controlled by adversarial agents [5], (vii) lack of cooperation among multiple decision units [6], (viii) lack of a common objective shared by multiple agents [6], and (ix) lack (or evolution) of trust among agents. These all lead to substantial degradation in performance and loss in efficiency if appropriate mechanisms are not put in place.

G. Norman and W. Sanders (Eds.): QEST 2014, LNCS 8657, pp. 5–8, 2014.

An appropriate framework for studying and analyzing multi-agent networked systems, and particularly those with adversarial intervention, is that of dynamic games [6]. We have the scenario of multiple agents (decision makers or players) picking policies (decision laws or strategies), which lead to actions that evolve over time. Policies are constructed based on information provided by the communication network (active as well as passive) and guided by individual utility or cost functions over the decision making horizon. The very special case of a single agent falls in the domain of stochastic control, and the special case of a single objective for all agents (and no adversaries) fall in the domain of stochastic teams [20]. The more general case of multiple agents and multiple objectives falls in the domain of zero-sum or nonzero-sum stochastic dynamic games. The former is the setting of stochastic teams operating in an environment with adversarial intervention, in which case the adversary or adversaries can be seen as members of a directly opposing party (and hence the name *zero-sum*, where the solution concept is that of *saddle point*); security games fall into that class [2], [14], and so do problems of jamming where the communication links between collaborating agents are jammed by a team of adversaries [1], [9], [10]. The most general framework is that of nonzero-sum games, where the solution concept adopted, in a noncooperative setting, is generally *Nash equilibrium*. Here the players (agents) are coupled through information exchange as well as their individual objective functions, and hence individual optimization by each player regardless of what other players do is not a possible approach; players have to anticipate the moves or actions of other players (at least in their neighborhoods of spatially or informationally connected agents) before they take their decisions. This necessitates the development of distributed algorithms and learning rules for the decision process to converge to the fixed point characterized by the Nash equilibrium [22]. The complexity created by the coupling among players can be somewhat alleviated in the large population regime (that is when the game has asymptotically an infinite number of players), in which case each player views the aggregate of other players as a *cloud* that is not affected by unilateral decisions or actions of individual players; this belongs to the domain of *mean field games*, which is an active area of research; see [7], [18], and references therein.

The complexity of obtaining an equilibrium solution in a stochastic dynamic game also depends on the nature of the information exchange among the players (that is, the structure of the information network). Roughly speaking, if the answer to the question "Is the quality of active and relevant information received by a player affected by actions of other players?" is *no*, then the underlying game is relatively easy to solve, in the sense that at least methodologies and computational algorithms exist for such games. If the answer is, however, *yes*, then the stochastic game (and as a special case, stochastic dynamic team) is generally very complex and is "difficult" to solve [20]. Such information structures or networks are known as *non-classical*, and they arise because of limitation on memory, presence of delay, or non-sharing of actionable information [3], [19], [20]. This interaction of agents on a non-classical information network is a complex one because, in addition to helping himself in performance improvement, each collaborating agent could also help neighboring agents to achieve an overall better performance through a judicious use of control/action signals that would improve the information content of transmitted messages. The presence of such a *dual role* makes the derivation of optimal solutions even in stochastic teams generally quite a challenging task. In an adversarial context, the same dual role emerges, this time not to help with improvement of performance but to degrade it. For some recent work on this topic, and additional references, see [4] and[16].

One class of problems that feature non-classical information are those where agents operate under strict limitations on the number or frequency of their actions, be it control (as in a closed-loop system) or transmission of information (as in a sensor network), with the restriction arising because of limitation on resources (such as energy). In traditional decision problems, the issue of interest has been *what to send*, or equivalently *how to shape the information/sensor and control signals* so as to meet targeted objectives. The resource limitation now brings in a second issue or question, one of *when to send*, given some constraints on the number of transmissions (which could include sensor signals, control signals, or communication between agents) and also given the unreliability of the transmission medium where the information on whether the signals sent have reached their intended destinations or not (due to random failures or adversarial intervention) is at best only partially available. The agents here are faced with a dynamic decision making process that trades off using more of a resource now (and therefore receiving instantaneous return in performance at the expense of not so good performance in the future) against holding on to more of the resource for a future use (and thus possibly obtaining an improved performance in the future at the expense of poor instantaneous return)–all this taking into account the possibility that the transmission medium could be lossy (with some statistical description of failures). Solutions to these problems involve *threshold-type policies* with online dynamic scheduling and offline computation; for some representative work, see [12] and [17].

References

1. Akyol, E., Rose, K., Başar, T.: Gaussian sensor networks with adversarial nodes. In: Proc. 2013 IEEE International Symposium on Information Theory (ISIT 2013), Istanbul, Turkey, July 8-12 (2013)
2. Alpcan, T., Başar, T.: Network Security: A Decision and Game Theoretic Approach. Cambridge University Press (2011)
3. Başar, T.: Variations on the theme of the Witsenhausen counterexample. In: Proc. 47th IEEE Conf. Decision and Control, pp. 1614–1619 (2008)
4. Başar, T.: Stochastic differential games and intricacy of information structures. In: Haunschmied, J., Veliov, V.M., Wrzaczek, S. (eds.) Dynamic Games in Economics, pp. 23–49. Springer (2014)
5. Başar, T., Bernhard, P.: H∞ Optimal Control and Related Minimax Design Problems: A Dynamic Game Approach. Systems & Control: Foundations and Applications Series. Birkhäuser, Boston (1995)
6. Başar, T., Olsder, G.J.: Dynamic Non-Cooperative Game Theory. SIAM Series in Classics in Applied Mathematics, Philadelphia (1999)

7. Bauso, D., Başar, T.: Large networks of dynamic agents: Consensus under adversarial disturbances. In: Proc. 46th Asilomar Conference on Signals, Systems and Computers, Pacific Grove, CA, November 4-7 (2012)

8. Bhattacharya, S., Başar, T.: Optimal strategies to evade jamming in heterogeneous mobile networks. In: Proc. 2010 IEEE International Conf. Robotics and Automation (Workshop on Search and Pursuit/Evasion in the Physical World: Efficiency, Scalability, and Guarantees; SPE-ICRA 2010), Anchorage, Alaska, May 3-8 (2010)

9. Bhattacharya, S., Başar, T.: Multi-layer hierarchical approach to double sided jamming games among teams of mobile agents. In: Proc. 51st IEEE Conference on Decision and Control (CDC 2012), Maui, Hawaii, December 10-13, pp. 5774–5779 (2012)

10. Bhattacharya, S., Başar, T.: Differential game-theoretic approach to a spatial jamming problem. In: Advances in Dynamic Game Theory and Applications, Annals of Dynamic Games, vol. 12, pp. 245–268. Birkhäuser (2012)

11. Gharesifard, B., Başar, T.: Resilience in consensus dynamics via competitive interconnections. In: Proc. 3rd IFAC Workshop on Estimation and Control of Networked Systems (NecSys 2012), Santa Barbara, CA, September 14-15 (2012)

12. Imer, O.C., Başar, T.: Optimal estimation with limited measurements. International J. Systems, Control and Communications, Special Issue on Information Processing and Decision Making in Distributed Control Systems 2(1/2/3), 5–29 (2010)

13. Imer, O.C., Yüksel, S., Başar, T.: Optimal control of LTI systems over unreliable communication links. Automatica 42(9), 1429–1440 (2006)

14. Manshaei, M.H., Zhu, Q., Alpcan, T., Başar, T., Hubaux, J.-P.: Game theory meets network security and privacy. ACM Computing Survey 45(3), 25:1–25:39 (2013)

15. Moon, J., Başar, T.: Control over lossy networks: a dynamic game approach. In: Proc. 2014 American Control Conference (ACC 2014), Portland, OR, June 4-6, pp. 5379–5384 (2014)

16. Nayyar, A., Başar, T.: Dynamic stochastic games with asymmetric information. In: Proc. 51st IEEE Conf. Decision and Control, Maui, Hawaii, pp. 7145–7150 (2012)

17. Nayyar, A., Başar, T., Teneketzis, D., Veeravalli, V.V.: Optimal strategies for communication and remote estimation with an energy harvesting sensor. IEEE Transactions on Automatic Control 58(9), 2246–2260 (2013)

18. Tembine, H., Zhu, Q., Başar, T.: Risk-sensitive mean-field games. IEEE Transactions on Automatic Control 59(4), 835–850 (2014)

19. Witsenhausen, H.: A counterexample in stochastic optimum control. SIAM J. Contr. 6, 131–147 (1968)

20. Yüksel, S., Başar, T.: Stochastic Networked Control Systems: Stabilization and Optimization under Information Constraints. Systems & Control: Foundations and Applications Series. Birkhäuser, Boston (2013)

21. Zhu, Q., Başar, T.: Game-theoretic methods for robustness, security and resilience of cyber-physical control systems: Games-in-games principle for optimal cross-layer resilient control systems. IEEE Control Systems Magazine (to appear, December 2014)

22. Zhu, Q., Tembine, H., Başar, T.: Hybrid learning in stochastic games and its applications in network security. In: Lewis, F.L., Liu, D. (eds.) Reinforcement Learning and Approximate Dynamic Programming for Feedback Control. Series on Computational Intelligence, ch. 14, pp. 305–329. IEEE Press/Wiley (2013)

A Structured Solution Approach
for Markov Regenerative Processes

Elvio Gilberto Amparore[1], Peter Buchholz[2], and Susanna Donatelli[1]

[1] Dipartimento di Informatica, Università degli Studi di Torino, Italy
{amparore,susi}@di.unito.it
[2] Department of Computer Science, TU Dortmund, Germany
peter.buchholz@udo.edu

Abstract. Two different methods have been introduced in the past for the numerical analysis of Markov Regenerative Processes. The first one generates the embedded Markov chain explicitly and solves afterwards the often dense system of linear equations. The second method avoids computation of the embedded Markov chain by performing a transient analysis in each step. This method is called "matrix free" and it is often more efficient in memory and time. In this paper we go one step further by even avoiding the storage of the generator matrices required by the matrix-free method, thanks to the use of a Kronecker representation.

1 Introduction and Previous Work

In many models of real systems the hypothesis that all durations have an exponential distribution may not be adequate (for example in case of timeouts or hard deadlines). Including general distribution may indeed lead to more realistic models, but with unfeasible solution, so that often simulation has to be used. Markov Regenerative Processes (MRPs) [9] constitute a natural model for whose systems in which the exponential choice is not realistic, but in whose behavior it is possible to identify a renewal sequence. The class of MRP we consider (MRP for short, from now on) is the subclass identified by the additional restriction that the system evolution between regeneration points can be described by a Continuous Time Markov Chain (CTMC), a condition which is true when, in any state, at most one non-exponential activity can take place. This (strong) modelling restriction is balanced by the fact that the solution of such an MRP can still be computed for "reasonably large" state spaces. This paper presents a technique that takes the MRP solution to even larger state spaces. This paper builds on two research lines: numerical MRP solutions and structured representation of state spaces. The high-level formalisms of reference are Stochastic Petri Nets (SPN), both Non-Markovian SPN (NMSPN) [16] and Deterministic SPN (DSPN) [10], from which an MRP is generated.

The MRP solution we build upon is that proposed by Chiola et al. in [1]. The MRP is characterized by three generator matrices $\mathbf{Q}, \bar{\mathbf{Q}}, \boldsymbol{\Delta}$, typically (very) sparse and its solution requires the construction of the MRP embedded Markov chain whose generator $\mathbf{P}$ is typically (very) dense, as well as the computation of

G. Norman and W. Sanders (Eds.): QEST 2014, LNCS 8657, pp. 9–24, 2014.
© Springer International Publishing Switzerland 2014

the matrix of sojourn times $\mathbf{C}$, of similar characteristics. The presence of dense matrices limits the solution to systems with a few thousand states. The solution of MRPs presented by German in [12] works directly with the generators matrices, without building, and storing, $\mathbf{P}$ and $\mathbf{C}$. This method was called "iterative" by the authors (later referred to as "matrix-free" or "implicit") and it trades-off memory saving for increased execution time, although memory saving can sometimes be so large so as to induce also better execution times. The method was extended in [3] to deal with projection methods (like GMRES [14]), and it has been made available through the DSPN-tool software [2].

The state space explosion problem of (stochastic) Petri nets (and similar formalisms) has been extensively studied in the past with the objective of limiting the memory required to store the underlying CTMC. In this paper we build on what is known as the "Kronecker approach" [13]: the system is built out of components and the CTMC infinitesimal generator of the whole system is defined by a Kronecker expression of matrices computed on the state space of the components. The saving in time can be huge (from multiplicative down to additive with respect to the state space of the components), and the overhead in execution time is typically at most linear in the number of components. Another technique for saving space is to use a decision diagram data structure to store the CTMC [11]. In this paper we consider a structured approach based on a Kronecker description, and we leave the use of decision diagram to a future work. This results in the following contribution: a steady-state solution of MRPs which combines the matrix-free technique of [12] and [3] with the Kronecker approach, leveraging on an extension of the Kronecker approach that builds a hierarchical representation of the state space, as defined in [6].

Section 2 presents the background material for the hierarchical structuring of CTMCs, Section 3 extends the structure to MRPs, and Section 4 defines a numerical solution algorithm that combines the structured approach and the implicit steady state solution of MRPs. Finally, Section 5 assesses experimentally the proposed solution technique and concludes the paper.

2   Background Material

In this section we shall introduce the running example and the necessary background material on the Kronecker solution of (exponential only) SPN.

The running example, shown in Figure 1, is a *moving server system* model (or polling system) with two stations. In the figure, white boxes are exponential transitions, black boxes are general transitions. Inter-arrival and service times are exponentially distributed and are realized by transitions t_{j1} and t_{j2} for component $j = 1, 2$. The capacity of the queues equals m_j. If the queue is empty, the server waits (transition t_{j3}) at the queue. When a customer arrives (t_{j1}) while the server is waiting, service starts again (transition t_{j3}), otherwise the server may move (t_{j4}) to the other queue. Waiting and moving times are generally distributed. For the rest of *this* section we assume that all transitions are exponential, to introduce the basic solution approach.

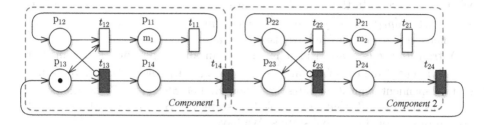

Fig. 1. Petri net model of a simple moving server system with two stations

We assume that models are built from a set $\mathcal{J}$ of components, with $J = |\mathcal{J}|$. The example in Figure 1 has two components ($J = 2$), graphically identified by the dashed boxes. Components have finite state spaces and interact via common transitions. In Petri net terms a component is an SPN with finite reachability set, and components have disjoint places, but can share some transitions. A transition is *shared* if it is connected to places that belong to at least two components, *local* otherwise. In the example t_{14} and t_{24} are the only shared transitions. The reachability set of the components, taken in isolation, would be infinite, but the structural analysis of the net (computation of p-semiflows) reveals that places p_{13}, p_{14}, p_{23}, p_{24}, are safe (1-bounded), and that the sum of tokens in p_{j1} and p_{j2} is bounded by m_j, for $j \in \{1, 2\}$ so that a finite state space can be generated for each component. For practical reasons we assume that all immediate transitions, if any, are local.

For component $j \in \mathcal{J}$ let $\mathcal{S}^{(j)}$ be the finite state space of cardinality $n^{(j)}$ and let $s_0^{(j)} \in \mathcal{S}^{(j)}$ be the initial state of component j. Moreover let $\mathcal{T}$ be the set of transitions, and let λ_t, for each $t \in \mathcal{T}$ be the transition rate of the exponential distribution associated to t. The dynamic behavior of a component according to the transitions from $\mathcal{T}$ is described by $n^{(j)} \times n^{(j)}$ matrices $\mathbf{E}_t^{(j)}$. Each matrix $\mathbf{E}_t^{(j)}$ is non-negative and has row sum 0 (i.e., the transition is disabled by the component in the state corresponding to the row) or 1 (i.e, the transition is enabled by the component in the state corresponding to the row). If component j is not related to transition t, then $\mathbf{E}_t^{(j)} = \mathbf{I}_{n^{(j)}}$, the identity matrix of order $n^{(j)}$. State dependent transition rates can be considered as well, in this case the row sums of matrices $\mathbf{E}_t^{(j)}$ are between 0 and 1 and not just 0 or 1.

For our running example, matrix $\mathbf{E}_{t_{11}}^{(1)}$ describes the contribution of the first component to the firing of t_{11}: a matrix with zero in all rows corresponding to markings which do not enable t_{11}, and with a single entry 1 in all other rows. In each non-zero row the "1" contribution is in the column corresponding to the change of marking realized by t_{11}. Matrix $\mathbf{E}_{t_{11}}^{(2)}$ is instead the identity matrix, since the state of the second component does not influence, and it is not influenced by, the firing of t_{11}.

Let $\mathcal{S}$ be the global state space of the model which contains the joint state of all components. Thus, global states are described by vector $\mathbf{s} = (\mathbf{s}(1), \ldots, \mathbf{s}(J))$

where $\mathbf{s}(j) \in \mathcal{S}^{(j)}$. In the example, the initial marking is $\mathbf{s}_0 = (s_0^{(1)}, s_0^{(2)})$, where $s_0^{(1)} = m1 \cdot m[p_{11}] + 1 \cdot m[p_{13}]$ and $s_0^{(2)} = m2 \cdot m[p_{21}]$, $m[p]$ indicates the number of tokens in place p, and we have used the bag notation for vectors.

A transition t is enabled iff it is enabled in all components, i.e., transition $t \in \mathcal{T}$ is enabled in $\mathbf{s}$ iff for all $j \in \mathcal{J}$: $\sum_{s' \in \mathcal{S}^{(j)}} \mathbf{E}_t^{(j)}(\mathbf{s}(j), s') = 1$ or > 0 if state dependent transition rates are allowed. Let $\mathcal{E}^{(j)}(s) \subseteq \mathcal{T}$ be the set of all transitions $t \in \mathcal{T}$ that are enabled in state $s \in \mathcal{S}^{(j)}$. $\mathcal{E}(\mathbf{s})$ is the set of all transitions enabled in global state $\mathbf{s}$. We have

$$\mathcal{E}(\mathbf{s}) = \bigcap_{j \in \mathcal{J}} \mathcal{E}^{(j)}(\mathbf{s}(j)).$$

If transition t occurs in state $\mathbf{s}$, state $\mathbf{s}'$ is the successor state with probability

$$\prod_{j \in \mathcal{J}} \mathbf{E}_t^{(j)}(\mathbf{s}(j), \mathbf{s}'(j)) .$$

Local and shared transitions can be distinguished based on the corresponding $\mathbf{E}_t^{(j)}$ matrices. Transition $t \in \mathcal{T}$ is local if $\mathbf{E}_t^{(j)} \neq \mathbf{I}$ for exactly one $j \in \mathcal{J}$. Otherwise the transition is shared and its enabling depends on the state of at least two components or the state changes synchronously in at least two components. For our example transitions t_{j1} and t_{j2} are local to component j, while t_{j3} and t_{j4} are global. The matrices of local *exponential* transitions can be combined into a single matrix, as in [13]: this aggregation increases the efficiency, but is not required for correctness.

The overall state space $\mathcal{S}$ can be over-approximated by the product of the state space of the components (potential state space) or can be exactly computed through an efficient reachability analysis of composed models, as described in [8,11], starting from state $\mathbf{s}_0 = (s_0^{(1)}, \ldots, s_0^{(J)})$. In each component, the state space $\mathcal{S}^{(j)}$ is partitioned into subsets $\mathcal{S}^{(j)}[k]$ $(k = 1, \ldots, K)$ such that two states $s, s' \in \mathcal{S}^{(j)}$ are put in the same subset iff

$$\begin{aligned}(\mathbf{s}(1), \ldots, \mathbf{s}(j-1), s, \mathbf{s}(j+1), \ldots, \mathbf{s}(J)) \in \mathcal{S} \Leftrightarrow \\ (\mathbf{s}(1), \ldots, \mathbf{s}(j-1), s', \mathbf{s}(j+1), \ldots, \mathbf{s}(J)) \in \mathcal{S}\end{aligned} \quad (1)$$

An algorithm that computes the subsets can be found in [6]. Let $N^{(j)}$ be the number of subsets for component j, and let $\mathcal{S}^{(j)}[k]$ be the k-th subset. States in $\mathcal{S}^{(j)}$ can then be ordered according to the subsets, such that the states belonging to subset 1 come first, followed by the states from subset 2, and so on. With this ordering all matrices $\mathbf{E}_t^{(j)}$ can be structured into block matrices $\mathbf{E}_t^{(j)}[k, l]$ with $1 \leq k, l \leq N^{(j)}$ including transitions between states for $\mathcal{S}^{(j)}[k]$ and $\mathcal{S}^{(j)}[l]$. Observe that (1) implies that all states in a subset are reachable in combination with exactly the same states of the environment. This condition is necessary and sufficient to describe a state space as a union of subsets where each subset is built from the cross product of subsets of component state spaces [6].

The global state space is decomposed into subsets, such that $\mathcal{S} = \bigcup_{\mathbf{k}} \mathcal{S}[\mathbf{k}]$ and $\mathbf{k} = (\mathbf{k}(1), \ldots, \mathbf{k}(J))$, where $1 \leq \mathbf{k}(j) \leq N^{(j)}$. We then have $\mathcal{S}[\mathbf{k}] = \times_{j \in \mathcal{J}} \mathcal{S}^{(j)}[\mathbf{k}(j)]$.

To simplify the notation the subsets of the global state space are numbered consecutively from 1 to N. Each subset $k \in \{1, \ldots, N\}$ belongs to a vector $\mathbf{k}$ and k_j denotes $\mathbf{k}(j)$. These subsets $\mathbf{k}$ are also called macro-states.

For the example, since both components have an identical structure, their state spaces are decomposed in the same way. Each component state space is decomposed in two subsets, one set $\mathcal{S}^{(j)}[1]$ where one of the places p_{j3} or p_{j4} is marked, and one subset $\mathcal{S}^{(j)}[2]$ where both places are empty. The overall state space of the composed net is then given by $\{\mathcal{S}^{(1)}[1] \times \mathcal{S}^{(2)}[2]\} \cup \{\mathcal{S}^{(1)}[2] \times \mathcal{S}^{(2)}[1]\}$ with the two macro states (k, l) with $k, l \in \{1, 2\}$ and $k \neq l$.

Given the structuring of the state space, a Kronecker structure of the infinitesimal generator can be derived as $\mathbf{Q} = \sum_{t \in \mathcal{T}} (\mathbf{R}_t - \mathbf{D}_t)$ where $\mathbf{D}_t = diag(\mathbf{R}_t \mathbf{1})$. It is well known that the matrix-vector multiplication required by the steady-state solution can be performed without the need to compute and store all the entries of $\mathbf{Q}$. Matrices $\mathbf{R}_t$ and $\mathbf{D}_t$ are block structured according to the decomposition of $\mathcal{S}$. For matrix $\mathbf{R}_t$ we have the following representation of the blocks:

$$\mathbf{R}_t[k, l] = \lambda_t \bigotimes_{j \in \mathcal{J}} \mathbf{E}_t^{(j)}[k_j, l_j] \tag{2}$$

3   A Kronecker Structure for MRPs

We now extend the Kronecker structured approach to non-Markovian stochastic Petri nets (NMSPN) and the associated MRPs. A similar approach is feasible also for other modeling formalisms. We shall first define a Kronecker expression for the MRP generator matrices, while in the next section we show how this structured description can be combined with a matrix-free solution for the MRP.

Again, we assume that models are composed of components from a set $\mathcal{J}$ of cardinality J and a set of transitions $\mathcal{T}$ which can now be split into $\mathcal{T}_e$, the set of transitions with exponentially distributed event times, and $\mathcal{T}_g$, the set of transitions with generally distributed event times. Let T_e and T_g be the number of transitions in $\mathcal{T}_e$ and $\mathcal{T}_g$, respectively. We use t for exponential transitions and g for general. λ_t is the transition rate of $t \in \mathcal{T}_e$, while $f_g(x)$ is the density and $F_g(x)$ is the distribution function of the inter-event time distribution of transition $g \in \mathcal{T}_g$. $\mathcal{E}_g(\mathbf{s}) = \mathcal{E}(\mathbf{s}) \cap \mathcal{T}_g$ is the set of general transitions that are enabled in state $\mathbf{s}$.

Following the standard notation for MRPs, $\mathbf{Q}$, $\bar{\mathbf{Q}}$ and $\boldsymbol{\Delta}$ denote the matrices including, respectively, the rates of non-preemptive exponential transitions ($\mathbf{Q}$), the rates of preemptive exponential transitions ($\bar{\mathbf{Q}}$) and the general branching probabilities of general transitions ($\boldsymbol{\Delta}$). Preemptive means that the firing of $t \in \mathcal{T}_e$ disables a currently enabled transition $g \in \mathcal{T}_g$.

For deriving a Kronecker expression for $\mathbf{Q}$ and $\bar{\mathbf{Q}}$, we need to consider the contribution of each exponential transition t to the generators $\mathbf{Q}_t$ and $\bar{\mathbf{Q}}_t$, and to distinguish, for each of them, firings of t that happen in 1) states in which no general transition is enabled or in 2) states in which a general transition g is enabled. Moreover we need to distinguish, for the second case, whether 2a) the firing of t disables the general or 2b) does not disable it, since the contribution of 2a goes into $\bar{\mathbf{Q}}_t$ and the one of 2b goes into $\mathbf{Q}_t$. Note that this distinction is not required for $\boldsymbol{\Delta}$, although also in this case we shall consider the contribution of one transition (general by definition) at a time.

The need to distinguish states according to the enabling of general transitions should therefore be reflected also in the hierarchical structuring of the state space: the condition of (1) is then refined to distinguish partitions depending on the enabling of general transitions, leading to the following condition: two states $s, s' \in \mathcal{S}^{(j)}$ are put in the same subset iff

$$
\begin{aligned}
(\mathbf{s}(1), \ldots, \mathbf{s}(j-1), s, \mathbf{s}(j+1), \ldots, \mathbf{s}(J)) &\in \mathcal{S} \Leftrightarrow \\
(\mathbf{s}(1), \ldots, \mathbf{s}(j-1), s', \mathbf{s}(j+1), \ldots, \mathbf{s}(J)) &\in \mathcal{S} \\
\text{and} \quad \mathcal{E}^{(j)}(s) \cap \mathcal{T}_g = \mathcal{E}^{(j)}(s') \cap \mathcal{T}_g.&
\end{aligned}
\tag{3}
$$

Rule (3) ensures that in each partition the set of enabled general transitions remains the same. Recall that $N^{(j)}$ is the number of subsets for component j and denote by $\mathcal{S}^{(j)}[k]$ the k-th subset.

For the running example each component state space is now decomposed into 4 subsets, $\mathcal{S}^{(j)}[1], \ldots, \mathcal{S}^{(j)}[4]$, since we have to distinguish whether the general transitions t_{i3} and t_{i4} are enabled or not. The structure becomes finer than in the Markovian case. $\mathcal{S}^{(j)}[1]$ contains states in which the server is not at the component, therefore places p_{j3} and p_{j4} are both empty (for a total of $m_j + 1$ states). $\mathcal{S}^{(j)}[2]$ contains states in which the server is at the queue but t_{j3} and t_{j4} are not enabled, therefore p_{j4} is empty, p_{j3} contains a token and p_{j2} is not empty (for a total of m_j states). $\mathcal{S}^{(j)}[3]$ contains all states where t_{j3} is enabled which implies that p_{j3} contains a token and p_{j2} is empty, which actually corresponds to a single state. Finally, $\mathcal{S}^{(j)}[4]$ contains all states where t_{j4} is enabled, which means that p_{j4} contains a token and the tokens can be in either p_{j2} or p_{j1} (for a total of $m_j + 1$ states). The complete state space $\mathcal{S}$ can then be described by the following subsets: $\mathcal{S}[1,2], \mathcal{S}[2,1], \mathcal{S}[1,3], \mathcal{S}[1,4], \mathcal{S}[3,1], \mathcal{S}[4,1]$, and contains $(m1+1)m2 + m1(m2+1) + (m1+1) + (m1+1)(m2+1) + (m2+1) + (m1+1)(m2+1) = 4(m1+1)(m2+1)$ states, as in the exponential case. We number the vectors for the subsets or macro states in the order given above from 1 through 6. Usually, the number of macro states generated according to (1) or (3) is significantly smaller than the number of states. If this is not the case, another component structure should be defined to benefit from a structured solution approach.

Remember that an MRP can be analyzed numerically if at most one general transition is enabled in a state. This condition can be expressed as $|\mathcal{E}_g(\mathbf{s})| \leq 1$ for all $\mathbf{s} \in \mathcal{S}$. This condition holds for the running example, since the sum of tokens in places p_{13}, p_{14}, p_{23}, and p_{24} is equal to one. Since for $\mathbf{s}, \mathbf{s}' \in \mathcal{S}[k]$ the relation $\mathcal{E}_g(\mathbf{s}) = \mathcal{E}_g(\mathbf{s}')$ holds by definition of the subset, it is meaningful to define $\mathcal{E}_g(\mathbf{k})$, the set of global transitions that are enabled in states from subset $\mathcal{S}[\mathbf{k}]$. We shall also assume that the state space of the system is irreducible. An extension of the approach to reducible case can be made, similarly to [3].

To ease the description of the generator matrices, the subsets of the global states space are ordered so that the subsets $1, \ldots, K_0$ contain only states where no transition from $\mathcal{T}_g$ is enabled. We assume that transitions from $\mathcal{T}_g$ are numbered from 1 through T_g, such that the subsets $K_{g-1} + 1, \ldots, K_g$ contain the states where transition $g \in \mathcal{T}_g$ is enabled. Let $\mathcal{S}_g = \bigcup_{k=K_{g-1}+1}^{K_g} \mathcal{S}[k]$ be the set of all states where $g \in \mathcal{T}_g$ is enabled.

We now show how the different matrices required for the analysis of the MRP can be composed from the small matrices of the components. Let us start with the matrix $\mathbf{Q}$ which includes all rates of transitions from $\mathcal{T}_e$ that do not preempt a general transition. The diagonal elements of $\mathbf{Q}$ contain the sum of exponential rates independently on whether they disable a general transition or not. This representation is consistent with [3]. For each $t \in \mathcal{T}_e$ we define a matrix $\mathbf{Q}_t$ and represent this matrix by matrices $\mathbf{R}_t$ containing the non-diagonal part and $\mathbf{D}_t$ containing the diagonal part such that $\mathbf{Q} = \sum_{t \in \mathcal{T}_e} (\mathbf{R}_t - \mathbf{D}_t)$ and $\mathbf{D} = \sum_{t \in \mathcal{T}_e} \mathbf{D}_t$. Matrices $\mathbf{R}_t$ and $\mathbf{D}_t$ are block structured according to the decomposition of $\mathcal{S}$. For matrix $\mathbf{R}_t$ we distinguish two expressions for the blocks of $\mathbf{R}_t$ in Equation 4 due to the occurrence of t: one for states in which only exponential transitions are enabled (first case), and one for states in which a general transition g is enabled and the occurrence of t does not disable it (second case); blocks corresponding to states in which a general transition g is disabled by firing exponential transition t are not part of matrix $\mathbf{R}_t$ (third case). In the equations $\mathbf{E}_t^{(j)}[k_j, l_j]$ is used to describe the firing of t for rows in which only exponentials are enabled. For blocks in which, at the component level, a general g is locally enabled and the firing of t leads to a local state in which g is still locally enabled, we need to distinguish two cases. First, the firings of t that "preempts and immediately re-enables" g, and, second, the firings of t that do not affect g at all. The latter are collected in $\mathbf{E}_{t,g}^j[k_j, l_j]$ and contribute to $\mathbf{R}_t$.

$$\mathbf{R}_t[k, l] = \begin{cases} \lambda_t \bigotimes_{j \in \mathcal{J}} \mathbf{E}_t^{(j)}[k_j, l_j] & \text{if } k \le K_0, \\ \lambda_t \bigotimes_{j \in \mathcal{J}} \mathbf{E}_{t,g}^{(j)}[k_j, l_j] & \text{if } K_{g-1} < k, l \le K_g \text{ for } 1 \le g \le T_g, \\ 0 & \text{otherwise.} \end{cases} \quad (4)$$

In matrix $\mathbf{D}_t$ only the diagonal blocks are non-zero and they are built out of diagonal matrices. Since diagonal elements contain all the rates of exponential transitions of both $\mathbf{Q}$ and $\bar{\mathbf{Q}}$, we can write

$$\mathbf{D}_t[k, k] = \lambda_t \bigotimes_{j \in \mathcal{J}} diag\left(\sum_{l=1}^{N} \mathbf{E}_t^{(j)}[k_j, l_j]\mathbf{1}\right) = \lambda_t \bigotimes_{j \in \mathcal{J}} \mathbf{F}_t^{(j)}[k_j, k_j] \quad (5)$$

where $diag(\mathbf{a})$ is a diagonal matrix with $\mathbf{a}(i)$ in position (i, i) and $\mathbf{F}_t^{(j)}[k_j, k_j] = diag\left(\sum_{l=1}^{N} \mathbf{E}_t^{(j)}[k_j, l_j]\mathbf{1}\right)$.

Matrix $\bar{\mathbf{Q}}$ represents the contribution of exponential events that preempt a general transition. As before, it can be represented as $\bar{\mathbf{Q}} = \sum_{t \in \mathcal{T}_e} \bar{\mathbf{R}}_t$, and we have two distinguished cases: transition g is disabled after the firing of t (first case), and g is enabled before and after the occurrence of t but is disabled by

t. Since g is disabled if it is disabled by at least one component, the resulting matrix consists of the difference between the matrix with all transitions of t and the matrix with all transitions where g is not preempted.

$$\bar{\mathbf{R}}_t[k,l] = \begin{cases} \lambda_t \bigotimes_{j\in\mathcal{J}} \mathbf{E}_t^{(j)}[k_j, l_j] & \text{if } K_{g-1} < k \le K_g, K_{h-1} < l \le K_h, \\ & g \neq h, 1 \le g \le T_g, 0 \le h \le T_g \\ \lambda_t \left(\bigotimes_{j\in\mathcal{J}} \mathbf{E}_t^{(j)}[k_j, l_j] - \bigotimes_{j\in\mathcal{J}} \mathbf{E}_{t,g}^{(j)}[k_j, l_j] \right) & \text{if } K_{g-1} < k, l \le K_g, \\ & \text{for } 1 \le g \le T_g \\ 0 & \text{otherwise.} \end{cases} \quad (6)$$

Finally, $\boldsymbol{\Delta} = \sum_{g\in\mathcal{T}_q} \boldsymbol{\Delta}_g$, where matrices $\boldsymbol{\Delta}_g$ describe the transition probabilities of the states reached after the firing of each transition $g \in \mathcal{T}_g$:

$$\boldsymbol{\Delta}_g[k,l] = \begin{cases} \bigotimes_{j\in\mathcal{J}} \mathbf{E}_g^{(j)}[k_j, l_j] & \text{if } K_{g-1} < k \le K_g \\ 0 & \text{otherwise.} \end{cases} \quad (7)$$

In our running example transitions t_{14} and t_{24} cannot be preempted, while t_{13} and t_{23} can be preempted in states corresponding to blocks 3 and 5, leading to the presence of non-zero sub-matrices $\bar{\mathbf{Q}}[3,1]$ and $\bar{\mathbf{Q}}[5,2]$ in $\bar{\mathbf{Q}}$. Furthermore, the exponential transitions of each component are local, so for each component there is a single matrix that describes t_{j1} and t_{j2}.

4   Numerical Computation of the Stationary Vector

Our goal for the numerical analysis of an MRP is the computation of the stationary distribution π using the implicit (matrix-free) method [12]. In the following we first review the explicit method, and then proceed in extending the method to account for a structured representation of the MRP generator matrices.

4.1   Explicit Method for the Computation of the Stationary Vector

The stationary vector is computed as the stationary solution of the embedded discrete time Markov chain which is afterwards normalized according to the sojourn times in the different states during one regeneration cycle [12]. Following [3,12] we define $\mathbf{I}_g$ as a $|\mathcal{S}| \times |\mathcal{S}|$ matrix where the diagonal elements belonging to the rows from $\mathcal{S}[K_{g-1}+1], \ldots, \mathcal{S}[K_g]$ are 1 and all remaining elements are 0. Similarly, for $\mathbf{I}_e$ the diagonal elements belonging to the rows from $\mathcal{S}[1], \ldots, \mathcal{S}[K_0]$ are 1 and the remaining diagonal elements are 0. Due to the ordering of states, the matrices select a consecutive subset of rows when multiplied with a matrix from the left. For some vector $\mathbf{x}$ we define $\mathbf{x}_g = \mathbf{x}\mathbf{I}_g$, $\mathbf{x}_e = \mathbf{x}\mathbf{I}_e$ and $\mathbf{x}_G = \mathbf{x}\sum_{g\in\mathcal{T}_g} \mathbf{I}_g$. Then the matrix of the embedded Markov chain equals

$$\mathbf{P} = \mathbf{I}_e - \mathbf{I}_e (\mathbf{D})^{-1} \mathbf{Q} + \boldsymbol{\Omega}\boldsymbol{\Delta} + \boldsymbol{\Psi}\bar{\mathbf{Q}} \quad (8)$$

where $(\mathbf{D})^{-1}$ is a diagonal matrix with the diagonal elements $(\mathbf{D}(x,x))^{-1}$ if $\mathbf{D}(x,x) \neq 0$ and 0 otherwise. The terms $\boldsymbol{\Omega}$ and $\boldsymbol{\Psi}$ are

$$
\begin{aligned}
\boldsymbol{\Omega} &= \sum_{g \in \mathcal{T}_g} \boldsymbol{\Omega}_g = \sum_{g \in \mathcal{T}_g} \mathbf{I}_g \int_0^\infty e^{\mathbf{I}_g(\mathbf{R}-\mathbf{D})x} f_g(x)\, dx = \sum_{m=0}^\infty (\mathbf{U}_g)^m \, \alpha_g^f(m, \mu_g) \\
\boldsymbol{\Psi} &= \sum_{g \in \mathcal{T}_g} \boldsymbol{\Psi}_g = \sum_{g \in \mathcal{T}_g} \mathbf{I}_g \int_0^\infty e^{\mathbf{I}_g(\mathbf{R}-\mathbf{D})x} \bar{F}_g(x)\, dx = \sum_{m=0}^\infty (\mathbf{U}_g)^m \, \alpha_g^{\bar{F}}(m, \mu_g)
\end{aligned} \tag{9}
$$

where $\mathbf{U}_g = \mathbf{I}_g + \mathbf{I}_g(\mathbf{R}-\mathbf{D})/\mu_g$, $\mu_g \geq \max_{x \in \mathcal{S}_g} |\sum_{t \in \mathcal{T}_e} \mathbf{D}_t(x,x)|$, function $\bar{F}_g(x)$ is $(1 - F_g(x))$, and functions $\alpha_g^f(k, \mu)$ and $\alpha_g^{\bar{F}}(k, \mu)$ are

$$
\alpha_g^f(m, \mu) = \int_0^\infty e^{-\mu x} \frac{(\mu x)^m}{m!} f_g(x)\, dx, \quad \alpha_g^{\bar{F}}(m, \mu) = \int_0^\infty e^{-\mu x} \frac{(\mu x)^m}{m!} \bar{F}_g(x)\, dx \tag{10}
$$

with $m \in \mathbb{N}$. For the evaluation of the functions for different densities $f_g(x)$ we refer to [12]. The terms $\boldsymbol{\Omega}$ and $\boldsymbol{\Psi}$ are matrix exponentials and can be computed as truncated Taylor series.

Although the matrices $\mathbf{R}_t$ can be represented as Kronecker products of small component matrices, this does not help when $\boldsymbol{\Omega}_g$ and $\boldsymbol{\Psi}_g$ are built. The matrices have to be computed from a sparse matrix representation of matrices $\mathbf{R}_t$ and even worse, the resulting matrices often become dense such that the memory effort grows to $O(n^2)$ where n is the size of the state space. This implies that the computation of matrix $\mathbf{P}$ can only be done for models of a moderate size and is very time and memory consuming.

After matrix $\mathbf{P}$ is available from (8) the left eigenvector $\mathbf{x}\mathbf{P} = \mathbf{x}$ subject to $\mathbf{x}\mathbf{1} = 1$ can be computed. Alternatively, the system $\mathbf{x}(\mathbf{P} - \mathbf{I}) = \mathbf{0}$ with $\mathbf{x}\mathbf{1} = 1$ can be solved using standard numerical techniques for the computation of the solution vector of a linear equation system. Commonly used algorithms are the Power method, SOR or projection methods (like GMRES, BiCG-stab, see [15]). In the case of the Power method, a series of vectors $\mathbf{x}^m$ has to be computed according to the relation: $\mathbf{x}^m = \mathbf{x}^{m-1}\mathbf{P}$, with $\mathbf{x}^0$ an appropriate initial vector with $\mathbf{x}^0 \geq 0$, $\mathbf{x}^0\mathbf{1} = 1$. In the case of projection methods, a sequence of residual vectors in the form $\mathbf{r}^m = \mathbf{x}^m\mathbf{P} - \mathbf{x}^m$ is required (see [3]).

Let $\mathbf{x}$ the (approximate) solution of the linear equation system $\mathbf{x} = \mathbf{x}\mathbf{P}$. The stationary vector of the MRP π is computed as

$$
\hat{\pi} = \mathbf{x}\mathbf{C} = \mathbf{x}\left(\mathbf{I}_e\,(\mathbf{D})^{-1} + \boldsymbol{\Psi}\right) \text{ such that } \pi = \frac{\hat{\pi}}{\hat{\pi}\mathbf{1}}. \tag{11}
$$

where $\mathbf{C}$ is the *conversion factors* matrix, which is usually as dense as $\mathbf{P}$.

The outlined approach is denoted as explicit method because it first builds matrices $\mathbf{P}$ and $\mathbf{C}$, and then computes the linear solution.

4.2   The Proposed Structured Implicit Method

The bottleneck of the above mentioned explicit solution method is the computation of the matrices $\mathbf{P}$ and $\mathbf{C}$ or, more precisely, the computation of $\boldsymbol{\Omega}$ and

$\mathbf{\Psi}$. To avoid this computation one has to compute vector products $\mathbf{x}^m\mathbf{P}$ without having pre-computed and stored matrix $\mathbf{P}$ (hence the "matrix-free" name). The product $\mathbf{x}^m\mathbf{P}$ can be rewritten [12] as:

$$\mathbf{x}^m\mathbf{P} = \mathbf{x}^m \left(\mathbf{I}_e - \mathbf{I}_e\left(\mathbf{D}\right)^{-1}\mathbf{Q} + \mathbf{\Omega\Delta} + \mathbf{\Psi\bar{Q}} \right)$$

$$= \mathbf{x}_e^m - \mathbf{x}_e^m\left(\mathbf{D}\right)^{-1}\mathbf{Q} + \left(\sum_{g\in\mathcal{T}_g} \mathbf{x}_g^m\mathbf{\Omega}_g \right)\mathbf{\Delta} + \left(\sum_{g\in\mathcal{T}_g} \mathbf{x}_g^m\mathbf{\Psi}_g \right)\mathbf{\bar{Q}} \qquad (12)$$

$$= \mathbf{x}_e^m - \underbrace{\mathbf{x}_e^m\left(\mathbf{D}\right)^{-1}\mathbf{Q}}_{\mathbf{u}^m} + \underbrace{\mathbf{a}^m\mathbf{\Delta}}_{\mathbf{v}^m} + \underbrace{\mathbf{b}^m\mathbf{\bar{Q}}}_{\mathbf{w}^m}$$

In the implicit method vectors $\mathbf{b}^m$ and $\mathbf{a}^m$ are computed without generation of the matrices $\mathbf{\Omega}$ and $\mathbf{\Psi}$. In this paper we go one step further from [12] and avoid even the generation of the matrices $\mathbf{Q}$, $\mathbf{\bar{Q}}$ and $\mathbf{\Delta}$ by using the implicit Kronecker representation of these matrices in the $\mathbf{xP}$ products of Eq (12). In what follows we present, at the same time, the implicit method and its adaptation to deal with Kronecker expressions of the generators. We begin with the vectors $\mathbf{a}^m$ and $\mathbf{b}^m$, which can be computed by an iterative approach based on (9). Both vectors are computed from the repeated multiplication of vector $\mathbf{x}^m$ with matrices $\mathbf{U}_g$, for each $g \in \mathcal{T}_g$. This implies that the first rows, belonging to the subsets $\mathcal{S}[1],\ldots,\mathcal{S}[K_0]$ are zero and the corresponding elements of vector $\mathbf{x}^m$ are not needed. Let $\mathbf{y}^0 = \mathbf{x}^m$ and perform the following iterations.

$$\mathbf{y}^{l+1}[k] = \sum_{h=K_{g-1}+1}^{K_g} \mathbf{y}^l[h]\left(\mathbf{I} - \tfrac{1}{\mu_g}\mathbf{Q}[h,k] \right)$$

$$= \sum_{h=K_{g-1}+1}^{K_g} \mathbf{y}^l[h]\left(\mathbf{I} - \sum_{t\in\mathcal{T}_e} \tfrac{1}{\mu_g}\mathbf{R}_t[h,k] \right) - \mathbf{y}^l[k]\sum_{t\in\mathcal{T}_e} \tfrac{1}{\mu_g}\mathbf{D}_t[k,k]$$

$$= \sum_{h=K_{g-1}+1}^{K_g} \mathbf{y}^l[h]\left(\mathbf{I} - \sum_{t\in\mathcal{T}_e} \tfrac{\lambda_t}{\mu_g}\bigotimes_{j\in\mathcal{J}} \mathbf{E}_t^{(j)}[h_j,k_j] \right) \qquad (13)$$

$$\qquad - \mathbf{y}^l[k]\sum_{t\in\mathcal{T}_e} \tfrac{\lambda_t}{\mu_g}\bigotimes_{j\in\mathcal{J}} \mathbf{F}_t^{(j)}[k_j,k_j]$$

for $g \in \mathcal{T}_g$ and $K_{g-1} < k \le K_g$. Observe that we only have to consider the subsets $\mathcal{S}[K_{g-1}+1],\ldots,\mathcal{S}[K_g]$ if we analyze the behavior of exponential transitions during the enabling time of general transition $g \in \mathcal{T}_g$ because g is disabled in subsets $h \le K_{g-1}$ and $h > K_g$. All subvectors $\mathbf{y}^l[k]$ and therefore the whole vector $\mathbf{y}^l$ can be computed by repeated use of the standard procedure to perform the multiplication of a vector with a Kronecker product of matrices [15,7]. The terms $\mathbf{a}$ and $\mathbf{b}$ are rewritten as

$$\mathbf{a}^m[k] = \begin{cases} \sum_{l=L_g}^{R_g} \alpha_g^f(l,\mu_g)\cdot\mathbf{y}^l & \text{if } K_{g-1}+1 < k \le K_g \text{ for } g\in\mathcal{T}_g \\ \mathbf{0} & \text{otherwise} \end{cases}$$

$$\mathbf{b}^m[k] = \begin{cases} \sum_{l=L_g}^{R_g} \alpha_g^{\bar{F}}(l,\mu_g)\cdot\mathbf{y}^l & \text{if } K_{g-1}+1 < k \le K_g \text{ for } g\in\mathcal{T}_g \\ \mathbf{0} & \text{otherwise} \end{cases} \qquad (14)$$

where L_g and R_g are the truncation points for the computations of the Taylor's series of the α-factors of g.

The values of $\mathbf{a}^m[k]$ and $\mathbf{b}^m[k]$ can then be used to compute the vectors $\mathbf{u}^m$, $\mathbf{v}^m$ and $\mathbf{w}^m$ of (12) (always without generating $\mathbf{Q}$, $\bar{\mathbf{Q}}$ and $\mathbf{\Delta}$ explicitly). We begin with $\mathbf{u}^m$. Vector $\mathbf{u}^m[k] = \mathbf{0}$ for $k > K_0$, i.e., in states where a general transition is enabled. For the case of $k \leq K_0$ we derive

$$
\begin{aligned}
\mathbf{u}^m[k] &= \sum_{h=1}^{K_0} \mathbf{x}^m[h] \left(\mathbf{D}[h,h]\right)^{-1} \mathbf{Q}[h,k] = \\
&= \sum_{h=1}^{K_0} \mathbf{x}^m[h] \left(\mathbf{D}[h,h]\right)^{-1} \left(\sum_{t \in \mathcal{T}_e} \lambda_t \bigotimes_{j \in \mathcal{J}} \mathbf{E}_t^{(j)}[h_j, k_j] \right)
\end{aligned}
\tag{15}
$$

Matrices $\mathbf{D}[k,k]$ are not represented as Kronecker products, but are stored as sparse matrices. However, since these matrices are diagonal matrices only, it is sufficient to store a vector of length $\sum_{k=1}^{K_0} |\mathcal{S}[k]|$, whose size is usually less than the length of the $\mathbf{u}^m$ vector. Vector $\mathbf{v}^m[k]$ is computed as

$$
\mathbf{v}^m[k] = \sum_{g \in \mathcal{T}_g} \sum_{h=K_{g-1}+1}^{K_g} \mathbf{a}^m[h] \mathbf{\Delta}_g[h,k] = \sum_{g \in \mathcal{T}_g} \sum_{h=K_{g-1}+1}^{K_g} \mathbf{a}^m[h] \bigotimes_{j \in \mathcal{J}} \mathbf{E}_g^{(j)}[h_j, k_j]
\tag{16}
$$

for $k = K_0 + 1, \ldots, N$ and $\mathbf{0}$ for $k \leq K_0$. Vector $\mathbf{w}^m$ is given by

$$
\begin{aligned}
\mathbf{w}^m[k] &= \sum_{g \in \mathcal{T}_g} \sum_{h=K_{g-1}+1}^{K_g} \mathbf{b}^m[h] \bar{\mathbf{Q}}[h,k] \\
&= \sum_{g \in \mathcal{T}_g} \sum_{h=K_{g-1}+1}^{K_g} \left(\mathbf{b}^m[h] \sum_{t \in \mathcal{T}_e} \lambda_t \cdot \left(\bigotimes_{j \in \mathcal{J}} \mathbf{E}_t^{(j)}[h_j, k_j] \right. \right. \\
&\qquad \left. \left. -\delta(k \in \{K_{g-1}+1, \ldots, K_g\}) \bigotimes_{j \in \mathcal{J}} \mathbf{E}_{t,g}^{(j)}[h_j, k_j] \right) \right)
\end{aligned}
\tag{17}
$$

where $\delta(b) = 1$ for $b = true$ and 0 otherwise. Again $\mathbf{w}^m[k] = \mathbf{0}$ for $k \leq K_0$.

These equations allow one to compute the steady state solution of an MRP using the smaller component matrices $\mathbf{E}^{(j)}$ rather than the much larger matrices of the complete system, as the computation of the stationary vector $\mathbf{x}$ and of π with (11) can be done by substituting the terms $\mathbf{u}^m$, $\mathbf{v}^m$ and $\mathbf{w}^m$ from (12) with the ones derived in (15), (16) and (17).

5   Experimental Assessment and Conclusions

To experimentally assess the proposed technique we have implemented the structured implicit solution by merging the Kronecker-based solution of the nsolve [5] tool with the implicit MRP solution provided by DSPN-tool [2]. nsolve is a tool developed at the TU Dortmund that handles structured stochastic Petri nets (exponential only), while DSPN-Tool is a Petri net solver [2] for stochastic Petri nets (exponential and deterministic), which solves MRPs with both the implicit and explicit methods. The implemented tool is an extension of nsolve that includes the implicit structured solution of MRPs presented in this paper. In the new tool, general transitions are currently restricted to deterministic ones, but other distributions can be supported as well, with the proper generation of the α-factors of (10) for other distributions, as exemplified in one of the examples.

The evaluation considers three different models, a *Flexible Manufacturing System* (FMS) [4], the *Moving Server System* of the running example and a sequential program with a fork and join section.

Objective of the evaluation is to understand the advantages/disadvantages of the proposed method in terms of memory usage and execution time. Readers should recall that Kronecker approaches are meant to save space, with a known overhead in terms of time [7] which, however, strongly depends on the structure of the model.Normally, we cannot expect the proposed method to be better that the implicit method implemented in DSPN-tool, unless the memory occupation becomes so large to affect execution times due to memory access or caching effects. In particular, the Kronecker approach we use can be at most J times slower than the unstructured one, when matrices involved are very sparse (much less than one entry per row). In our experience this factor is usually around 2 or 3. Note that, although not reported, performance indicators have been computed as well, and they fully coincide between the two tools. All results are computed with a relative error ϵ of 10^{-8} on the same PC.

Flexible Manufacturing System. The net considered here has been presented in [4]. It consists of four machines $M_{1...4}$ that operate a set of N circulating objects. Machine M_2 can break during operation, but a set of Spares can be used to repair it immediately. Machine M_3 also may break, but it always requires a full repairing. A repairman checks and repairs the machines, if needed. All activities of the repairman are deterministic. Since there is a single repairman it is never the case that two deterministic are enabled in the same state. The net consists of 24 places, 8 immediate, 4 deterministic and 8 exponential transitions.

This example is used to compare the structured implicit solution of this paper with the standard implicit solution in terms of space and execution time for three different numerical methods: Power method [15], which directly implements (12), and two projection methods, GMRES and BiCG-Stabilized [3].

Table 1 reports, in the order, the number N of objects, the number of states, the number of macro states, the number of non-zero entries (nnz) in the Kronecker representation of the generators, the solution time and number of iterations of the three numerical methods for the proposed structured solution implemented in nsolve. The same set of data is reported in the last 7 columns for the unstructured case, as implemented in DSPN-tool; in this case nnz is the number of non-zeros in the generator matrices. Despite the large number of macro states, which indicates the fairly complex structure of the model, we observe a significant saving in space. Note the superior performance of GMRES and BiCGstab over Power method, especially in terms of iterations for reaching convergence. Moreover we have a counter-intuitive result for what concerns solution

Table 1. FMS model: steady state solution comparison (time and space)

			Structured implicit							Unstructured implicit					
				Power		GMRES		BiCGstab			Power		GMRES		BiCGstab
N	states	mst	nnz	Time	It	Time	It	Time	It	nnz	Time	It	Time	It	Time	It
5	4361	220	9442	0.76	90	0.44	45	0.37	40	18556	0.4	87	0.3	39	0.2	38
7	12744	306	20670	2.13	110	1.27	55	1.04	46	60120	1.6	110	0.9	48	0.8	43
10	42636	435	50472	7.37	140	4.14	69	3.34	58	219351	7.4	147	3.8	62	3.2	50
12	80899	521	81290	13.92	150	8.27	79	6.31	62	431535	14.8	152	8.9	75	7.1	59
14	140480	607	122716	25.93	170	17.07	100	11.44	68	769675	28.3	166	16.6	80	13.2	63
16	228123	693	176334	38.81	160	27.08	100	20.10	76	1275867	59.8	179	41.9	109	28.5	71
18	351500	779	243728	62.13	170	47.87	117	29.61	72	1998287	123.1	189	85.1	117	53.5	72
20	519211	865	326482	95.87	180	71.06	119	45.21	74	2991191	195.6	182	144.3	119	88.6	73

times, since the structured solution can even be faster than the unstructured one (for large values of the state space). As we shall see in later examples, this is a phenomenon that is always observed when memory occupation becomes large. This could be caused by the fact that a smaller number of non-zeros can lead to a better caching effect for the matrices.

Moving Server System. This is our running example (Figure 1) modified as to consider 4 stations (leading to 4 components), and to include general transitions (black and thick in the figure). Experiments are ran for a variable number m_i of requests arriving to the station's queues (same value for all queues). The number of macro states is not reported since it is constant and equal to 12. With this example we also compare the differences of a MRP solution with deterministic transitions and a CTMC solution where deterministic transitions are approximated by Erlang-k distributions.

Table 2. MSS: State space sizes and memory occupations

	Structured representation (nsolve)								Unstructured (DSPN-Tool)		
	General		Exp		Erl-2		Erl-3		Implicit	Explicit	
m_i	states	nnz	states	nnz	states	nnz	states	nnz	$Q, \bar{Q}, \Delta$	nnz P	nnz C
1	128	80	128	64	224	236	320	416	384	544	388
5	10368	240	10368	224	16416	684	22464	1056	44928	312384	262932
10	117128	440	117128	424	181016	1244	244904	1856	543048	7452116	6705612
15	524288	640	524288	624	802816	1804	1081344	2656	2490368	33127648	29974864
20	1555848	840	1555848	824	2370816	2364	3185784	3456	7482888	88318068	79771404
25	3655808	1040	3655808	1024	5554016	2924	7452224	4256	17716608	-	-
30	7388168	1240	7388168	1224	11201416	3484	15014664	5056	35987528	-	-

Table 2 summarizes the comparison in space: number of states and number of non-zeros for the four cases of structured representation (general, exponential, Erlang-2 and 3) and for the unstructured one (last three columns). Erlang is here implemented through expansion of the CTMC, which justifies the larger number of states and of non-zeros in the Kronecker representation. As expected, the explicit solution (last two columns) is not able to cope with large state spaces, while the implicit unstructured, although better than explicit, occupies much more space than the corresponding structured case proposed in this paper, as an example: 840 entries against more than 7 millions for the $m_i = 20$ case. Note also that, space wise, the (structured) approximation with an Erlang (columns 6 to 9) performs better than the implicit unstructured. If we compare the results of the deterministic versus the Erlang distributions for waiting and moving time, significant differences can be observed for different measures like the mean population of place p_{i2} (mean buffer population), or the probability that all m_i tokens or on p_{i2}, even if Erlang-3 distributions are used. This shows that the use of phase type distributions as approximation for deterministic distributions is either only a rough approximation (exponential case) or requires additional effort (Erlang-k distributions with a large k).

Table 3. MSS: steady state solution times, using the BiCG-Stabilized method

											Unstructured (DSPN-tool)			
		Structured implicit (nsolve)												
		General (MRP)				Phase expansion (CTMC)					Implicit		Explicit	
	Det		Erl-3		Exp		Erl-2		Erl-3		Det		Det	
m_i	Time	Iters	Time	Iters	Time	Iters	Time	Iters	Time	Iters	Time	Iters	Time	Iters
5	0.1	130	0.1	130	0.1	142	0.1	200	0.2	278	0.6	125	1.1	104
10	1.8	228	1.8	218	0.7	226	1.5	318	2.5	416	11.8	226	67.7	220
15	10.9	290	11.7	286	4.3	320	9.2	440	16.3	562	126.0	317	1488.0	320
20	44.2	384	47.5	382	16.8	398	35.4	532	64.3	698	606.8	424	12202.1	416
25	131.8	476	140.5	466	50.8	494	106.3	632	208.5	836	1687.8	509	-	-
30	341.6	588	369.3	584	151.7	632	331.5	772	613.4	960	4048.4	611	-	-

Table 3 reports the execution times for the steady-state solution of the same systems as in Table 2, in which we consider two types of general: deterministic and Erlang-3. Columns 2 to 11 show the time and number of iterations of nsolve for the four distribution variations, while the last four columns report time and iterations for the unstructured (implicit and explicit) solution of DSPN-tool. Results correspond to the BiCG-stabilized method, which has shown to be the best method for these models. When comparing the structured solution (column 2 and 3 against unstructured one (columns 12 and 13), the unexpected gain in time, which is even more striking than for the FMS case, makes even more explicit the need for a change in DSPN-tool.

Third example: The model is depicted in Figure 2, consists or two parts: a sequential part (upper part of the net) and a loop over a fork and join part. The fork consists of three branches, and each fork creates N threads in each branch. Computation of the fork and join is controlled by a time-out, while the number of loops is at most K (at most since there is a transition that eliminates tokens from the loop control place). Deterministic transitions are here used to have a more realistic model: the time-out is deterministic and the execution of the initial sequential part is made of activities with low variability. This low variability

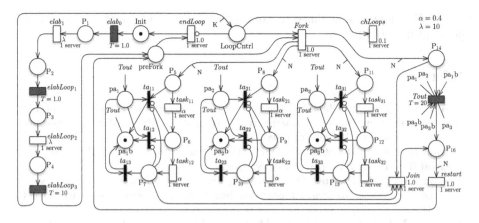

Fig. 2. The Petri net of a sequential Program with a parametric fork and join

has been realized with an alternation of a deterministic of duration 1 and of an
exponential of mean duration 0.1 (parameter λ set to 10).

Table 4. Third example: steady state solution comparison (time and space)

			Structured implicit				Unstructured implicit		
				BiCGstab				BiCGstab	
K	N	states	mst	nnz	Time	It	nnz	Time	It
10	2	2456	6	351	0.1	44	11962	0.1	49
10	4	46206	8	1409	4.4	42	278247	5.7	45
10	6	358336	10	3811	47.3	46	2382K	57.0	45
10	8	1736146	12	8133	282.2	44	12237K	298.2	45
20	8	3472286	12	8413	847.6	58	24648K	950.8	57
30	8	5208426	12	8693	1838.0	74	37059K	1745.2	73
40	8	6944566	12	8969	3306.0	88	49471K	2743.7	89

The experiments were conducted for varying N and K, and are reported in
Table 4 for the single case of BiCGstab, which was the faster solver. For these
models the number of macro states depends on N. For $N \geq 20$ only results for
$K = 8$ are shown. Also in this more realistic case of the use of deterministic
transitions, the saving in space for the storage of non-zeros is striking (8969
entries against more than 49 millions!) and execution times, and iterations, are
comparable.

The experiments reported here (as well as all the others that we have performed) show very clearly the advantage of the proposed technique and encourage to move on in this line on research, both to consolidate our numerical solvers
and to extend the technique to gain efficiency by exploiting symmetries or to
go towards approximate solutions. Another line that should be pursued is the
exploitation of decision diagram(DD) to understand how the DD encoding of

the CTMCs should be modified to work with MRPs, and what is the impact on the implicit method.

References

1. Ajmone Marsan, M., Chiola, G.: On Petri nets with deterministic and exponentially distributed firing times. In: Rozenberg, G. (ed.) APN 1987. LNCS, vol. 266, pp. 132–145. Springer, Heidelberg (1987)
2. Amparore, E.G., Donatelli, S.: DSPN-Tool: a new DSPN and GSPN solver for GreatSPN. In: QEST 2010, pp. 79–80 (2010)
3. Amparore, E.G., Donatelli, S.: Revisiting matrix-free solution of Markov regenerative processes. Numerical Linear Algebra with Applications 18(6), 1067–1083 (2011)
4. Balbo, G., Beccuti, M., De Pierro, M., Franceschinis, G.: First Passage Time Computation in Tagged GSPNs with Queue Places. The Computer Journal (2010)
5. Buchholz, P.: Markov matrix market, http://ls4-www.cs.tu-dortmund.de/download/buchholz/struct-matrix-market.html
6. Buchholz, P.: Hierarchical structuring of superposed GSPNs. IEEE Trans. Software Eng. 25(2), 166–181 (1999)
7. Buchholz, P., Ciardo, G., Donatelli, S., Kemper, P.: Complexity of memory-efficient Kronecker operations with applications to the solution of Markov models. INFORMS Journal on Computing 12(3), 203–222 (2000)
8. Buchholz, P., Kemper, P.: Hierarchical reachability graph generation for Petri nets. Formal Methods in System Design 21(3), 281–315 (2002)
9. Choi, H., Kulkarni, V.G., Trivedi, K.S.: Markov regenerative stochastic Petri nets. Performance Evaluation 20(1-3), 337–357 (1994)
10. Ciardo, G., Lindemann, C.: Analysis of Deterministic and Stochastic Petri Nets. In: PNPM 1993, pp. 160–169. IEEE Computer Society (1993)
11. Ciardo, G., Lüttgen, G., Siminiceanu, R.: Saturation: An efficient iteration strategy for symbolic state-space generation. In: Margaria, T., Yi, W. (eds.) TACAS 2001. LNCS, vol. 2031, pp. 328–342. Springer, Heidelberg (2001)
12. German, R.: Iterative analysis of Markov regenerative models. Perform. Eval. 44(1-4), 51–72 (2001)
13. Plateau, B., Fourneau, J.M.: A methodology for solving Markov models of parallel systems. J. Parallel Distrib. Comput. 12(4), 370–387 (1991)
14. Saad, Y., Schultz, M.H.: GMRES: a generalized minimal residual algorithm for solving nonsymmetric linear systems. SIAM Journal on Scientific and Statistical Computing 7(3), 856–869 (1986)
15. Stewart, W.J.: Introduction to the numerical solution of Markov chains. Princeton University Press (1994)
16. Vicario, E., Sassoli, L., Carnevali, L.: Using stochastic state classes in quantitative evaluation of dense-time reactive systems. IEEE Transactions on Software Engineering 35(5), 703–719 (2009)

Low-Rank Tensor Methods
for Communicating Markov Processes

Daniel Kressner[1] and Francisco Macedo[1,2]

[1] EPF Lausanne, SB-MATHICSE-ANCHP,
Station 8, CH-1015 Lausanne, Switzerland
{daniel.kressner,francisco.macedo}@epfl.ch
[2] IST, Alameda Campus, Av. Rovisco Pais, 1, 1049-001 Lisbon, Portugal

Abstract. Stochastic models that describe interacting processes, such as stochastic automata networks, feature a dimensionality that grows exponentially with the number of processes. This state space explosion severely impairs the use of standard methods for the numerical analysis of such Markov chains. In this work, we discuss the approximation of solutions by matrix product states or, equivalently, by tensor train decompositions. Two classes of algorithms based on this low-rank decomposition are proposed, using either iterative truncation or alternating optimization. Our approach significantly extends existing approaches based on product form solutions and can, in principle, attain arbitrarily high accuracy. Numerical experiments demonstrate that the newly proposed algorithms are particularly well suited to deal with pairwise neighbor interactions.

1 Introduction

Markov processes featuring high-dimensional state spaces regularly arise when modelling processes that interact with each other. Termed communicating Markov processes in [4], this class includes queuing networks, stochastic automata networks, and stochastic Petri nets. The need for considering the joint probability distribution for a network of stochastic processes is responsible for the exponential growth of the state space dimension, which severely impairs the numerical analysis of such Markov processes. For example, all standard iterative solvers [2] for addressing the linear system or, equivalently, the eigenvalue problem needed for determining the stationary probability distribution have a complexity that scales at least linearly with the state space dimension.

The high dimensionality of the state space is usually coped with by either performing model reduction or by exploiting the rich structure of the transition rate matrix in the numerical solution procedure. Product form solutions represent a particularly popular reduction technique and have been successfully used in a wide range of applications. The basic idea of this reduction is to yield a system for which the stationary distribution factorizes into a product of distributions for the individual processes. This reduced system then allows for a much less expensive numerical treatment. General techniques for arriving at product form

G. Norman and W. Sanders (Eds.): QEST 2014, LNCS 8657, pp. 25–40, 2014.

solutions are described, e.g., in [18, Ch. 6]. Extensive work has been done on finding conditions under which such a product form approach applies; see [10,9] for some recent results. However, its practical range of applicability is still limited to specific subclasses. A rather different approach is based on the observation that the transition rate matrix of a communicating Markov process can often be represented by a short sum of Kronecker products [25]. This property can then be exploited when performing matrix-vector multiplications or constructing preconditioners [19,20] to reduce the cost of iterative solvers significantly. Still, the complexity scales linearly with the state space dimension.

The approach proposed in this paper can be viewed as a combination of the two approaches above, performing (nonlinear) model reduction along with the iterative solution. For this purpose, we exploit the fact that a vector containing joint probabilities can be naturally rearranged into a tensor. This then allows us to use established low-rank tensor approximation techniques [12]. Such an approach has already been considered by Buchholz [6], using the so called CANDECOMP/PARAFAC (CP) decomposition. As explained below, this decomposition may not always be the best choice as it does not exploit the topology of interactions. We therefore propose the use of matrix product states (MPS) or, equivalently, tensor train (TT) decompositions. Note that MPS and related low-rank tensor decompositions have already been used for simulating stochastic systems [14,15]. The novelty of our contribution consists of explicitly targeting the computation of the stationary distribution for a finite-dimensional communicating Markov process, developing and comparing different algorithmic approaches.

The rest of this work is organized as follows. In Section 2, we will introduce the low-rank tensor decompositions that will subsequently, in Section 3, be used to develop efficient algorithms for computing stationary distributions. Section 4 compares the performance of these algorithms for two popular examples.

2   Low-Rank Decompositions of Tensors

In this section, we discuss low-rank decompositions for compressing vectors. When considering, for example, a network of d communicating finite state Markov processes, the vector π containing the stationary distribution has length $n_1 n_2 \cdots n_d$, where n_μ denotes the number of states in the μth process for $\mu = 1, \ldots, d$. Quite naturally, the entries of this vector can be rearranged into an $n_1 \times \cdots \times n_d$ array, defining a dth order tensor $\mathcal{X} \in \mathbb{R}^{n_1 \times \cdots \times n_d}$. The entries of $\mathcal{X}$ are denoted by

$$\mathcal{X}_{i_1, i_2, \ldots, i_d}, \quad 1 \leq i_\mu \leq n_\mu, \quad \mu = 1, \ldots, d.$$

The opposite operation is denoted by $\mathrm{vec}(\mathcal{X})$, which stacks the entries of $\mathcal{X}$ back into a long vector, so that the indices of $\mathcal{X}$ are sorted in lexicographical order.

For $d = 2$, $\mathcal{X}$ becomes a matrix and there is a unique notion of rank, which can be computed by the singular value decomposition (SVD) [11]. The extension of this concept to $d > 2$ is by no means unique, and several different notions of low rank decompositions for tensors have been developed; see [16] for an overview.

The *CP decomposition* takes the form

$$\text{vec}(\mathcal{X}) = \sum_{r=1}^{R} u_r^{(1)} \otimes u_r^{(2)} \otimes \cdots \otimes u_r^{(d)} = \sum_{r=1}^{R} \bigotimes_{\mu=1}^{d} u_r^{(\mu)}, \tag{1}$$

where each $u_r^{(\mu)}$ is a vector of length n_μ, and $\otimes$ denotes the usual Kronecker product. The tensor rank of $\mathcal{X}$ is the smallest R admitting such a decomposition. The individual entries of (1) are given by

$$\mathcal{X}_{i_1,i_2,\ldots,i_d} = \sum_{r=1}^{R} u_{r,i_1}^{(1)} u_{r,i_2}^{(2)} \cdots u_{r,i_d}^{(d)}, \quad 1 \le i_\mu \le n_\mu, \quad \mu = 1,\ldots,d.$$

This reveals that tensors of rank $R = 1$ are closely related to the concept of product form solutions.

The approximation of stationary distributions with tensors of tensor rank $R > 1$ has been proposed by Buchholz [6], using a combination of greedy low-rank and alternating optimization schemes. Despite certain theoretical drawbacks [16], the CP decomposition has been observed to perform fairly well in practice. On the other hand, this decomposition aims at a simultaneous separation of all d processes. This ignores the topology of interactions between processes and may result in relatively high ranks.

2.1 TT Decomposition / Matrix Product States

Low-rank decompositions that benefit from the locality of interactions are well established in computational physics, in particular for simulating quantum systems [28,30]. A *matrix product state* (MPS) takes the form

$$\mathcal{X}_{i_1,\ldots,i_d} = G_1(i_1) \cdot G_2(i_2) \cdots G_d(i_d), \quad G_\mu(i_\mu) \in \mathbb{R}^{r_{\mu-1} \times r_\mu}, \tag{2}$$

where $r_0 = r_d = 1$. In the numerical analysis community, this decomposition was proposed in [22,24] and termed *tensor train* (TT) decomposition. During the last few years, MPS/TT have been used in a wide range of applications; see [12] for a literature survey.

To give a concrete example, consider a vector of the form $\text{vec}(\mathcal{X}) = x^{(1)} \otimes y^{(1)} \otimes y^{(1)} \otimes y^{(1)} + y^{(2)} \otimes x^{(2)} \otimes y^{(2)} \otimes y^{(2)} + y^{(3)} \otimes y^{(3)} \otimes x^{(3)} \otimes y^{(3)} + y^{(4)} \otimes y^{(4)} \otimes y^{(4)} \otimes x^{(4)}$ for arbitrary vectors $x^{(\mu)}, y^{(\mu)} \in \mathbb{R}^n$. The coefficients in the decomposition (2) of $\mathcal{X}$ are then given by

$$G_1(i_1) = \begin{bmatrix} x^{(1)}(i_1) & y^{(1)}(i_1) \end{bmatrix} \in \mathbb{R}^{1 \times 2}, \quad G_2(i_2) = \begin{bmatrix} y^{(2)}(i_2) & 0 \\ x^{(2)}(i_2) & y^{(2)}(i_2) \end{bmatrix} \in \mathbb{R}^{2 \times 2},$$

$$G_3(i_3) = \begin{bmatrix} y^{(3)}(i_3) & 0 \\ x^{(3)}(i_3) & y^{(3)}(i_3) \end{bmatrix} \in \mathbb{R}^{2 \times 2}, \quad G_4(i_4) = \begin{bmatrix} y^{(4)}(i_4) \\ x^{(4)}(i_4) \end{bmatrix} \in \mathbb{R}^{2 \times 1}.$$

The TT decomposition (2) is closely connected to certain matricizations of $\mathcal{X}$. Let $X^{(1,\ldots,\mu)}$ denote the $(n_1 \cdots n_\mu) \times (n_{\mu+1} \cdots n_d)$ matrix obtained by reshaping the entries of $\mathcal{X}$ such that the indices $(i_1,\ldots,i_\mu)$ become row indices and $(i_{\mu+1}, \ldots, i_d)$ become column indices, sorted, again, in lexicographical order. This operation has an important interpretation for a network of d stochastic automata: Merging the automata $1, \ldots, \mu$ into one subsystem and the automata $\mu+1, \ldots, d$ into another subsystem yields a network with only two (aggregated) automata, see Figure 1. The tensor corresponding to this network has order 2

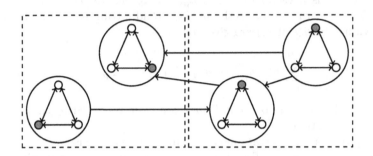

Fig. 1. Network of 4 interacting automata. The aggregation of automata into two disjoint subsystems (indicated by the dashed lines) corresponds to a matricization of the tensor.

and coincides with the matrix $X^{(1,\ldots,\mu)}$. The following result is well known; we include its proof to illustrate the use of matricizations.

Lemma 1. *Given a TT decomposition* (2), *the matricization* $X^{(1,\ldots,\mu)}$ *of* $\mathcal{X}$ *satisfies* $\mathrm{rank}\big(X^{(1,\ldots,\mu)}\big) \leq r_\mu$ *for every* $\mu = 1, \ldots, d$.

Proof. Let us define the so called interface matrices

$$\begin{aligned}
\mathbf{G}_{\leq\mu} &= \big[G_1(i_1)\cdots G_\mu(i_\mu)\big] \in \mathbb{R}^{(n_1\cdots n_\mu)\times r_\mu} \\
\mathbf{G}_{\geq\mu+1} &= \big[G_{\mu+1}(i_{\mu+1})\cdots G_d(i_d)\big]^T \in \mathbb{R}^{(n_{\mu+1}\cdots n_d)\times r_\mu}
\end{aligned} \tag{3}$$

Then, by definition (2), we have $X^{(1,\ldots,\mu)} = \mathbf{G}_{\leq\mu}\mathbf{G}_{\geq\mu+1}^T$, which implies the statement of the lemma since each of the factors has only r_μ columns. □

Motivated by Lemma 1, the tuple $\big(\mathrm{rank}\big(X^{(1,\ldots,\mu)}\big)\big)_{\mu=1,\ldots,d-1}$ is called the *TT rank* of $\mathcal{X}$. Following the proof of Lemma 1, successive low-rank factorizations can be used to compute the TT decomposition (2) of a given tensor with TT rank $\mathbf{r} = (r_1, \ldots, r_{d-1})$. More importantly, by using truncated SVD, we can truncate any tensor $\mathcal{X}$ to a tensor $\mathcal{X}_\mathbf{r}$ of rank $\mathbf{r}$ verifying

$$\|\mathcal{X} - \mathcal{X}_\mathbf{r}\|_2^2 \leq \sum_{\mu=1}^{d-1} \sum_{j=r_\mu+1}^{n_\mu} \sigma_j\big(X^{(1,\ldots,\mu)}\big)^2, \tag{4}$$

where the 2-norm of a tensor is defined via its vectorization and $\sigma_j(\cdot)$ denotes the jth largest singular value of a matrix. In our algorithms, we will use a related procedure described in [22] to truncate a tensor in TT decomposition to lower TT rank. Properly implemented, this procedure takes $O(dNR^3)$ operations, where R and N are upper bounds on the TT ranks and the sizes of the tensor $\mathcal{X}$, respectively.

2.2   Example

One fundamental assumption in all low-rank techniques is that the data (in our case, the stationary probability distribution) can actually be well approximated by a low-rank tensor. According to (4), this can be quantified by considering the singular values of the matricizations – good accuracy can only be expected when these singular values decay sufficiently fast.

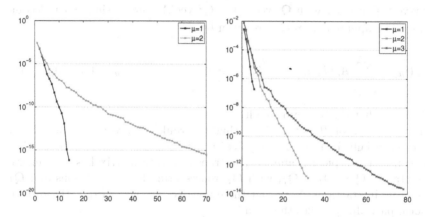

Fig. 2. Singular values of $X^{(1,\ldots,\mu)}$ for the stationary distribution of the large overflow model, for $d = 4$ and $n_\mu = 20$ (left plot) as well as $d = 6$ and $n_\mu = 6$ (right plot). Each graph shows the singular values of the corresponding matricization sorted in non-increasing order.

Figure 2 displays the singular values for the relevant matricizations of two instances of the large overflow model, one of the queuing networks studied in Section 4. Note that only the first half of the matricizations are considered, as the singular values of the second half display similar behavior. It turns out that the singular values have a very fast decay, showing that this model can be well approximated with very low TT ranks.

2.3   Operator TT Decomposition / Matrix Product Operator

Our algorithms require the repeated application of the transition matrix $\mathbf{Q} \in \mathbb{R}^{(n_1\cdots n_d)\times(n_1\cdots n_d)}$ to a vector in TT decomposition. It is therefore important to represent $\mathbf{Q}$ in a form that allows to perform this operation efficiently. In principle, a sum of T Kronecker products

$$\mathbf{Q} = \sum_{i=1}^{T} \bigotimes_{\mu=1}^{d} Q_\mu^{(i)} \tag{5}$$

would be suitable for this purpose. However, the application of such a $\mathbf{Q}$ will increase all TT ranks by the factor T. Taking into account that low TT ranks are decisive to attain efficiency, we prefer to work with a more suitable representation of $\mathbf{Q}$.

An *operator TT decomposition*, also called *matrix product operator* (MPO), takes the form

$$\mathbf{Q}_{(i_1,\ldots,i_d),(j_1,\ldots,j_d)} = A_1(i_1,j_1) \cdot A_2(i_2,j_2) \cdots A_d(i_d,j_d), \tag{6}$$

where $A_\mu(i_\mu,j_\mu) \in \mathbb{R}^{t_{\mu-1} \times t_\mu}$, and $t_0 = t_d = 1$. This is simply the TT decomposition (2) applied to $\mathrm{vec}(\mathbf{Q})$, merging each index pair (i_μ, j_μ) in lexicographical order into a single index ranging from 1 to n_μ^2. The tensor $\widetilde{\mathcal{X}}$ resulting from a matrix-vector product with $\mathbf{Q}$, $\mathrm{vec}(\widetilde{\mathcal{X}}) = \mathbf{Q}^T \mathrm{vec}(\mathcal{X})$, has a simple TT decomposition. The updated cores $\widetilde{G}_\mu$ are given by

$$\widetilde{G}_\mu(i_\mu) = \sum_{j_\mu=1}^{n_\mu} A_\mu(j_\mu,i_\mu) \otimes G_\mu(j_\mu) \in \mathbb{R}^{r_{\mu-1}t_{\mu-1} \times r_\mu t_\mu}, \quad i_\mu = 1,\ldots,n_\mu, \tag{7}$$

which shows that the TT ranks multiply.

The advantage of (6) over (5) is that the ranks t_μ are often much lower than T, especially in the case of pairwise neighbor interactions. To see this, let us consider a typical example. The transition rate matrix has the general representation $\mathbf{Q} = \mathbf{Q}_L + \mathbf{Q}_I$, with $\mathbf{Q}_L$ representing local transitions and $\mathbf{Q}_I$ representing interactions. Note that these matrices are never explicitly formed. The local part always takes the form

$$\mathbf{Q}_L = \sum_{\mu=1}^{d} I_{n_1} \otimes \cdots \otimes I_{n_{\mu-1}} \otimes L_\mu \otimes I_{n_{\mu+1}} \otimes \cdots \otimes I_{n_d},$$

where $L_\mu \in \mathbb{R}^{n_\mu \times n_\mu}$ contains the local transitions in the μth system and I_{n_μ} denotes the identity matrix of size n_μ. For the large overflow model considered in Section 4.1, L_μ contains arrival and departure rates. In this example, the matrix $\mathbf{Q}_I$ is given by

$$\sum_{1 \leq \mu_1 < \mu_2 \leq d} I_d \otimes \cdots \otimes I_{\mu_2+1} \otimes B_{\mu_2} \otimes C_{\mu_2-1} \otimes \cdots \otimes C_{\mu_1+1} \otimes D_{\mu_1} \otimes I_{\mu_1-1} \otimes \cdots \otimes I_1$$

for some matrices $B_\mu, C_\mu, D_\mu \in \mathbb{R}^{n_\mu \times n_\mu}$ directly obtained from the description of the model. Hence, the CP-like decomposition (5) of the operator $\mathbf{Q}$ requires $T = \frac{d(d+1)}{2}$ terms. On the other hand, by direct calculation, it can be shown that $\mathbf{Q}$ admits an operator TT decomposition (6) with the cores

$$A_1(i_1,j_1) = \begin{bmatrix} L_1(i_1,j_1) & B_1(i_1,j_1) & I_1(i_1,j_1) \end{bmatrix}, \quad A_d(i_d,j_d) = \begin{bmatrix} I_d(i_d,j_d) \\ D_d(i_d,j_d) \\ L_d(i_d,j_d) \end{bmatrix},$$

and

$$A_\mu(i_\mu, j_\mu) = \begin{bmatrix} I_\mu(i_\mu, j_\mu) & 0 & 0 \\ C_\mu(i_\mu, j_\mu) & B_\mu(i_\mu, j_\mu) & 0 \\ L_\mu(i_\mu, j_\mu) & D_\mu(i_\mu, j_\mu) & I_\mu(i_\mu, j_\mu) \end{bmatrix}, \quad \mu = 2, \ldots, d-1.$$

In particular, the corresponding TT ranks t_μ are 3, for *any* value of d.

3   Low-Rank Algorithms

In terms of the transition rate matrix $\mathbf{Q}$, the computation of the stationary distribution $\mathbf{x}$ requires the solution of the problem

$$\mathbf{Q}^T\mathbf{x} = 0, \quad \mathbf{e}^T\mathbf{x} = 1, \tag{8}$$

where $\mathbf{e}$ is the vector of all ones. Clearly, a solution of (8) also solves the equivalent constrained least-squares problem

$$\min_{\mathbf{x}}\{\|\mathbf{Q}^T\mathbf{x}\|_2^2 : \mathbf{e}^T\mathbf{x} = 1\}. \tag{9}$$

Remark 1. For an irreducible ergodic Markov Chain, each of the problems (8) and (9) admits a unique solution [26, Ch. 4]. Reducible Markov Chains appear quite frequently, e.g., this is the case for the Kanban control model considered in Section 4.2. In this example, the states are combinations from 0 to the maximum capacity of the queue for all types of customers. Some states are not well-defined due to restrictions on the total number of customers. To address this, one can eliminate combinations not verifying the restriction by ordering and filtering the states in a specific way described in [5].

3.1   Truncated Power Method

A time discretization with step size $\Delta t > 0$ results in the matrix

$$\mathbf{P} = I + \Delta t \mathbf{Q},$$

and (8) becomes equivalent to the eigenvalue problem

$$\mathbf{P}^T\mathbf{x} = \mathbf{x}, \quad \mathbf{e}^T\mathbf{x} = 1. \tag{10}$$

As explained in [29, Ch. 1], $\Delta t > 0$ needs to be chosen sufficiently small to ensure that $\mathbf{P}$ is a non-periodic stochastic matrix. More specifically, it is required that $0 < \Delta t < |q_{ii}|^{-1}$ holds for every diagonal entry of $\mathbf{Q}$. In our experiments, we chose $\Delta t = 0.9999 \times (\max_i |q_{ii}|)^{-1}$. Note that $\mathbf{Q}$ is only given implicitly, in terms of the low-rank Kronecker representations (5) or (6) and it may therefore not be feasible to evaluate all diagonal entries of $\mathbf{Q}$. In such cases, we use an upper bound on $|q_{ii}|$ instead. For example, when a Kronecker product representation of the form (5) is available, an inexpensive upper bound is given by

$$\prod_{\mu=1}^{d} \max_{i=1,\ldots,n_\mu} |[Q_\mu^{(1)}]_{ii}| + \cdots + \prod_{\mu=1}^{d} \max_{i=1,\ldots,n_\mu} |[Q_\mu^{(T)}]_{ii}|.$$

Originally proposed in [3], in combination with the CP decomposition, the truncated power method is probably the simplest low-rank tensor method for solving the eigenvalue problem (10). Starting with a random vector $\mathbf{x}_0$ of rank 1, the jth iteration consists of computing

$$\widetilde{\mathbf{x}}_{j+1} = \mathsf{trunc}\big(\mathbf{x}_j + \mathsf{trunc}\big(\Delta t \mathbf{Q}^T \mathbf{x}_j\big)\big), \quad \mathbf{x}_{j+1} = \widetilde{\mathbf{x}}_{j+1}/(\mathbf{e}^T \widetilde{\mathbf{x}}_{j+1}).$$

Note that all iterates $\mathbf{x}_j$ are represented in terms of their TT decomposition. Since both the application of $\mathbf{Q}$ and the addition of two tensors increase the TT ranks, it is mandatory to repeatedly use the operation trunc, which truncates the tensor back to lower TT ranks within a specified tolerance; see also Section 2.1. This truncation destroys the property $\mathbf{e}^T \widetilde{\mathbf{x}}_{j+1} = 1$, which would otherwise be preserved by the power method. The required inner product $\mathbf{e}^T \widetilde{\mathbf{x}}_{j+1}$ to restore this normalization is inexpensive since $\mathbf{e}$ corresponds to a tensor of rank 1 [22].

3.2   Alternating Least-Squares (ALS)

A rather different approach consists of constraining the optimization problem (8) further to the set $\mathcal{M}_{\mathbf{r}}$ of tensors having fixed TT ranks $\mathbf{r} = (r_1, \ldots, r_{\mu-1})$:

$$\min_{\mathbf{x} \in \mathcal{M}_{\mathbf{r}}} \{\|\mathbf{Q}^T \mathbf{x}\|_2^2 : \mathbf{e}^T \mathbf{x} = 1\}. \tag{11}$$

Due to the nonlinear nature of $\mathcal{M}_{\mathbf{r}}$, the solution of this optimization problem is by no means simple. On the other hand, each individual core $G_\mu(\cdot)$ enters the TT decomposition (2) linearly and therefore the optimization with respect to a single core (while keeping all other cores fixed) should pose no problem. To discuss the resulting alternating least-squares (ALS) method, we require some additional notation.

Recalling the definition (3) of interface matrices for a given TT decomposition, let

$$\mathbf{G}_{\neq\mu} = \mathbf{G}_{\leq\mu-1} \otimes I_{n_\mu} \otimes \mathbf{G}_{\geq\mu+1}.$$

Then

$$\mathbf{x} = \mathbf{G}_{\neq\mu} \cdot \mathbf{g}_\mu, \quad \text{where} \quad \mathbf{g}_\mu = \mathrm{vec}\left([G_\mu(1), \ldots, G_\mu(n_\mu)]\right) \in \mathbb{R}^{n_\mu r_{\mu-1} r_\mu}.$$

We additionally assume that the columns of $\mathbf{G}_{\leq\mu-1}$ and $\mathbf{G}_{\geq\mu+1}$ are orthonormal, which can always be achieved by the orthogonalization procedure described in [22]. In turn, the optimization of the μth core in ALS is performed by minimizing

$$\|\mathbf{Q}^T \mathbf{x}\|_2^2 = \left\|\mathbf{Q}^T\big(\mathbf{G}_{\neq\mu} \cdot \mathbf{g}_\mu\big)\right\|_2^2 = \mathbf{g}_\mu^T \big(\mathbf{G}_{\neq\mu}^T \mathbf{Q}\mathbf{Q}^T \mathbf{G}_{\neq\mu}\big)\mathbf{g}_\mu$$

among all $\mathbf{g}_\mu$ satisfying $\mathbf{e}^T \mathbf{G}_{\neq\mu} \mathbf{g}_\mu = (\mathbf{G}_{\neq\mu}^T \mathbf{e})^T \mathbf{g}_\mu = 1$. This problem can be addressed by solving the linear system

$$\begin{bmatrix} \mathbf{G}_{\neq\mu}^T \mathbf{Q}\mathbf{Q}^T \mathbf{G}_{\neq\mu} & \widetilde{\mathbf{e}} \\ \widetilde{\mathbf{e}}^T & 0 \end{bmatrix} \begin{bmatrix} \mathbf{g}_\mu \\ \lambda \end{bmatrix} = \begin{bmatrix} \mathbf{0} \\ 1 \end{bmatrix} \qquad (12)$$

where $\widetilde{\mathbf{e}} = \mathbf{G}_{\neq\mu}^T \mathbf{e}$ is computed by performing contractions.

After (12) having been solved, the μth core of $\mathbf{x}$ is updated with $\mathbf{g}_\mu$. The ALS method continues with a left-orthonormalization of this core, followed by the optimization of the $(\mu + 1)$th core. A half sweep of ALS consists of processing all cores from the left to the right until reaching $\mu = d$. Similarly, the second half sweep of ALS consists of processing all cores from the right to the left until reaching $\mu = 1$. Two subsequent half sweeps constitute a full sweep of ALS.

Remark 2. For small ranks, the reduced problem (12) is solved by explicitly forming the matrix $\mathbf{L} = \mathbf{G}_{\neq\mu}^T \mathbf{Q}\mathbf{Q}^T \mathbf{G}_{\neq\mu}$ of size $n_\mu r_{\mu-1} r_\mu$ and utilizing a direct solver. When this becomes infeasible due to large TT ranks, we should resort to an iterative solver. As explained in [17, Sec. 3.3] the matrix $\mathbf{L}$ can be represented by a short sum of Kronecker products which greatly speeds up matrix-vector multiplications with $\mathbf{L}$. Moreover, by a standard manipulation [21, Ch. 16], the symmetric indefinite linear system (12) can be transformed into a symmetric positive definite linear system. This allows us to apply the conjugate gradient (CG) method to (12). In our experiments, we sometimes observed the CG method to suffer from slow convergence. However, the construction of effective preconditioners to accelerate convergence for such reduced problems appears to be a rather challenging task.

3.3   Local Enrichment with Residuals

An obvious drawback of ALS is that all TT ranks of $\mathbf{x}$ need to be chosen *a priori*. This can be addressed by starting with very low ranks, say $r_\mu \equiv 1$, and subsequently increasing the ranks by enriching the cores with additional information. For the optimization problem (11), we can mimic gradient descent by incorporating information from the residual $\mathbf{Q}^T \mathbf{x}$. Such an approach has been suggested by White [31] for eigenvalue problems, and extended to linear systems by Dolgov and Savostyanov [7,8].

As explained in Section 2.3, the tensor corresponding to $\mathbf{Q}^T \mathbf{x}$ is again in TT decomposition, with the updated cores $\widehat{G}_\mu(\cdot)$ defined in (7). The basic idea of enrichment is to augment the μth core of $\mathbf{x}$ with $\widetilde{G}_\mu(\cdot)$. However, augmenting all cores in each microstep of ALS would increase the TT ranks way too quickly and result in an inefficient algorithm. (Note that one microstep corresponds to the optimization of one core in the TT decomposition.) To avoid this, this procedure is modified as follows:

1. Before the enrichment, the residual is truncated to low TT ranks, within a specified accuracy.

2. In the μth microstep of the first half sweep, only the cores $G_\mu(\cdot)$ and $G_{\mu+1}(\cdot)$ are augmented. In the μth microstep of the second half sweep, only the cores $G_\mu(\cdot)$ and $G_{\mu-1}(\cdot)$ are augmented. This local enrichment benefits the next microstep while it avoids that all TT ranks increase simultaneously. The rank of the enrichment is fixed to 3.

We refrain from a more detailed discussion of these points and refer to [8,17] for more details. Following [8], the resulting variant of ALS that allows for rank adaptation is called AMEn.

4   Numerical Experiments

We have implemented all methods described in Section 3 – truncated power method, ALS, and AMEn – in MATLAB version 2013b. These methods are based on the TT decomposition and we make use of functionality from the TT-Toolbox [23]. For reference, we have also implemented the ALS method for the CP decomposition [6], based on functionality from the tensor toolbox [1]. To distinguish between the two different ALS methods, we will denote them by ALS-TT and ALS-CP.

We nearly always used a direct solver for addressing the reduced problems (12) in ALS-TT and AMEn. Only for one example, we applied the CG method with an adaptive stopping criterion. The CG method is terminated when the residual norm has been decreased by a factor of 100 relative to the current residual norm or when the number of iterations exceeds 1000, which was frequently the case.

The truncated power method and AMEn rely on repeated low-rank truncations to prevent excessive rank growth. If not stated otherwise, the tolerance for performing this truncation is set adaptively to $\|\mathbf{r}\|_2$ for the power method and to $\|\mathbf{r}\|_2/100$ for AMEn. The TT ranks used in ALS-TT are all set to the same value, the maximum of the TT ranks of the approximate solution produced by AMEn. In ALS-CP, we extensively tried different tensor ranks, choosing the one exhibiting the best performance.

All computations were performed on a 12-core Intel Xeon CPU X5675 3.07 GHz with 192 GB RAM, under 64-Bit Linux version 2.6.32.

4.1   Large Overflow Model

This model was used to test ALS-CP in [6]. It consists of a queuing network where customers can arrive in each of $d = 6$ ordered queues according to independent Poisson processes and, in case the queue of arrival is full, they get blocked and try to enter the next one, until finding one that is not full. If all queues turn out to be full, the customer gets lost. The corresponding interactions are thus associated with functional transitions, where the arrival rates of the queues depend on the state of the previous ones (being full or not). Services follow an exponential distribution with mean 1. The arrival rates for queues 1 to 6 are given by 1.2, 1.1, 1.0, 0.9, 0.8 and 0.7, respectively. Choosing a maximum capacity of 10 for each queue leads to a total of $11^6 = 1771561$ states. As the capacity of each system is not infinite and the arrival rates depend on the states of other queues, this model does not constitute a Jackson network [13] and therefore the product-form approach cannot be applied.

Table 1. Obtained execution times (in seconds) and corresponding ranks for the large overflow model with respect to different accuracies

Accuracy	1×10^{-5}		4.6×10^{-6}		2.2×10^{-6}		1×10^{-6}	
	time	rank	time	rank	time	rank	time	rank
ALS-CP	169.0	120	430.8	180	1330.3	240	3341.4	300
Power method	35.9	12.03	47.5	13.67	61.6	15.09	78.8	17.32
ALS-TT	10.7	13.92	11.4	13.92	22.8	15.67	35.9	16.84
AMEn	4.9	12.96	4.9	12.96	9.3	14.85	12.5	15.85
AMEn (CG)	266.5	11.65	–		–		–	

We have applied ALS-CP to the large overflow model for tensor ranks ranging from 120 to 300. The obtained accuracy (in terms of the residual norm) is displayed in the first row of Table 1. We then iterate the truncated power method, ALS-TT, and AMEn until the same accuracy is reached. The obtained results are shown in rows 4–7 of Table 1. For TT decompositions, the rank refers to the effective rank r_{eff}. Following [27], r_{eff} is determined as the solution of the equation

$$\text{memory}(r_1, \ldots, r_{d-1}) = \text{memory}(r_{\text{eff}}, \ldots, r_{\text{eff}}),$$

where $\text{memory}(r_1, \ldots, r_{d-1})$ is the amount of storage needed by a TT decomposition with TT ranks $r_1, \ldots, r_{d-1}$. All algorithms attained the target accuracy except for AMEn (CG), which refers to AMEn with the reduced problems solved by the CG method. In this case, convergence stagnated and AMEn was stopped after 10 full sweeps, reaching an accuracy of 3.622×10^{-5}. Due to the lack of an effective preconditioner for CG, the reduced problems could not be solved to sufficient accuracy. This renders the use of the CG method unattractive and we therefore do not consider it in the following experiments.

The execution times of ALS-CP are much higher compared to the times of the TT-based algorithms. This is mainly due to the fact that the number of terms in the Kronecker product representation (5) of the operator is $T = \frac{d(d+1)}{2} = 21$, while the TT ranks of the operator TT decomposition (6) are 3, see Section 2.3.

Among the TT-based algorithms, AMEn is clearly the best and the power method appears to offer the poorest performance. This picture changes, however, when demanding higher accuracies. This results in higher TT ranks and consequently makes the solution of the reduced problems in AMEn and ALS-TT more expensive. For example, when demanding an accuracy of 10^{-8}, the truncated power method requires 275 seconds, ALS-TT 1120 seconds, and AMEn 367 seconds.

To give some additional indication on the accuracy of the different methods, Table 1 displays two quantities that can be easily extracted from the approximate solution: (1) the mean population of the last queue, and (2) the probability that the last queue is full. The reference values for these two quantities given in [6] are (1) 3.949 and (2) 7.66×10^{-2}, respectively. Note that the inaccuracy of AMEn (CG) is due to convergence failures of the CG method.

Table 2. Mean population ('mean') and probability of being full ('prob') for the last queue with respect to the different accuracies used in Table 1

Accuracy	1×10^{-5}		4.6×10^{-6}		2.2×10^{-6}		1×10^{-6}	
	mean	prob	mean	prob	mean	prob	mean	prob
ALS-CP	3.951	7.64×10^{-2}	3.956	7.68×10^{-2}	3.950	7.67×10^{-2}	3.950	7.66×10^{-2}
Power method	3.854	7.33×10^{-2}	3.904	7.51×10^{-2}	3.927	7.59×10^{-2}	3.939	7.63×10^{-2}
ALS-TT	3.887	7.44×10^{-2}	3.879	7.43×10^{-2}	3.912	7.56×10^{-2}	3.936	7.62×10^{-2}
AMEn	3.879	7.42×10^{-2}	3.879	7.42×10^{-2}	3.939	7.63×10^{-2}	3.939	7.63×10^{-2}
AMEn (CG)	2.990	3.87×10^{-2}	–		–		–	

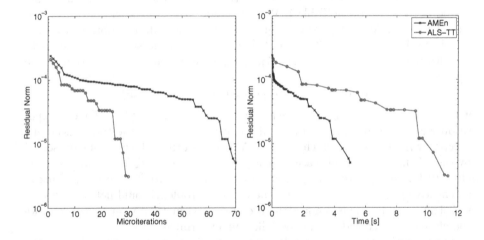

Fig. 3. Evolution of the residual norm with respect to the number of microsteps and with respect to the accumulated execution time (in seconds) for ALS-TT and AMEn applied to the large overflow model

The left plot of Figure 3 shows how the residual norm evolves for AMEn and ALS-TT during the microsteps. Not surprisingly, ALS-TT converges faster as it uses the maximal TT ranks right from the first sweep. This picture changes significantly when considering the evolution with respect to the accumulated execution time in the right plot of Figure 3. AMEn operates with much smaller TT ranks during the first sweeps, making them less expensive and resulting in a smaller total execution time despite the fact that the total number of sweeps is higher.

Reduced arrival rates. We now divide the arrival rates by 2. This significantly decreases the execution times. To attain an accuracy of about 10^{-8}, the truncated power method now requires only 14 seconds, ALS-TT requires 11 seconds, and AMEn 4.9 seconds. With 32 seconds, ALS-CP is still the slowest method.

Exploring AMEn. Since the experiments above clearly reveal the advantages of AMEn, we investigate its performance for high-dimensional problems in more detail. For this purpose, we reduce the maximum capacity to 2 customers in each queue, that is $n_\mu \equiv 3$, and target an accuracy of 10^{-5}. We vary the number of queues from $d = 7$ to $d = 24$. Service rates are 1 for all queues while the arrival rates have been adjusted to $\frac{12-0.1\times i}{8}$ for the ith queue. To avoid convergence problems, the tolerance for performing low-rank truncations has been decreased to $\|\mathbf{r}\|_2/1000$.

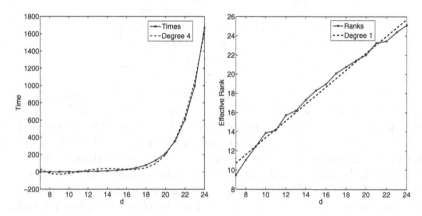

Fig. 4. Execution time in seconds (left plot) and effective rank (right plot) for AMEn applied to the large overflow model with $d = 7, 8, \ldots, 24$ queues

Figure 4 reveals how the execution time and the effective TT rank grow as d increases. The TT rank appears to grow less than linearly, while the execution time seems to grow proportionally with d^4. This is due to the need for solving the reduced problems, which require $O(r^6)$ operations when using a direct solver. Note that the largest Markov chain is associated with a total of $3^{24} \approx 2.82 \times 10^{11}$ states, which is clearly infeasible for standard solvers. In contrast, AMEn requires less than 2000 seconds to obtain a good approximation of the solution.

4.2 Kanban Control Model

We now consider the Kanban control model [5], where customers arrive in the first queue, being then served in sequence until the last queue, leaving the network afterwards. We assume an infinite source – there is automatically a new arrival when the first queue is not full. A customer that finishes the service and experiences that the next queue is full needs to wait in the current queue.

We choose $d = 12$ queues, each with a maximum capacity of one customer. Services follow an exponential distribution with mean 1. The time spent traveling from one queue to the next is also exponential, with expected value $\frac{1}{10}$. For the queues 2 to 11, one needs to distinguish the type of customer (already served, waiting to move to the next queue, or no customer), so that there are 3 possible states. We therefore have, in total, $2 \times 3^{10} \times 2 = 236196$ states. The resulting matrix operator has TT rank 4 and CP rank $3d - 2 = 34$. The tolerance for performing low-rank truncations is again decreased to $\|\mathbf{r}\|_2/1000$ for AMEn.

Table 3. Obtained execution times (in seconds) and corresponding ranks for the Kanban control model

	Accuracy	time	rank
ALS-CP	2×10^{-3}	3024.7	200
Power method	1×10^{-5}	314.5	20.95
ALS-TT	1×10^{-5}	624.9	39.51
AMEn	1×10^{-5}	200.8	28.02

Table 3 shows that the execution times for the Kanban control model are clearly higher than the ones for the large overflow model, see Table 1, despite the fact that there is a smaller total number of states. For ALS-CP, we needed to stop the algorithm before reaching the target accuracy of 10^{-5}, to avoid an excessive consumption of computational resources. In contrast to the large overflow model, this model features interactions between non-consecutive queues, as a customer has to go through all queues before leaving the system. This non-neighbor interaction can be expected to lead to the observed higher TT ranks. The largest TT rank of the approximate solution for this model is 47, while the TT ranks for the large overflow model never exceed 30, even for $d = 24$.

5   Conclusions and Future Work

We have proposed three algorithms for approximating stationary distributions by low-rank TT decompositions: the truncated power method, ALS and AMEn. Preliminary numerical experiments with stochastic automata networks that feature a fairly simple topology of interactions demonstrate that these methods, in particular AMEn, can perform remarkably well for very high-dimensional problems. They clearly outperform an existing approach based on CP decompositions.

Having obtained approximate stationary distributions in low-rank TT decomposition, it is comparably cheap to compute statistics of the solution. For example, the marginal probabilities of a particular queue are obtained by partial contractions with the vector of all ones, involving a cost that is negligible compared to the rest of the computation.

A bottleneck of ALS-TT and AMEn is that they use a direct solver for the reduced problems, which becomes rather expensive for larger TT ranks. Our future work will therefore focus on the further development of preconditioned iterative methods to address this issue. Being able to deal with larger TT ranks will then also allow us to study networks with a more complicated topology of interactions.

While our algorithms are designed to guarantee that the entries of the approximate solution sum up to one, the preservation of nonnegativity is another crucial aspect that remains to be addressed.

Acknowledgments. We thank Peter Buchholz for helpful discussions and the reviewers for their constructive feedback.

References

1. Bader, B.W., Kolda, T.G., et al.: Matlab tensor toolbox version 2.5 (January 2012), http://www.sandia.gov/~tgkolda/TensorToolbox/
2. Barrett, R., Berry, M., Chan, T.F., Demmel, J.W., Donato, J., Dongarra, J.J., Eijkhout, V., Pozo, R., Romine, C., van der Vorst, H.: Templates for the Solution of Linear Systems: Building Blocks for Iterative Methods. SIAM, Philadelphia (1994)
3. Beylkin, G., Mohlenkamp, M.J.: Algorithms for numerical analysis in high dimensions. SIAM J. Sci. Comput. 26(6), 2133–2159 (2005)
4. Buchholz, P.: A framework for the hierarchical analysis of discrete event dynamic systems, Habilitationsschrift, Fachbereich Informatik, Universität Dortmund (1996)
5. Buchholz, P.: An adaptive decomposition approach for the analysis of stochastic petri nets. In: Proceedings of the 2002 International Conference on Dependable Systems and Networks, DSN 2002, pp. 647–656. IEEE Computer Society, Washington, DC (2002)
6. Buchholz, P.: Product form approximations for communicating Markov processes. Performance Evaluation 67(9), 797–815 (2010)
7. Dolgov, S.V., Savostyanov, D.V.: Alternating minimal energy methods for linear systems in higher dimensions. Part I: SPD systems. arXiv preprint arXiv:1301.6068 (2013)
8. Dolgov, S.V., Savostyanov, D.V.: Alternating minimal energy methods for linear systems in higher dimensions. Part II: Faster algorithm and application to nonsymmetric systems. arXiv preprint arXiv:1304.1222 (2013)
9. Fourneau, J.-M.: Product form steady-state distribution for stochastic automata networks with domino synchronizations. In: Thomas, N., Juiz, C. (eds.) EPEW 2008. LNCS, vol. 5261, pp. 110–124. Springer, Heidelberg (2008)
10. Fourneau, J.M., Plateau, B., Stewart, W.J.: An algebraic condition for product form in stochastic automata networks without synchronizations. Perform. Eval. 65(11-12), 854–868 (2008)
11. Golub, G.H., Van Loan, C.F.: Matrix Computations, 3rd edn. Johns Hopkins University Press, Baltimore (1996)
12. Grasedyck, L., Kressner, D., Tobler, C.: A literature survey of low-rank tensor approximation techniques. GAMM-Mitt. 36(1), 53–78 (2013)

13. Jackson, J.R.: Jobshop-like queueing systems. Manage. Sci. 50(12 suppl.), 1796–1802 (2004)
14. Johnson, T.H., Clark, S.R., Jaksch, D.: Dynamical simulations of classical stochastic systems using matrix product states. Phys. Rev. E 82, 036702 (2010)
15. Kazeev, V., Khammash, M., Nip, M., Schwab, C.: Direct solution of the chemical master equation using quantized tensor trains. Technical Report 2013-04, Seminar for Applied Mathematics, ETH Zürich (2013)
16. Kolda, T.G., Bader, B.W.: Tensor decompositions and applications. SIAM Review 51(3), 455–500 (2009)
17. Kressner, D., Steinlechner, M., Uschmajew, A.: Low-rank tensor methods with subspace correction for symmetric eigenvalue problems. MATHICSE preprint 40.2013, EPF Lausanne, Switzerland (2013)
18. Kulkarni, V.G.: Introduction to modeling and analysis of stochastic systems, 2nd edn. Springer Texts in Statistics. Springer, New York (2011)
19. Langville, A.N., Stewart, W.J.: The Kronecker product and stochastic automata networks. J. Comput. Appl. Math. 167(2), 429–447 (2004)
20. Langville, A.N., Stewart, W.J.: A Kronecker product approximate preconditioner for SANs. Numer. Linear Algebra Appl. 11(8-9), 723–752 (2004)
21. Nocedal, J., Wright, S.J.: Numerical Optimization, 2nd edn. Series in Operations Research. Springer (2006)
22. Oseledets, I.V.: Tensor-train decomposition. SIAM J. Sci. Comput. 33(5), 2295–2317 (2011)
23. Oseledets, I.V.: TT-Toolbox Version 2.2 (2012), `https://github.com/oseledets/TT-Toolbox`
24. Oseledets, I.V., Tyrtyshnikov, E.E.: Breaking the curse of dimensionality, or how to use SVD in many dimensions. SIAM J. Sci. Comput. 31(5), 3744–3759 (2009)
25. Plateau, B., Fourneau, J.-M., Lee, K.-H.: PEPS: A package for solving complex Markov models of parallel systems. In: Puigjaner, R., Potier, D. (eds.) Modeling Techniques and Tools for Computer Performance Evaluation, pp. 291–305. Springer US (1989)
26. Ross, S.M.: Introduction to probability models., 7th edn. Harcourt/Academic Press, Burlington (2000)
27. Savostyanov, D.V.: QTT-rank-one vectors with QTT-rank-one and full-rank Fourier images. Linear Algebra Appl. 436(9), 3215–3224 (2012)
28. Schollwöck, U.: The density-matrix renormalization group in the age of matrix product states. Ann. Physics 326, 96–192 (2011)
29. Stewart, W.J.: Introduction to the Numerical Solution of Markov Chains. Princeton University Press, Princeton (1994)
30. White, S.R.: Density matrix formulation for quantum renormalization groups. Phys. Rev. Lett. 69, 2863–2866 (1992)
31. White, S.R.: Density matrix renormalization group algorithms with a single center site. Phys. Rev. B 72, 180403 (2005)

A Statistical Approach
for Computing Reachability of Non-linear
and Stochastic Dynamical Systems*

Luca Bortolussi[1] and Guido Sanguinetti[2,3]

[1] DMG, University of Trieste, and CNR/ISTI, Pisa, Italy
[2] School of Informatics, University of Edinburgh, UK
[3] SynthSys, Centre for Synthetic and Systems Biology, University of Edinburgh, UK

Abstract. We present a novel approach to compute reachable sets of dynamical systems with uncertain initial conditions or parameters, leveraging state-of-the-art statistical techniques. From a small set of samples of the true reachable function of the system, expressed as a function of initial conditions or parameters, we emulate such function using a Bayesian method based on Gaussian Processes. Uncertainty in the reconstruction is reflected in confidence bounds which, when combined with template polyhedra ad optimised, allow us to bound the reachable set with a given statistical confidence. We show how this method works straightforwardly also to do reachability computations for uncertain stochastic models.

1 Introduction

Reachability computation for dynamical systems is a core topic in theoretical computer science, with important implications for the analysis and control of applications such as embedded systems or cyber-physical systems [17]. In most applications, dynamical systems are described by a set of (nonlinear) differential equations or differential inclusions. Reachability computation for this class of systems is intimately connected with understanding the *impact of uncertainty* on the dynamics: the interest normally lies in understanding how uncertainty on the initial conditions, or on model parameters, or on the dynamics itself (e.g. by external bounded perturbations) propagates during the time evolution of the system. Usually, one is interested in computing if the system remains stable even in presence of uncertainty, or how the dynamics is modified. Uncertainty is treated in this context *non-deterministically*, i.e. in the least committal way, as no assumption is made on the distribution of initial conditions or of parameter values. This problem is relevant for many complex systems in areas as diverse as systems biology, economics, computer networks. All those systems show a non-linear dynamics depending on parameters which are seldom known exactly, hence uncertainty is the norm rather than the exception.

* G.S. acknowledges support from the ERC under grant MLCS 306999. L.B. acknowledges partial support from EU-FET project QUANTICOL (nr. 600708) and by FRA-UniTS.

G. Norman and W. Sanders (Eds.): QEST 2014, LNCS 8657, pp. 41–56, 2014.

There are many approaches in the literature that tackle the reachability problem for uncertain dynamical systems, which can be roughly divided in two classes: exact over-approximation methods and simulation-based methods [26]. The first class of methods manipulates directly sets of states, finitely represented, for instance, as polytopes [17], ellipsoids [24], or zonotopes [22]. The dynamics of the system is lifted at the set level, so that one computes the evolution of the reachable set under the action of the dynamics. As exact computations may be too expensive or unfeasible, one is usually satisfied with an over-approximation of the real reachable set, which is formally guaranteed to contain all reachable points. Most of the methods of this class deal with linear systems, which are common in many engineering applications. For this class of systems, efficient methods exist [22], scaling up to several hundred dimensions, which are implemented in tools like SpaceEx [20]. Over-approximation methods for non-linear systems, instead, are much less developed, mainly because the problem becomes much more difficult. Here we recall hybridization [15], which is based on the idea of splitting the state space into many regions, and linearising the dynamics in each of them, with formal guarantees on the introduced error. Other methods work for restricted classes of non-linear systems, like multi-affine [4] and polynomial systems [14,29]. However, all these methods suffer from the curse of dimensionality.

An alternative approach is offered by simulation methods, which simulate some trajectories of the system, for fixed parameter values and initial conditions, and try to infer the reachable set from a finite number of them. An approach based on bisimulation metrics is [23], while [19] uses a barrier certificate apporach. The method of [18], instead, leverages the sensitivity of the system with respect to initial conditions to (approximatively) compute how a covering of the set of uncertain initial conditions is enlarged or reduced during the dynamics, obtaining a covering of the reachable set at a given time T. Other methods rely on the law of large numbers to give asymptotic guarantees; for example, the procedure of [5,13,16] constructs an under-approximation of the reachable set, relying on a Monte-Carlo sampling method which is guaranteed to cover the whole reachable set as the number of sampled points goes to infinity.

The method we propose in this paper is also *simulation-based* and it is not exact. Our idea, however, is to tackle the problem from a statistical perspective, trying to *statistically control the error* introduced by considering only a finite number of sample points belonging to the reachable set. To this end, we leverage recent non-parametric Bayesian methods [28] developed in the context of machine learning to treat the error in a sound way. Hence, we will produce a statistical over-approximation: the true reachable set will be contained in the output set of our procedure with a given confidence.

In general, we advocate the use of statistical methods in this context: machine learning is concerned with reconstructing plausible models/ functions from few data points, hence it is natural that techniques developed in this context can be used to infer the reachable set from few simulations. More specifically, we will consider Gaussian Process emulation [28], which has been previously used as a tool for sensitivity analysis of complex simulated models [27]. The appeal of these methods lies in their ability to quickly provide good, statistically sound, and analytically computable approximations of functions that can only be observed, possibly with noise, in few points. To our knowledge, this is the first work in which these methods are used to compute reachable sets of dynamical systems.

However, our approach has further advantages. First of all, it features a sound treatment of noise. This means that the samples used to construct our approximation do not have to be exact observations of the true reachability function (up to numerical error), but they can be noisy estimates. A classical scenario when this happens is if we are interested in understanding the impact of uncertainty in stochastic models, for instance on their average or on any other statistics. Our approach applies *tout-court* also to this class of problems, for stochastic processes as diverse as Continuous Time Markov Chains or Stochastic Differential Equations [21]. Furthermore, the treatment of noise allows us to keep consistently into account the numerical error incurred during simulation. Third, this method does not make any assumption about the nature of the dynamics: it works for arbitrary (smooth) non-linear systems. We stress that our approach applied to stochastic systems estimates the impact of uncertainty on the average or other statistics, but it does not compute the probability of reaching a certain set of states as a function of model parameters, although similar ideas can be used for this problem, see [9]. In this respect, this is work is different from e.g. [10], which uses bayesian statistics to compute the reachability probability.

The paper is organised as follows: in Section 2, we formalise the problem we are tackling and introduce the SIR model from epidemiology [1], which will serve as the running example throughout the paper. In Section 3, we discuss our method from a high-level perspective, postponing the technical details to Section 4. We then present some experimental results on the epidemic spreading model in Section 5, comparing the performance of our approach with the state-of-the-art. We conclude the paper in Section 6 by discussing the merits and limitations of the proposed method, as well as a number of possible extensions to the work.

2   Problem Definition

We consider a deterministic (dynamical) system with values in $\mathbb{R}^n$:

$$\boldsymbol{x} : \mathbb{R}_{\geq 0} \times D \times D_0 \to \mathbb{R}^n, \quad \boldsymbol{x}(t, \boldsymbol{p}, \boldsymbol{x_0}) \in \mathbb{R}^n.$$

The dynamics of $\boldsymbol{x}$ depends on a tuple of d parameters $\boldsymbol{p} \in D$ and on a tuple of initial conditions $\boldsymbol{x_0} \in D_0$. Here D is a compact subset of $\mathbb{R}^{d_p}$, usually a hyper-rectangle of the form $D = \prod_{i=1}^{d_p} [\underline{p}_i, \overline{p}^i]$, while D_0 is a compact subset of $\mathbb{R}^{d_n}$, where $d_n \leq n$. The idea is that D and D_0 contain only the parameters and initial conditions that we consider in the uncertainty analysis.

Furthermore, we assume to have a simulation algorithm which can provide us with an observation $\boldsymbol{y}^i$ of the process $\boldsymbol{x}$ at specific points $\boldsymbol{z}^i = (t^i, \boldsymbol{p}^i, \boldsymbol{x_0}^i) \in \mathbb{R}_{\geq 0} \times D \times D_0$. Such observations can be *noisy*, i.e. we assume

$$\boldsymbol{y}^i = \boldsymbol{x}(t^i, \boldsymbol{p}^i, \boldsymbol{x_0}^i) + \epsilon^i,$$

where ϵ^i is a multivariate Gaussian random variable with mean zero and covariance matrix $C = \sigma^i I_n$, with σ^i being a vector of input-dependent standard deviations. This is a very general setting, which encompasses the two cases in which we are interested:

1. $x(t, p, x_0)$ is the flow of a differential equation defined by the vector field $F(x, t, p) \in \mathbb{R}^n$, i.e. the solution of the initial value problem

$$\frac{d}{dt}x(t, p, x_0) = F(x(t, p, x_0), t, p), \quad x(0, p, x_0) = x_0.$$

In this case the observations y^i are essentially noise free, being produced by a numerical integration algorithm. However, we will set σ^i equal to the absolute tolerance of the numerical integrator [11], so that we can keep numerical imprecisions into account. In particular, we assume that this error has a Gaussian distribution, which can therefore take larger values than the absolute tolerance. This is a common assumption, which simplifies the statistical treatment [27].

2. $x(t, p, x_0)$ is the average or another statistic (e.g. the variance, or a quantile of the distribution) of a stochastic process $X : \mathbb{R}_{\geq 0} \to Dist(\mathbb{R}^n)$, like a Continuous Time Markov Chain or a stochastic differential equation [21]. In this case, we assume that x has been estimated by a batch of N simulations, and we assume the error term ϵ^i to be the standard deviation of the estimator, either computed analytically or by bootstrapping. Here the central limit theorem guarantees the Gaussian assumption about the noise to be approximately correct.

In this paper, we will consider two bounded variants of the reachability problem for $x(t, p, x_0)$. We assume to have uncertainty in both (some) parameters and (some) initial conditions. We fix a finite time horizon $T > 0$.

Reachability at time T: given noisy observations $y^1, \ldots, y^m$ at points $z^1 = (T, p^1, x_0^1), \ldots, z^m = (T, p^m, x_0^m)$, compute a set $\hat{R}^T$ such that the set $R^T = \{x(T, p, x_0) \mid p \in D, x_0 \in D_0\}$ is contained in $\hat{R}^T$. In this case, the input space is $\mathcal{D} = \{T\} \times D \times D_0$ and has dimension $d = d_p + d_n$.

Reachability between times T_1 and T_2 ($[T_1, T_2] \subseteq [0, T]$, $T_1 < T_2$): given noisy observations $y^1, \ldots, y^m$ at points $z^1 = (t_1, p^1, x_0^1), \ldots, z^m = (t_m, p^m, x_0^m)$, with $t_i \in [T_1, T_2]$ $\forall i$, compute a set $\hat{R}^{[T_1, T_2]}$ such that the set $R^{[T_1, T_2]} = \{x(t, p, x_0) \mid t \in [T_1, T_2], p \in D, x_0 \in D_0\}$ is contained in $\hat{R}^{[T_1, T_2]}$. In this case, the input space is $\mathcal{D} = [\mathcal{T}_\infty, \mathcal{T}_\in] \times \mathcal{D} \times \mathcal{D}$, and has dimension $d = d_p + d_n + 1$.

More precisely, in this paper we will be concerned with a statistical variant of the above reachability problem:

Definition 1 (Statistical Reachability). *Given noisy observations $y^1, \ldots, y^m$, compute a set $\hat{R}$ that over-approximates (as tightly as possible) the exact reachable set R with a given confidence level α.*[1]

---

[1] This means such that if we repeat the reachability algorithm, R is contained in $\hat{R}$ approximatively α percent of the times.

2.1 A Working Example: Epidemics in a LAN Computer Network

We consider a simple variation of the classical SIR infection model [1], in which an initial population of N susceptible nodes can be infected either from outside the network (e.g. by receiving an infected email message) or by the active spread of the virus by infected computers in the network. Infected nodes can be patched, and become immune to the worm for some time, after which they are susceptible again (for instance, to a new version of the worm).

This system is modelled as a set of ODEs or as a population CTMC. In both cases, the state space is described by a vector $\boldsymbol{X}$ of three variables, returning the number of nodes that are in the susceptible (X_S), infected (X_I), and patched state (X_R). The dynamics can be described by a list of transitions, in the population CTMC style (see e.g. [7]). Each transition is defined by an update vector ν, indicating how each variable is modified, and by a rate function. From this description, it is straightforward to construct a CTMC or an ODE [7].

External infection: $\nu_e = (-1, 1, 0)$, with rate function $r_e(\boldsymbol{X}) = k_e X_S$;
Internal infection: $\nu_i = (-1, 1, 0)$, with rate function $r_i(\boldsymbol{X}) = k_i X_S X_I$;
Patching: $\nu_r = (0, -1, 1)$, with rate function $r_r(\boldsymbol{X}) = k_r X_I$;
Immunity loss: $\nu_e = (1, 0, -1)$, with rate function $r_s(\boldsymbol{X}) = k_s X_R$;

The vector field of the ODEs is obtained as $F(\boldsymbol{x}) = \sum \nu_j r_j(\boldsymbol{x})$. In this case, we can further rely on the conservation law $x_S + x_I + x_R = N$ to get rid of one variable, obtaining

$$\begin{cases} \frac{d}{dt} x_S = -k_e x_S - k_i x_S x_I + k_s (N - x_S - x_I) \\ \frac{d}{dt} x_I = k_e x_S + k_i x_S x_I - k_r x_I \end{cases}$$

3 Statistical Computation of Reachability

We introduce now a high level overview of the approach to solve the statistical reachability problem, postponing the presentation of the actual statistical tools we are using to the next section. We will start by introducing few key concepts and then discussing the algorithm to compute reachability. The method will turn out to be the same for both the bounded reachability problems we discussed in the previous section.

3.1 Bayesian Emulation and Optimisation

Bayesian emulation is a method to reconstruct a function $f : \mathcal{D} \to \mathbb{R}$, given some noisy observations of its values at a certain set of points. This approach is similar to regression [6], but is based on assigning an a-priori distribution $p(g)$ to functions g of a properly chosen class $\mathcal{F}$, which is assumed to approximate well the real function we are observing. Given (noisy) observations $\boldsymbol{y} = \boldsymbol{y}^1, \ldots, \boldsymbol{y}^m$ of function values at points $\boldsymbol{Z} = \boldsymbol{z}^1, \ldots, \boldsymbol{z}^m$, one can in principle compute a posterior distribution on the class of functions $\mathcal{F}$, $p(g|\boldsymbol{Z}, \boldsymbol{y})$ through Bayes theorem. In practice, this is seldom possible analytically, except in the important case when the prior distribution over functions is a *Gaussian Process* (see Section 4.1).

While approximating an unknown input/ output relation is the domain of regression, one is often interested in also optimizing the response function, assuming a smooth dependence on some tunable parameters. The availability of a probability distribution over the response function offers considerable advantages for optimisation, as it enables a principled trade-off between *exploitation*, i.e. local search near high scoring input values, and *exploration*, i.e. selection of new input points in areas where the function values is very uncertain. Bayesian optimization refers to a class of algorithms that leverage these concepts for optimisation: at the core is the definition of an *acquisition function*, a rule which dictates which new input points to query, considering the emulated function and uncertainty. In Section 4.4 we discuss one such algorithm, which has recently been shown to be provably globally convergent [30].

3.2 Templates

The second ingredient we need for our reachability algorithm are *polytope templates*, similarly to [12,29]. The idea is simply to consider a polytope defined as $Wz \leq a$, W a $h \times n$ matrix, i.e. as the intersection of h linear hyperplanes of the form $w_j z \leq a$, where w_j is the vector of coefficients of the normal to the hyperplane (and the jth row of W). The idea behind a polytope template is that it defines a family of polytopes of $\mathbb{R}^d$ as a function of the coefficients a, and the goal of the reachability algorithm becomes to find the optimal vector a such that the polytope over-approximates the true reachable set as tightly as possible. This can be rephrased as the following optimisation problem:

$$\forall j = 1, \ldots, h, \ a_j = \max_{z \in \mathcal{D}} \{w_j x(z)\}.$$

Note that the matrix W can be interpreted as defining the linear observables $Wx(t, p, x_0)$ of the dynamical system x. Templates can be chosen according to different heuristics. For instance, in [12], the authors take the polytope defining the convex hull of (a subset of) the sample points. In [29], instead, the authors start from a polytope bounding the initial conditions and refine it dynamically taking into account rotational effects of the vector field.

3.3 Reachability Algorithm

We consider a template polytope defined by the matrix W, and try to upper bound as tightly as possible the linear observables $Wx(t, p, x_0)$. The function $x(t, p, x_0)$ is unknown, and so are the h observables $r_j(t, p, x_0) = w_j x(t, p, x_0)$. The basic idea of our method is to emulate r_j from (noisy) observations $w_j y^1, \ldots, w_j y^m$ at points $z^1, \ldots, z^m \in \mathcal{D}$. To compute the over-approximation of the reachable set, instead of finding a tight upper-bound of the prediction function $\hat{r}_j$, we work with the upper confidence bound $\hat{r}_j(z) + \varphi_\alpha \sigma_{\hat{r}_j}(z)$ for a given confidence level α. From a high level perspective, our algorithm works as follows:

- Let W be a template polytope, $z^1, \ldots, z^m$ be the input points, $y^1, \ldots, y^m$ be the noisy observations of $x(z^i)$, and α be the confidence level.
- For each $j = 1, \ldots, h$
    1. Construct the emulation $\hat{r}_j$ of $w_j x(z)$ and obtain its maximum $\hat{\rho}_j$ (with associated uncertainty $\hat{\sigma}_{\hat{\rho}_j}$) via Bayesian optimisation.
    2. Compute $\hat{a}_j = \hat{\rho}_j + \varphi_\alpha \hat{\sigma}_{\hat{\rho}_j}$ with φ_α a constant chosen such that the true maximum value is below $\hat{a}_j$ with confidence α.
- The over-approximation $\hat{R}_\alpha$ with confidence α of the reachable set is the polytope $Wz \leq \hat{a}$

We note here that the confidence level α is computed *conditionally* on the Bayesian optimisation algorithm having obtained a true maximum. While (some) Bayesian optimisation algorithms come with strong convergence guarantees, these generally only imply convergence in probability (where the probability of convergence can be tuned to be as close as desired to 1). Therefore, our algorithm, as all statistical algorithms, only returns an approximation which is with high probability correct.

4   Statistical Methods

We now explain in more detail the statistical tools that we use to construct the emulation function and to solve the global optimisation problem. We first introduce Gaussian Processes (GP) and GP-regression, then comment on the (hyper)parameters of the GP kernel, and finally discuss a GP-based Bayesian optimisation.

4.1   Gaussian Processes

Gaussian Processes (GPs) are a natural extension of the multivariate normal distribution to infinite dimensional spaces of functions [28]. A GP is a probability measure over the space of continuous functions (over a suitable input space $\mathcal{D}$) such that the random vector obtained by evaluating a sample function at *any* finite set of points $z^1, \ldots, z^N \in \mathcal{D}$ follows a multivariate normal distribution. A GP is uniquely defined by its *mean* and *covariance* (or *kernel*) functions, denoted by $\mu(z)$ and $k(z, z')$. By definition, we have that for every finite set of points

$$f \sim \mathcal{GP}(\mu, k) \leftrightarrow \mathbf{f} = \left(f(z^1), \ldots, f(z^N) \right) \sim \mathcal{N}(\boldsymbol{\mu}, K) \qquad (1)$$

where $\boldsymbol{\mu}$ is the vector obtained evaluating the mean function μ at every point, and K is the matrix obtained by evaluating the covariance function k at every pair of points. The prior mean function is often taken to be identically zero (a non-zero mean can be added post-hoc to the predictions w.l.o.g., cf Section 4.3).

The choice of covariance function is an important modelling decision, as it essentially determines the type of functions which can be sampled from a GP (more precisely, it can assign prior probability zero to large subsets of the space of continuous functions). In this paper, we consider the *ARD Gaussian covariance* (ARD-G)

$$k(z, z') = \gamma \exp \left[-\sum_{i=1}^{n} \frac{(z_i - z_i')^2}{\lambda_i^2} \right]. \tag{2}$$

The main motivation for this choice is that sample functions from a GP with ARD-G covariance are with probability one infinitely differentiable functions [9]. For more details, we refer the interested reader to [28].

GPs depend on a number of hyperparameters through their mean and co-variance functions: the ARD-G covariance depends on the *amplitude* γ and a *lengthscale* λ_i for each coordinate. For the mean, we will model it as a linear combination of basis functions $\mu(x) = \boldsymbol{\beta}^T \boldsymbol{\varphi}(x)$, with the coefficients constituting additional hyperparameters.

4.2   GP Regression and Prediction

Suppose now that we are given a set of noisy observations $\mathbf{y}$ of the function value at input values $\mathbf{Z} = z^1, \dots, z^N$, distributed around an unknown true value $f(\mathbf{z})$ with spherical Gaussian noise of variance $\sigma^2(z)$. We are interested in determining how these observations influence our belief over the function value at a further input value z^* where the function value is unobserved.

By using the basic rules of probability and matrix algebra, we have that the predictive distribution at z^* is again Gaussian with mean μ^* and variance k^*:

$$\mu^* = \boldsymbol{k}(z^*) \hat{K}_N^{-1} \mathbf{y} \tag{3}$$

$$k^* = k(z^*, z^*) - \boldsymbol{k}(z^*) \hat{K}_N^{-1} \boldsymbol{k}(z^*)^T. \tag{4}$$

where $\boldsymbol{k}(z^*) = \left(k(z^*, z^1), \dots, k(z^*, z^N) \right)$, and $\hat{K}_N$ is the *Gram matrix*, obtained by evaluating the covariance function at each pair of training points and adding the diagonal matrix $diag(\sigma^2(z^1), \dots, \sigma^2(z^N))$. Notice that the first term on the r.h.s of equation (4) is the prior variance at the new input point; therefore, we see that the observations lead to a *reduction* of the uncertainty over the function value at the new point. The variance however returns to the prior variance when the new point becomes very far from the observation points.

Equation (3) warrants two important observations: first, as a function of the new point z^*, μ^* is a linear combination of a finite number of *basis* functions $k(z^*, z)$ centred at the observation points. Secondly, the posterior mean at a fixed z^* is a linear combination of the observed values, with weights determined by the specific covariance function used. For the ARD-G covariance, input points further from the new point z^* are penalised exponentially, hence contribute less to the predicted value.

4.3   Hyperparameters

GP emulation depends in a complex way on the hyperparameters. The length-scales, in particular, play a delicate role, because they control the Lipschitz constants of the predictive function along each coordinate, and as such they govern the roughness of the predictive function and the quality of the emulation. Furthermore, amplitude is also crucial as it controls the magnitude of the variance, and hence the quality and correctness of the over-approximation. In order to automatically choose the value of hyperparameters, we can estimate them from the set of samples.

Lengthscale Optimisation. In order to choose the lengthscales, we can treat the estimation of hyperparameters as a model selection problem, which can be tackled in a maximum-likelihood perspective by optimising the model evidence

$$p(y|Z, \lambda) = \int p(y|\hat{f}, Z)p(\hat{f}|Z, \lambda)d\hat{f}.$$

$p(y|Z, \lambda)$ is the marginal likelihood of the observed data; its dependence on the hyperparameters can be calculated analytically [28]; optimisation of the marginal likelihood is then both fast and theoretically well founded. Optimisation of model evidence is done relying on a simple global optimisation scheme, running several times a Newton-Raphson local optimisation algorithm from random starting points [11]. Experimentally, we found that the model evidence tends to behave quite well, with a global optimum having a large basin of attraction, a phenomenon often observed in practice [28], so that few runs of the optimisation routine suffice.

Amplitude Optimisation. If we fix the lengthscales, then the amplitude and the mean coefficients can be estimated in a fully Bayesian way, following the approach of [27]. Assuming a (non-informative) normal inverse gamma prior on the coefficients β and the amplitude γ, one obtains that the posterior GP-process follows pointwise a t-student distribution with $m - 2$ degrees of freedom (where m is the number of observation points), with mean $m^*(z)$ in a generic point z equal to

$$m^*(z) = h(z)\hat{\beta} + k(z)^t \hat{K}_N^{-1}(y - H\hat{\beta}),$$

and covariance equal to

$$k^*(z, z') = \hat{\alpha} \Big[k(z, z') - k(z)^t \hat{K}_N^{-1} k(z) +$$

$$(h(z) - k(z)^t \hat{K}_N^{-1} H)(H^t \hat{K}_N^{-1} H)(h(z') - k(z')^t \hat{K}_N^{-1} H)^t \Big],$$

where $\hat{K}^N$ is the Gram matrix and $k(z)$ is the kernel evaluated between the point z and all the observation points (see Section 4.2), , $H^t = (h(z_1)^t, \ldots, h(z_m)^t)$ is matrix of the basis function of the prior mean evaluated at observation points, $\hat{\beta} = V^*(H^t \hat{K}_N^{-1} y)$, $V^* = (H^t \hat{K}_N^{-1} H)$, and the predictive posterior amplitude $\hat{\gamma}$ is equal to

$$\hat{\gamma} = \Big[y^t \hat{K}_N^{-1} y - \hat{\beta}^t (V^*)^{-1} \hat{\beta} \Big] / (m - 2).$$

4.4 Adaptive Bayesian Optimisation

The main computational challenge behind the algorithm sketched in Section 3.3 is to find the upper confidence bound (UCB) for the maximum of the reachability function. However, if the variance of the GP-emulation is large, this can result in a very poor over-approximation. To refine the method, we would like to force the UCB of the maximum to be as close as possible to the maximum of the real function $w_i x(z)$. Staten otherwise, we would like to reduce the variance to a minimum around the true maximum, and control the variance in other regions of the state space to guarantee that the maximum we have identified is close to the true one. This suggests an adaptive sampling scheme, which is driven by the information contained in the samples already collected. Indeed, this is achieved by state-of-the-art Bayesian optimisation algorithms, which realise a trade-off between exploration of the state space and exploitation of the search around the more promising areas. This is achieved by observing that the new regions to explore are those with high variance (as we have few information about them), while the regions to exploit are those with high value of the predictive function. This reflectes in an adaptive optimisation scheme of the UCB, which provably converges to the global optimum of the true function [30]. The GP-UCB optimisation algorithm works as follows:

- Sample an initial set of input points $z^1, \ldots, z^m$ with a good coverage of the input space (e.g., from a regular grid or by Latin Hypercube Sampling strategies [31]), and compute the (noisy) value of the function to optimise.
- Iterate until the convergence criterion is met:
  1. Construct the GP emulator $\hat{f}$ and its standard deviation $\sigma_{\hat{f}}$ from the current set of t input points.
  2. Find $z^{t+1} = \mathrm{argmax}_{z \in \mathcal{D}} \hat{f}(z) + \beta_t \sigma_{\hat{f}}(z)$ using e.g. a multi-start gradient-based optimisation strategy [11].[2]
  3. Compute the true function y^{t+1} at z^{t+1} and add this to the input points.

Convergence is usually declared when no improvement in the maximum value is observed in few attempts. β_t has to be taken as a logarithmically divergent sequence of positive values to enforce convergence [30].

5   Experimental Results

Consider the computer network epidemics model discussed in Section 2.1, fixing the number of nodes to $N = 100$. Model parameters are set to $k_e = 0.01$, $k_i = 0.1/N$, $k_r = 0.05$, $k_s = 0.005$, and initial conditions to $x_S^0 = N$, $x_I^0 = 0$. We will first focus on the differential equation model, and investigate two scenarios:

1. uncertainty in the internal infection parameter k_i, with $k_i \in [0.05/N, 0.15/N]$;

---

[2] Practically, we evaluate the UCB function on a random grid, choose the best k points, and run a local optimisation starting from these points.

2. uncertainty in the initial conditions, assuming $(x_S^0, x_I^0) \in [0.8 \cdot N, 0.9 \cdot N] \times [0, 0.1 \cdot N]$

In case 1, we will investigate the reachability problem in $[0, T]$ fixing $T = 300$, a time sufficient for the ODE system to reach equilibrium for the whole range of the parameter k_i considered. In case 2, instead, we will investigate the reachability problem at given times $T \in \{3, 6, \ldots, 300\}$. We will then consider a stochastic variant of the network epidemics model, interpreted as a population CTMC [7]. Here we will investigate the following scenario:

3. uncertainty in infection and recovery parameters, with $k_i \in [0.0005, 0.0015]$ and $k_r \in [0.01, 0.1]$.

In particular, we will be interested in the impact of the uncertainty on the average of the steady state distribution, and in bounding the region of the (x_S, x_I) plane in which the steady state is contained with 95% probability for any possible value of (k_i, k_r).

Draft software to recreate the experiments is available from the authors on request; we are currently working on a distributable and more efficient implementation, which will be released soon.

Template Selection. Our approach depends crucially on a choice of a polytope template. We will combine basic information about the system with a statistical analysis of the set of points based on Principal Component Analysis (PCA, [6]). The idea is to include in the template the following hyperplanes:

$$W^t = \begin{pmatrix} 1 & -1 & 0 & 0 & 1 & -1 \\ 0 & 0 & 1 & -1 & 1 & -1 \end{pmatrix} w_a{}^t w_b{}^t \end{pmatrix}$$

The first four columns correspond to upper and lower bounds on the variables x_S and x_I, respectively. The fifth and sixth columns upper and lower bound the quantity $x_S + x_I$, which must always be less than N. Finally, the last two columns w_a and w_b correspond to the two principal components (PC) of the sample points of $x(t, p, x^0) = (x_S(t, p, x^0), x_I(t, p, x^0))$. In this way, we bound along the two orthogonal directions better explaining the variation of data.

Case 1: Reachability in $[0, T]$. To approximate the reachable set in $[0, T]$, we start by computing the solution $(x_S(t, p, x^0), x_I(t, p, x^0))$ in an equi-spaced grid of 5 points for k_i and 50 points for the time interval $[0, T]$. Then we ran a hyperparameter optimisation for the lengthscale, fixing the amplitude to 1, and estimate the amplitude using the Bayesian approach of the previous section. As for the prior mean, we considered a set of polynomial basis functions of degree 2, i.e. $h(z) = (1, z_1, z_2, z_1^2, z_1 z_2, z_2^2)$. We defined a template from the PCA, using the samples up to time 150, when the system has almost stabilised on the steady state (more points would bias the PCA by placing more and more weight on the steady state value). In Figure 1 left, we see the method applied to the whole set of sample points at once, using the GP-UCB algorithm, initialised with the grid

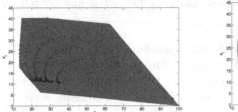
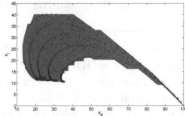

Fig. 1. Reachability in $[0, T]$. Left: overapproximation by the PCA polytope for the whole set of reachable points in $[0, T]$, varying k_i. Blue dots are sample points from the true function, for different values of time and k_i. Right: over-approximation by the PCA polytope, splitting the computation of the reachable set in several blocks.

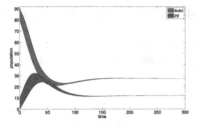
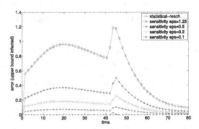

Fig. 2. Reachability at time T, starting from uncertain initial conditions. Left: upper and lower bound for variables x_S and x_T, as a function of time. Right: Comparison between our method and the sensitivity-based one of [18], for different grid sizes.

of 5×50 initial samples. Convergence is reached after few additional function evaluations.

In Figure 1 right, instead, we split as in [12] the reachability problem into blocks, to increase the precision of the overapproximation. More specifically, we decompose the interval $[0, 300]$ in the following sub-intervals: $[10 * i, 10 * (i + 1)]$, for $i = 0, \ldots, 9$, $[100, 150]$, and $[150, 300]$. For each subproblem, we sample from an equi-spaced grid of 30 time points and 5 k_i points. We use the PCA procedure to define the template and run the reachability algorithm. The bound obtained is quite tight. Furthermore, we experimentally tested the quality of the overapproximation, simulating many points for different values of time and k_i, and checking how many of them are contained in the polytope. In this case, 100% of all the tested points (more than 20000) was inside. In general, however, we may expect to find some points outside, due to the statistical nature of the method.

Case 2: Reachability at Time T. In this second scenario , we looked at the reachability at a given time instant T, taking T from $\{0, 3, 6 \ldots, 300\}$. Here we varied the initial conditions, keeping parameters fixed to their nominal value. In Figure 2 left, we plot the upper and lower bounds for the two variables x_S and

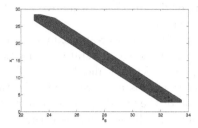 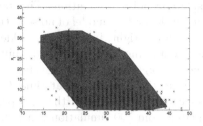

Fig. 3. Left: over-approximation of the steady-state average of the CTMC interpretation of the epidemics model, with uncertainty on $k_i - k_r$. Right: over-approximation of the region containing 95% of the steady state probability mass, again varying $k_i - k_r$.

x_I as a function of time. As we can see, we correctly recover the fact that all the trajectories of the model converge to its unique, globally attracting, steady state. Here we used GP-UCB initialised with a grid of 25 points.

In Figure 2 right, instead, we present a comparison of our approach with the method of [18], which uses sensitivity analysis to estimate how much a neighbourhood of the initial condition x_0^i is expanded or compressed during the dynamics. This allows us to cover the reachable set at time T by a collection of balls centered at points $x(T, p, x_0^i)$. We used squared balls (w.r.t the infinity norm), choosing the same covering grid as the one with which we initialised our method, in which balls have a radius of $\epsilon = 1.25$, and denser grids for $\epsilon = 0.5$, $\epsilon = 0.2$, and $\epsilon = 0.1$. Then we upper and lower bounded the reachable sets for x_I at the time instants $\{0, 1.5, 3 \ldots, 150\}$, and compared the results against our method. In particular, in Figure 2 right we show the absolute error in the estimate of the upper bound of infected individuals.[3] As we can see, our approach consistently produces tighter bounds. Computationally, our method (implemented in Matlab) took about 6.4 seconds, while the cost of a comparable Matlab implementation of [18] varies between 1.5 seconds ($\epsilon = 1.25$) to 225 seconds ($\epsilon = 0.1$). However, the computational cost of the method of [18] (for the precision level of $\epsilon = 0.1$) can be reduced closer to that of our algorithm using an adaptive strategy to generate the grid, as discussed in [18].

Case 3: Reachability for Stochastic Systems. We consider now the reachability problem for the stochastic version of the network epidemics model. Each transition is interpreted as a random event taking an exponentially distributed amount of time to happen, obtaining a population CTMC model (see e.g. [7]). In Figure 3 left, we try to over-approximate the steady state average, for $k_i \in [0.0005, 0.0015]$ and $k_r \in [0.01, 0.1]$. We took the distribution at time $T = 300$ as a proxy of the steady state distribution. For this problem, the use of GP-UCB optimisation is crucial to reduce the number of function evaluations, which are costly. In fact, the average is estimated at each sampled point from 200 simulation runs, resulting in

---

[3] The true upper bound has been estimated exploiting monotonicity properties of the model. In particular, it is defined by two trajectories for extreme values of $x_S(0)$ and $x_I(0)$. The error peak of Figure 2 right happens at their crossing point.

considerably noisy observations. The standard deviation of the estimate is used for the input dependent model of noise in the GP-emulation.

In Figure 3 right, instead, we tackle the problem of over-approximating the region containing 95% of the probability mass of the steady state distribution in the (x_S, x_I)-plane, again varying $k_i \in [0.0005, 0.0015]$ and $k_r \in [0.01, 0.1]$ (and using the distribution at time $T = 300$). To this end, we maximised the upper confidence bound of the 97.5 percentile of the steady state distribution, estimated from 500 simulation runs, and symmetrically we minimised the lower confidence bound of the 2.5 percentile. We used the GP-UCB algorithm, starting from a grid of 25 initial samples. The optimisation procedure converges quickly, after only 5-10 additional function evaluations. To asses the quality of the over-approximation, we sampled 25000 points from the distribution at time $T = 300$ and computed the fraction contained in the over-approximating polytope. It turned out that such a fraction was equal to 0.976, which is close to the expected value of 0.95.

6   Discussion

We presented a novel statistical approach for reachability computation in non-linear dynamical systems, as part of a research programme deploying advanced machine learning tool to solve approximation problems in theoretical computer science [8,9,2]. The method leverages recent advances in statistical machine learning which provide convergent algorithms to optimise unknown functions from noisy pointwise observations [30]. We then adopt a template-based approach [12,29] to construct an over-approximation to the reachable set in terms of linear observables, and determine a tight approximation to the reachable set by optimising the linear observable from finite observations. Experimental evaluations suggest that the method can have high accuracy and it is computationally competitive with state-of-the art reachability computation methods [18].

In our view, a major advantage of our method is its ability to handle noise in a structural way, through a Bayesian treatment that combines the system uncertainty and the observation noise in the final predictions. As such, the method can be directly used to study the impact of uncertainty on the statistics of a (non-linear) stochastic process. We showed on a nontrivial example that our method provides accurate and fast reconstructions of reachable set both for deterministic and stochastic systems, and can be easily applied both to problems of reachability at a fixed time and reachability within an interval. Furthermore, Bayesian optimisation provides an effective method to efficiently solve the maximisation problems involved, but also to adaptively check if an unsafe set has been reached (using the distance from the unsafe set as a non-linear observable), keeping the number of samples to a minimum. This is relevant in high-dimensional problems and for stochastic models, for which it is expensive to compute statistics of the process.

While we believe this approach introduces a new idea for reachability computation, there are plenty of potential improvements/ outstanding challenges to be addressed. A straightforward enhancement, which could further improve accuracy, would be the use of derivative information in the definition of the GP kernel, in the spirit of the sensitivity-based reachability computations of [18]. Derivative kernels can be constructed for GPs [28] and have been used in a biological modelling scenario in e.g. [25]. Another important problem is related to the choice of the template, which can impact the quality of results. In the paper, we advocated a statistical approach and used hyperplanes identified by PCA. In general, the idea of learning templates from sample data should be investigated with more detail, and possibly combined with the use of non-linear templates, such as ellipsoids [24] or other polynomial curves. Since the GP emulation procedure does not depend on the details of the emulated function, our method should be directly applicable to this case as well.

Two major challenges for the emulation based approach are scaling to higher dimensions, and handling discontinuous reachability functions. Large spaces of uncertain parameters are problematic due to the curse of dimensionality: exploring the reachability function to a sufficient degree to start the GP emulation procedure involves a number of function evaluations which is exponential in the number of uncertain parameters. This is further compounded by the cubic scaling of GP emulation with the number of input points (due to the necessity of inverting the Gram matrix). This can be circumvented by using sparse approximations [28] but it will inevitably introduce a trade-off between scalability and accuracy. Discontinuous functions, such as the ones encountered for bi-stable systems or bifurcations, violate the smoothness condition implied by the use of GPs. This condition can be relaxed to a piecewise smoothness condition by using decision trees to partition the input space [3]; however, such methods have not been previously used in the context of Bayesian emulation or optimisation.

References

1. Andersson, H., Britton, T.: Stochastic Epidemic Models and Their Statistical Analysis. Springer (2000)
2. Bartocci, E., Bortolussi, L., Nenzi, L., Sanguinetti, G.: On the robustness of temporal properties for stochastic models. In: Proc. of HSB. EPTCS, vol. 125, pp. 3–19 (2013)
3. Becker, W., Worden, K., Rowson, J.: Bayesian sensitivity analysis of bifurcating nonlinear models. Mechanical Systems and Signal Processing 34(1-2), 57–75 (2013)
4. Belta, C., Habets, L.C.: Controlling a class of nonlinear systems on rectangles. IEEE Trans. on Automatic Control 51(11), 1749–1759 (2006)
5. Bhatia, A., Frazzoli, E.: Incremental search methods for reachability analysis of continuous and hybrid systems. In: Alur, R., Pappas, G.J. (eds.) HSCC 2004. LNCS, vol. 2993, pp. 142–156. Springer, Heidelberg (2004)
6. Bishop, C.M.: Pattern recognition and machine learning. Springer, NY (2009)
7. Bortolussi, L., Hillston, J., Latella, D., Massink, M.: Continuous approximation of collective systems behaviour: a tutorial. Performance Evaluation (2013)
8. Bortolussi, L., Sanguinetti, G.: Learning and designing stochastic processes from logical constraints. In: Joshi, K., Siegle, M., Stoelinga, M., D'Argenio, P.R. (eds.) QEST 2013. LNCS, vol. 8054, pp. 89–105. Springer, Heidelberg (2013)
9. Bortolussi, L., Sanguinetti, G.: Smoothed model checking for uncertain Continuous Time Markov Chains. arXiv preprint arXiv:1402.1450 (2014)

10. Bujorianu, L.M.: A statistical inference method for the stochastic reachability analysis. In: Proceedings of IEEE CDC 2005 (2005)
11. Burden, R.L., Faires, J.D.: Numerical analysis. Brooks/Cole, Cengage Learning, Boston (2011)
12. Chutinan, A., Krogh, B.H.: Computing polyhedral approximations to flow pipes for dynamic systems. In: Proc. of IEEE CDC, vol. 2 (1998)
13. Dang, T., Dreossi, T.: Falsifying oscillation properties of parametric biological models. In: Proc. of HSB. EPTCS, vol. 125 (2013)
14. Dang, T., Gawlitza, T.M.: Template-based unbounded time verification of affine hybrid automata. In: Yang, H. (ed.) APLAS 2011. LNCS, vol. 7078, pp. 34–49. Springer, Heidelberg (2011)
15. Dang, T., Le Guernic, C., Maler, O.: Computing reachable states for nonlinear biological models. Theor. Comput. Sci. 412(21) (2011)
16. Dang, T., Nahhal, T.: Coverage-guided test generation for continuous and hybrid systems. Formal Methods in System Design 34(2) (2009)
17. De Schutter, B., Heemels, W.P., Lunze, J., Prieur, C.: Survey of modeling, analysis, and control of hybrid systems. In: Handbook of Hybrid Systems Control–Theory, Tools, Applications, pp. 31–55 (2009)
18. Donzé, A., Maler, O.: Systematic simulation using sensitivity analysis. In: Bemporad, A., Bicchi, A., Buttazzo, G. (eds.) HSCC 2007. LNCS, vol. 4416, pp. 174–189. Springer, Heidelberg (2007)
19. Duggirala, P.S., Mitra, S., Viswanathan, M.: Verification of annotated models from executions. In: Proc. of ACM EMSOFT 2013 (2013)
20. Frehse, G., et al.: SpaceEx: Scalable verification of hybrid systems. In: Gopalakrishnan, G., Qadeer, S. (eds.) CAV 2011. LNCS, vol. 6806, pp. 379–395. Springer, Heidelberg (2011)
21. Gardiner, C.W.: Stochastic methods. Springer (2009)
22. Girard, A., Le Guernic, C., Maler, O.: Efficient computation of reachable sets of linear time-invariant systems with inputs. In: Hespanha, J.P., Tiwari, A. (eds.) HSCC 2006. LNCS, vol. 3927, pp. 257–271. Springer, Heidelberg (2006)
23. Girard, A., Pappas, G.J.: Verification using simulation. In: Hespanha, J.P., Tiwari, A. (eds.) HSCC 2006. LNCS, vol. 3927, pp. 272–286. Springer, Heidelberg (2006)
24. Kurzhanski, A.B., Varaiya, P.: On ellipsoidal techniques for reachability analysis. Optimization Methods and Software 17(2) (2002)
25. Lawrence, N.D., Sanguinetti, G., Rattray, M.: Modelling transcriptional regulation using gaussian processes. In: NIPS, pp. 785–792. MIT Press (2006)
26. Maler, O.: Computing reachable sets: an introduction. Technical report (2008), http://www-verimag.imag.fr/maler/Papers/reach-intro.pdf
27. Oakley, J.E., O'Hagan, A.: Probabilistic sensitivity analysis of complex models: a bayesian approach. J. of the Royal Statistical Society B 66(3), 751–769 (2004)
28. Rasmussen, C.E., Williams, C.K.I.: Gaussian processes for machine learning. MIT Press, Cambridge (2006)
29. Ben Sassi, M.A., Testylier, R., Dang, T., Girard, A.: Reachability analysis of polynomial systems using linear programming relaxations. In: Chakraborty, S., Mukund, M. (eds.) ATVA 2012. LNCS, vol. 7561, pp. 137–151. Springer, Heidelberg (2012)
30. Srinivas, N., Krause, A., Kakade, S., Seeger, M.: Information-theoretic regret bounds for Gaussian process optimisation in the bandit setting. IEEE Trans. Inf. Th. 58(5), 3250–3265 (2012)
31. Tang, B.: Orthogonal array-based latin hypercubes. Journal of the American Statistical Association 88(424), 1392–1397 (1993)

Formal Synthesis and Validation of Inhomogeneous Thermostatically Controlled Loads*

Sadegh Esmaeil Zadeh Soudjani[1], Sebastian Gerwinn[2], Christian Ellen[2], Martin Fränzle[2], and Alessandro Abate[3,1]

[1] Delft Center for Systems and Control, TU Delft, The Netherlands
[2] OFFIS, Universität Oldenburg, Germany
[3] Department of Computer Science, University of Oxford, United Kingdom
S.EsmaeilZadehSoudjani@tudelft.nl

Abstract. This work discusses the construction of a finite-space stochastic dynamical model as the aggregation of the continuous temperature dynamics of an inhomogeneous population of thermostatically controlled loads (TCLs). The temperature dynamics of a TCL is characterized by a differential equation in which the TCL status (ON, OFF) is controlled by a thresholding mechanism, and which displays inhomogeneity as its thermal resistance changes in time according to a Poisson process. In the aggregation procedure, each TCL model in the population is formally abstracted as a Markov chain, and the cross product of these Markov chains is lumped into its coarsest (exact) probabilistic bisimulation. Quite importantly, the abstraction procedure allows for the quantification of the induced error. Assuming that the TCLs explicitly depend on a control input, the contribution investigates the problem of population-level power reference tracking and load balancing. Furthermore, for the corresponding closed-loop control scheme we show how the worst case performance can be lower bounded statistically, thereby guaranteeing robustness versus power-tracking when the underlying assumption on the inhomogeneity term is relaxed.

Keywords: Thermostatically controlled load, Markov chain, Poisson process, Formal abstraction, Probabilistic bisimulation, Stochastic optimal control, Noisy optimization.

1 Introduction and Background

Household appliances such as water boilers/heaters, air conditioners and electric heaters – all referred to as thermostatically controlled loads (TCLs) – can store energy due to their thermal mass. These appliances generally operate within a dead-band around a temperature set-point. Control of the aggregate power consumption of a population of TCLs can provide a variety of benefits to the

* This work was supported by the European Commission STREP project MoVeS 257005 and by the European Commission IAPP project AMBI 324432.

G. Norman and W. Sanders (Eds.): QEST 2014, LNCS 8657, pp. 57–73, 2014.

electricity grid. First, ancillary service requests can be partially addressed locally, which reduces the need for additional transmission line capacity. Second, controlling a large population of TCLs may improve robustness, since even if a few TCLs fail to provide the required service the consequence on the population as a whole would be small. Such benefits highlight the importance of precise modeling and quantitative control of TCL populations.

Modeling efforts over populations of TCLs and applications to load control arguably initiate with the work in [9]. A discrete-time stochastic model for a TCL is studied in [25], where a simulation model is developed based on a Markov chain approximation of the discrete-time dynamics. A diffusion approximation framework is introduced in [21] to model the dynamics of the electric demand of large aggregates of TCLs by a system of coupled ordinary and partial differential equations. These equations are further studied in [8], where a linear time-invariant dynamical model is derived for the population.

A range of recent contributions [4,19,22,24] employ a partitioning of the TCL temperature range to obtain an aggregate state-space model for the TCL population. Matrices and parameters of the aggregate model are computed either analytically or via system identification techniques. Additional recent efforts have targeted the application of this approach towards higher-order dynamical models [29,30] and the problem of energy arbitrage [23]. The main limitation of these approaches is the lack of a quantitative measure on the accuracy of the constructed aggregated model. Motivated by this drawback, [15,18] have looked at the problem from the perspective of formal abstractions: in contrast to all related approaches in the literature, stymied by the lack of control on the introduced aggregation error, [15,18] have introduced a formal abstraction procedure that provides an upper bound on the error, which can be precisely tuned to match a desired level before computing the actual aggregation.

The purpose of this work is to focus on a new inhomogeneous model for the TCL dynamics, where the inhomogeneity enters the model through a thermal resistance capturing the effect of the opening/closing of windows, of people entering/leaving the room, and so on. The inhomogeneity enters randomly via a Poisson process with a fixed arrival rate, which changes the value of the thermal resistance within a given finite set. We show that the TCL dynamics can be equivalently represented by a discrete-time Markov process, of which we explicitly compute its stochastic kernel. Next, we employ the mentioned abstraction techniques in [1,13] to formally approximate it with a Markov chain. Thus, the aggregated behavior of a *population* of Markov chains can be modeled as a stochastic difference equation [14,15], which is then used for state estimation and closed-loop control of the total power consumption for tracking a load profile.

A crucial assumption in our work underpinning the construction of the stochastic abstraction and the synthesis of the corresponding control scheme is the homogeneity in the *parameters* of the population of TCLs. This assumption might be practically violated. Additionally, some variables such as the initial state of the population and the desired load profile to be tracked are potentially not known in advance. Nevertheless, we would like to guarantee a desired performance of a control scheme. For example, a desirable property could be that the control scheme is able to follow a load profile up to a given deviation while keeping the temperature of the individual households within a certain range. Once a property is specified, guaranteeing such performance amounts to solving a stochastic satisfiability problem. More precisely, one has to choose initial states and load profiles pessimistically to minimize the expect value of the associated cost function. Mathematically this problem can be formulated using stochastic satisfiability modulo theory [17]. Unfortunately, most tools for solving this kind of satisfiability problems are not suited to handle continuous non-determinism. However, there is a tight connection between noisy optimization and stochastic satisfiability [10], rendering the problem suitable for dedicated methods such as [20,28]. We employ these techniques to investigate the robustness of the overall control scheme.

The article is organized as follows. Section 2 introduces the dynamics of a TCL in continuous time, along with the inhomogeneity injected via the Poisson process. Modeling of the TCL dynamics as a discrete-time Markov process and computation of the associated stochastic kernel are presented in Section 3. Abstraction and aggregate modeling of a population of TCLs are then discussed in Section 4. Power reference tracking through closed-loop control of the population is described in Section 5. Finally in Section 6, robustness of the performance of the synthesized controller is validated a-posteriori, against violations on the assumptions on the model.

Throughout this article we use the following notation: $\mathbb{N} = \{1, 2, \ldots\}$ for the natural numbers, $\mathbb{N}_0 = \mathbb{N} \cup \{0\}$, and $\mathbb{N}_n = \{1, 2, \ldots, n\}$ for $n \in \mathbb{N}$.

2   Model of a Thermostatically Controlled Load

Consider the temperature $\theta(t)$ of a TCL evolving in continuous time according to the equation

$$d\theta(t) = \frac{dt}{R(t)C}(\theta_a \pm m(t)R(t)P_{rate} - \theta(t)) \ , \tag{1}$$

where θ_a is the ambient temperature, C indicates the thermal capacitance, and P_{rate} is the rate of energy transfer. In equation (1) a $+$ sign is used for a heating TCL, whereas a $-$ sign is used for a cooling TCL. Inhomogeneity enters via the thermal resistance $R(\cdot) : \mathbb{R}^{\geq 0} \to \{R_0, R_1\}$, which is a function of time and switches between two different values (R_0, R_1), where the switching times are distributed according to the (homogeneous) Poisson process $N(\cdot)$, namely

$$R(t) = \begin{cases} R_0 & \text{if } N(t) \equiv 0 \pmod{2} \\ R_1 & \text{if } N(t) \equiv 1 \pmod{2} \ . \end{cases} \tag{2}$$

The Poisson process accounts for the number of switches and their occurrence time within a given time interval. $N(\cdot)$ is characterized by a specified rate parameter λ, so that the number of switches within the time interval $(t, t + \tau]$ follows a Poisson distribution with

$$\mathbb{P}\{N(t+\tau) - N(t) = n\} = \frac{e^{-\lambda\tau}(\lambda\tau)^n}{n!} \quad \forall n \in \mathbb{N}_0 \ .$$

In equation (1), the quantity $m(\cdot)$ represents the status of the thermostat, namely $m : \mathbb{R}^{\geq 0} \to \{0,1\}$, where $m(t) = 1$ represents the ON mode and $m(t) = 0$ the OFF mode. For the sake of simplicity, we assume that the Poisson process $N(\cdot)$ is initialized probabilistically according to

$$N(0) = \begin{cases} 0 & \text{with probability } q = 1 - p \\ 1 & \text{with probability } p \ . \end{cases}$$

Moreover, we select $p = q = 1/2$, which means $\mathbb{P}\{R(t) = R_0\} = \mathbb{P}\{R(t) = R_1\} = 1/2$, for all $t \in \mathbb{R}^{\geq 0}$. This simplifying assumption on the initialization of the Poisson process and on the special selection of parameter p can be easily relaxed by including the thermal resistance within the discrete state of the TCL. Without loss of generality, we further assume that $R_0 > R_1$.

With focus on a cooling TCL ($-$ sign in equation (1)), the temperature of the load is regulated by a digital controller $m(t + \tau) = f(m(t), \theta(t))$ that is based on a binary switching mechanism, as follows:

$$f(m, \theta) = \begin{cases} 0 & \text{if } \theta < \theta_s - \delta_d/2 \doteq \theta_- \\ 1 & \text{if } \theta > \theta_s + \delta_d/2 \doteq \theta_+ \\ m & \text{else }, \end{cases} \quad (3)$$

where θ_s and δ_d denote the temperature set-point and the dead-band width, respectively, and together characterize the operating temperature range. Note that the switching control signal is applied only at discrete time instants $\{k\tau, k \in \mathbb{N}_0\}$: the mode $m(t)$ may change only at these times, and is fixed in between any two time instants $k\tau$ and $(k+1)\tau$, during which the temperature evolves based on equations (1)-(2) with a fixed $m(k\tau)$. In other words, the operational frequency of the digital controller is $\frac{1}{\tau}$.

The power consumption of the single TCL at time t is equal to $\frac{1}{\eta}m(t)P_{rate}$, which is then equal to zero in the OFF mode and is positive in the ON mode, and where the parameter η is the coefficient of performance (COF). The constant $\frac{1}{\eta}P_{rate}$, namely the power consumed by a single TCL when it is in the ON mode, will be shortened as P_{ON} in the sequel.

In the next section we show that the power consumption of the TCL can be modeled as a Markov process in discrete time, of which we compute the stochastic kernel.

3   Discrete-Time Markov Process Associated to the TCL

We consider a discrete-time Markov process (dtMP) $\{s_k, k \in \mathbb{N}_0\}$, defined over a general (e.g., continuous or hybrid) state space [2]. The model is denoted by the pair $\mathfrak{S} = (\mathcal{S}, T_s)$ in which $\mathcal{S}$ is an uncountable state space. We denote by $\mathcal{B}(\mathcal{S})$ the associated sigma algebra and refer the reader to [2,5] for details on measurability and topological considerations. The stochastic kernel $T_s : \mathcal{B}(\mathcal{S}) \times \mathcal{S} \to [0,1]$ assigns to each state $s \in \mathcal{S}$ a probability measure $T_s(\cdot|s)$, so that for any set $A \in \mathcal{B}(\mathcal{S}), k \in \mathbb{N}_0, \mathbb{P}\{s_{k+1} \in A|s_k = s\} = T_s(A|s)$. We assume that the stochastic kernel T_s admits a representation by its conditional density function $t_s : \mathcal{S} \times \mathcal{S} \to \mathbb{R}^{\geq 0}$, namely $T_s(d\bar{s}|s) = t_s(\bar{s}|s)d\bar{s}$, for any $s, \bar{s} \in \mathcal{S}$.

The digital controller of Section 2 ensures that the power consumption of the TCL is a piecewise-constant signal, where the jumps can happen only at the sampling times $\{k\tau, k \in \mathbb{N}_0\}$. We define $\theta_k = \theta(k\tau), m_k = m(k\tau)$ as the values of the random processes $\theta(\cdot), m(\cdot)$ at the sampling time $k\tau$. Despite the fact that the temperature evolves stochastically in between sampling times, we show that the dynamics of the temperature and of the mode at the sampling times (that is, in discrete time with sampling constant τ) can be modeled as a dtMP and we compute the corresponding conditional density function. In other words, the goal of this section is to compute the density function of the process state at the next discrete time step, conditioned on the state at the current time step.

Define the hybrid state of the dtMP $\mathfrak{S}$ as $s_k = (m_k, \theta_k) \in \mathcal{S} \doteq \{0,1\} \times \mathbb{R}$. The evolution of the mode is given by the deterministic equation $m_{k+1} = f(m_k, \theta_k)$, while the temperature evolves stochastically and depends on the conditional density function $t_\theta(\theta_{k+1}|\theta_k, m_k)$. Then for all $s_k = (m_k, \theta_k), s_{k+1} = (m_{k+1}, \theta_{k+1}) \in \mathcal{S}$,

$$t_s(s_{k+1}|s_k) = \delta\left[m_{k+1} - f(m_k, \theta_k)\right] t_\theta(\theta_{k+1}|\theta_k, m_k) ,$$

where $\delta[\cdot]$ is the discrete Kronecker delta function. The rest of this section is dedicated to the computation of t_θ, and to the study of its dependency on characteristic parameters. We focus on the explicit computation of t_θ for the case where the mode of the current state is OFF, namely $t_\theta(\theta_{k+1}|\theta_k, m_k = 0)$. The case of $t_\theta(\theta_{k+1}|\theta_k, m_k = 1)$ is similar and thus discussed at the end of this section.

Lemma 1. *Suppose the TCL is in the OFF mode, $m(t) = 0$, during the interval $t \in [t_1, t_2]$. The value of the temperature at the end of the interval solely depends on the relative time the temperature evolves with either of the two resistance values $\{R_0, R_1\}$, and is independent of the actual order or number of occurrence of the two values.*

Lemma 1 states that the distribution of the temperature at the next time step θ_{k+1}, conditioned on the current temperature value θ_k, depends exclusively on the relative time duration that the temperature evolves with any of the two resistances, within the interval: the number or the order of switchings between the resistances is not important, thus the corresponding time can be simply accumulated. Since the resistance changes value based on the jumps of a Poisson process, we define the sum of the length of the sub-intervals of $[k\tau, (k + 1)\tau]$ in which the temperature evolves with R_0 (resp. R_1) as the random variable w_0 (resp. w_1). Let us now compute the density functions of these two random variables. Despite the fact that ω_0, ω_1 are defined with respect to the particularly chosen interval $[k\tau, (k + 1)\tau]$, next lemma shows that their density functions are independent of k and solely depend on the length of the interval τ.

Lemma 2. *The density function of ω_0 can be expressed as*

$$f_{\omega_0}(x) = e^{-\lambda\tau}\delta(\tau - x) + \lambda e^{-\lambda\tau}\left[I_0\left(2\lambda\sqrt{x(\tau - x)}\right) + \sqrt{\tfrac{x}{\tau-x}}I_1\left(2\lambda\sqrt{x(\tau - x)}\right)\right], \quad (4)$$

which is parametrized by λ, τ, and is independent of k. $I_0(\cdot), I_1(\cdot)$ are modified Bessel functions of the first kind [3, Chapter 9] and $\delta(\cdot)$ is the Dirac delta function. The density function of ω_1 is $f_{\omega_1}(x) = f_{\omega_0}(\tau - x)$.

Lemma 2 indicates that the random variable ω_0 (resp. ω_1) is of mixed type, including a probability density function with the interval $[0, \tau]$ as its support, and a probability mass at $x = \tau$ (resp. $x = 0$). Once we know the distributions of ω_0, ω_1, we can compute $t_\theta(\theta_{k+1}|\theta_k, m_k)$ in the OFF mode.

Theorem 1. *The conditional density function $t_\theta(\theta_{k+1}|\theta_k, m_k = 0)$ is of the form*

$$t_\theta(\theta_{k+1}|\theta_k, m_k = 0) = \frac{1}{2}t_0(\theta_{k+1}|\theta_k, R_0) + \frac{1}{2}t_1(\theta_{k+1}|\theta_k, R_1) , \quad (5)$$

where the functions t_0, t_1 are computed based on the density functions of ω_0, ω_1:

$$t_i(\theta_{k+1}|\theta_k, R_i) = \frac{1}{|(\theta_{k+1} - \theta_a)\gamma|}f_{\omega_i}\left(\frac{1}{\gamma}\left[\ln\frac{\theta_{k+1} - \theta_a}{\theta_k - \theta_a} + \tau_1\tau\right]\right) \quad i \in \{0, 1\} , \quad (6)$$

and where $\tau_0 = \frac{1}{R_0 C}$, $\tau_1 = \frac{1}{R_1 C}$, and $\gamma = \tau_1 - \tau_0 > 0$.

We emphasize that the conditional density function t_θ is independent of k, which results in a time-homogeneous dtMP. It has the support $[\theta_{min}, \theta_{max}]$ with

$$\theta_{min} \doteq \theta_a + (\theta_k - \theta_a)e^{-\tau_0\tau} , \quad \theta_{max} \doteq \theta_a + (\theta_k - \theta_a)e^{-\tau_1\tau} ,$$

and includes two Dirac delta functions at the boundaries of its support

$$\frac{1}{2}e^{-\lambda\tau}\left[\delta\left(\theta_{k+1} - \theta_{min}\right) + \delta\left(\theta_{k+1} - \theta_{max}\right)\right] .$$

Moreover it is discontinuous at the boundaries of its support with the following discontinuities:

$$t_\theta(\theta_{min}|\theta_k, m_k = 0) = \frac{\lambda e^{-\lambda\tau}(2+\lambda\tau)}{2|(\theta_k-\theta_a)\gamma|e^{-\tau_0\tau}} , \quad t_\theta(\theta_{max}|\theta_k, m_k = 0) = \frac{\lambda e^{-\lambda\tau}(2+\lambda\tau)}{2|(\theta_k-\theta_a)\gamma|e^{-\tau_1\tau}} . \quad (7)$$

All the above derivations for the density function t_θ conditioned in the OFF mode can be likewise obtained for that in the ON mode: $t_\theta(\theta_{k+1}|\theta_k, m_k = 1)$ is formulated exactly as $t_\theta(\theta_{k+1}|\theta_k, m_k = 0)$ where the quantity θ_a is replaced by the steady-state value of the temperature in the ON mode. The only required assumption is that the temperature trajectories are steered toward the same steady-state value regardless of the thermal resistance. Such a steady-state value is $\theta_\infty = \theta_a - m(t)R(t)P_{rate}$, which is a function of $R(t)$ in the ON mode. This assumption technically allows us to swap the order of the intervals in which the temperature evolves with different resistances and leads to being able to simplify the computations and to obtain the conditional density functions.

4 Abstraction of a Population of Inhomogeneous TCLs

4.1 Abstraction of a TCL as a Markov Chain

The interpretation of the Poisson-driven TCL model as a dtMP allows leveraging an abstraction technique, proposed in [1] and extended in [12,13], aimed at reducing an uncountable state-space dtMP into a (discrete-time) finite-state Markov chain. This abstraction is based on a state-space partitioning procedure as follows. Consider an arbitrary, finite partition of the continuous domain $\mathbb{R} = \cup_{i=1}^{n}\Theta_i$, and arbitrary representative points within the partitioning regions denoted by $\{\bar{\theta}_i \in \Theta_i, i \in \mathbb{N}_n\}$. Introduce a finite-state Markov chain $\mathcal{M}$, characterized by $2n$ states $s_{im} = (m, \bar{\theta}_i), m \in \{0, 1\}, i \in \mathbb{N}_n$. The transition probability matrix of $\mathcal{M}$ is made up of the entries

$$\mathbb{P}(s_{im}, s_{i'm'}) = \int_{\Theta_{i'}} t_s\left(m', \theta'|m, \bar{\theta}_i\right) d\theta' \quad \forall m, m' \in \{0, 1\}, \; i, i' \in \mathbb{N}_n \; . \tag{8}$$

The initial probability mass of $\mathcal{M}$ is obtained as $p_0(s_{im}) = \int_{\Theta_i} \pi_0(m, \theta)d\theta$, where $\pi_0 : \mathcal{S} \to \mathbb{R}^{\geq 0}$ is the density function of the initial hybrid state of the TCL. For simplicity of notation we rename the states of $\mathcal{M}$ by the bijective map $\ell(s_{im}) = mn + i, m \in \{0, 1\}, i \in \mathbb{N}_n$, and accordingly we introduce the new notation

$$P_{ij} = \mathbb{P}(\ell^{-1}(i), \ell^{-1}(j)) \; , \quad p_{0i} = p_0(\ell^{-1}(i)) \quad \forall i, j \in \mathbb{N}_{2n} \; .$$

Notice that the conditional density function of the stochastic system capturing the dynamics of a TCL is discontinuous, due to the presence of equation (3). Further, the density function $t_\theta(\theta_{k+1}|\theta_k, m_k)$ is the summation of two Dirac delta functions and of a piecewise-Lipschitz continuous part. The existence of Dirac delta functions produces technical difficulties in the analysis of properties of interest [16]. Despite these irregularities, we show in the next section that we can compute an upper bound on the error of Markov chain abstraction, based on [15,18]. The abstraction error is composed of three terms related to the Dirac delta functions, the discontinuity at the boundaries, and the state-space discretization.

4.2 Error Computation of the Markov Chain Abstraction

We compute the abstraction error based on [13, pp. 933-934], which gives an upper bound for the abstraction error via the constant $\mathcal{H}$ satisfying the inequality

$$\int_{\mathbb{R}} |t_\theta(\theta_{k+1}|\theta_k, m_k) - t_\theta(\theta_{k+1}|\theta'_k, m_k)|d\theta_{k+1} \leq \mathcal{H}|\theta_k - \theta'_k| \quad \forall \theta_k, \theta'_k, m_k \; . \tag{9}$$

Notice that the integration in the left-hand side is with respect to the next state θ_{k+1}, while the changes are applied to the current state θ_k. In order to compute the constant $\mathcal{H}$, we first establish the Lipschitz continuity of the continuous part of t_θ in Theorem 2 which is founded on the Lipschitz continuity of the bounded part of the functions $f_{\omega_0}(\cdot), f_{\omega_1}(\cdot)$ presented in Lemma 3.

Lemma 3. *Functions* $g_{\omega_0}(x) \doteq f_{\omega_0}(x) - e^{-\lambda\tau}\delta(\tau - x)$ *and* $g_{\omega_1}(x) \doteq f_{\omega_1}(x) - e^{-\lambda\tau}\delta(x)$, *representing the bounded part of functions* $f_{\omega_0}, f_{\omega_1}$, *satisfy the Lipschitz condition*

$$|g_{\omega_i}(x) - g_{\omega_i}(x')| \le h|x - x'| \quad \forall x, x' \in [0, \tau], \, i \in \{0, 1\} \ ,$$

where $h = \lambda^2 e^{-\lambda\tau} M(2\lambda\tau)$ *and* $M(a) = \max_{u \in [0,1]} |\zeta(u, a)|$. *The function* ζ *is defined as*

$$\zeta(u, a) = I_0\left(a\sqrt{u(1-u)}\right) + \frac{1-2u}{\sqrt{u(1-u)}} I_1\left(a\sqrt{u(1-u)}\right) - \frac{u}{1-u} I_2\left(a\sqrt{u(1-u)}\right) \ ,$$

where $I_0(\cdot), I_1(\cdot), I_2(\cdot)$ *are modified Bessel functions of the first kind [3].*

Theorem 2. *The density function* $t_\theta(\theta_{k+1}|\theta_k, m_k)$ *satisfies the Lipschitz condition*

$$|t_\theta(\theta_{k+1}|\theta_k, m_k) - t_\theta(\theta_{k+1}|\theta_k', m_k)| \le \kappa|\theta_k - \theta_k'| \quad \forall\theta_k, \theta_k' \in [\theta_-, \theta_+], \, m_k \in \{0, 1\} \ ,$$

for all θ_{k+1} *in the intersection of the supports of* $t_\theta(\cdot|\theta_k, m_k), t_\theta(\cdot|\theta_k', m_k)$, *with* $\kappa = \dfrac{h}{\varrho\gamma^2}$, *and the constant* $\varrho = \min\left\{(\theta - \theta_\infty)^2, \theta \in [\theta_-, \theta_+]\right\}$.

Recall that the density function t_θ is discontinuous at the boundaries of its support, with jumps quantified in (7). The value of these jumps appear directly in the left-hand side of inequality (9). Then we have to establish the Lipschitz continuity of the jumps of t_θ with respect to the current state, which is done in the next Lemma.

Lemma 4. *The jumps of the density functions* $t_\theta(\theta_{min}|\theta_k, m_k), t_\theta(\theta_{max}|\theta_k, m_k)$, *as in (7), are Lipschitz continuous with respect to the current temperature* θ_k, *with the following constants:*

$$\kappa_0 = \frac{\lambda}{2\varrho|\gamma|}(2 + \lambda\tau)e^{-\lambda\tau}e^{\tau_0\tau} \ , \quad \kappa_1 = \frac{\lambda}{2\varrho|\gamma|}(2 + \lambda\tau)e^{-\lambda\tau}e^{\tau_1\tau} \ .$$

Finally, the transition probabilities of the Markov chain are computed by eliminating the Dirac delta functions and integrating over the partition sets. The total abstraction error is formulated in Theorem 3 using the constants of Theorem 2 and Lemma 4, and leveraging results of [13].

Theorem 3. *If we partition the temperature range with the diameter* δ_p, *the one-step abstraction error is* $\varepsilon = 2(e^{-\lambda\tau} + \kappa_0 + \kappa_1 + \kappa\mathcal{L})\delta_p$, *where* $\mathcal{L}$ *is the Lebesgue measure of the temperature range.*

The error computed in Theorem 3 is useful towards two different purposes. First, it provides a measure on the distance between the power consumption of the model in (1)-(3) and that of the abstracted model [15], namely

$$|\mathbb{E}\left[m(k\tau) - m_{abs}(k\tau)|\, \theta(0), m(0)\right]| \le k\varepsilon \ .$$

Second, it can be used to check Bounded Linear Temporal Logic (BLTL) specifications over the abstracted Markov chain, providing a guarantee on the specification for the original population model. The error caused by the abstraction for checking any BLTL specification is $N\varepsilon$, where N is the horizon of the specification [26]. Notice that the error can be tuned by proper selection of the partition diameter, time-step, and arrival rate. For instance the error is $\varepsilon = 1.3 \times 10^7 \delta_p$ for the physical parameters of Table 1 (left) that are widely used in the literature, and for the values of Table 1 (right) specifically selected for the model in this study. The large constant in the expression of ε is mainly due to the Lipschitz constant of the density function, which can be reduced by selection of the discretization time step τ.

Table 1. Physical parameters of a residential air conditioner as a TCL [8] (left) and selected parameters for the model in (1)-(3) (right)

Parameter	Interpretation	Value
θ_s	set-point	$20\,[^\circ C]$
δ_d	dead-band width	$0.5\,[^\circ C]$
θ_a	ambient temperature	$32\,[^\circ C]$
C	thermal capacitance	$10\,[kWh/^\circ C]$
P_{rate}	power	$14\,[kW]$
η	COF	2.5

Parameter	Interpretation	Value
R_0	thermal resistance	$1.5\,[^\circ C/kW]$
R_1	thermal resistance	$2.5\,[^\circ C/kW]$
τ	time step	$10\,[sec]$
λ	arrival rate	$1\,[sec^{-1}]$

4.3   Aggregate Model of a Population of TCLs

In the previous section we described how a TCL model is formally abstracted as a Markov chain. In this section we develop a *stochastic* difference equation (SDE) as the aggregate model of the TCL population, which is to be later employed in the state estimation and closed-loop control. Our modeling approach generalizes the results of [7,6] for our setting in which convergence to a *deterministic* difference equation for large populations of Markov chains is investigated in the context of mean field limits. To construct the SDE model, we first take the cross product of all the abstracted Markov chains, to obtain a (admittedly large) Markov chain $\mathcal{Z}$ with finitely many states characterizing the behavior of the population. As a second stage we assign labels $X \in \mathbb{R}^{2n}$ to the states of $\mathcal{Z}$ in which the ith entry of $X(X(i), i \in \mathbb{N}_{2n})$ indicates the proportion of individual Markov chains within the ith state. Given such a labeling, we define an equivalence relation over the labeled Markov chain $\mathcal{Z}$, which relates all states with the same label. This equivalence relation is in fact an exact probabilistic bisimulation [15] and makes it possible to consider equivalence classes as lumped states and thus reduce the state space of $\mathcal{Z}$. In order for this to be an exact probabilistic bisimulation, however, all transition matrices have to be the same, which reduces to TCL characterized by equal parameters, unlike a different ambient temperature (see Section 6). Let us remark that in practice the lumped chain can be obtained directly, with no need to go through the construction of $\mathcal{Z}$. The dynamics over the labels in the reduced Markov chain can be modeled by the following SDE

$$X_{k+1} = P^T X_k + W_k \ , \tag{10}$$

where $X_k \in \mathbb{R}^{2n}$ is the value of the label of the reduced Markov chain at sample time $k\tau$, and its ith entry represents the portion of TCLs with mode and temperature inside the partition set associated to the representative point $\ell^{-1}(i)$ (cf. Section 4.1). Matrix $P = [P_{ij}]_{i,j \in \mathbb{N}_{2n}}$ is the transition probability matrix of the Markov chain $\mathcal{M}$, and W_k is a state-dependent process noise with $\mathbb{E}[W_k] = 0$ and $Cov(W_k) = [\Sigma_{ij}(X_k)]_{i,j \in \mathbb{N}_{2n}} \in \mathbb{R}^{2n \times 2n}$, where

$$\Sigma_{ii}(X_k) = \frac{1}{n_p} \sum_{r=1}^{2n} X_k(r) P_{ri}(1 - P_{ri}) \ , \quad \Sigma_{ij}(X_k) = -\frac{1}{n_p} \sum_{r=1}^{2n} X_k(r) P_{ri} P_{rj} \ ,$$

for all $i, j \in \mathbb{N}_{2n}$, $i \neq j$. The process noise W_k converges in distribution to a multivariate normal random vector for large population sizes n_p [15]. Moreover, the transition probability matrix P in (10) depends on the set-point θ_s, which is utilized in the next section for power tracking.

The total power consumption obtained from the aggregation of the original models in (1)-(3), with variables $(m^j, \theta^j)(t), j \in \mathbb{N}_{n_p}$, is

$$y(t) = \sum_{j=1}^{n_p} m^j(t) P_{ON} \ ,$$

which is piecewise-constant in time due to the presence of the digital controller updating the modes of the TCLs. Then the total power consumption can be represented as

$$y(t) = \sum_{k=0}^{\infty} y_k \mathbb{I}_{[k\tau,(k+1)\tau)}(t) \ , \quad y_k = \sum_{j=1}^{n_p} m_k^j P_{ON} \ , \tag{11}$$

where $\mathbb{I}_A(\cdot)$ is the indicator function of a given set A, i.e. $\mathbb{I}_A(x) = 1$ for $x \in A$ and zero otherwise. With focus on the abstract model (10), the power consumption of the model is also piecewise-constant in time with the representation

$$y_{abs}(t) = \sum_{k=0}^{\infty} y_k^{abs} \mathbb{I}_{[k\tau,(k+1)\tau)}(t) \ , \quad y_k^{abs} = H X_k \ , \quad H = n_p P_{ON}[0_n, 1_n] \ , \tag{12}$$

where 0_n and 1_n are n-dimensional row vectors with all the entries equal to zero and one, respectively.

The performance and precision of the aggregated model in (10) is displayed in Figure 1 by comparing its normalized output with the normalized power consumption of the TCL population (namely, by comparing $y_{abs}(t), y(t)$ divided by $n_p P_{ON}$). In these simulations, the initial temperatures of the TCLs are distributed uniformly over the dead-band and the modes are obtained as samples of Bernoulli trials with a success probability of 0.5. Two different population sizes $n_p = 100$ (top panel) and $n_p = 1000$ (bottom panel) are considered. The oscillations ranges are $[0.35, 0.58]$ and $[0.40, 0.52]$, respectively, which indicates that the oscillations amplitude depends on the population size: it decreases as the population size increases. The number of introduced abstract states in both cases is $2n = 860$. The transition probability matrix P is computed within 8.3 seconds and is sparse with 0.3% non-zero entries.

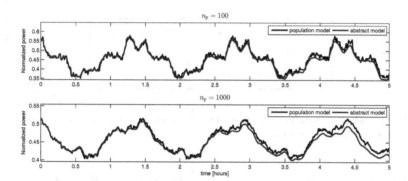

Fig. 1. Comparison of the total power consumption of the population with that of the aggregated model, with population sizes 100 (top) and 1000 (bottom) respectively. The initial conditions have been distributed uniformly.

5   Closed-Loop Control of the TCL Population

As recently investigated in related work [15,18], we consider the set-point θ_s as the control input and regulate the value of this control uniformly over all TCLs. The application of this control scheme practically leads to a change in the position of the non-zero entries of the transition matrix P derived from the dynamics of the single TCL. We discretize the domain of allowable set-point control input by the same partition diameter δ_p used for the temperature range: this allows retaining the definition of the states in X_k.

We employ the stochastic Model Predictive Control (MPC) framework used in [15] over the controlled model in order to track a reference signal for the total power consumption. More precisely, we optimize the following cost function at each time step to track the reference power signal $y_{ref}(\cdot)$:

$$\min_{\theta_s(k\tau) \in [\theta_-, \theta_+]} |\mathbb{E}[y_{abs}(k\tau + \tau)|X_k, \theta_s(k\tau)] - y_{ref}(k\tau + \tau)| . \tag{13}$$

A Kalman filter with state-dependent process noise is employed for state estimation when the information of states X_k is not available and only the total power consumption of the population $y(k\tau)$ is measured. Figure 2 presents the closed-loop control scheme for the power reference tracking problem.

The performance of the closed-loop control of Figure 2 is illustrated in Figure 3 (left). A population size $n_p = 100$ is selected and the number of states $2n = 1000$

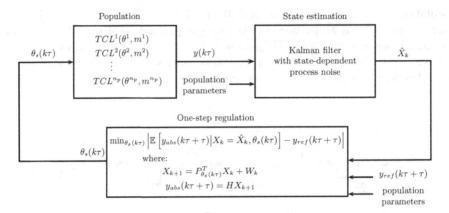

Fig. 2. State estimation and one-step regulation architecture for the closed-loop control of the total power consumption

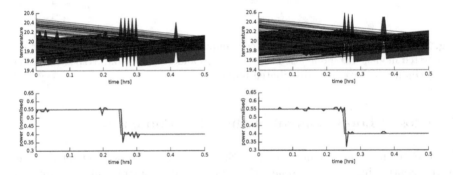

Fig. 3. Illustration of the controller based on a Kalman filter (ref. Figure 2), applied to a population of 100 TCLs, modeled with equations (1)-(3), for either a homogeneous (left) or heterogeneous (right) population. Each TCL has its own initial condition. The upper panels show the evolution of the temperatures across the population in black. The blue region indicates the applied set-point $\pm$ the δ_d dead-band. The lower panels plot in red the desired load profile. The blue lines show the actual load, as achieved when applying the control scheme. For the heterogeneous case the ambient temperature of each TCL is allowed to vary randomly with ± 2 C$^\circ$ around the average ambient temperature of the homogeneous case.

is considered. The lower panel presents the reference signal and the normalized power consumption of the population, while the upper panel shows the synthesized temperature dead-band $[\theta_s(k) - \delta_d/2, \theta_s(k) + \delta_d/2]$. The simulations indicate that the population can accurately track the desired power reference signal. In order to examine the robustness of the control scheme against heterogeneity in the parameters of the population, we run simulations for the case where the ambient temperature of each TCL is allowed to vary randomly with ± 2 C$^\circ$ around the average ambient temperature of the population. The result is presented in Figure 3 (right), which shows that the control scheme is qualitatively robust against violation of this homogeneity assumption. In the next section we quantify more formally the robustness of the closed-loop control scheme.

6   Validation of the Robustness of the Synthesized Controller

As claimed theoretically and further suggested by Figure 3, the control scheme devised on the aggregated population in the previous section can accurately perform tracking of a given load profile for the total power consumption. This requires an assumption on the parameter homogeneity of the TCL population. The simulation shown in Figure 3, however, further suggests that the presented control scheme is robust against heterogeneity, which is to be investigated in this section. Although the following approach could be optimally applied in a closed-loop setting in order to automatically validate the actions taken by the control scheme, we restrict ourselves to an a-posteriori validation due to the computational complexity that is discussed below. We have investigated how well the control scheme based on the Kalman-filter is able to track power consumption in a system that switches modes instantaneously once the temperature hits given thresholds. To apply this control in practice we would like to know how safely the control scheme performs in the worst case, that is across possible unknown variables such as the initial conditions and the desired load profiles. With safe here we mean that the achieved load is within a given range of the desired load profile. As the underlying system is stochastic our goal is to bound the probability of violating such a safety target. Mathematically, we can state such a problem as a noisy optimization problem, representing the probability as an expectation over a binary function:

$$\max_{x \in \mathcal{X}} \mathbb{E}_{y(t), T_0 < t < T}[\phi_{rob}(y)|x] \ , \tag{14}$$

where ϕ_{rob} is an indicator function characterizing whether a trajectory $y(t)$ is in close proximity of the desired load profile. Specifically, we use

$$\phi_{rob}(y) = \mathbb{I}_{|y - y_{ref}| \le \Delta_l}(y) = \begin{cases} 1 & \text{if } \forall t \in [T_0, T] : |y(t) - y_{ref}(t)| \le \Delta_l \\ 0 & \text{else ,} \end{cases}$$

where Δ_l is a parameter controlling the desired degree of load tracking. The vector x represents all variables (in a given space, to be discussed shortly) over which we would like to optimize safety, including the desired load profile (y_{ref}), the initial conditions $(\theta^1(0), m^1(0), \ldots, \theta^{n_p}(0), m^{n_p}(0))$ of TCLs, and the ambient temperatures $\theta_a^1, \ldots \theta_a^{n_p}$. Parameters T_0 and T are used to select the relevant time period over which robustness is validated. Formally, equation (14) falls into the same class of problems as equation (13) and therefore could be solved similarly. As we are interested in robustness properties against heterogeneity, the corresponding optimization space ($\mathcal{X}$) comprises as many dimensions as individual TCLs (x contains all initial conditions of individual TCLs, see above). Formulating the problem as a maximization over an expectation can be interpreted as assigning a probability value to a SSMT formula, comprised of an existentially quantified variable followed by a randomized quantified variable, ϕ_{rob} being an atomic SMT formula (see [17,10] for more details). As formal approaches to solve such problems depend critically on the dimension of the state space, we follow a statistical approach instead, thereby relying only on simulations of the TCL population. To solve such a noisy optimization problem, we are adopting statistical methods, presented in [20,28] to obtain *probably approximate near optimizers*.

Suppose that $g : Y \to \mathbb{R}$, that $\mathbb{P}_Y$ is a given probability measure on Y, and that $\alpha, \epsilon > 0$ are given numbers. A number $g_0 \in \mathbb{R}$ is said to be a Type 3 near minimum of g to level α, or a probably approximate near minimum of g to accuracy ϵ and level α, if $g_0 \geq \min_{y \in Y} g(y) - \epsilon$, and in addition: $\mathbb{P}_Y\{y \in Y : g(y) < g_0 - \epsilon\} \leq \alpha$.

If the probability measure $\mathbb{P}_Y$ is chosen to be the uniform distribution, the probably approximate near minimum is equivalent to the notion of approximate domain optimizer with value imprecision ϵ and residual domain α [20]. These notions of approximate near optimizers can further be extended to hold only with a given confidence ρ, if the approximate near optimizers g_0 have an additional dependence on further random variables [20]. We present a corresponding algorithm to obtain such an optimizer based on uniform sampling.

As mentioned, we aim at a statistical solution to the noisy optimization problem of equation (14). To this end, we use the following simple algorithm [20,28]. The algorithm first samples the parameters over which we would like to optimize the probability of satisfying the robustness property. For each such parameter (in particular containing the initial conditions), the behavior of each TCL is sampled under the closed-loop control from the previous section. Using these samples the probability of satisfying the robustness property can be estimated. The necessary number of samples to achieve the desired accuracy can be calculated in advance (N and M in the above algorithm). For this algorithm it can be shown that the output is a probably approximate near minimum to accuracy ϵ and level α with probability at least ρ, see [27,20,28].

Algorithm 1. Randomized Black-Box optimization algorithm

---

function RANDOPT(satProperty, α, ϵ, ρ)

    $N \leftarrow \frac{\log \frac{2}{1-\rho}}{\log \frac{1}{1-\alpha}}$; $M \leftarrow \frac{1}{2\epsilon^2} \log \frac{4N}{1-\rho}$

    $x^1, \ldots, x^N \leftarrow$ SAMPLEOPTIMIZATIONPARAMS($\mathcal{X}$)   ▷ sample conditions/profiles

    for $n = 1, \ldots, N$ **do**

        $y_n^1, \ldots, y_n^M \leftarrow$ SAMPLETRAJECTORIES(x^n)   ▷ Sample power for condition x^n

        for $m = 1 \ldots, M$ **do**

            $g_{n,m} \leftarrow$ CHECKROBUSTNESSONSAMPLE($\phi_{rob}(y_n^m)$)

        end for

    end for

    $\hat{g}_n \leftarrow$ AVERAGE($g_{n,m}$ across m)   ▷ Estimate robustness for initial conditions

    return $\min_n \hat{g}_n$

end function

Table 2. Parameters for a posteriori verification of the closed-loop control scheme

Δ_l	0.15	Population size n_p	100
Range within ambient temp.	$\pm\,2°$	Level α	0.8
Accuracy ϵ	0.1	Confidence $1-\delta$	0.8
Duration between load switches	15 min	Time horizon T_0, T	25 min, 60 min

We are interested in estimating the worst case performance of the control scheme of Section 5. In such a setting, controllers are typically designed for performance, say to follow a predefined load profile as closely as possible (see equation (13)). Safety is usually not considered during the design phase. Having a system speci-fication for which one can simulate an already designed controller also allows for verifying safety with a statistical procedure such as the one presented in the previ-ous section. If we define safety for a given controller by a maximal allowed deviation Δ_l of the actual load from the desired load profile, we can quantify the number of simulations needed (in terms of precision, level and confidence) in order to guaran-tee the safety of a given controller for a worst case scenario. To simulate, we have to additionally assume a finite time horizon: we have chosen 4 times the period needed for updating the desired load profile, in order to cover the relevant changes between desired values (see Table 2). Using Algorithm 1 we can now verify that such a controller achieves a desired load profile robustly across a heterogeneous parameter within the ambient temperature. More precisely, using the parameters in Table 2, we could verify that the worst-case probability of resulting in a non-safe system using the closed-loop control scheme is guaranteed to be bounded by ϵ (accuracy=0.1) up to residual α level and confidence, as given in Table 2. The number of simulations for such parameters is $N = 738$ and the objective is to have the total power consumption within ±0.15 of the desired reference load. The worst case were determined over all possible initial conditions (temperatures and modes of TCLs) as well as over a set of desired load profiles. For the set of load profiles we considered step-functions, which change the desired load every 15 minutes and have a height $\in [0.4, 0.6]$. Due to the independence between different simulations, Algorithm 1 can be parallelized efficiently. Nevertheless, the necessary number of samples quickly increases with the different parameters for accuracy using this pro-cedure, therefore, it still needs considerable computational effort. Although such a procedure could in principle also be used to verify safety or solve the control problem in a closed-loop setting, we did not investigate such scheme due to the computational effort. To illustrate the feasibility of the approach and due to the high computational load, we set α, or residual domain to 0.8 thereby allowing 80 percent of the optimization domain to have a potentially worse robustness level.

7   Conclusions

In this work the problem of aggregate modeling and control of a population of TCLs has been addressed. The temperature evolution is modeled in continuous time and combined with a digital ON/OFF switching controller. The TCLs are allowed to have different thermal resistances which change values in time based on a Poisson arrival process, and thus induce inhomogeneity in the dynamics. The power consumption of each TCL is modeled as a Markov process and formally abstracted to a Markov chain, which is then used to develop an aggregate model for the total power consumption of the population. Finally, a control scheme is proposed to track a power reference signal and its robustness is examined against violation of the homogeneity assumption by use of statistical techniques.

References

1. Abate, A., Katoen, J.-P., Lygeros, J., Prandini, M.: Approximate model checking of stochastic hybrid systems. European Journal of Control 6, 624–641 (2010)
2. Abate, A., Prandini, M., Lygeros, J., Sastry, S.: Probabilistic reachability and safety for controlled discrete time stochastic hybrid systems. Automatica 44(11), 2724–2734 (2008)
3. Abramowitz, M., Stegun, I.A.: Handbook of mathematical functions with formulas, graphs, and mathematical tables. Dover, New York (1964), ninth dover printing, tenth gpo printing edition
4. Bashash, S., Fathy, H.K.: Modeling and control insights into demand-side energy management through setpoint control of thermostatic loads. In: Proceedings of the 2011 American Control Conference, pp. 4546–4553 (June 2011)
5. Bertsekas, D.P., Shreve, S.E.: Stochastic optimal control: the discrete-time case. Athena Scientific (1996)
6. Bortolussi, L., Hillston, J., Latella, D., Massink, M.: Continuous approximation of collective system behaviour: A tutorial. Performance Evaluation 70(5), 317–349 (2013)
7. Le Boudec, J.-Y., McDonald, D., Mundinger, J.: A generic mean field convergence result for systems of interacting objects. In: 4th International Conference on the Quantitative Evaluation of SysTems (QEST), pp. 3–18 (2007)
8. Callaway, D.S.: Tapping the energy storage potential in electric loads to deliver load following and regulation, with application to wind energy. Energy Conversion and Management 50(5), 1389–1400 (2009)
9. Chong, C.-Y., Debs, A.S.: Statistical synthesis of power system functional load models. In: 18th IEEE Conference on Decision and Control including the Symposium on Adaptive Processes, vol. 18, pp. 264–269 (1979)
10. Ellen, C., Gerwinn, S., Fränzle, M.: Confidence bounds for statistical model checking of probabilistic hybrid systems. In: Jurdziński, M., Ničković, D. (eds.) FORMATS 2012. LNCS, vol. 7595, pp. 123–138. Springer, Heidelberg (2012)
11. Esmaeil Zadeh Soudjani, S., Abate, A.: Adaptive gridding for abstraction and verification of stochastic hybrid systems. In: Proceedings of the 8th International Conference on Quantitative Evaluation of Systems, pp. 59–69 (September 2011)
12. Esmaeil Zadeh Soudjani, S., Abate, A.: Higher-order approximations for verification of stochastic hybrid systems. In: Chakraborty, S., Mukund, M. (eds.) ATVA 2012. LNCS, vol. 7561, pp. 416–434. Springer, Heidelberg (2012)
13. Esmaeil Zadeh Soudjani, S., Abate, A.: Adaptive and sequential gridding procedures for the abstraction and verification of stochastic processes. SIAM Journal on Applied Dynamical Systems 12(2), 921–956 (2013)
14. Esmaeil Zadeh Soudjani, S., Abate, A.: Aggregation of thermostatically controlled loads by formal abstractions. In: European Control Conference (ECC), Zurich, Switzerland, pp. 4232–4237 (July 2013)

15. Esmaeil Zadeh Soudjani, S., Abate, A.: Aggregation and control of populations of thermostatically controlled loads by formal abstractions. IEEE Transactions on Control Systems Technology (accepted, 2014)

16. Esmaeil Zadeh Soudjani, S., Abate, A.: Probabilistic reach-avoid computation for partially-degenerate stochastic processes. IEEE Transactions on Automatic Control 59(2), 528–534 (2014)

17. Fränzle, M., Hermanns, H., Teige, T.: Stochastic satisfiability modulo theory: A novel technique for the analysis of probabilistic hybrid systems. In: Egerstedt, M., Mishra, B. (eds.) HSCC 2008. LNCS, vol. 4981, pp. 172–186. Springer, Heidelberg (2008)

18. Kamgarpour, M., Ellen, C., Esmaeil Zadeh Soudjani, S., Gerwinn, S., Mathieu, J.L., Mullner, N., Abate, A., Callaway, D.S., Franzle, M., Lygeros, J.: Modeling options for demand side participation of thermostatically controlled loads. In: International Conference on Bulk Power System Dynamics and Control (IREP), pp. 1–15 (August 2013)

19. Koch, S., Mathieu, J.L., Callaway, D.S.: Modeling and control of aggregated heterogeneous thermostatically controlled loads for ancillary services. In: 17th Power Systems Computation Conference, Stockholm, Sweden (August 2011)

20. Lecchini Visintini, A., Lygeros, J., Maciejowski, J.: Stochastic optimization on continuous domains with finite-time guarantees by Markov chain Monte Carlo methods. IEEE Transactions on Automatic Control 55(12), 2858–2863 (2010)

21. Malhame, R., Chong, C.-Y.: Electric load model synthesis by diffusion approximation of a high-order hybrid-state stochastic system. IEEE Transactions on Automatic Control 30(9), 854–860 (1985)

22. Mathieu, J.L., Callaway, D.S.: State estimation and control of heterogeneous thermostatically controlled loads for load following. In: Hawaii International Conference on System Sciences, Hawaii, USA, pp. 2002–2011 (2012)

23. Mathieu, J.L., Kamgarpour, M., Lygeros, J., Callaway, D.S.: Energy arbitrage with thermostatically controlled loads. In: European Control Conference (ECC), pp. 2519–2526 (2013)

24. Mathieu, J.L., Koch, S., Callaway, D.S.: State estimation and control of electric loads to manage real-time energy imbalance. IEEE Transactions on Power Systems 28(1), 430–440 (2013)

25. Mortensen, R.E., Haggerty, K.P.: A stochastic computer model for heating and cooling loads. IEEE Transactions on Power Systems 3(3), 1213–1219 (1988)

26. Tkachev, I., Abate, A.: Formula-free finite abstractions for linear temporal verification of stochastic hybrid systems. In: Proceedings of the 16th International Conference on Hybrid Systems: Computation and Control, Philadelphia, PA, pp. 283–292 (April 2013)

27. Vidyasagar, M.: Learning and generalization: with applications to neural networks. Springer (1997)

28. Vidyasagar, M.: Randomized algorithms for robust controller synthesis using statistical learning theory. Automatica 37, 1–2 (2001)

29. Zhang, W., Kalsi, K., Fuller, J., Elizondo, M., Chassin, D.: Aggregate model for heterogeneous thermostatically controlled loads with demand response. In: IEEE PES General Meeting, San Diego, CA (July 2012)

30. Zhang, W., Lian, J., Chang, C.-Y., Kalsi, K.: Aggregated modeling and control of air conditioning loads for demand response. IEEE Transactions on Power Systems 28(4), 4655–4664 (2013)

Finite Abstractions
of Stochastic Max-Plus-Linear Systems[*]

Dieky Adzkiya[1], Sadegh Esmaeil Zadeh Soudjani[1], and Alessandro Abate[2,1]

[1] Delft Center for Systems and Control
TU Delft, Delft University of Technology, The Netherlands
{D.Adzkiya,S.EsmaeilZadehSoudjani}@tudelft.nl
[2] Department of Computer Science
University of Oxford, United Kingdom
Alessandro.Abate@cs.ox.ac.uk

Abstract. This work investigates the use of finite abstractions to study the finite-horizon probabilistic invariance problem over Stochastic Max-Plus-Linear (SMPL) systems. SMPL systems are probabilistic extensions of discrete-event MPL systems that are widely employed in the engineering practice for timing and synchronisation studies. We construct finite abstractions by re-formulating the SMPL system as a discrete-time Markov process, then tailoring formal abstraction techniques in the literature to generate a finite-state Markov Chain (MC), together with precise guarantees on the level of the introduced approximation. This finally allows to probabilistically model check the obtained MC against the finite-horizon probabilistic invariance specification. The approach is practically implemented via a dedicated software, and elucidated in this work over numerical examples.

Keywords: Max-plus-linear systems, Max-plus algebra, Discrete-time stochastic processes, Continuous state spaces, Abstractions, Approximate probabilistic bisimulations.

1 Introduction

Max-Plus-Linear (MPL) systems are a class of discrete-event systems [1, 2] with a continuous state space characterising the timing of the underlying sequential discrete events. MPL systems are predisposed to describe the timing synchronisation between interleaved processes, under the assumption that timing events are dependent linearly (within the max-plus algebra) on previous event occurrences. MPL systems are widely employed in the analysis and scheduling of infrastructure networks, such as communication and railway systems [3], production and manufacturing lines [4, 5], or biological systems [6].

---

[*] The first two authors have equally contributed to this work. This work has been supported by the European Commission via STREP project MoVeS 257005, Marie Curie grant MANTRAS 249295, and IAPP project AMBI 324432.

G. Norman and W. Sanders (Eds.): QEST 2014, LNCS 8657, pp. 74–89, 2014.

Stochastic Max-Plus-Linear (SMPL) systems [7–9] are MPL systems where the delays between successive events (in the examples above, the processing or transportation times) are now characterised by random quantities. In practical applications SMPL systems are more realistic than simple MPL ones: for instance in a model for a railway network, train running times depend on driver behaviour, on weather conditions, and on passenger numbers at stations: they can arguably be more suitably modelled by random variables.

Only a few approaches have been developed in the literature to study the steady-state behaviour of SMPL systems, for example employing Lyapunov exponents and asymptotic growth rates [10–15]. The Lyapunov exponent of an SMPL system is analogous to the max-plus eigenvalue for an autonomous MPL system. The Lyapunov exponent of SMPL systems under some assumptions has been studied in [10], and later extended to approximate computations under other technical assumptions in [11, p. 251]. The application of model predictive control and system identification to SMPL systems is studied in [16, 17]. In contrast, our work focuses on one-step properties of SMPL systems and is based on developing finite-state abstractions: this is parallel to the approach in [18] for (deterministic) MPL systems. To the best of our knowledge, this contribution represents the first work on finite-state abstractions of SMPL systems.

Verification techniques and tools for deterministic, discrete-time, finite-state systems have been widely investigated and developed in the past decades [19]. The application of formal methods to stochastic models is typically limited to discrete-state structures, either in continuous or in discrete time [20, 21]. Continuous-space models on the other hand require the use of finite abstractions, as it is classically done for example with finite bisimulations of timed automata, which can be computed via the known region construction [22]. With focus on stochastic models, numerical schemes based on Markov Chain (MC) approximations of stochastic systems have been introduced in [23, 24], and applied to the approximate study of probabilistic reachability or invariance in [25, 26], however these finite abstractions do not come with explicit error bounds. On the contrary in [27], a technique has been introduced to instead provide formal abstractions of discrete-time, continuous-space Markov models [29], with the objective of investigating their probabilistic invariance by employing probabilistic model checking over a finite MC. In view of scalability and of generality, the approach has been improved and optimised in [30]. Interestingly the procedure has been shown [31] to introduce an approximate probabilistic bisimulation of the concrete model [32].

The aim of this work is to characterise and to compute the approximate solution of the finite-time invariance problem over SMPL systems: more precisely, for any allowable initial event time, we determine the probability that the time associated to the occurrence of N consecutive events will remain close to a given deterministic N-step schedule (cf. Section 2.2). The probabilistic invariance problem can be regarded as the dual of a reachability problem [29], and can be computed by constructing finite abstractions of the SMPL system, which are quantifiably close to the concrete model [27]. More precisely, our approach works as follows. We first formulate the given SMPL system as a discrete-time Markov process, as suggested by [8, 9]. Then we adapt the techniques in [27, 30] to the structure of the SMPL system, in order to generate a finite-state MC, together with guarantees on the level of approximation introduced in the process. The invariance property over the obtained MC can then be analysed via probabilistic model checking [20] and computed by existing software [33, 34]. The result obtained from the model checking software is then combined with the approximation guarantees, in order to provide an overall assessment of the probability that the concrete SMPL system satisfies the given property.

The article is structured as follows. Initially, Section 2.1 introduces the SMPL formalism, whereas Section 2.2 presents the probabilistic invariance problem. Section 3 discusses the formal abstraction of an SMPL system as an MC. Furthermore, with focus on the probabilistic invariance problem, the quantification of the abstraction error and some numerical examples are presented in Section 4. Finally, Section 5 concludes the presentation of this work.

2 Preliminaries

This section introduces the basics of max-plus algebra and of autonomous SMPL systems, and discusses the probabilistic invariance problem, which is to be further elaborated throughout the paper.

2.1 Modelling: Stochastic Max-Plus-Linear Systems

The notations $\mathbb{N}$ and $\mathbb{N}_n$ represent the whole positive integers $\{1, 2, \ldots\}$ and the first n positive integers $\{1, 2, \ldots, n\}$, respectively. We use the bold letters for vectors and usual letters with the same name and index for the elements of the vector, for instance $\boldsymbol{x} = [x_1, \ldots, x_n]^T$. Furthermore we define $\mathbb{R}_\varepsilon$, ε and e respectively as $\mathbb{R} \cup \{\varepsilon\}$, $-\infty$ and 0. For $\alpha, \beta \in \mathbb{R}_\varepsilon$, introduce the two operations

$$\alpha \oplus \beta = \max\{\alpha, \beta\} \qquad \text{and} \qquad \alpha \otimes \beta = \alpha + \beta ,$$

where the element ε is considered to be absorbing w.r.t. $\otimes$ [12, Definition 3.4], namely $\alpha \otimes \varepsilon = \varepsilon$ for all $\alpha \in \mathbb{R}_\varepsilon$. The rules for the order of evaluation of the max-algebraic operators correspond to those in the conventional algebra: max-algebraic multiplication has a higher precedence than max-algebraic addition [12, Sect. 3.1].

The basic max-algebraic operations are extended to matrices as follows. If $A, B \in \mathbb{R}_\varepsilon^{m \times n}$; $C \in \mathbb{R}_\varepsilon^{m \times p}$; $D \in \mathbb{R}_\varepsilon^{p \times n}$; and $\alpha \in \mathbb{R}_\varepsilon$, then

$$[\alpha \otimes A]_{ij} = \alpha \otimes A_{ij} , \quad [A \oplus B]_{ij} = A_{ij} \oplus B_{ij} , \quad [C \otimes D]_{ij} = \bigoplus_{k=1}^{p} C_{ik} \otimes D_{kj} ,$$

for each $i \in \mathbb{N}_m$ and $j \in \mathbb{N}_n$. Notice the analogy between $\oplus$, $\otimes$ and respectively $+$, $\times$ for matrix and vector operations in the conventional algebra. In this paper, the following notation is adopted for reasons of convenience. A vector with each component being equal to 0 (resp., $-\infty$) is also denoted by e (resp., ε). Furthermore, for practical reasons, the state space is taken to be $\mathbb{R}^n$ (rather than $\mathbb{R}_\varepsilon^n$).

An autonomous SMPL system is defined as:

$$x(k+1) = A(k) \otimes x(k) \ , \tag{1}$$

where $x(k) = [x_1(k), \ldots, x_n(k)]^T \in \mathbb{R}^n$; $\{A_{ij}(\cdot)\}$ are discrete-time stationary random processes[1] taking values in $\mathbb{R}_\varepsilon$; further $A_{ij}(k)$ are independent for all $k \in \mathbb{N} \cup \{0\}$ and $i, j \in \mathbb{N}_n$. We assume each random variable has fixed support [7, Definition 1.4.1], i.e. the probability of ε is either 0 or 1. The random sequence $\{A_{ij}(\cdot)\}$ is then characterised by a given density function $t_{ij}(\cdot)$ and corresponding distribution function $T_{ij}(\cdot)$ (cf. Theorem 1). The independent variable k denotes an increasing occurrence index, whereas the state variable $x(k)$ defines the (continuous) time of the k-th occurrence of the discrete events. The state component $x_i(k)$ denotes the time of the k-th occurrence of the i-th event. Since this article is based exclusively on autonomous (that is, not non-deterministic) SMPL systems, the adjective will be dropped for simplicity.

Example 1. Consider the following SMPL system representing a simple railway network between two connected stations. The state variables $x_i(k)$ for $i = 1, 2$ denote the time of the k-th departure at station i:

$$x(k+1) = A(k) \otimes x(k), \quad A(k) = \begin{bmatrix} 2 + e_{11}(k) & 5 + e_{12}(k) \\ 3 + e_{21}(k) & 3 + e_{22}(k) \end{bmatrix} \quad \text{or equivalently,}$$

$$\begin{bmatrix} x_1(k+1) \\ x_2(k+1) \end{bmatrix} = \begin{bmatrix} \max\{2 + e_{11}(k) + x_1(k), 5 + e_{12}(k) + x_2(k)\} \\ \max\{3 + e_{21}(k) + x_1(k), 3 + e_{22}(k) + x_2(k)\} \end{bmatrix} ,$$

where we have assumed that $e_{11}(\cdot) \sim Exp(1)$, $e_{12}(\cdot) \sim Exp(2/5)$, $e_{21}(\cdot) \sim Exp(2/3)$, and $e_{22}(\cdot) \sim Exp(2/3)$, and $Exp(\lambda)$ represents the exponential distribution with rate λ. Notice that $A_{ij}(\cdot)$ denotes the traveling time from station j to station i and amounts to a deterministic constant plus a delay modelled by the random variable $e_{ij}(\cdot)$. A few sample trajectories of the SMPL system, initialised at $x(0) = [1, 0]^T$, are displayed in Figure 1. Note that when all random delays are assumed to be equal to zero, the above deterministic system admits the unique solution $x(k) = x(0) + dk = [1 + 4k, 4k]^T$, where $d = 4$ is the max-plus eigenvalue of matrix A, and $[1, 0]^T$ is the corresponding eigenvector of the deterministic MPL system [12, Sect. 3.7]. Such a periodic trajectory can be used as a regular schedule for the train departures (cf. Section 2.2). □

---

[1] Notice that, for deterministic MPL systems, matrix A is instead given and time-invariant.

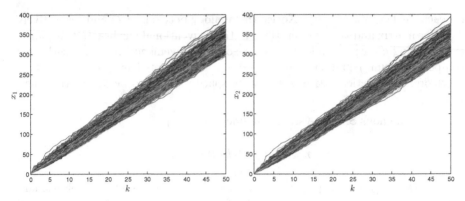

Fig. 1. Sample trajectories of the SMPL system in Example 1 for 50 discrete steps (horizontal axis) and both coordinates (vertical axis)

2.2   Problem: Probabilistic Invariance

Let us consider events that are scheduled to occur regularly, that is let us select a time between consecutive events that is a positive given constant, say d. We call this a *regular schedule* and assume that it does not affect the time of occurrence of all events, e.g. any event may occur ahead of the regular schedule. In this work, we consider an N-step finite-horizon probabilistic invariance problem w.r.t. a regular schedule: more specifically, for each possible time of initial occurrence of all events $(x_i(0), i \in \mathbb{N}_n)$, we are interested in determining the probability that the time of k-th occurrence of all events $(\boldsymbol{x}(k))$ remains close to the corresponding time of the regular schedule, for $k \in \mathbb{N}_N \cup \{0\}$. For instance, we may want to determine the probability that the time of occurrence of all events is at least 5 time units ahead of the given regular schedule, as well as at most 5 time units behind it. The safe set is then defined as the desired time of occurrence of all events w.r.t. the regular schedule.

The techniques in [27, 30], developed to provide the characterisation and the computation of the quantity of interest over general Markov processes, can be directly applied to the SMPL system (1). However, in order to prevent the growth of the safe set as the event horizon N increases (which in general leads to a decrease in computational performance), we reformulate the SMPL system based on the given regular schedule, so that a fixed safe set is obtained. Since we are interested in the delay of event occurrences with respect to the given schedule, we introduce new variables defined as the difference between the states of the original SMPL system and the regular schedule. More precisely, first we define a vector $\boldsymbol{s}(\cdot)$ that characterises the regular schedule. The dynamics of $\boldsymbol{s}(\cdot)$ are determined by the time duration $d \in \mathbb{R}$ between consecutive events[2] and the arbitrary initial condition $\boldsymbol{s}(0) \in \mathbb{R}^n$, i.e. $\boldsymbol{s}(k+1) = d \otimes \boldsymbol{s}(k)$. As mentioned, new

---

[2] Our results can be generalised to event-dependent time durations. In this case the Markov process becomes inhomogeneous, which will greatly increase the computational complexity of the procedure.

states $z(\cdot)$ are defined as the difference between the states of the original SMPL system (1) and the regular schedule $s(\cdot)$, i.e. $z(k) = x(k) - s(k)$ for $k \in \mathbb{N} \cup \{0\}$. The dynamics of the newly introduced SMPL system are then given by

$$z(k+1) = [A(k) + D] \otimes z(k) ,   \qquad (2)$$

where $D = [d_{ij}]_{i,j} \in \mathbb{R}^{n \times n}$ (i.e. d_{ij} is the entry of matrix D at row i and column j), $d_{ij} = s_j(0) - s_i(0) - d$, and $z(k) = [z_1(k), \ldots, z_n(k)]^T \in \mathbb{R}^n$. Notice that $A_{ij}(k) \otimes d_{ij}$ are independent for all $k \in \mathbb{N} \cup \{0\}$ and $i, j \in \mathbb{N}_n$. The density (resp., distribution) function of $A_{ij}(k) \otimes d_{ij}$ corresponds to the density (resp., distribution) function of $A_{ij}(k)$ shifted forward of d_{ij} units. The independent variable k again denotes an increasing occurrence index, whereas the state variable $z(k)$ defines the delay w.r.t. the schedule of k-th occurrence of all events: in particular the state component $z_i(k)$ denotes the delay w.r.t. the schedule of k-th occurrence of the i-th event. Notice that if the delay is negative then the event occurs ahead of schedule, whereas if the delay is positive then the event occurs behind schedule. The next theorem shows that, much like the original model in (1), the new SMPL system can be described as a discrete-time homogeneous Markov process.

Theorem 1. *The SMPL system in (2) is fully characterised by the following conditional density function*

$$t_z(\bar{z}|z) = \prod_{i=1}^{n} t_i(\bar{z}_i|z) \quad where$$

$$t_i(\bar{z}_i|z) = \sum_{j=1}^{n} \left[t_{ij}(\bar{z}_i - d_{ij} - z_j) \prod_{k=1, k \neq j}^{n} T_{ik}(\bar{z}_i - d_{ik} - z_k) \right] \quad i \in \mathbb{N}_n . \qquad (3)$$

Employing the introduced SMPL system (2), the problem can be formulated as the following N-step invariance probability

$$P_{z_0}(\mathcal{A}) = \Pr\{z(k) \in \mathcal{A} \text{ for all } k \in \mathbb{N}_N \cup \{0\} | z(0) = z_0\} ,$$

where $\mathcal{A}$ is called the safe set and is assumed to be Borel measurable. The next proposition provides a theoretical framework to study the problem.

Proposition 1 ([29, Lemma 1]). *Consider value functions $V_k : \mathbb{R}^n \to [0, 1]$, for $k \in \mathbb{N}_N \cup \{0\}$, computed through the following backward recursion:*

$$V_k(z) = \mathbb{1}_{\mathcal{A}}(z) \int_{\mathcal{A}} V_{k+1}(\bar{z}) t_z(\bar{z}|z) d\bar{z} \quad for all z \in \mathbb{R}^n ,$$

initialised with $V_N(z) = \mathbb{1}_{\mathcal{A}}(z)$ for all $z \in \mathbb{R}^n$. Then $P_{z_0}(\mathcal{A}) = V_0(z_0)$.

For any $k \in \mathbb{N}_N \cup \{0\}$, notice that $V_k(z)$ represents the probability that an execution of the SMPL system (2) remains within the safe set $\mathcal{A}$ over the residual event horizon $\{k, \ldots, N\}$, starting from z at event step k. This result characterises the finite-horizon probabilistic invariance problem as a dynamic programming problem. Since an explicit analytical solution to the problem is generally impossible to be found, we leverage the techniques developed in [27, 30] to provide a numerical computation with exact associated error bounds. This is elaborated in the next section.

3   Abstraction by a Finite State Markov Chain

We tailor the abstraction procedure presented in [27, Sect. 3.1] towards the goal of generating a finite-state MC $(\mathcal{P}, T_p)$ from a given SMPL system and a safe set $\mathcal{A}$, and employ it to approximately compute the probabilistic invariance of interest.

Let $\mathcal{P} = \{\phi_1, \ldots, \phi_{m+1}\}$ be a set of finitely many discrete states and $T_p : \mathcal{P} \times \mathcal{P} \to [0, 1]$ a related transition probability matrix, such that $T_p(\phi_i, \phi_j)$ characterises the probability of transitioning from state ϕ_i to state ϕ_j and thus induces a conditional discrete probability distribution over the finite space $\mathcal{P}$. Given a safe set $\mathcal{A}$, Algorithm 1 provides a procedure to abstract an SMPL system by a finite-state MC. The set $\mathcal{A}_p = \{\phi_1, \ldots, \phi_m\}$ denotes the discrete safe set. In Algorithm 1, $\Xi : \mathcal{A}_p \to 2^{\mathcal{A}}$ represents the concretisation function, i.e. a set-valued map that associates to any discrete state (point) $\phi_i \in \mathcal{A}_p$ the corresponding continuous partition set $\mathcal{A}_i \subset \mathcal{A}$. Furthermore the abstraction function $\xi : \mathcal{A} \to \mathcal{A}_p$ associates to any point $z \in \mathcal{A}$ on the SMPL state space, the corresponding discrete state in $\mathcal{A}_p$. Additionally, notice that an absorbing discrete state ϕ_{m+1} is added to the state space of the MC in order to render the transition probability matrix T_p stochastic: the absorbing discrete state ϕ_{m+1} represents the complement of the safe set $\mathcal{A}$ for the SMPL system, namely $\mathbb{R}^n \backslash \mathcal{A}$, and accounts for the associated dynamics.

Algorithm 1. Generation of a finite-state MC from an SMPL system and a safe set

---

Input: An SMPL system in (2) and a safe set $\mathcal{A}$
Output: A finite-state MC $(\mathcal{P}, T_p)$

1. Select a finite partition of set $\mathcal{A}$ of cardinality m, as $\mathcal{A} = \cup_{i=1}^{m} \mathcal{A}_i$
2. For each $\mathcal{A}_i$, select a single representative point $z_i \in \mathcal{A}_i$
3. Define $\mathcal{A}_p = \{\phi_i, i \in \mathbb{N}_m\}$ and take $\mathcal{P} = \mathcal{A}_p \cup \{\phi_{m+1}\}$ as the finite state-space of the MC (ϕ_{m+1} is an absorbing state, as explained in the text)
4. Compute the transition probability matrix T_p as

$$
T_p(\phi_i, \phi_j) = \begin{cases} \int_{\Xi(\phi_j)} t_z(\bar{z}|z_i)d\bar{z} \ , & \text{if } 1 \le j \le m \text{ and } 1 \le i \le m \ , \\ 1 - \sum_{\bar{\phi} \in \mathcal{A}_p} \int_{\Xi(\bar{\phi})} t_z(\bar{z}|z_i)d\bar{z} \ , & \text{if } j = m+1 \text{ and } 1 \le i \le m \ , \\ 1 \ , & \text{if } j = i = m+1 \ , \\ 0 \ , & \text{if } 1 \le j \le m \text{ and } i = m+1 \ , \end{cases}
$$

Remark 1. The bottleneck of Algorithm 1 lies in the computation of transition probability matrix T_p, due to the integration of kernel t_z. This integration can be circumvented if the distribution functions $T_{ij}(\cdot)$ for all $i, j \in \mathbb{N}_n$ have explicit analytical form, e.g. an exponential distribution.

The procedure in Algorithm 1 has been shown [31] to introduce an approximate probabilistic bisimulation of the concrete model [32].

Algorithm 1 can be applied to abstract an SMPL system as a finite-state MC, regardless of the particular safe set $\mathcal{A}$. However the quantification of the abstraction error in Section 4 requires that the safe set $\mathcal{A}$ is bounded.   □

Considering the obtained finite-state, discrete-time MC $(\mathcal{P}, T_p)$ and the discretised safe set $\mathcal{A}_p \subset \mathcal{P}$, the probabilistic invariance problem amounts to evaluating the probability that a finite execution associated with the initial condition $\phi_0 \in \mathcal{P}$ remains within the discrete safe set $\mathcal{A}_p$ during the given event horizon. This can be stated as following probability:

$$p_{\phi_0}(\mathcal{A}_p) = \Pr\{\phi(k) \in \mathcal{A}_p \text{ for all } k \in \mathbb{N}_N \cup \{0\} | \phi(0) = \phi_0\} \ ,$$

where $\phi(k)$ denotes the discrete state of the MC at step k.

The solution of this finite-horizon probabilistic invariance problem over the MC abstraction can be determined via a discrete version of Proposition 1.

Proposition 2. *Consider value functions $V_k^p : \mathcal{P} \to [0, 1]$, for $k \in \mathbb{N}_N \cup \{0\}$, computed through the following backward recursion:*

$$V_k^p(\phi) = \mathbb{1}_{\mathcal{A}_p}(\phi) \sum_{\bar{\phi} \in \mathcal{P}} V_{k+1}^p(\bar{\phi}) T_p(\phi, \bar{\phi}) \quad \text{for all } \phi \in \mathcal{P} \ ,$$

initialised with $V_N^p(\phi) = \mathbb{1}_{\mathcal{A}_p}(\phi)$ for all $\phi \in \mathcal{P}$. Then $p_{\phi_0}(\mathcal{A}_p) = V_0^p(\phi_0)$.

For any $k \in \mathbb{N}_N \cup \{0\}$, notice that $V_k^p(\phi)$ represents the probability that an execution of the finite-state MC remains within the discrete safe set $\mathcal{A}_p$ over the residual event horizon $\{k, \ldots, N\}$, starting from ϕ at event step k. The quantities in Proposition 2 can be easily computed via linear algebra. It is of interest to provide a quantitative comparison between the discrete outcome obtained by Proposition 2 and the continuous solution that results from Proposition 1: in other words, we are interested in deriving bounds on the abstraction error. The following section accomplishes this goal.

4   Quantification of the Abstraction Error

This section starts by precisely defining the error related to the abstraction procedure, which is due to the approximation of a continuous concrete model with a finite discrete one. Then a bound of the approximation error in [30] is recalled, and applied to the probabilistic invariance problem under some structural assumptions, namely in the case of Lipschitz continuous density functions, or alternatively of piecewise Lipschitz continuous density functions.

The approximation error is defined as the maximum difference between the outcomes obtained by Propositions 1 and 2 for any pair of initial conditions $z_0 \in \mathcal{A}$ and $\xi(z_0) \in \mathcal{A}_p$. Since an exact computation of this error is not possible in general, we resort to determining an upper bound of the approximation error, which is denoted as E. More formally, we are interested in quantifying E that satisfies

$$|P_{z_0}(\mathcal{A}) - p_{\xi(z_0)}(\mathcal{A}_p)| \leq E \quad \text{for all } z_0 \in \mathcal{A} \ . \tag{4}$$

We raise the following assumption on the SMPL system. Recall that the density function of $A_{ij}(k) \otimes d_{ij}$ in (2) corresponds to the density function of $A_{ij}(k)$ in (1) shifted d_{ij} units forward.

Assumption 3. *The density functions $t_{ij}(\cdot)$ for $i, j \in \mathbb{N}_n$ are bounded:*

$$t_{ij}(z) \leq M_{ij} \quad \text{for all } z \in \mathbb{R} .$$

Assumption 3 implies the distribution functions $T_{ij}(\cdot)$ for $i, j \in \mathbb{N}_n$ are Lipschitz continuous. Recall that the (global) Lipschitz constant of a one-dimensional function can be computed as the maximum of the absolute value of the first derivative of the function. Thus

$$|T_{ij}(z) - T_{ij}(z')| \leq M_{ij}|z - z'| \quad \text{for all } z, z' \in \mathbb{R} .$$

For computation of the bound on approximation error, we use the following result based on [30], which has inspired most of this work.

Proposition 4 ([30, pp. 933-934]). *Suppose Assumption 3 holds and the density function $t_z(\bar{z}|z)$ satisfies the condition*

$$\int_{\mathcal{A}} |t_z(\bar{z}|z) - t_z(\bar{z}|z')| d\bar{z} \leq H \|z - z'\| \quad \text{for all } z, z' \in \mathcal{A} ,$$

then an upper bound on the approximation error in (4) is $E = NH\delta$, where N is the event horizon and δ is the partition diameter.

The partition diameter δ is defined in [27, Sect. 3.1]. We first determine the constant H for Lipschitz continuous density functions, then generalise the result to piecewise Lipschitz continuous density functions.

4.1 Lipschitz Continuous Density Functions

Assumption 5. *The density functions $t_{ij}(\cdot)$ for $i, j \in \mathbb{N}_n$ are Lipschitz continuous, namely there exist finite and positive constants h_{ij}, such that*

$$|t_{ij}(z) - t_{ij}(z')| \leq h_{ij}|z - z'| \quad \text{for all } z, z' \in \mathbb{R} .$$

Under Assumptions 3 and 5, the conditional density function $t_z(\bar{z}|z)$ is Lipschitz continuous. This opens up the application of the results in [27, 30] for the approximate solution of the probabilistic invariance problem. Notice that the Lipschitz constant of $t_z(\bar{z}|z)$ may be large, which implies a rather conservative upper bound on the approximation error. To improve this bound, we can instead directly use Proposition 4 presented before – an option also discussed in [30]. In particular we present three technical lemmas that are essential for the computation of the constant H. After the derivation of the improved bound, the obtained results are applied to a numerical example.

Lemma 1. *Any one-dimensional continuous distribution function $T(\cdot)$ satisfies the inequality*

$$\int_{\mathbb{R}} |T(\bar{z} - z) - T(\bar{z} - z')| d\bar{z} \leq |z - z'| \quad \text{for all } z, z' \in \mathbb{R} \ .$$

Lemma 2. *Suppose the random vector $\bar{z}$ can be organised as $\bar{z} = [\bar{z}_1^T, \bar{z}_2^T]^T$, so that its conditional density function is the multiplication of conditional density functions of $\bar{z}_1, \bar{z}_2$ as:*

$$f(\bar{z}|z) = f_1(\bar{z}_1|z) f_2(\bar{z}_2|z) \ .$$

Then for a given set $\mathcal{A} \in \mathcal{B}(\mathbb{R}^n)$ it holds that

$$\int_{\mathcal{A}} |f(\bar{z}|z) - f(\bar{z}|z')| d\bar{z} \leq \sum_{i=1}^{2} \int_{\Pi_i(\mathcal{A})} |f_i(\bar{z}_i|z) - f_i(\bar{z}_i|z')| d\bar{z}_i \ ,$$

where $\Pi_i(\cdot)$ represents the projection operator on the i-th axis.

Lemma 3. *Suppose the vector z can be organised as $z = [z_1^T, z_2^T]^T$, and that the density function of the conditional random variable $(\bar{z}|z)$ is of the form*

$$f(\bar{z}|z) = f_1(\bar{z}, z_1) f_2(\bar{z}, z_2) \ ,$$

where $f_1(\bar{z}, z_1), f_2(\bar{z}, z_2)$ are bounded non-negative functions with $M_1 = \sup f_1(\bar{z}, z_1)$ and $M_2 = \sup f_2(\bar{z}, z_2)$. Then for a given set $\mathcal{C} \in \mathcal{B}(\mathbb{R})$:

$$\int_{\mathcal{C}} |f(\bar{z}|z_1, z_2) - f(\bar{z}|z_1', z_2')| d\bar{z}$$

$$\leq M_2 \int_{\mathcal{C}} |f_1(\bar{z}, z_1) - f_1(\bar{z}, z_1')| d\bar{z} + M_1 \int_{\mathcal{C}} |f_2(\bar{z}, z_2) - f_2(\bar{z}, z_2')| d\bar{z} \ .$$

Theorem 2. *Under Assumptions 3 and 5, the constant H in Proposition 4 is*

$$H = \sum_{i,j=1}^{n} H_{ij} + (n-1) M_{ij} \ ,$$

where $H_{ij} = \mathcal{L}_i h_{ij}$, and where the constant $\mathcal{L}_i = \mathcal{L}(\Pi_i(\mathcal{A}))$ is the Lebesgue measure of the projection of the safe set onto the i-th axis.

We now elucidate the above results on a case study, and select a beta distribution to characterise delays. A motivation for employing a beta distribution is that its density function has bounded support. Thus by scaling and shifting the density function, we can construct a distribution taking positive real values within an interval. Recall that this distribution is used to model processing or transportation time, and as such it can only take positive values. Furthermore, the beta distribution can be used to approximate the normal distribution with arbitrary accuracy.

Definition 1 (Beta Distribution). *The general formula for the density function of the beta distribution is*

$$t(x; \alpha, \beta, a, b) = \frac{(x-a)^{\alpha-1}(b-x)^{\beta-1}}{B(\alpha, \beta)(b-a)^{\alpha+\beta-1}} \qquad \text{if } a \leq x \leq b ,$$

and 0 otherwise, where $\alpha, \beta > 0$ *are the shape parameters;* $[a, b]$ *is the support of the density function; and* $B(\cdot, \cdot)$ *is the beta function. A random variable* X *characterised by this distribution is denoted by* $X \sim Beta(\alpha, \beta, a, b)$.

The case where $a = 0$ and $b = 1$ is called the standard beta distribution. Let us remark that the density function of the beta distribution is unbounded if any of the shape parameters belongs to the interval $(1, 2)$. We remark that if the shape parameters are positive integers, the beta distribution has a piecewise polynomial density function, which has been used for system identification of SMPL systems in [17, Sect. 4.3].

Example 2. We apply the results in Theorem 2 to the following two-dimensional SMPL system (1), where $A_{ij}(\cdot) \sim Beta(\alpha_{ij}, \beta_{ij}, a_{ij}, b_{ij})$,

$$\begin{bmatrix} \alpha_{11} & \alpha_{12} \\ \alpha_{21} & \alpha_{22} \end{bmatrix} = \begin{bmatrix} 2 & 4 \\ 2 & 2 \end{bmatrix}, \begin{bmatrix} \beta_{11} & \beta_{12} \\ \beta_{21} & \beta_{22} \end{bmatrix} = \begin{bmatrix} 5 & 2 \\ 2 & 4 \end{bmatrix}, \begin{bmatrix} a_{11} & a_{12} \\ a_{21} & a_{22} \end{bmatrix} = \begin{bmatrix} 0 & 2 \\ 2 & 0 \end{bmatrix}, \begin{bmatrix} b_{11} & b_{12} \\ b_{21} & b_{22} \end{bmatrix} = \begin{bmatrix} 7 & 6.5 \\ 4 & 9 \end{bmatrix}.$$

Skipping the details of the direct calculations, the supremum and the Lipschitz constant of the density functions are respectively

$$\begin{bmatrix} M_{11} & M_{12} \\ M_{21} & M_{22} \end{bmatrix} = \begin{bmatrix} 1536/4375 & 15/32 \\ 3/4 & 15/64 \end{bmatrix}, \quad \begin{bmatrix} h_{11} & h_{12} \\ h_{21} & h_{22} \end{bmatrix} = \begin{bmatrix} 30/49 & 80/81 \\ 3/2 & 20/81 \end{bmatrix}.$$

Considering a regular schedule with $s(0) = [0, 0]^T$ and $d = 4$, selecting safe set $\mathcal{A} = [-5, 5]^2$, and event horizon $N = 5$, according to Theorem 2 we obtain an error $E = 176.4\delta$. In order to obtain an approximation error bounded by $E = 0.1$, we would need to discretise set $\mathcal{A}$ uniformly with 24942 bins per each dimension (step 1 of Algorithm 1). The representative points have been selected at the centre of the squares obtained by uniform discretisation (step 2). The obtained finite-state MC has $24942^2 + 1$ discrete states (step 3). The procedure to construct transition probability matrix (step 4) works as follows. For each $i, j \in \{1, \ldots, 24942^2 + 1\}$, we compute $T_p(\phi_i, \phi_j)$ which consists of four possible cases. If $1 \leq i, j \leq 24942^2$, then $T_p(\phi_i, \phi_j)$ is defined as the probability of transitioning from the i-th representative point z_i to the j-th partition set $\mathcal{A}_j$. If $1 \leq i \leq 24942^2$ and $j = 24942^2 + 1$, then $T_p(\phi_i, \phi_j)$ is defined as the probability of transitioning from the i-th representative point z_i to the complement of the safe set $\mathbb{R}^n \setminus \mathcal{A}$. Since the discrete state ϕ_{24942^2+1} is absorbing, then $T_p(\phi_{24942^2+1}, \phi_j) = 1$ if $j = 24942^2 + 1$, and is equal to 0 otherwise. The solution of the invariance problem obtained over the abstract model (cf. Proposition 2) is computed via the software tool FAUST^2 [35] and is depicted in Figure 2 (left panel). □

4.2   Piecewise Lipschitz Continuous Density Functions

It is clear that the structural assumptions raised in the previous section pose limitations on the applicability of the ensuing results. For the sake of generality, we now extend the previous results to the more general case encompassed by the following requirement.

Assumption 6. *The density functions $t_{ij}(\cdot)$ for $i, j \in \mathbb{N}_n$ are piecewise Lipschitz continuous, namely there exist partitions $\mathbb{R} = \cup_{k=1}^{m_{ij}} D_{ij}^k$ and corresponding finite and positive constants h_{ij}^k, such that*

$$t_{ij}(z) = \sum_{k=1}^{m_{ij}} t_{ij}^k(z) \mathbb{1}_{D_{ij}^k}(z) \qquad \text{for all } z \in \mathbb{R} \;,$$

$$|t_{ij}^k(z) - t_{ij}^k(z')| \le h_{ij}^k |z - z'| \qquad \text{for all } k \in \mathbb{N}_{m_{ij}} \text{ and } z, z' \in D_{ij}^k \;.$$

The notation k used in Assumption 6 is not a power and is not an occurrence index (1), but it denotes the index of a set in the partition of cardinality $\sum_{i,j} m_{ij}$. Notice that if Assumption 6 holds and the density functions are Lipschitz continuous, then Assumption 5 is automatically satisfied with $h_{ij} = \max_k h_{ij}^k$. In other words, with Assumption 6 we allow relaxing Assumption 5 to hold only within arbitrary sets partitioning the state space of the SMPL system. In fact, we could limit the assumptions to the safe set.

Under Assumptions 3 and 6, we now present a result extending Theorem 2 for the computation of the constant H.

Theorem 3. *Under Assumptions 3 and 6, the constant H in Proposition 4 is*

$$H = \sum_{i,j=1}^{n} H_{ij} + (n-1)M_{ij} \;,$$

where $H_{ij} = \mathcal{L}_i \max_k h_{ij}^k + \sum_k |J_{ij}^k|$ and $\mathcal{L}_i = \mathcal{L}(\Pi_i(\mathcal{A}))$. The notation $J_{ij}^k = \lim_{z \downarrow c_{ij}^k} t_{ij}(z) - \lim_{z \uparrow c_{ij}^k} t_{ij}(z)$ denotes the jump distance of the density function $t_{ij}(\cdot)$ at the k-th discontinuity point c_{ij}^k.

The constants H_{ij} in Theorem 3 are chosen for the satisfaction of the following inequalities

$$\int_{\Pi_i(\mathcal{A})} |t_{ij}(\bar{z}_i - d_{ij} - z_j) - t_{ij}(\bar{z}_i - d_{ij} - z_j')| d\bar{z}_i \le H_{ij} |z_j - z_j'| \;. \qquad (5)$$

In some cases, it is possible to obtain a smaller value for H_{ij} by substituting the density function directly into the inequality in (5). Furthermore H_{ij} may be independent of the size of the safe set. For instance, if the delay is modelled by an exponential distribution as in Example 1, then $A_{ij}(\cdot)$ for all $i, j \in \mathbb{N}_n$ follows a shifted exponential distribution, i.e. $A_{ij}(\cdot) \sim SExp(\lambda_{ij}, \varsigma_{ij})$. In this case, $H_{ij} = \lambda_{ij} + \lambda_{ij}^2 \mathcal{L}_i$, as per Theorem 3. However if we compute directly the left-hand side of (5), we get the quantity $H_{ij} = 2\lambda_{ij}$, which is independent of the shape of the safe set. This fact is now proven in general, for a class of distribution functions, in Theorem 4. Let us first introduce the following definition.

Definition 2 (Shifted Exponential Distribution). *The density function of an exponential distribution shifted by* ς *is given by*

$$t(x; \lambda, \varsigma) = \lambda \exp\{-\lambda(x - \varsigma)\}\theta(x - \varsigma) \;,$$

where $\theta(\cdot)$ *is the unit step function. A random variable* X *characterised by this distribution is denoted by* $X \sim SExp(\lambda, \varsigma)$.

Theorem 4. *Any random sequence* $A_{ij}(\cdot) \sim SExp(\lambda_{ij}, \varsigma_{ij})$ *satisfies inequality* (5) *with* $H_{ij} = 2\lambda_{ij}$.

Given the previous result, the bound related to the invariance-related abstraction error over SMPL systems with $A_{ij}(\cdot) \sim SExp(\lambda_{ij}, \varsigma_{ij})$ can be improved and explicitly shown as follows. The maximum value of the density function $t_{ij}(\cdot)$ equals λ_{ij}, i.e. $M_{ij} = \lambda_{ij}$ for all $i, j \in \mathbb{N}_n$. By Theorem 3 and Proposition 4, the bound of the approximation error is then

$$E = N\delta(n + 1) \sum_{i,j} \lambda_{ij} \;.$$

Let us go back to Example 2 and adapt according to Definition 2 and Theorem 4.

Example 3. Consider the following two-dimensional SMPL system (1), where $A_{ij}(\cdot) \sim SExp(\lambda_{ij}, \varsigma_{ij})$ and

$$\begin{bmatrix} \lambda_{11} & \lambda_{12} \\ \lambda_{21} & \lambda_{22} \end{bmatrix} = \begin{bmatrix} 1/2 & 1/3 \\ 1 & 1/3 \end{bmatrix} , \quad \begin{bmatrix} \varsigma_{11} & \varsigma_{12} \\ \varsigma_{21} & \varsigma_{22} \end{bmatrix} = \begin{bmatrix} 0 & 2 \\ 2 & 0 \end{bmatrix} .$$

Considering a regular schedule with $s(0) = [0,0]^T$ and $d = 4$, selecting safe set $\mathcal{A} = [-5,5]^2$, and event horizon $N = 5$, we get $E = 32.5\delta$. In order to obtain a desired error $E = 0.1$, we need to use 4597 bins per dimension on a uniform discretisation of the set $\mathcal{A}$. The solution of the invariance problem over the abstract model is presented in Figure 2 (right panel).

Let us now validate this outcome. We have computed 1000 sample trajectories, with an initial condition that has been uniformly generated from the level set corresponding to the probability 0.3, namely within the set $\{z : P_z(\mathcal{A}) \geq 0.3\}$. Practically, this means we have sampled the initial condition on points corresponding to colours warmer than the "orange line." Given the error bound $E = 0.1$, we would expect that the trajectories are invariant with a likelihood greater than 0.2. Among the cohort, we have found that 374 trajectories stay inside the safe set for the given 5 steps, which is aligned with the guarantee we have derived.

Furthermore we have compared the approximate solution against the following empirical approach: for each representative point, we generate 1000 sample trajectories starting from it and compute ratio of the number of trajectories that stay in the safe set for 5 steps to the total number of trajectories (1000).

The maximum absolute difference between the approximate solution and the empirical approach for all representative points is 0.0565, which aligns with the error bound of 0.1.

We have also done these two comparisons for the SMPL system in Example 2. The results are quite analogous to the ones obtained in this example.   □

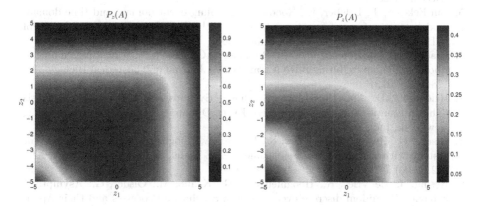

Fig. 2. The left and right plots show solution of the finite-horizon probabilistic invariance problem for two-dimensional SMPL systems with beta (Example 2) and exponential (Example 3) distributions, respectively. The plots have been obtained by computing the problem over finite abstractions obtained by uniform discretisation of the set of interest and selection of central representative points.

5   Conclusions and Future Work

This work has employed finite abstractions to study the finite-horizon probabilistic invariance problem over Stochastic Max-Plus-Linear (SMPL) systems. We have assumed that each random variable has a fixed support, which implies that the topology of the SMPL system is fixed over time. Along this line, we are interested to relax this assumption in order to obtain results that are robust against small topological changes. Furthermore, we are interested in considering extensions of the probabilistic invariance problem. Computationally, we are interested in improving the software and integrating it with FAUST^2 [35]. Finally, we have been exploring the existence of distributions associated to an analytical solution to the finite-horizon probabilistic invariance problem.

References

1. Baccelli, F., Cohen, G., Gaujal, B.: Recursive equations and basic properties of timed Petri nets. Discrete Event Dynamic Systems: Theory and Applications 1(4), 415–439 (1992)
2. Hillion, H., Proth, J.: Performance evaluation of job-shop systems using timed event graphs. IEEE Trans. Autom. Control 34(1), 3–9 (1989)

3. Heidergott, B., Olsder, G., van der Woude, J.: Max Plus at Work–Modeling and Analysis of Synchronized Systems: A Course on Max-Plus Algebra and Its Applications. Princeton University Press (2006)
4. Roset, B., Nijmeijer, H., van Eekelen, J., Lefeber, E., Rooda, J.: Event driven manufacturing systems as time domain control systems. In: Proc. 44th IEEE Conf. Decision and Control and European Control Conf. (CDC-ECC 2005), pp. 446–451 (December 2005)
5. van Eekelen, J., Lefeber, E., Rooda, J.: Coupling event domain and time domain models of manufacturing systems. In: Proc. 45th IEEE Conf. Decision and Control (CDC 2006), pp. 6068–6073 (December 2006)
6. Brackley, C.A., Broomhead, D.S., Romano, M.C., Thiel, M.: A max-plus model of ribosome dynamics during mRNA translation. Journal of Theoretical Biology 303, 128–140 (2012)
7. Heidergott, B.: Max-Plus Linear Stochastic Systems and Perturbation Analysis (The International Series on Discrete Event Dynamic Systems). Springer-Verlag New York, Inc., Secaucus (2006)
8. Olsder, G., Resing, J., De Vries, R., Keane, M., Hooghiemstra, G.: Discrete event systems with stochastic processing times. IEEE Trans. Autom. Control 35(3), 299–302 (1990)
9. Resing, J., de Vries, R., Hooghiemstra, G., Keane, M., Olsder, G.: Asymptotic behavior of random discrete event systems. Stochastic Processes and their Applications 36(2), 195–216 (1990)
10. van der Woude, J.W., Heidergott, B.: Asymptotic growth rate of stochastic max-plus systems that with a positive probability have a sunflower-like support. In: Proc. 8th Int. Workshop Discrete Event Systems, pp. 451–456 (July 2006)
11. Goverde, R., Heidergott, B., Merlet, G.: A coupling approach to estimating the Lyapunov exponent of stochastic max-plus linear systems. European Journal of Operational Research 210(2), 249–257 (2011)
12. Baccelli, F., Cohen, G., Olsder, G., Quadrat, J.P.: Synchronization and Linearity, An Algebra for Discrete Event Systems. John Wiley and Sons (1992)
13. Baccelli, F., Hong, D.: Analytic expansions of max-plus Lyapunov exponents. The Annals of Applied Probability 10(3), 779–827 (2000)
14. Gaubert, S., Hong, D.: Series expansions of Lyapunov exponents and forgetful monoids. Technical Report RR-3971, INRIA (July 2000)
15. Merlet, G.: Cycle time of stochastic max-plus linear systems. Electronic Journal of Probability 13(12), 322–340 (2008)
16. Farahani, S., van den Boom, T., van der Weide, H., De Schutter, B.: An approximation approach for model predictive control of stochastic max-plus linear systems. In: Proc. 10th Int. Workshop Discrete Event Systems, Berlin, DE, pp. 386–391 (August/September 2010)
17. Farahani, S., van den Boom, T., De Schutter, B.: Exact and approximate approaches to the identification of stochastic max-plus-linear systems. In: Discrete Event Dynamic Systems: Theory and Applications, pp. 1–25 (April 2013)
18. Adzkiya, D., De Schutter, B., Abate, A.: Finite abstractions of max-plus-linear systems. IEEE Trans. Autom. Control 58(12), 3039–3053 (2013)
19. Kurshan, R.P.: Computer-Aided Verification of Coordinating Processes: The Automata-Theoretic Approach. Princeton Series in Computer Science. Princeton University Press (1994)
20. Baier, C., Katoen, J.-P., Hermanns, H.: Approximate symbolic model checking of continuous-time Markov chains. In: Baeten, J.C.M., Mauw, S. (eds.) CONCUR 1999. LNCS, vol. 1664, pp. 146–162. Springer, Heidelberg (1999)

21. Kwiatkowska, M., Norman, G., Segala, R., Sproston, J.: Verifying quantitative properties of continuous probabilistic timed automata. In: Palamidessi, C. (ed.) CONCUR 2000. LNCS, vol. 1877, pp. 123–137. Springer, Heidelberg (2000)

22. Alur, R., Dill, D.: A theory of timed automata. Theoretical Computer Science 126(2), 183–235 (1994)

23. Kushner, H.J., Dupuis, P.: Numerical Methods for Stochastic Control Problems in Continuous Time. Springer, New York (2001)

24. Chaput, P., Danos, V., Panangaden, P., Plotkin, G.: Approximating Markov processes by averaging. Journal of the ACM 61(1), 5:1–5:45 (2014)

25. Koutsoukos, X., Riley, D.: Computational methods for reachability analysis of stochastic hybrid systems. In: Hespanha, J.P., Tiwari, A. (eds.) HSCC 2006. LNCS, vol. 3927, pp. 377–391. Springer, Heidelberg (2006)

26. Prandini, M., Hu, J.: Stochastic reachability: Theory and numerical approximation. In: Cassandras, C., Lygeros, J. (eds.) Stochastic Hybrid Systems. Automation and Control Engineering Series, vol. 24, pp. 107–138. Taylor & Francis Group/CRC Press (2006)

27. Abate, A., Katoen, J.P., Lygeros, J., Prandini, M.: Approximate model checking of stochastic hybrid systems. European Journal of Control 16(6), 624–641 (2010)

28. Esmaeil Zadeh Soudjani, S., Abate, A.: Adaptive gridding for abstraction and verification of stochastic hybrid systems. In: Proceedings of the 8th International Conference on Quantitative Evaluation of Systems, pp. 59–69 (September 2011)

29. Abate, A., Prandini, M., Lygeros, J., Sastry, S.: Probabilistic reachability and safety for controlled discrete time stochastic hybrid systems. Automatica 44(11), 2724–2734 (2008)

30. Esmaeil Zadeh Soudjani, S., Abate, A.: Adaptive and sequential gridding procedures for the abstraction and verification of stochastic processes. SIAM Journal on Applied Dynamical Systems 12(2), 921–956 (2013)

31. Abate, A.: Approximation metrics based on probabilistic bisimulations for general state-space Markov processes: a survey. Electronic Notes in Theoretical Computer Sciences, 3–25 (2014)

32. Desharnais, J., Edalat, A., Panangaden, P.: Bisimulation for labelled Markov processes. Information and Computation 179(2), 163–193 (2002)

33. Kwiatkowska, M., Norman, G., Parker, D.: PRISM 4.0: Verification of probabilistic real-time systems. In: Gopalakrishnan, G., Qadeer, S. (eds.) CAV 2011. LNCS, vol. 6806, pp. 585–591. Springer, Heidelberg (2011)

34. Katoen, J.P., Khattri, M., Zapreev, I.S.: A Markov reward model checker. In: Proc. 2nd Int. Conf. Quantitative Evaluation of Systems (QEST 2005), pp. 243–244. IEEE Computer Society, Los Alamos (2005)

35. Esmaeil Zadeh Soudjani, S., Gevaerts, C., Abate, A.: FAUST^2: Formal Abstractions of Uncountable-STate STochastic processes arXiv:1403.3286 (2014)

Mean Field for Performance Models
with Generally-Distributed Timed Transitions

Richard A. Hayden[1], Illés Horváth[2], and Miklós Telek[3,4]

[1] Department of Computing, Imperial College London, United Kingdom
[2] MTA-BME Information Systems Research Group
[3] Budapest University of Technology and Economics
Department of Networked Systems and Services
[4] Inter-University Center of Telecommunications and Informatics, Debrecen

Abstract. In this paper we extend the mean-field limit of a class of stochastic models with exponential and deterministic delays to include exponential and generally-distributed delays. Our main focus is the rigorous proof of the mean-field limit.

Keywords: mean-field limit, generally-distributed delay, delayed differential equation.

1   Introduction

In this paper, we further develop the new mean-field methodology introduced in [16] for a class of massively-parallel *generalised semi-Markov processes (GSMPs)* [20,14,15]. We focus on population models where individuals can enable both Markovian and generally-timed transitions, which are crucial for the accurate modelling of many real-world computer and networking protocols. We encode such models in a low-level formalism, the *population generalised semi-Markov process (PGSMP)*.

The motivation for the mean-field approach is the same as in the continuous-time Markov chain (CTMC) case — unsurprisingly, GSMP models with many components also become computationally intractable to explicit state techniques [7,9] rapidly as a result of the familiar state-space explosion problem. Our approach is based on the derivation of *delay differential equations (DDEs)* from PGSMP models and generalises the traditional mean-field approach as applied to CTMC models based on ordinary differential equations (ODEs) [1,13,4,17].

The class of models to which our approach applies is very broad — the only significant restriction we make is that at most one generally-timed transition may be enabled by each individual in any given local state. However, globally, there is no restriction on the concurrent enabling of generally-timed transitions by different individuals.

As in the CTMC case, the size of the system of DDEs is equal to the number of physical local states that components in the model can be in. Therefore

G. Norman and W. Sanders (Eds.): QEST 2014, LNCS 8657, pp. 90–105, 2014.

this approach represents a significant improvement with respect to both accuracy and efficiency when compared with the traditional CTMC mean-field approach where generally-timed transitions are approximated using phase-type distributions. The mean-field approach based on DDEs presented here captures generally-timed distributions directly without the need for additional physical states or for approximation of the generally-timed distribution itself.

This paper focuses on the non-racing case, that is, under the assumption that generally-timed transitions do not compete locally with exponential transitions (hence the term *delay-only* PGSMPs). The main contribution is to show how systems of coupled DDEs can be derived directly from PGSMP models with generally-timed transitions, and to give a full proof of transient mean-field convergence. The paper is quite proof-heavy; for a worked example, see Section 4, and for more examples, we refer to [16] and [5].

1.1   Related Work

Related work can be found in the biology and chemistry literature. Systems of DDEs have been derived to approximate stochastic models of reaction networks where deterministic delays are possible after reactions occur [3,6,21]. However, these models differ from those considered here in a number of critical ways; most importantly, the presentation in this paper lacks the severe rigidity of models encountered in biology and chemistry, making it suitable for a much larger class of population models.

Closest related work is due to [16] and [5] which both deal with deterministic delay-only PGSMPs in different ways; our presentation is closest in spirit to [16], but the upgrade from deterministic delays to generally-timed delays calls for a careful and involved analysis.

The approach in [5] highlights the connection to ODE approximations of DDEs [19] which is directly analogous to the Erlang approximation of the delay in the PGSMP. The approach in the present paper, however, avoids any Erlang approximations whatsoever, proving the mean-field limit directly via probability concentration theorems.

2   Population Generalised Semi-Markov Processes

2.1   Definition of PGSMPs

A PGSMP model consists of many interacting components each inhabiting one of a finite set of *local states* $\mathcal{S}$. The global state space, say $\mathcal{X}$, of a PGSMP model then consists of elements $\mathbf{x} = (x_s)_{s \in \mathcal{S}}$ where each $x_s \in \mathbf{Z}^+$ tracks the number of components currently in the local state s.

Exponential transitions are specified by a finite set of *Markovian transitions* $\mathcal{C}$. Each $c \in \mathcal{C}$ specifies a finite *change multiset* L_c, which consists of tuples $(s, s') \in \mathcal{S} \times \mathcal{S}$ each of which specifies that as part of a c-transition, a distinct, uniformly randomly selected component currently in local state s moves to local state s'. Write also $\mathbf{l}^c = (l_s^c)_{s \in \mathcal{S}}$ where each $l_s^c := |\{(s', s) \in L_c\}| - |\{(s, s') \in L_c\}|$, which represents the total change in components that are in local state s when a c-transition occurs. The aggregate rate of c-transitions is given by a *rate function* $r_c : \mathcal{X} \to \mathbf{R}^+$. We assume that the rate function is defined such that it is zero whenever a transition is not possible due to there not being enough distinct components in the required local states.

Generally-timed transitions are specified by *event clocks* in a similar fashion to standard GSMPs [14]. Specifically, we assume a finite set of event clocks $\mathcal{E}$. Each event clock $e \in \mathcal{E}$ is specified by its *set of active states* $\mathcal{A}_e \subseteq \mathcal{S}$, its *event transition probability function* $p_e : \mathcal{S} \times \mathcal{S} \to [0, 1]$ and the *clock time distribution* given by a cumulative distribution function (CDF) F_e used to the set the clock.

When a component enters a state s in $\mathcal{A}_e$ for the first time, the clock is initialized according to the CDF F_e. After the clock time has elapsed, it moves immediately to a new local state by sampling from the discrete probability distribution $p_e(s, \cdot)$. The clock is disabled when the component leaves s and is reset by resampling from the distribution if it later returns to the set of active states.

As mentioned above, the key restriction we make for all PGSMP models considered in this paper is that at most one event clock may be active in any local state. That is, for each $s \in \mathcal{S}$, $|\{e \in \mathcal{E} : s \in \mathcal{A}_e\}| \leq 1$. We will see that this restriction is necessary for the mean-field analyses presented in the sequel. This restriction also means that, with probability one, it is not possible for two transitions (Markovian or generally-timed) to occur simultaneously within a single component. Finally, we write $\mathbf{x}^0$ for the initial state of the model.

2.2 Delay-Only PGSMPs

We will focus on a class of PGSMPs with the structural restriction that, within a given component, generally-timed transitions may not be enabled concurrently with Markovian ones. We refer to such models as *delay only* since the general transitions in the constituent components then serve only to introduce generally-distributed delays between periods of otherwise Markovian behaviour.

Formally, the class of delay-only PGSMPs is specified by two restrictions: for all $e \in \mathcal{E}$; if $s \in \mathcal{A}_e$ then there can be no $c \in \mathcal{C}$ with $(s, s') \in L_c$; and for each $s \in \mathcal{A}_e$, there must exist some $s' \in \mathcal{S}$ such that $s' \notin \mathcal{A}_{e'}$ for any $e' \in \mathcal{E}$ and $p_e(s, s') = 1$. The first restriction guarantees that no Markovian transitions are enabled concurrently with general transitions, as above. The second restriction guarantees, firstly, that after any general transition completes, the component jumps into a unique state.[1] Secondly, it also ensures that the completion of a general transition cannot immediately enable another.

---

[1] This is a technical but not, in fact, a modelling restriction, as the state space may be reconfigured so the general transition is followed by a Markovian transition sampling from any discrete probability distribution p_e.

2.3 Construction of Delay-Only PGSMPs in Terms of Poisson Processes

In this section we give a construction of the population processes of a delay-only PGSMP in terms of Poisson processes. Write $\mathbf{x}(t) \in \mathbf{Z}^{+|\mathcal{S}|}$ for the underlying population process of a delay-only PGSMP, where $x_s(t) \in \mathbf{Z}^+$ tracks the number of components currently in the local state s.

Now let $\{P_c : c \in \mathcal{C}\}$ be a set of mutually-independent rate-1 Poisson processes and, for each $e \in \mathcal{E}$, let $\{T_i^e\}_{i=1}^{\infty}$ be mutually independent sequences of identically-distributed random variables distributed according to F_e, all also mutually independent of the Poisson processes. Then we may write:

$$
\begin{aligned}
x_s(t) = {} & x_s^0 + \sum_{c \in \mathcal{C}} l_s^c P_c \left(\int_0^t r_c(\mathbf{x}(u))\, du \right) \\
& + \sum_{e \in \mathcal{E}} \sum_{c \in \mathcal{C}} \left(\sum_{s' \in \mathcal{A}_e} p_e(s',s) l_{s'}^c - \mathbf{1}_{\{s \in \mathcal{A}_e\}} l_s^c \right) \\
& \times \int_{z=0}^t \mathbf{1}_{\left\{ T_{P_c(\int_0^z r_c(\mathbf{x}(u))\, du)}^e \le t-z \right\}} \, dP_c \left(\int_0^z r_c(\mathbf{x}(u))\, du \right)
\end{aligned}
\qquad (1)
$$

This is similar to the direct definition of [16]; the extra variables T_k^e are necessary due to the more complicated nature of the process.

3 Mean-Field Approximation of Delay-Only PGSMPs

For each local state $s \in \mathcal{S}$, we write $v_s(t)$ for the mean-field approximation to the number of components in state s at time $t \in \mathbf{R}^+$ and we also let $\mathbf{v}(t) = (v_s(t))_{s \in \mathcal{S}}$. The mean-field approximations satisfy the following system of integral equations:

$$
\begin{aligned}
v_s(t) = {} & v_s^0 + \sum_{c \in \mathcal{C}} l_s^c \int_0^t r_c(\mathbf{v}(u))\, du \\
& + \sum_{e \in \mathcal{E}} \sum_{c \in \mathcal{C}} \left(\sum_{s' \in \mathcal{A}_e} p_e(s',s) l_{s'}^c - \mathbf{1}_{\{s \in \mathcal{A}_e\}} l_s^c \right) \times \int_0^t F_e(t-u) r_c(\mathbf{v}(u))\, du
\end{aligned}
\qquad (2)
$$

3.1 Transient Mean-Field Convergence

In this section of the paper, we prove transient mean-field convergence for delay-only PGSMPs. We begin by constructing a sequence of delay-only PGSMP models indexed by $N \in \mathbf{Z}^+$ with increasing total component population size. Their underlying stochastic processes are denoted $\{\mathbf{x}^N(t) \in \mathbf{R}^{+|\mathcal{S}|}\}_{N \in \mathbf{Z}^+}$, where $\mathbf{x}^N(t) = (x_s^N(t))_{s \in \mathcal{S}}$ and $x_s^N(t) \in \mathbf{Z}^+$ tracks the number of components currently in the local state s for the Nth model.

We assume that the set of local states $\mathcal{S}$; the set of transitions $\mathcal{C}$ and the change multisets L_c; the sets of event clocks $\mathcal{E}$, the sets of active states $\mathcal{A}_e$, the transition probability functions p_e and the delay CDFs F_e are all fixed for all elements of the sequence. The rate functions r_c^N are allowed to vary with N and the initial conditions for the Nth model in the sequence are given by $N\mathbf{x}^0$ for some $\mathbf{x}^0 \in \mathbf{Z}^{+|\mathcal{S}|}$. Write $\mathcal{X}^N \subseteq \mathbf{Z}^{+|\mathcal{S}|}$ for the reachable state space of the Nth model. Note that following Section 2.3, we may write, in terms of a single set of Poisson processes $\{P_c : c \in \mathcal{C}\}$ and delay variables $\{T_i^e : e \in \mathcal{E}\}_{i=1}^\infty$:

$$x_s^N(t) = x_s^N(0) + \sum_{c \in \mathcal{C}} l_s^c P_c \left(\int_0^t r_c^N(\mathbf{x}^N(u))\, du \right)$$

$$+ \sum_{e \in \mathcal{E}} \sum_{c \in \mathcal{C}} \left(\sum_{s' \in \mathcal{A}_e} p_e(s',s) l_{s'}^c - \mathbf{1}_{\{s \in \mathcal{A}_e\}} l_s^c \right)$$

$$\times \int_0^t \mathbf{1}_{\left\{ T_{P_c(\int_0^z r_c^N(\mathbf{x}^N(u))\, du)}^e \leq t-z \right\}}\, dP_c \left(\int_0^z r_c^N(\mathbf{x}^N(u))\, du \right) \quad (3)$$

Similarly to the case of *density-dependent Markov chains* [12,18], we assume that we may define $r_c(\mathbf{x}) := (1/N) r_c^N(N\mathbf{x})$ for all $\mathbf{x} \in \mathbf{R}^{+|\mathcal{S}|}$ independently of N. Furthermore, we assume that r_c satisfies a local Lipschitz condition on $\mathbf{R}^{+|\mathcal{S}|}$ and that for all $c \in \mathcal{C}$, $r_c^N(\mathbf{x}) \leq R\|\mathbf{x}\|$ for all $\mathbf{x} \in \mathcal{X}^N$ where $R \in \mathbf{R}^+$ is independent of N. Define the rescaled processes $\bar{\mathbf{x}}^N(t) := (1/N)\mathbf{x}^N(t)$ that thus satisfy $r_c(\bar{\mathbf{x}}^N(t)) = r_c^N(\mathbf{x}^N(t))$.

We assume that initially, the system is concentrated on the non-active states $\mathcal{C} \setminus \cup_{e \in \mathcal{E}} \mathcal{A}_e$, in which case no initialization is necessary for the non-Markovian clocks. (For a discussion of the issue of initialization in the deterministic delay case, see [5]). We also assume

$$\|\mathbf{v}^0 - \bar{\mathbf{x}}^N(0)\| \to 0.$$

Note that in most applications, it is perfectly natural to set a deterministic initial condition, but we may also allow $\bar{\mathbf{x}}^N(0)$ to be random; in this case, assume

$$\mathbf{P}(\|\mathbf{v}^0 - \bar{\mathbf{x}}^N(0)\| > \epsilon) \to 0.$$

Theorem 1. *Under the assumptions and setup given above, we have, for any $T > 0$ and $\epsilon > 0$:*

$$\lim_{N \to \infty} \mathbf{P} \left\{ \sup_{t \in [0,T]} \|\bar{\mathbf{x}}^N(t) - \mathbf{v}(t)\| > \epsilon \right\} = 0$$

Remark. Actually, assuming the initial condition converges almost surely, we have almost sure convergence in the theorem, which is stronger than convergence in probability, but, since the probabilistic coupling presented in (3) is a technical issue with no underlying deeper connection, there is not much of a difference.

Proof. Define the auxiliary process

$$y_s^N(t) := v_s^0 + \sum_{c\in\mathcal{C}} l_s^c \int_0^t r_c(\bar{\mathbf{x}}^N(u))\,du$$

$$+ \sum_{e\in\mathcal{E}}\sum_{c\in\mathcal{C}}\left(\sum_{s'\in\mathcal{A}_e} p_e(s',s)l_{s'}^c - \mathbf{1}_{\{s\in\mathcal{A}_e\}}l_s^c\right) \times \int_0^t F_e(t-u)r_c(\bar{\mathbf{x}}^N(u))\,du$$

Then

$$|\bar{x}_s^N(t) - v_s(t)| \le |\bar{x}_s^N(t) - y_s^N(t)| + |y_s^N(t) - v_s(t)|.$$

Denote

$$D_s^N(T) = \sup_{t\in[0,T]} |\bar{x}_s^N(t) - y_s^N(t)|$$

We estimate $\mathbf{y}^N(t) - \mathbf{v}(t)$ by

$$|y_s^N(t) - v_s(t)| \le \sum_{c\in\mathcal{C}} |l_s^c| \int_0^t |r_c^N(\mathbf{x}^N(u)) - r_c(\mathbf{v}(u))|\,du$$

$$+ \sum_{e\in\mathcal{E}}\sum_{c\in\mathcal{C}}\left|\sum_{s'\in\mathcal{A}_e} p_e(s',s)l_{s'}^c - \mathbf{1}_{\{s\in\mathcal{A}_e\}}l_s^c\right| \times \int_0^t F_e(t-u)|r_c(\bar{\mathbf{x}}^N(u)) - r_c(\mathbf{v}(u))|\,du$$

$$\le ZR \int_0^t \|\mathbf{x}^N(u) - \mathbf{v}(u)\|\,du$$

where

$$Z := \sum_{c\in\mathcal{C}} |l_{k,s}^c| + \sum_{e\in\mathcal{E}_k}\sum_{c\in\mathcal{C}}\left|\sum_{s'\in\mathcal{A}_e} p_e(s',s)l_{k,s'}^c - \mathbf{1}_{\{s\in\mathcal{A}_e\}}l_{k,s}^c\right|$$

and $\|.\|$ is the maximum norm on $\mathbb{R}^{|\mathcal{S}|}$. We aim to show that $D_s^N(T) \to 0$ in probability for each $s \in \mathcal{S}$; once we have that, we have

$$\|\bar{\mathbf{x}}^N(t) - \mathbf{v}(t)\| \le \max_{s\in\mathcal{S}} D_s^N(t) + ZR \int_0^t \|\bar{\mathbf{x}}^N(u) - \mathbf{v}(u)\|\,du \qquad (4)$$

and an application of Grönwall's lemma ([12], page 498) readily yields

$$\|\bar{\mathbf{x}}^N(t) - \mathbf{v}(t)\| \le \max_{s\in\mathcal{S}} D_s^N(T)\exp(ZRT),$$

proving the theorem.

It now remains to show that for each $s \in \mathcal{S}$, $D_s^N(T) \to 0$ in probability as $N \to \infty$. To see this note that:

$$D_s^N(T) \le |v_s^0 - \bar{x}_s^N(0)| + \sum_{c\in\mathcal{C}} \frac{|l_s^c|}{N} \sup_{t\in[0,T]}\left|P_c\left(\int_0^t r_c(\bar{\mathbf{x}}^N(u))\,du\right) - \int_0^t r_c(\bar{\mathbf{x}}(u))\,du\right|$$

$$+ \sum_{c\in\mathcal{C}}\sum_{e\in\mathcal{E}} \frac{Y_{c,e}}{N} \sup_{t\in[0,T]}\left|\int_0^t \mathbf{1}_{\left\{T_{J_c^N(u)}^e \le t-u\right\}}\,dJ_c^N(u) - \int_0^t F_e(t-u)r_c^N(\bar{\mathbf{x}}(u))\,du\right|$$

where $Y_{c,e} := \left| \sum_{s' \in \mathcal{A}_e} p_e(s',s) l^c_{s'} - 1_{\{s \in \mathcal{A}_e\}} l^c_s \right|$, using the shorthand $J^N_c(u) :=$ $P_c \left(\int_0^u r^N_c(\mathbf{x}^N(z)) \, dz \right)$.

The first term converges per our assumptions; we argue that the second and third terms on the right-hand side converge almost surely. The second term is handled in the following lemma.

Lemma 1. *For any $c \in \mathcal{C}$*

$$\frac{1}{N} \sup_{t \in [0,T]} \left| P_c \left(\int_0^t r^N_c(\mathbf{x}^N(u)) du \right) - \int_0^t r^N_c(\mathbf{x}^N(u)) du \right| \to 0$$

almost surely as $N \to \infty$.

Proof. By the Lipschitz-condition, $0 \le \int_0^t r^N_c(\mathbf{x}^N(u)) du \le RTN$ and thus

$$\frac{1}{N} \sup_{t \in [0,T]} \left| P_c \left(\int_0^t r^N_c(\mathbf{x}^N(u)) \, du \right) - \int_0^t r^N_c(\mathbf{x}^N(u)) \, du \right| \le$$

$$\le \frac{1}{N} \sup_{s \in [0,RT]} |P_c(Ns) - Ns|,$$

which goes to 0 almost surely by the functional strong law of large numbers (FSLLN) for the Poisson process ([22], Section 3.2).

To handle the third term we note that:

$$\left| \int_0^t 1_{\left\{ T^e_{J^N_c(u)} \le t-u \right\}} \, dJ^N_c(u) - \int_0^t F_e(t-u) r^N_c(\mathbf{x}^N(u)) \, du \right| \le$$

$$\left| \int_0^t 1_{\left\{ T^e_{J^N_c(u)} \le t-u \right\}} \, dJ^N_c(u) - \int_0^t F_e(t-u) \, dJ^N_c(u) \right|$$

$$+ \left| \int_0^t F_e(t-u) \, dJ^N_c(u) - \int_0^t F_e(t-u) r^N_c(\mathbf{x}^N(u)) \, du \right|$$

($Y_{c,e}$'s are constants not depending on N and t, and as such, there is no need to carry them around.) The two terms on the right hand side require tools different enough to separate them into Lemmas 2 and 3. The proof of Lemma 2 is essentially a consequence of the FSLLN for the Poisson process, while the heart of the proof of Lemma 3 is a probability concentration (or large deviation) theorem (*Azuma's inequality*).

We have some more preparations first. We already have that

$$\sup_{t \in [0,T]} \frac{1}{N} \left| P_c \left(\int_0^t r^N_c(\mathbf{x}^N(u)) du \right) - \int_0^t r^N_c(\mathbf{x}^N(u)) du \right| \to 0$$

almost surely as $N \to \infty$. As a direct consequence of this, we also have

$$\sup_{s,t \in [0,T]} \frac{1}{N} \left| P_c \left(\int_s^t r^N_c(\mathbf{x}^N(u)) du \right) - \int_s^t r^N_c(\mathbf{x}^N(u)) du \right| \to 0$$

almost surely since

$$\sup_{s,t\in[0,T]} \left|\int_s^t \cdot\right| = \sup_{s,t\in[0,T]} \left|\int_0^t \cdot - \int_0^s \cdot\right| \le 2 \sup_{t\in[0,T]} \left|\int_0^t \cdot\right|$$

Also as a preparation, we have

$$\sup_{t\in[0,T]} \frac{1}{N} \int_0^t r_c^N(\mathbf{x}^N(u))du \le \sup_{t\in[0,T]} \frac{1}{N} \int_0^t R\|\mathbf{x}^N(u)\|du \le \frac{1}{N} \sup_{t\in[0,T]} NRt = RT$$

independent of N, again using $\|\mathbf{x}^N\| \le N$ and $r_c^N(\mathbf{x}) \le R\|\mathbf{x}\| \ \forall \mathbf{x} \in \mathcal{X}^N$. Lemma 1 then also implies $\frac{1}{N}\int_0^t dJ_c^N(u) \le RT + \varepsilon_N$, where $\varepsilon_N \to 0$ almost surely as $N \to \infty$.

Lemma 2

$$\sup_{t\in[0,T]} \frac{1}{N} \left|\int_0^t F_e(t-u)dJ_c^N(u) - \int_0^t F_e(t-u)r_c^N(\mathbf{x}^N(u))du\right| \to 0$$

almost surely as $N \to \infty$.

Proof. Let $\varepsilon > 0$ be fixed. Write

$$F_e(t-u) = g_{e,t,\varepsilon}(u) + h_{e,t,\varepsilon}(u),$$

where $g = g_{e,t,\varepsilon}$ is a piecewise constant function with $0 \le g(u) \le 1$ and $\|h\|_\infty \le \varepsilon$. Their exact definition is as follows. Take the $\varepsilon, 2\varepsilon, \dots$ quantiles of $F_e(t-u)$ (recall $F_e(t-u)$ is a nonincreasing function between 0 and 1); that is, let $u_k = \inf\{u : F(t-u) \le k\varepsilon\}$. Some of these u_k's may be equal if F has discontinuities. The number of distinct quantiles is certainly no more than $\lceil \varepsilon^{-1} \rceil$, independent of N and t.

Let g be the piecewise constant function

$$g(u) = F_e(t-u_k) \quad \text{if} \quad u \in (u_{k-1}, u_k],$$

so $g(u) \le F_e(t-u)$. The choice of u_k's guarantees that $h(u) = F_e(t-u) - g(u) \le \varepsilon$. Then we can write

$$\frac{1}{N} \left|\int_0^t F_e(t-u)dJ_c^N(u) - \int_0^t F_e(t-u)r_c^N(\mathbf{x}^N(u))du\right| \le$$
$$\frac{1}{N} \left|\int_0^t g(u)dJ_c^N(u) - \int_0^t g(u)r_c^N(\mathbf{x}^N(u))du\right| +$$
$$\frac{1}{N} \left|\int_0^t h(u)dJ_c^N(u) - \int_0^t h(u)r_c^N(\mathbf{x}^N(u))du\right|$$

Since g is piecewise constant,

$$\frac{1}{N}\left|\int_0^t g \mathrm{d}J_c^N(u) - \int_0^t g(u)r_c^N(\mathbf{x}^N(u))\mathrm{d}u\right| =$$

$$\frac{1}{N}\left|\sum_{k=1}^{\lceil\varepsilon^{-1}\rceil} g(u_k)\left(J_c^N(u_k) - J_c^N(u_{k-1}) - \int_{u_{k-1}}^{u_k} r_c^N(\mathbf{x}^N(u))\mathrm{d}u\right)\right| \leq$$

$$\frac{1}{N}\sum_{k=1}^{\lceil\varepsilon^{-1}\rceil} |g(u_k)|\left|\left(J_c^N(u_k) - J_c^N(u_{k-1}) - \int_{u_{k-1}}^{u_k} r_c^N(\mathbf{x}^N(u))\mathrm{d}u\right)\right| \leq$$

$$\frac{1}{N}\sum_{k=1}^{\lceil\varepsilon^{-1}\rceil} \left|J_c^N(u_k) - J_c^N(u_{k-1}) - \int_{u_{k-1}}^{u_k} r_c^N(\mathbf{x}^N(u))\mathrm{d}u\right| \leq$$

$$\sum_{k=1}^{\lceil\varepsilon^{-1}\rceil} \sup_{s,t\in[0,T]} \frac{1}{N}\left|P_c\left(\int_s^t r_c^N(\mathbf{x}^N(u))\mathrm{d}u\right) - \int_s^t r_c^N(\mathbf{x}^N(u))\right| =$$

$$\lceil\varepsilon^{-1}\rceil \cdot \sup_{s,t\in[0,T]} \frac{1}{N}\left|P_c\left(\int_s^t r_c^N(\mathbf{x}^N(u))\mathrm{d}u\right) - \int_s^t r_c^N(\mathbf{x}^N(u))\mathrm{d}u\right| \to 0$$

almost surely as $N \to \infty$ since ε is independent of N.

Since $\|h\|_\infty \leq \varepsilon$, we have

$$\frac{1}{N}\left|\int_0^t h(u)\mathrm{d}J_c^N(u) - \int_0^t h(u)r_c^N(\mathbf{x}^N(u))\mathrm{d}u\right| \leq$$

$$\frac{1}{N}\left|\int_0^t h(u)\mathrm{d}J_c^N(u)\right| + \frac{1}{N}\left|\int_0^t h(u)r_c^N(\mathbf{x}^N(u))\mathrm{d}u\right| \leq$$

$$\frac{\varepsilon}{N}\left|\int_0^t \mathrm{d}J_c^N(u)\right| + \frac{\varepsilon}{N}\left|\int_0^t r_c^N(\mathbf{x}^N(u))\mathrm{d}u\right| \leq \varepsilon(2RT + \varepsilon_N),$$

independent of t (with $\varepsilon_N \to 0$ almost surely as $N \to \infty$).

Letting $\varepsilon \to 0$ proves

$$\sup_{t\in[0,T]}\left|\int_0^t F_e(t-u)\mathrm{d}J_c^N(u) - \int_0^t F_e(t-u)r_c^N(\mathbf{x}^N(u))\mathrm{d}u\right| \to 0$$

almost surely as $N \to \infty$.

Lemma 3

$$\sup_{t\in[0,T]} \frac{1}{N}\left|\int_0^t \mathbf{1}_{\left\{T_{J_c^N(u)}^e \leq t-u\right\}} \mathrm{d}J_c^N(u) - \int_0^t F_e(t-u)\mathrm{d}J_c^N(u)\right| \to 0$$

almost surely as $N \to \infty$.

Proof. Let ε be fixed. Also fix t for now. We want to prove

$$\mathbf{P}\left(\left|\int_0^t \mathbf{1}_{\left\{T^e_{J^N_c(u)} \leq t-u\right\}} \mathrm{d}J^N_c(u) - \int_0^t F_e(t-u)\mathrm{d}J^N_c(u)\right| > \varepsilon\right)$$

is exponentially small in N via Azuma's inequality [8,2]. Once we have that, we can apply Borel–Cantelli lemma (see e.g. [10] Chapter 2.3) to conclude that for any fixed ϵ, the above event happens only finitely many times, which is equivalent to almost sure convergence to 0. To apply Azuma, we need to write the above integral as a martingale with bounded increments. The measure $\mathrm{d}J^N_c(u)$ is concentrated on points u where P_c has an arrival at $\int_0^u r^N_c(\mathbf{x}^N(z))\mathrm{d}z$. Let we denote these points by $u_1, u_2, \ldots$. The integral only has contributions from these points; it is natural to write (using a slightly different notation)

$$S_l := \left(\mathbf{1}_{\{T^e_1 \leq t-u_1\}} - F_e(t-u_1)\right) + \cdots + \left(\mathbf{1}_{\{T^e_l \leq t-u_l\}} - F_e(t-u_l)\right)$$

$$M_N := P_c\left(\int_0^t r^N_c(\mathbf{x}^N(z))\mathrm{d}z\right)$$

so that

$$\int_0^t \mathbf{1}_{\left\{T^e_{J^N_c(u)} \leq t-u\right\}} \mathrm{d}J^N_c(u) - \int_0^t F_e(t-u)\mathrm{d}J^N_c(u) = S_{M_N}.$$

We first resolve the difficulty that M_N is in fact random.

$$\mathbf{P}\left(\frac{1}{N}\left|\int_0^t \mathbf{1}_{\left\{T^e_{J^N_c(u)} \leq t-u\right\}} \mathrm{d}J^N_c(u) - \int_0^t F_e(t-u)\mathrm{d}J^N_c(u)\right| > \varepsilon\right) =$$

$$\mathbf{P}\left(\left|\frac{S_{M_N}}{N}\right| > \varepsilon\right) = \sum_{l=0}^\infty \mathbf{P}\left(\left|\frac{S_l}{N}\right| > \varepsilon, M_N = l\right) \leq$$

$$\sum_{l=0}^{2RTN} \mathbf{P}\left(\left|\frac{S_l}{N}\right| > \varepsilon\right) + \sum_{2RTN+1}^\infty \mathbf{P}(M_N = l) =$$

$$\sum_{l=0}^{2RTN} \mathbf{P}\left(\left|\frac{S_l}{N}\right| > \varepsilon\right) + \mathbf{P}(M_N > 2RTN).$$

The sum was cut at $2RTN$ because M_N is stochastically dominated by a Poisson distribution with parameter RTN, so $\mathbf{P}(M_N > 2RTN)$ is exponentially small due to Cramér's large deviation theorem (see e.g. Theorem II.4.1 in [11]):

$$\mathbf{P}(M_N > 2RTN) \leq e^{-RTN(2\ln 2 - 1)}.$$

(The Cramér rate function of the Poisson-distribution with parameter λ is $I(x) = x\ln(x/\lambda) - x + \lambda$.)

To apply Azuma to each of the terms $\mathbf{P}\left(\left|\frac{S_l}{N}\right| > \varepsilon\right)$, we also need to check that S_l is indeed a martingale with bounded increments. To set it up properly as a martingale, note that $\{u_l\}$ is an increasing sequence of stopping times, so the filtration $\{\mathcal{F}_l\}$ is well-defined; $\mathcal{F}_l$ contains all the information known up to time u_l, including the values of all of the non-Markovian clocks that started by the time u_l.

S_l has bounded increments, since

$$\left|\mathbf{1}_{\{T_l^e \leq t - u_l\}} - F_e(t - u_l)\right| \leq 1.$$

The last step to apply Azuma is that we need to check that S_l is a martingale with respect to $\mathcal{F}_l$. It is clearly adapted, and

$$\mathbf{E}(\mathbf{1}_{\{T_{l+1}^e \leq t - u_{l+1}\}}|\mathcal{F}_l) = \mathbf{E}(\mathbf{E}(\mathbf{1}_{\{T_{l+1}^e \leq t - u_{l+1}\}}|\mathcal{F}_l, u_{l+1})|\mathcal{F}_l) =$$
$$\mathbf{E}(\mathbf{P}(T_{l+1}^e \leq t - u_{l+1})|\mathcal{F}_l, u_{l+1})|\mathcal{F}_l) = \mathbf{E}(F_e(t - u_{l+1})|\mathcal{F}_l)$$

shows that it is a martingale as well. (In the last step, we used the fact that u_{l+1} is measurable with respect to $\sigma\{\mathcal{F}_l \cup \{u_{l+1}\}\}$ while T_{l+1}^e is independent from it.)

We have everything assembled to apply Azuma's inequality:

$$\sum_{l=0}^{2RTN} \mathbf{P}\left(\left|\frac{S_l - \mathbf{E}(S_l)}{N}\right| > \varepsilon\right) \leq \sum_{l=0}^{2RTN} 2e^{-\frac{2\varepsilon^2 N^2}{l}} \leq$$
$$\leq 2RTN \cdot 2e^{-\frac{2\varepsilon^2 N^2}{2RTN}} = 4RTNe^{-\frac{\varepsilon^2 N}{RT}}.$$

In the last inequality, we estimated each term in the sum by the largest one, which is for $l = 2RN$.

The estimate obtained is

$$\mathbf{P}\left(\frac{1}{N}\left|\int_0^t \mathbf{1}_{\left\{T_{J_c^N(u)}^e \leq t - u\right\}} dJ_c^N(u) - \int_0^t F_e(t - u)dJ_c^N(u)\right| > \varepsilon\right) \leq$$
$$4RTNe^{-\frac{\varepsilon^2 N}{RT}} + e^{-RTN(2\ln 2 - 1)}.$$

Remember that t was fixed; we need to upgrade this estimate into an estimate that is valid for $\sup_{t \in [0,T]}(.)$ before applying Borel–Cantelli lemma. We do this by partitioning the interval $[0, T]$ into N subintervals uniformly, and then controlling what happens *at* the partition points and *between* the partition points separately. For the former, we apply the previous estimate. Let

$$t_i := \frac{iT}{N}, \ i = 0, 1, \ldots N,$$

then

$$\mathbf{P}\left(\max_{0 \leq i \leq N} \frac{1}{N}\left|\int_0^{t_i} \mathbf{1}_{\left\{T_{J_c^N(u)}^e \leq t - u\right\}} dJ_c^N(u) - \int_0^{t_i} F_e(t - u)dJ_c^N(u)\right| > \varepsilon\right) \leq$$
$$(N + 1)\left(4RTNe^{-\frac{\varepsilon^2 N}{RT}} + e^{-RTN(2\ln 2 - 1)}\right),$$

which is still summable.

Now we turn our attention to the intervals $[t_i, t_{i+1}]$. Since

$$\int_0^t 1_{\left\{T^e_{J^N_c(u) \leq t-u}\right\}} dJ^N_c(u) \quad \text{and} \quad \int_0^t F_e(t-u)dJ^N_c(u)$$

are both increasing in t, we only have to check that neither of them increases by more than εN over an interval $[t_i, t_{i+1}]$.

Let i be fixed. We handle the two integrals separately. First, for

$$\int_0^t F_e(t-u)dJ^N_c(u),$$

we have

$$\int_0^{t_{i+1}} F_e(t_{i+1} - u)dJ^N_c(u) - \int_0^{t_i} F_e(t_i - u)dJ^N_c(u) =$$

$$\int_0^{t_i} F_e(t_{i+1} - u) - F_e(t_i - u)dJ^N_c(u) + \int_{t_i}^{t_{i+1}} F_e(t_{i+1} - u)dJ^N_c(u) \leq$$

$$\int_0^{t_i} F_e(t_{i+1} - u) - F_e(t_i - u)dJ^N_c(u) + \int_{t_i}^{t_{i+1}} 1dJ^N_c(u).$$

The second term is equal to $J^N_c(t_{i+1}) - J^N_c(t_i)$, e.g. the number of arrivals of P_c in the interval $[t_i, t_{i+1}]$. By the Lipschitz-condition, this is stochastically dominated from above by $Z \sim \text{Poisson}(RT)$ given that the length of the interval is T/N, and thus

$$\mathbf{P}\left(\frac{1}{N}\int_{t_i}^{t_{i+1}} F_e(t_{i+1} - u)dJ^N_c(u) > \varepsilon\right) \leq \mathbf{P}\left(\frac{Z}{N} > \varepsilon\right) = \mathbf{P}\left(\frac{Z}{\varepsilon} > N\right).$$

Note that the right hand side is summable in N, its sum being equal to the expectation of $\lceil \frac{Z}{\varepsilon} \rceil$.

To estimate the other term, note that

$$u \in [t_{l-1}, t_l] \implies F_e(t_{i+1} - u) - F_e(t_i - u) \leq F_e(t_{i+1} - t_{l-1}) - F_e(t_i - t_l) =$$
$$F_e(t_{i+1} - t_{l-1}) - F_e(t_{i+1} - t_l) + F_e(t_{i+1} - t_l) - F_e(t_i - t_l),$$

which gives

$$\int_0^{t_i} F_e(t_{i+1} - u) - F_e(t_i - u)dJ^N_c(u) =$$

$$\sum_{l=1}^i \int_{t_{l-1}}^{t_l} F_e(t_{i+1} - u) - F_e(t_i - u)dJ^N_c(u) \leq$$

$$\sum_{l=1}^i \int_{t_{l-1}}^{t_l} F_e(t_{i+1} - t_{l-1}) - F_e(t_i - t_l)dJ^N_c(u) =$$

$$\sum_{l=1}^i (F_e(t_{i+1} - t_{l-1}) - F_e(t_i - t_l))(J^N_c(t_l) - J^N_c(t_{l-1})).$$

We use two things here: the fact that $(J_c^N(t_l) - J_c^N(t_{l-1}))$ is stochastically dominated by Poisson(RT) and the fact that the sum

$$\sum_{l=1}^{i}(F_e(t_{i+1} - t_{l-1}) - F_e(t_i - t_l)) = \sum_{l=1}^{i} F_e(t_{i-l+2}) - F_e(t_{i-l}) =$$
$$F_e(t_{i+1}) + F_e(t_i) - F_e(1) - F_e(0) \leq 2$$

is telescopic. This means that the whole sum can be stochastically dominated by Poisson$(2RT)$ (note that the number of clocks starting at each interval is not independent, but because of the Lipschitz-condition, we may still use independent Poisson variables when stochastically dominating the sum). Using the notation $Z \sim$ Poisson(RT) again, we get that

$$\sum_{N=1}^{\infty} \mathbf{P}\left(\frac{2Z}{N} > \varepsilon\right) = \sum_{N=1}^{\infty} \mathbf{P}\left(\frac{2Z}{\varepsilon} > N\right) \leq \frac{2RT}{\varepsilon} + 1.$$

(In fact, $\mathbf{P}\left(\frac{2Z}{\varepsilon} > N\right)$ goes to 0 superexponentially in N.)

The last term to estimate is the increment of

$$\int_0^t \mathbf{1}_{\left\{T^e_{J_c^N(u)} \leq t-u\right\}} \mathrm{d}J_c^N(u).$$

between t_i and t_{i+1}, e.g. the number of clocks expiring between t_i and t_{i+1}.

Partition the clocks according to which interval $[t_{l-1}, t_l]$ they started in. The number of clocks starting in $[t_{l-1}, t_l]$ is stochastically dominated by $Z \sim$ Poisson(RT) by the Lipschitz-condition, and for each such clock, the probability that it goes off in $[t_i, t_{i+1}]$ is less than or equal to $F_e(t_{i+1}) - F_e(t_{l-1})$. This implies that the number of the clocks starting in $[t_{l-1}, t_l]$ and going off in $[t_i, t_{i+1}]$ is stochastically dominated by $W_{i,l} \sim$ Poisson$(RT(F_e(t_{i+1}) - F_e(t_{l-1})))$. The total number of clocks going off in $[t_i, t_{i+1}]$ is stochastically dominated by Poisson$(RT \sum_{l=1}^{i}(F_e(t_{i+1}) - F_e(t_{l-1})))$, where the familiar telescopic sum appears in the parameter. (Once again, the Lipschitz-condition was used implicitly.) So the total number of clocks going off in $[t_i, t_{i+1}]$ is stochastically dominated by Poisson$(2RT)$, which means we arrive at the also familiar $\mathbf{P}\left(\frac{2Z}{\varepsilon} > N\right)$ value, which we already examined and proved to be summable in N.

Putting it altogether, we get that

$$\mathbf{P}\left(\sup_{t \in [0,T]} \frac{1}{N}\left|\int_0^t \mathbf{1}_{\left\{T^e_{J_c^N(u)} \leq t-u\right\}} \mathrm{d}J_c^N(u) - \int_0^t F_e(t-u)\mathrm{d}J_c^N(u)\right| > \varepsilon\right) \leq C_{N,\varepsilon}$$

where

$$\sum_{N=1}^{\infty} C_{N,\varepsilon} < \infty,$$

so the Borel–Cantelli lemma gives almost sure convergence as $N \to \infty$.

With Lemmas 1-3 finished, the proof of Theorem 1 is complete.

Theorem 1 proves mean-field convergence in the transient case. The question of stationary regime is quite different; for some remarks on the stationary regime, we refer to Section 5 of [16].

4   Example

In this section, we derive the system of DDEs as defined in the previous section for a simple example model of a peer-to-peer software update process. For a more detailed discussion of a peer-to-peer update example, we refer to [16], where essentially the same model was introduced, albeit with deterministic delays instead of general ones.

We consider two general types of nodes in this model which we term *old* and *updated*. Old nodes are those running an old software version and new nodes are those which have been updated to a new version. Nodes alternate between being *on* and *off*; when an old node turns on, it searches for an update in peer-to-peer fashion, with the probability of successfully finding an update being proportional to the number of nodes already updated. If it does not find an update, it gives up after a timeout. After that, it stays on for some time and then eventually turns off. New nodes do not search for updates, just alternate between on and off. We assume that the off time of a node is random and Pareto-distributed. So, nodes have five possible local states: updated nodes can be on and off, which we denote by a and b, respectively. Old nodes can be on (c), off (e) or in a state representing an old node which is on but has given up seeking updates (d). In the notation of Section 2.1, the set of local states is thus $\mathcal{S} := \{a, b, c, d, e\}$. The local behaviour of a node is depicted in Figure 1.

In this example, we consider all transitions to be Markovian except for the transitions bringing nodes from their off state into their on state, which have density function $f(s)$. Formally, there are two event clocks t_0 and t_1 with $\mathcal{A}_{t_0} := \{e\}$, $d_{t_0} := \eta$, $p_{t_0}(e, c) = 1$, $\mathcal{A}_{t_1} := \{b\}$. $d_{t_1} := \eta$ and $p_{t_1}(b, a) := 1$.

The DDEs corresponding to this model are:

$$\dot{v}_a(t) = -\rho v_a(t) + \beta v_c(t) v_a(t) + \rho \int_0^t v_a(t-s) f(s)\,\mathrm{d}s$$

$$\dot{v}_b(t) = -\rho \int_0^t v_a(t-s) f(s)\,\mathrm{d}s + \rho v_a(t)$$

$$\dot{v}_c(t) = -\rho v_c(t) - \beta v_c(t) v_a(t) - \kappa v_c(t)$$
$$+ \rho \int_0^t (v_d(t-s) + v_c(t-s)) f(s)\,\mathrm{d}s$$

$$\dot{v}_d(t) = -\rho v_d(t) + \kappa v_c(t)$$

$$\dot{v}_e(t) = -\rho \int_0^t (v_d(t-s) + v_c(t-s)) f(s)\,\mathrm{d}s + \rho v_d(t) + \rho v_c(t)$$

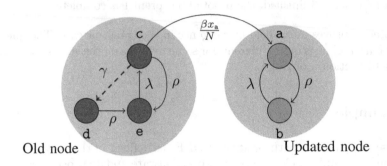

Fig. 1. Representation of the behaviour of a single node in the delay-only software update model

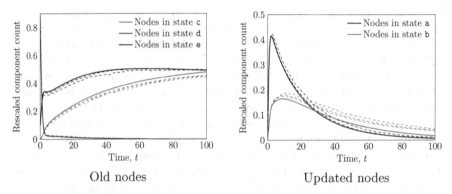

Fig. 2. Delay-only software update model rescaled DDE approximation (solid lines) compared with rescaled actual means for $N = 20$, 50 and 200 (dashed lines). Initial component proportions are $(0.1, 0, 0.9, 0, 0)$ and parameters are $\beta = 2.0$, $\rho = 0.1$, $\kappa = 0.67$ with $f(s)$ a Pareto density with scale parameter 1.5 and shape parameter 0.9.

This system of DDEs can be integrated numerically by adapting existing ODE solvers or specialised DDE routines such as the `dde23` routine in MATLAB®. The solution of these DDEs for one set of parameters is shown in Figure 2 compared with the corresponding rescaled component-count expectations as computed by many stochastic simulation replications. We observe that the means do appear to converge to the mean-field solutions in line with Theorem 1.

Acknowledgment. This work was partially supported by the OTKA K101150 project and by the TAMOP-4.2.2C-11/1/KONV-2012-0001 project.

References

1. Anselmi, J., Verloop, M.: Energy-aware capacity scaling in virtualized environments with performance guarantees. Perf. Eval. 68(11), 1207–1221 (2011)
2. Azuma, K.: Weighted sums of certain dependent random variables. Tohoku Mathematical Journal 19(3), 357–367 (1967)

3. Barbuti, R., Caravagna, G., Maggiolo-Schettini, A., Milazzo, P.: Delay stochastic simulation of biological systems: A purely delayed approach. In: Priami, C., Back, R.-J., Petre, I., de Vink, E. (eds.) Transactions on Computational Systems Biology XIII. LNCS, vol. 6575, pp. 61–84. Springer, Heidelberg (2011)
4. Benaïm, M., Le Boudec, J.-Y.: A class of mean field interaction models for computer and communication systems. Performance Evaluation 65(11-12), 823–838 (2008)
5. Bortolussi, L., Hillston, J.: Fluid approximation of ctmc with deterministic delays. In: Int. Conf. on Quantitative Evaluation of Systems, pp. 53–62 (2012)
6. Caravagna, G., Hillston, J.: Bio-PEPAd: A non-Markovian extension of Bio-PEPA. Theoretical Computer Science 419, 26–49 (2012)
7. Choi, H., Kulkarni, V., Trivedi, K., Marsan, M.A.: Transient analysis of deterministic and stochastic petri nets. In: Ajmone Marsan, M. (ed.) ICATPN 1993. LNCS, vol. 691, pp. 166–185. Springer, Heidelberg (1993)
8. Chung, F., Lu, L.: Concentration inequalities and martingale inequalities: A survey. Internet Mathematics 3(1), 79–127 (2006)
9. Cox, D.R.: The analysis of non-Markovian stochastic processes by the inclusion of supplementary variables. Mathematical Proceedings of the Cambridge Philosophical Society 51(03), 433–441 (1955)
10. Durrett, R.: Probability: Theory and Examples. Cambridge series on statistical and probabilistic mathematic. Cambridge University Press (2010)
11. Ellis, R.: Entropy, Large Deviations, and Statistical Mechanics. Classics in Mathematics. Springer (2005)
12. Ethier, S.N., Kurtz, T.G.: Markov Processes: Characterization and Convergence. Wiley (2005)
13. Gast, N., Bruno, G.: A mean field model of work stealing in large-scale systems. SIGMETRICS Perform. Eval. Rev. 38(1), 13–24 (2010)
14. Glynn, P.W.: A GSMP formalism for discrete event systems. Proceedings of the IEEE 77(1), 14–23 (1989)
15. Harrison, P.G., Strulo, B.: SPADES - a process algebra for discrete event simulation. Journal of Logic and Computation 10(1), 3–42 (2000)
16. Hayden, R.A.: Mean field for performance models with deterministically-timed transitions. In: 2012 Ninth International Conference on Quantitative Evaluation of Systems (QEST), pp. 63–73 (September 2012)
17. Hayden, R.A., Bradley, J.T.: A fluid analysis framework for a markovian process algebra. Theoretical Computer Science 411, 2260–2297 (2010)
18. Kurtz, T.G.: Strong approximation theorems for density dependent Markov chains. Stochastic Processes and their Applications 6(3), 223–240 (1978)
19. Maset, S.: Numerical solution of retarded functional differential equations as abstract cauchy problems. J. Comput. Appl. Math. 161(2), 259–282 (2003)
20. Matthes, K.: Zur theorie der bedienungsprozesse. In: 3rd Prague Conf. on Inf. Theory, Statistical Decision Functions and Random Processes, pp. 512–528 (1962)
21. Schlicht, R., Winkler, G.: A delay stochastic process with applications in molecular biology. Journal of Mathematical Biology 57(5), 613–648 (2008)
22. Whitt, W.: Internet supplement to Stochastic-Process Limits (2002)

Mean-Field Approximation
and Quasi-Equilibrium Reduction
of Markov Population Models[*]

Luca Bortolussi[1,2] and Rytis Paškauskas[2]

[1] DMG, University of Trieste, Italy
[2] ISTI Area della Ricerca CNR, via G. Moruzzi 1, 56124 Pisa, Italy

Abstract. Markov Population Model is a commonly used framework to describe stochastic systems. Their exact analysis is unfeasible in most cases because of the state space explosion. Approximations are usually sought, often with the goal of reducing the number of variables. Among them, the mean field limit and the quasi-equilibrium approximations stand out. We view them as techniques that are rooted in independent basic principles. At the basis of the mean field limit is the law of large numbers. The principle of the quasi-equilibrium reduction is the separation of temporal scales. It is common practice to apply both limits to an MPM yielding a fully reduced model. Although the two limits should be viewed as completely independent options, they are applied almost invariably in a fixed sequence: MF limit first, QE reduction second. We present a framework that makes explicit the distinction of the two reductions, and allows an arbitrary order of their application. By inverting the sequence, we show that the double limit does not commute in general: the mean field limit of a time-scale reduced model is not the same as the time-scale reduced limit of a mean field model. An example is provided to demonstrate this phenomenon. Sufficient conditions for the two operations to be freely exchangeable are also provided.

1 Introduction

Many complex systems whose dynamics is the result of the interaction of populations of indistinguishable agents can be described by Markov Population Models (MPM, [9,17]). This is the case, for instance, for biological systems and computer systems like queuing networks. Quantitative formal methods offer a powerful framework to describe and analyse them, using tools from verification and model checking. However, formal analysis of the Continuous Time Markov Chain (CTMC) that underlies an MPM is extremely challenging due to its usually large state space. Approximation techniques are therefore extremely useful, as they can lead to considerable simplifications of the analysis phase.

In this paper, we discuss two such methods. The first one is the fluid or mean-field approximation [9,12,1]. It has received considerable attention in the

---

[*] We acknowledge partial support from EU-FET project QUANTICOL (nr. 600708) and FRA-UniTS.

G. Norman and W. Sanders (Eds.): QEST 2014, LNCS 8657, pp. 106–121, 2014.

quantitative formal methods community in the past years with applications also to passage time computations [16] and stochastic model checking [5,10]. This method is based on a version of the law of large numbers for stochastic processes, known as Kurtz' theorem [12] which guarantees that, for large populations, an MPM is close to a (deterministic) ordinary differential equation (called fluid or mean-field ODE), converging to the latter in the limit of infinite population. This approximation holds for the transient behaviour under mild conditions on rate functions and model transitions, and can be extended to the steady state behaviour under additional assumptions on the limit ODEs [3,2].

Multiple time scale reduction, on the other hand, is based on a common intrinsic property of multi-dimensional dynamical systems to equilibrate unevenly. Several dimensions can be removed from a model if certain degrees of freedom equilibrate much faster than the rest. This is achieved by identifying the fast components and approximating them with the conditional equilibrium distribution. A rigorous definition involves singular perturbation [28]: A model that is a singular perturbation of another model is a multi-scale model. But this definition is too restrictive for real life situations (models with numerical rate constants of the same order may be multi-scale). Several methodologies and criteria to detect multiple time scales have been developed over the years: quasi-equilibrium and quasi-stationary state [18], computational singular perturbation[19], intrinsic low dimensional manifold [22] etc. Most of them have originated in chemistry [18,22,19] (ODE models) and [24,23,11,29] (stochastic models) and quite often the impression is that ODE and stochastic reductions are based on different assumptions. In the present article we propose a framework that eliminates this prejudice for a special, but important class of so-called quasi-equilibrium reductions [6]. Our contributions can be summarised as follows:

- we provide a consistent and constructive definition of the quasi-equilibrium reduction for MPMs. In particular, we treat uniformly mean field equations and stochastic processes by constructing reductions at a level of the MPM formalism [17]. For the stochastic case, we also formally prove the convergence of the full model to the reduced one when fast and slow time scales diverge.
- by examining the relationship between the QE reduction of the MF limit of a population process and the MF limit of the QE-reduced stochastic system we give sufficient conditions for the mean-field limit of the reduced stochastic system to exist and to be equal to the reduced mean field model. We also show that this is not true in general, and discuss scenarios where, application of the two limits in different order results in non-equivalent approximations.

The paper is organised as follows: In Section 2, we introduce Markov Population Models, while in Section 3 we review the mean-field approximation. Section 4, instead, is devoted to the presentation of the Quasi-Equilibrium reduction, both for differential equations and for MPMs. Section 5 contains our results about the relationship between mean-field and quasi-equilibrium, while in Section 6 we draw the final conclusions.

2   Markov Population Models

A Markov Population Model [17,9] is a simple formalism to describe models of populations of interacting agents based on Continuous-Time Markov Chains (CTMC). The formalism is inspired by chemical reaction networks [15], and is formally characterised by a tuple $\mathcal{X} = (\mathbf{X}, \mathcal{M}, \mathcal{T}, \mathbf{X}_0)$ where

1. $\mathbf{X} = (X_1, \ldots, X_n)^\mathsf{T}$ is a (column) vector of variables describing the n species of the model.
2. $\mathcal{M}$ is the domain of $\mathbf{X}$. Usually X_i counts the number of elements in a population of a species, therefore we assume $X_i \in \mathbb{N}$ and $\mathcal{M} \subseteq \mathbb{N}^n$.
3. $\mathcal{T} = \{\tau_1, \ldots, \tau_r\}$ is the set of r *transitions*, of the form $\tau = (\boldsymbol{\nu}, W)$, where:
   (a) $\boldsymbol{\nu} = (\nu_1, \ldots, \nu_n)^\mathsf{T} \in \mathcal{M}$ is a (column) *update vector*. This vector determines the stoichiometry of a transition, i.e. its elements equal the net change of the corresponding variable due to the transition.
   (b) $W : \mathcal{M} \mapsto \mathbb{R}_{\geq 0}$ is the *rate function*. We impose that all rate functions satisfy $W(\mathbf{X}) \geq 0$ and $W(\mathbf{X}) = 0$ if $\mathbf{X} + \boldsymbol{\nu} \notin \mathcal{M}$.
4. $\mathbf{X}_0 \in \mathcal{M}$ is the initial state: the process starts in $\mathbf{X}_0$ with probability one.

An MPM describes a Markovian stochastic process $\mathbf{X}(t)$ with r competing Poissonian (memoryless) transitions $\mathbf{X} \longrightarrow \mathbf{X} + \boldsymbol{\nu}_j$, with rates $W_j(\mathbf{X})$. Its analytic formulation is a 'master equation' for the probability mass $P(\mathbf{X}; t)$:

$$\partial_t P(\mathbf{X}; t) = \sum_{i=1}^{r} \{ W_i(\mathbf{X} - \boldsymbol{\nu}_i) P(\mathbf{X} - \boldsymbol{\nu}_i; t) - W_i(\mathbf{X}) P(\mathbf{X}; t) \}. \tag{1}$$

2.1   A Self-repressing Gene Network

We introduce now a simple 'running' example to illustrate the main concepts of the paper. Specifically, we consider the simplest gene network, composed of a single gene repressing its own expression. Despite its simplicity, this system is ubiquitously present in the genome [21]. We model it by a PCTMC $\mathcal{X} = (\mathbf{X}, \mathcal{M}, \mathcal{T}, \mathbf{X}_0)$ with three variables, $\mathbf{X} = (X_1, X_2, X_3)$, counting the amounts of, respectively, the repressed gene (X_1); the active, protein-producing gene (X_2); and the protein (X_3). The transcription-translation is lumped in one single step. The state space is $\mathcal{M} = \{0, \ldots, N\} \times \{0, \ldots, N\} \times \mathbb{N}$, where N is the number of copies of the gene in the system (cf. also the discussion at the end of Section 3). The model is specified by the following four transitions:

- $\tau_1 = (\boldsymbol{\nu}_1 = (\ 0, 0,\ \ 1)^\mathsf{T}, W_1(\mathbf{X}) = \varepsilon k_p X_2)$ – protein production;
- $\tau_2 = (\boldsymbol{\nu}_2 = (\ 0, 0, -1)^\mathsf{T}, W_2(\mathbf{X}) = \varepsilon k_d X_3)$ – protein degradation;
- $\tau_3 = (\boldsymbol{\nu}_3 = (-1, 1,\ \ 0)^\mathsf{T}, W_3(\mathbf{X}) = k_b X_2 X_3 / N)$ – repression, caused by the protein binding to a gene;
- $\tau_4 = (\boldsymbol{\nu}_4 = (1, -1, 0)^\mathsf{T}, W_4(\mathbf{X}) = k_u X_1)$ – the unbinding event.

Two remarks are in order: first, we do not remove a protein from the system when it bounds to the repressor. This is a minor tweak that simplifies the following discussion. Secondly, as typical for bimolecular reactions [15], we rescale the binding rate by the volume N, which for simplicity we assume here to equal the total amount of genes. In this way, the (copy number) concentration of the gene is between zero and one.

3   The Mean Field Limit of a MPM

Consider a MPM for a fixed *system size* N. The system size is usually interpreted as either the total population (typical of ecology and queueing networks' applications), or volume (chemical reaction networks). We can easily define a normalised MPM, by dividing variables by N, $\mathbf{X}^N = \mathbf{X}/N$, and expressing rates and updates with respect to these new variables. We call $\mathcal{M}^N$ the normalised state space, and further assume that the normalised state space satisfies $\bigcup_{N \in \mathbb{N}} \mathcal{M}^N \subseteq E$ for some open set $E \subseteq \mathbb{R}^n$. We call $W_j^N : E \mapsto \mathbb{R}_{\geq 0}$ the normalised rate functions for system size N, and assume $W_j^N(\mathbf{x})$ is defined for each $\mathbf{x} \in E$ (as usually the case).

Assumption 1. We require that:

(a) For each $j = 1, \ldots, r$, uniformly for $\mathbf{x} \in E$ it holds that

$$w_j(\mathbf{x}) = \lim_{N \to \infty} \frac{W_j^N(\mathbf{x})}{N}. \tag{2}$$

(b) Smoothness of functions $w_j(\mathbf{x})$, at least locally Lipschitz continuous.
(c) The normalised initial conditions converge: $\mathbf{X}_0^N \to \mathbf{x}_0 \in E$.

Under this assumption, the sequence of MPM $\mathbf{X}^N(t)$ converges (in probability, for any finite time horizon) to the solution $\mathbf{x}(t) = \mathbf{x}(t, \mathbf{x}_0)$ of the initial value problem

$$\frac{d\mathbf{x}}{dt}(t) = F(\mathbf{x}(t)), \quad \mathbf{x}(0) = \mathbf{x}_0, \quad F(\mathbf{x}) = \sum_{i=1}^{r} \nu_i w_i(\mathbf{x}), \tag{3}$$

where $F(\mathbf{x})$ is the (mean field) *drift* of the MPM. More formally, the following theorem holds [12]:

Theorem 1. *Under conditions a, b, and c above, for any $T < \infty$ and $\varepsilon > 0$,*

$$\lim_{N \to \infty} \mathbb{P} \left\{ \sup_{t \leq T} \left\| \mathbf{X}^N(t) - \mathbf{x}(t) \right\| > \varepsilon \right\} = 0.$$

We stress that Theorem 1 holds for any finite time window but it does not address the important question of steady state behaviour ($T = \infty$). Here the phenomenology is much wilder, and few things are known with certainty. However, if the mean field ODE (3) has a unique, globally attracting steady state $\mathbf{x}(\infty)$, i.e. for each $\mathbf{x}_0 \in E$, $\lim_{t \to \infty} \mathbf{x}(t, \mathbf{x}_0) = \mathbf{x}(\infty)$, then we have [3,2,9]:

Theorem 2. *Under the conditions of Theorem 1, if $\mathbf{X}^N(t)$ is ergodic and $\mathbf{x}(t, \mathbf{x}_0)$ has a unique globally attracting steady state, then*

$$\lim_{N \to \infty} \mathbf{X}^N(\infty) = \delta_{\mathbf{x}(\infty)} \quad \text{in probability,}$$

where $\delta_{\mathbf{x}(\infty)}$ is the point-wise mass probability at $\mathbf{x}(\infty)$.

Running Example. The mean field equations for the simple gene model are

$$\frac{dx_2}{dt}(t) = -\frac{dx_1}{dt}(t) = k_b x_2 x_3 - k_u x_1, \qquad \frac{dx_3}{dt}(t) = \varepsilon k_p x_2 - \varepsilon k_d x_3.$$

Theorem 1 asserts that a solution of these ODEs is exactly equivalent to the corresponding MPM in the limit $N = \infty$. The important question is whether this ODE is an acceptable approximation when $N < \infty$, as is always the case in practice. Intuitively, if there are many (paralogue) copies of the gene, so that transcription can happen concurrently, this ODE may be expected to be an excellent approximation to the MPM with a finite, but large N. If the number of gene copies remains small and constant with respect to N, we can still construct a hybrid limit, see [4]. For a discussion about the accuracy of mean field approximation, see [9].

4   Quasi-Equilibrium Reduction

In this section we provide formal definitions of the *Quasi-Equilibrium* framework with two objectives in mind. Firstly, we aim at generalising the 'canonical' setting where the fast and slow components of a model are decoupled by premise. We assume that they could be entangled, paying the price of a little extra formality. The second goal is to present a formal guideline of reducibility in the form of a list of easily verifiable conditions. This is achieved in Assumption 2 of section 4.3. However, we start by recalling two key ingredients of the reduction, *coordinate transforms* and *stoichiometric invariants* applied to MPMs.

4.1   Image of a MPM under a Change of Coordinates

A linear operator L acting on a finite dimensional vector space $\mathcal{M}$ is equivalent to matrix multiplication. We would like describe the L-action on an MPM. Define $L_A(\mathbf{x}) = A^\intercal \cdot \mathbf{x}$, where A is a real $n \times m$ matrix, and $\mathbf{x} \in \mathcal{M}$. If, in addition, $\mathbf{y} = L_A(\mathbf{x})$ is invertible (A is a square, invertible matrix) then the inverse, denoted by $\mathbf{x} = L_A^{-1}(\mathbf{y})$, is unique and $L_A^{-1} = L_{A^{-1}}$.

Fix such an invertible L and consider an MPM $\mathcal{X} = (\mathbf{X}, \mathcal{M}, \mathcal{T}, \mathbf{X}_0)$. The L-image of $\mathcal{X}$ is defined as $\mathcal{X}_L := L \circ \mathcal{X} = (\mathbf{Y}, \mathcal{N}, \mathcal{T}, \mathbf{Y}_0)$, where

- $\mathbf{Y} = L(\mathbf{X})$, $\mathbf{Y}_0 = L(\mathbf{X}_0)$, and $\mathcal{N} = L(\mathcal{M})$;
- Each transition $\tau = (\boldsymbol{\nu}, W)$ of $\mathcal{X}$ becomes the transition $\tau' = (\boldsymbol{\mu}, W')$, where $\boldsymbol{\mu} = L(\boldsymbol{\nu})$ and $W'(\mathbf{y}) = W(L^{-1}(\mathbf{y}))$.

It is obvious that, as L preserves all the update rules $\mathbf{X} \longrightarrow \mathbf{X} + \boldsymbol{\nu}_j$, $\mathcal{X}_L$ is equivalent to $\mathcal{X}$, in a sense that $\mathcal{X}_L$ represents the same stochastic process as $\mathcal{X}$, viewed in transformed coordinates $\mathbf{Y} = L(\mathbf{X})$.

4.2 Image of a MPM under a Stoichiometry Reduction

The $n \times r$ *stoichiometry matrix* S of a MPM $\mathcal{X}$ is a matrix composed from all the state change vectors $\boldsymbol{\nu}$, arranged as columns:

$$S_{\mathcal{X}} = (\boldsymbol{\nu}_1, \ldots, \boldsymbol{\nu}_r) \,. \tag{4}$$

Two important characteristics of S are, the rank $rank\,(S)$, and the co-dimension

$$codim\,(S) := n - rank\,(S) \geq \max\{0, n - r\} \,. \tag{5}$$

A MPM $\mathcal{X}$ is called *(stoichiometry) reducible* iff $m_{\mathcal{X}} := codim\,(S_{\mathcal{X}}) > 0$. By definition, there exist $m_{\mathcal{X}}$ linearly independent vectors $\mathbf{c}_1, \ldots, \mathbf{c}_{m_{\mathcal{X}}}$ such that $L_{\mathbf{c}_i}(\boldsymbol{\nu}_j) = 0$ for all i, j. This implies, for each $Y_i = L_{\mathbf{c}_i}(\mathbf{X})$, a transition $Y_i \longrightarrow Y_i + L_{\mathbf{c}_i}(\boldsymbol{\nu}_j) = Y_i$. Therefore, the vector $\mathbf{Y} = (Y_1, \ldots, Y_{m_{\mathcal{X}}})$ is conserved by dynamics. Its components are called *p-invariants*. They maintain constant values throughout dynamics therefore they can be made into parameters, rather than remaining independent variables. To achieve this, fix additional $n - m_{\mathcal{X}}$ vectors $\mathbf{k}_1, \ldots, \mathbf{k}_{n-m_{\mathcal{X}}}$, requiring that $\{\mathbf{c}_i\}$ and $\{\mathbf{k}_j\}$ should span $\mathcal{M}$. We arrange those vectors in two matrices $\mathsf{C} = (\mathbf{c}_1, \ldots, \mathbf{c}_{m_{\mathcal{X}}})$ and $\mathsf{K} = (\mathbf{k}_1, \ldots, \mathbf{k}_{n-m_{\mathcal{X}}})$. The matrix (C, K) is then invertible by definition. The *(stoichiometry) reduced* image of $\mathcal{X}$ is defined as $\mathcal{X}_{\mathsf{C},\mathsf{K}} := (\mathbf{Z}, \mathcal{K}, \mathcal{T}, \{\mathbf{Z}_0, \mathbf{Y}_0\})$ where

- $\mathbf{Z} = L_{\mathsf{K}}(\mathbf{X})$, $\mathcal{K} = L_{\mathsf{K}}(\mathcal{M})$, $\mathbf{Y}_0 = L_{\mathsf{C}}(\mathbf{X}_0)$, and $\mathbf{Z} = L_{\mathsf{K}}(\mathbf{X}_0)$;
- Each transition $\tau = (\boldsymbol{\nu}, W)$ of $\mathcal{X}$ becomes the transition $\tau = (\boldsymbol{\sigma}, \overline{W}_{\mathbf{Y}_0})$, where $\boldsymbol{\sigma} = L_{\mathsf{K}}(\boldsymbol{\nu})$ and $\overline{W}_{\mathbf{Y}_0}(\mathbf{Z}) = W\big(L^{-1}_{(\mathsf{C},\mathsf{K})}(\mathbf{Y}_0, \mathbf{Z})\big)$.

Running Example. Going back to the example of section 2.1, we have

$$S = (\boldsymbol{\nu}_1, \boldsymbol{\nu}_2, \boldsymbol{\nu}_3, \boldsymbol{\nu}_4) = \begin{pmatrix} 0 & 0 & -1 & 1 \\ 0 & 0 & 1 & -1 \\ 1 & -1 & 0 & 0 \end{pmatrix} \,.$$

We may recognise that $\mathbf{c} = (1, 1, 0)^{\mathsf{T}}$ is a p-invariant of the system. Letting $\mathbf{k}_1 = (0, 1, 0)^{\mathsf{T}}$, $\mathbf{k}_2 = (0, 0, 1)^{\mathsf{T}}$, we obtain the following reduced PCTMC model:

- $\mathbf{Z} = (Z_1, Z_2) = (X_2, X_3)$, $Y_0 = N$, $\mathcal{K} = \{0, \ldots, N\} \times \mathbb{N}$;
- State changes $\boldsymbol{\sigma}_i$ are obtained from the corresponding $\boldsymbol{\nu}_i$s by crossing out the first element. Rates $\tilde{W}$ are equal to Ws expressed in the new variables $\mathbf{Z}$. The rate of the 'unbind' transition becomes $\tilde{W}_4(Z_1, Z_2) = k_u(N - Z_1)$.

4.3 Fast-Slow Rate and Variable Decomposition of a PCTMC

We are now in position to describe the quasi-equilibrium reduction.

Assumption 2. Consider an MPM $\mathcal{X} = (\mathbf{X}, \mathcal{M}, \mathcal{T}, \mathbf{X}_0)$ such that

(a) There exist two parameters $T^{\text{slow}} > T^{\text{fast}} > 0$ and an integer s, $1 < s < r$, such that the ordering of all rate functions

$$\underbrace{W_1(\mathbf{X}) \leq \cdots \leq W_s(\mathbf{X})}_{\text{slow transitions}} \leq \frac{N}{T^{\text{slow}}} < \frac{N}{T^{\text{fast}}} \leq \underbrace{W_{s+1}(\mathbf{X}),\ldots,W_r(\mathbf{X})}_{\text{fast transitions}} \qquad (6)$$

is valid for all $\mathbf{X}$ in a sufficiently large subspace of $\mathcal{M}$, containing the initial condition $\mathbf{X}_0$. This condition is equivalent to requiring that rate functions behave with respect to dimensionless parameter $\varepsilon = \frac{T^{\text{fast}}}{T^{\text{slow}}}$ as follows

$$W_i(\mathbf{X};\varepsilon) \underset{\varepsilon \to 0}{\sim} \varepsilon W_{0,i}(\mathbf{X}) + O(\varepsilon^2), \quad i = 1,\ldots,s \qquad (7)$$

$$W_i(\mathbf{X};\varepsilon) \underset{\varepsilon \to 0}{\sim} W_{0,i}(\mathbf{X}) + O(\varepsilon), \quad i = s+1,\ldots,r \qquad (8)$$

where $W_{0,i}(\mathbf{X})$ are functions that do not depend on ε.

The set of transitions $\mathcal{T}$ is thus partitioned into slow transitions $\mathcal{T}^{\text{slow}} = \{\tau_1,\ldots,\tau_s\}$, and fast transitions $\mathcal{T}^{\text{fast}} = \{\tau_{s+1},\ldots,\tau_r\}$.

(b) $\mathcal{X}$ restricted to $\mathcal{T}^{\text{fast}}$ is stoichiometry reducible according to section 4.2, i.e.

$$m := codim\,(\boldsymbol{\nu}_{s+1},\ldots,\boldsymbol{\nu}_r) > 0. \qquad (9)$$

If both these conditions are satisfied, we may separate slow and fast components of $\mathbf{X}$, such separation being the basis of the subsequent dimensional reduction. Matrices C and K can be identified such that, following section 4.1, (C,K) is invertible and $L_{\mathsf{C}}(\boldsymbol{\nu}_i) = 0$, but only for $i = s+1,\ldots,r$. Define

$$\underbrace{\mathbf{Y} = (Y_1,\ldots,Y_m) = L_{\mathsf{C}}(\mathbf{X})}_{\text{slow variables}} \quad \underbrace{\mathbf{Z} = (Z_1,\ldots,Z_{n-m}) = L_{\mathsf{K}}(\mathbf{X})}_{\text{fast variables}} \qquad (10)$$

Note that $\mathbf{Y}$, owing to its definition in terms of fast transitions rather than all transitions, is not a p-invariant. This means that some transitions of the slow variable will occur, given by the updated vectors $\boldsymbol{\mu}$, defined as follows

$$L_{\mathsf{C}}(\boldsymbol{\nu}_1,\ldots,\boldsymbol{\nu}_s,\boldsymbol{\nu}_{s+1},\ldots,\boldsymbol{\nu}_r) = (\boldsymbol{\mu}_1,\ldots,\boldsymbol{\mu}_s,0,\ldots,0).$$

The fast subspace update vectors are similarly defined: $\boldsymbol{\sigma}_i = L_{\mathsf{K}}(\boldsymbol{\nu}_i)$.

Running Example. We assume that $\varepsilon \ll 1$ is a small dimensionless parameter. This assumption implies the partition $\mathcal{T}^{\text{slow}} = \{\tau_1,\tau_2\}$ and $\mathcal{T}^{\text{fast}} = \{\tau_3,\tau_4\}$. If all other parameters are $O(1)$, then there is a large gap between T^{fast} and T^{slow}, guaranteed by the smallness of ε, which we leave as the scale separation parameter. The procedure of stoichiometry reduction, applied to $\mathcal{T}^{\text{fast}}$, provides

$$m = codim \begin{pmatrix} 1 & -1 \\ 0 & 0 \end{pmatrix} = 2 - 1 = 1.$$

Since $m > 0$, this model is QE-reducible and indeed, $\boldsymbol{c} = (0,1)^{\mathsf{T}}$ is a p-invariant of $\mathcal{T}^{\text{fast}}$. Complementing the basis with $\boldsymbol{k} = (1,0)^{\mathsf{T}}$ we conclude that the slow variable is $Y = L_{\boldsymbol{c}}(\mathbf{X}) = X_3$ (protein) and the fast variable is $Z = L_{\boldsymbol{k}}(\mathbf{X}) = X_2$ (active gene). The ε-rescaled rates $W_{0,i}$, expressed in the slow-fast variables, are

$$W_{0,1} = k_p Z, \quad W_{0,2} = k_d Y, \quad W_{0,3} = k_b Y Z/N, \quad W_{0,4} = k_u(N - Z). \qquad (11)$$

4.4   Quasi-Equilibrium Reduction of the Mean-Field Model

As a demonstration of utility of our formalism, we will obtain the canonical equations of the singular perturbation theory [28] from the standard quasi-equilibrium approximation of ODEs.

Recall from Section 3 the definition of limit rate functions (2) and that of the limit drift vector $\boldsymbol{F}(\mathbf{x}) = \sum_{i=1}^{r} \boldsymbol{\nu}_i w_i(\mathbf{x}; \varepsilon)$, where we made explicit the dependence on a small parameter ε. If the MPM satisfies Assumption 2, then the asymptotic $\varepsilon \to 0$ dependence of the rate functions is $w_i(\mathbf{x}; \varepsilon) \sim \varepsilon w_{0,i}(\mathbf{x}) + O(\varepsilon^2)$ for $i = 1, \ldots, s$, and $w_i(\mathbf{x}; \varepsilon) \sim w_{0,i}(\mathbf{x}) + O(\varepsilon)$ for $j = s+1, \ldots, r$ and $1 < s < r$. Define the *slow variables* $\mathbf{y} = L_C(\mathbf{x})$, the *fast variables* $\mathbf{z} = L_K(\mathbf{x})$, and the *slow time* $\tau = \varepsilon t$. It is then straightforward to demonstrate that the mean field limit equations are equivalent to

$$\frac{d\mathbf{y}}{d\tau} = G(\mathbf{y}, \mathbf{z}) + O(\varepsilon), \quad \varepsilon \frac{d\mathbf{z}}{d\tau} = H(\mathbf{y}, \mathbf{z}) + O(\varepsilon) \tag{12}$$

where

$$G = \sum_{i=1}^{s} L_C(\boldsymbol{\nu}_i) w_{0,i}(L_{(C,K)}^{-1}(\mathbf{y}, \mathbf{z})), \quad H = \sum_{j=s+1}^{r} L_K(\boldsymbol{\nu}_j) w_{0,j}(L_{(C,K)}^{-1}(\mathbf{y}, \mathbf{z})).$$

Since ε multiplies the highest order derivative in (12) (right), the perturbation in ε is singular [28]. The construction of a reduced model from equations (12) is governed by further assumptions provided by the *Tikhonov theorem* [28, Theorem 8.1].

Assumption 3. Consider the initial value problem (12) for $\tau \geq 0$, with $\mathbf{y}(0) = \mathbf{y}_0$, $\mathbf{z}(0) = \mathbf{z}_0$. We further require:

(a) the drifts $G(\mathbf{y}, \mathbf{z})$ and $H(\mathbf{y}, \mathbf{z})$ are sufficiently smooth functions of their arguments.
(b) a unique solution $\mathbf{y}^\varepsilon(\tau)$, $\mathbf{z}^\varepsilon(\tau)$ of the initial value problem (12) exists;
(c) a unique solution $\overline{\mathbf{y}}(\tau)$, $\overline{\mathbf{z}}(t)$ of the reduced initial value problem exists; the reduced problem being defined by

$$d\mathbf{y}/d\tau = G(\mathbf{y}, \mathbf{z}), \quad \mathbf{y}(0) = \mathbf{y}_0, \quad 0 = H(\mathbf{y}, \mathbf{z}),$$

(d) equation $0 = H(\mathbf{y}, \mathbf{z})$ is solved by $\mathbf{z} = \phi(\mathbf{y})$ where ϕ is continuous, and it is an isolated root;
(e) $\mathbf{z} = \phi(\mathbf{y})$ is an asymptotically stable solution of $d\mathbf{z}/dt = H(\mathbf{y}, \mathbf{z})$ uniformly in $\mathbf{y}(\tau)$, considered as a (fixed) parameter;
(f) $\mathbf{z}(0)$ is contained in an interior subset of the domain of attraction of $\mathbf{z} = \phi(\mathbf{y})$ for $\mathbf{y} = \mathbf{y}(0)$.

The previous conditions guarantee that the solution of the reduced problem (defined in (c) above) is actually the $\varepsilon \to 0$ limit of the original system, as proved in the following:

Theorem 3 (Tikhonov (1958)). *Under conditions (a)–(f) above, $\forall T < \infty$*

$$\lim_{\varepsilon \to 0} \mathbf{y}^\varepsilon(\tau) = \overline{\mathbf{y}}(\tau), \quad \lim_{\varepsilon \to 0} \mathbf{z}^\varepsilon(\tau) = \overline{\mathbf{z}}(\tau), \quad 0 < \tau \leq T \tag{13}$$

Running Example. In our example, the slow variable y is the protein concentration, the fast variable z is the active gene concentration. They satisfy (12) in the slow time variable $\tau = \varepsilon t$, with $G(y, z) = k_p z - k_d y$ and $H(y, z) = k_b y z - k_u(1 - z)$. Solving $H(y, z) = 0$ for z, we get $\phi(y) = \frac{k_u}{k_u + k_b y}$, hence finding the classic ODE for lumped gene transcription:

$$\frac{d\bar{y}}{d\tau} = \frac{k_p k_u}{k_u + k_b \bar{y}} - k_d \bar{y}$$

4.5 The Quasi-Equilibrium Reduction of an MPM

Let the Assumption 2 hold for (1) (the rate functions and the variable are decomposable into fast and slow subsets). Inserting the decomposition, described in section 4.3, into (1), yields

$$\partial_t P(\mathbf{Y}, \mathbf{Z}; t) = \sum_{i=1}^{r} \Big\{ W_i(\mathbf{Y} - \boldsymbol{\mu}_i, \mathbf{Z} - \boldsymbol{\sigma}_i) P(\mathbf{Y} - \boldsymbol{\mu}_i, \mathbf{Z} - \boldsymbol{\sigma}_i; t)$$
$$-W_i(\mathbf{Y}, \mathbf{Z}) P(\mathbf{Y}, \mathbf{Z}; t) \Big\}, \quad P(\mathbf{Y}, \mathbf{Z}; 0) = P_0(\mathbf{Y}, \mathbf{Z}). \quad (14)$$

In addition to requiring that a corresponding MPM satisfies Assumption 2, we further require

Assumption 4. (Ergodicity)

(a) The full process $\mathbf{X}(t) = (\mathbf{Y}, \mathbf{Z})(t)$ is ergodic;
(b) The stochastic process $\mathbf{Z_Y}(t)$ describing the fast subsystem is ergodic for each fixed $\mathbf{Y}$.

Under these further requirements, the master equation of the reduced system is

$$\partial_\tau P(\mathbf{Y}; \tau) = \sum_{i=1}^{s} \Big\{ \widetilde{W}_{0,i}^\infty(\mathbf{Y} - \boldsymbol{\mu}_i) P(\mathbf{Y} - \boldsymbol{\mu}_i; \tau) - \widetilde{W}_{0,i}^\infty(\mathbf{Y}) P(\mathbf{Y}; \tau) \Big\} \quad (15)$$

$$\widetilde{W}_{0,i}^\infty = \mathbb{E}_{\mathbf{Z_Y}(\infty)}(W_{0,i}(\mathbf{Y}, \mathbf{Z})) = \sum_{\mathbf{Z}} W_{0,i}(\mathbf{Y}, \mathbf{Z}) \overline{P}_{\mathbf{Y}}(\mathbf{Z}), \quad i = 1, \ldots, s \quad (16)$$

and $\mathbf{Z_Y}(\infty)$ is the unique steady state measure of the fast process $\mathbf{Z_Y}(t)$ (due to 4.b), with $\overline{P}_{\mathbf{Y}}(\mathbf{Z})$ being the steady state probability of the master equation

$$\partial_t P_{\mathbf{Y}}(\mathbf{Z}; t) = \sum_{j=s+1}^{r} \Big\{ W_{0,j}(\mathbf{Y}, \mathbf{Z} - \boldsymbol{\sigma}_j) P_{\mathbf{Y}}(\mathbf{Z} - \boldsymbol{\sigma}_j; t) - W_{0,j}(\mathbf{Y}, \mathbf{Z}) P_{\mathbf{Y}}(\mathbf{Z}; t) \Big\} \quad (17)$$

The slow process $\widetilde{\mathbf{Y}}(\tau)$ defined by the master equation (15) is indeed the limit of the full process for $\varepsilon \to 0$ (see [7] for the proof):

Theorem 4 (Quasi-equilibrium reduction). *Under assumptions 4.(a)–(b),*

$$\lim_{\varepsilon \to 0} \sum_{\mathbf{Z}} P(\mathbf{Y}, \mathbf{Z}; \tau/\varepsilon) = P(\mathbf{Y}; \tau) \tag{18}$$

for all $T > 0$ and $0 \le \tau \le T$. □

We can now lift Theorem 4 to the MPM level. Consider an MPM $\mathcal{X} = (\mathbf{X}, \mathcal{M}, \mathcal{T}, \mathbf{X}_0)$, that is QE reducible (see section 4.3). The quasi-equilibrium image of $\mathcal{X}$ is defined as $\mathcal{X}^{\text{qe}} = (\mathbf{Y}, \mathcal{N}, \mathcal{T}^{\text{slow}}, \{\mathbf{Y}_0, \mathbf{Z}_0\})$, where

- $\mathbf{Y} = L_{\mathsf{C}}(\mathbf{X})$, $\mathcal{N} = L_{\mathsf{C}}(\mathcal{M})$, $\mathbf{Y}_0 = L_{\mathsf{C}}(\mathbf{X}_0)$, and $\mathbf{Z}_0 = L_{\mathsf{K}}(\mathbf{X}_0)$;
- Each *slow* transition $\tau = (a, \boldsymbol{\nu}, W) \in \mathcal{T}^{\text{slow}}$ of $\mathcal{X}$ becomes the transition $\tau = (a, \boldsymbol{\mu}, \widetilde{W}_0^\infty)$, where $\boldsymbol{\mu} = L_{\mathsf{C}}(\boldsymbol{\nu})$ and $\widetilde{W}_0^\infty(\mathbf{Y})$ is defined by (16).

It is also straightforward to define a *family* of MPMs for the fast subsystem, parameterised by the slow variable $\mathbf{Y}$, described by the master equation (17).

Running Example. The most important new information are the expressions for the averaged slow rates, given by the definition (16). We find

$$\widetilde{W}_{0,1}(Y) = \sum_Z k_p Z \overline{P}_Y(Z) = k_p \langle Z \rangle_{Z_Y(\infty)}, \quad \widetilde{W}_{0,2}(Y) = k_p Y. \tag{19}$$

These rates, together with the state change vectors $\mu_1 = 1$, $\mu_2 = -1$, complete the definition of a reduced MPM, which is easily seen to describe a birth-death process. The rates of this process are given by (19). In this simple case, the fast process, conditional on Y, is also a birth-death process, hence owing to the linearity of (11), we get $\langle Z \rangle_{Z_Y(\infty)} = \frac{N k_u}{k_u + k_b Y/N}$, which gives an explicit expression for the rates of Y. We emphasise that in general this is not true, as the stationary distribution of Z_Y may not be known explicitly, so that one has still to rely on numerical methods, like simulation [29].

5 Comparing Mean Field and Quasi-Equilibrium

Consider an MPM $\mathcal{X}$, and, as in Section 3, let $\mathbf{X}^N(t)$ be the normalised model with respect to system size N. In this paper we have introduced two possible model simplification strategies: the mean field approximation and the QE reduction. To fix notation in the rest of this section, we will refer to the former by the operator $\mathbb{M}$, and to the latter by the operator $\mathbb{Q}$. Hence, $\mathbb{M}(\mathbf{X}^N(t)) = \mathbf{x}(t)$ is the mean field limit of $\mathbf{X}^N(t)$, and $\mathbb{Q}(\mathbf{X}^N(t)) = \widetilde{\mathbf{Y}}^N(\tau)$ is the QE reduction of $\mathbf{X}^N(t)$, whereas $\mathbb{Q}(\mathbf{x}(t)) = \widetilde{\mathbf{y}}(\tau) = \mathbb{Q}(\mathbb{M}(\mathbf{X}^N(t)))$ the QE reduction of $\mathbf{x}(t)$.

The issue we wish to address in this section is how these two procedures are related. In particular, it is natural to ask if the two operators $\mathbb{M}$ and $\mathbb{Q}$ commute, as shown in Fig. 1. The diagram illustrates the following two possibilities. We could either construct the QE reduction upon the mean field limit of a MPM $\mathbf{X}^N(t)$ and obtain a deterministic process $\widetilde{\mathbf{y}}(\tau)$, or we could first apply the QE reduction to $\mathbf{X}^N(t)$, and then attempt to construct the mean field limit of $\widetilde{\mathbf{Y}}^N(t)$. Two questions arise naturally

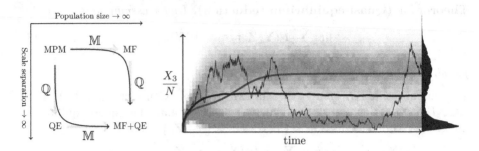

Fig. 1. Left: commutation diagram. The curved paths illustrate two distinct limiting procedures to arrive from an MPM to a fully reduced model. Right: toggle switch counter example to Theorem 5. The blue curve is the solution of the reduced mean field ODE, while the solid black curve is the average of the reduced stochastic process, which is bistable (cf. the empirical distribution on the right).

1. Does $\mathbb{M}(\mathbb{Q}(\mathbf{X}^N(t)))$, i.e. the mean field limit of $\widetilde{\mathbf{Y}}^N(\tau)$ exist?
2. If so, is it the same as $\widetilde{\mathbf{y}}(\tau) = \mathbb{Q}(\mathbb{M}(\mathbf{X}^N(t)))$, i.e. does the diagram in Figure 1 commutes?

We show that the answer is 'yes to both questions' only if some additional requirements for the fast subsystem are fulfilled. We will demonstrate that the answer to question 2 is 'no in general', and that even question 1 may have a negative answer. The problem is intimately connected with the extension of Theorem 1 to the steady state, hence with Theorem 2. In fact, when we construct the QE reduction $\mathbb{Q}(\mathbf{X}^N(t))$ of $\mathbf{X}^N(t)$, we need to average the slow rates with respect to the steady state distribution $\mathbf{Z}_{\mathbf{Y}}^N(\infty)$ of the fast subsystem $\mathbf{Z}_{\mathbf{Y}}^N$. Assumptions 4.(a) and 4.(b) enforce ergodicity, hence existence and uniqueness of such a steady state distribution $\mathbf{Z}_{\mathbf{Y}}^N(\infty)$ for each N and $\mathbf{Y}$. However, to construct the mean field limit of $\mathbb{Q}(\mathbf{X}^N(t))$, we also need to know how such sequence behaves as N goes to infinity. Essentially, we need to know if it has a limit, and what such limit is. Unfortunately, this is one of the most delicate points of mean-field approximation theory: Little is known about the limiting behaviour of the steady state, except from Theorem 2. Hence, we can provide a positive answer to questions 1 and 2 only if we place ourselves in the conditions of such a theorem. This leads to the following

Assumption 5. The solution $\mathbf{z} = \phi(\mathbf{y})$ of $0 = H(\mathbf{y}, \mathbf{z})$ is unique, i.e. the mean field limit of the fast subsystem $\bar{\mathbf{z}}(t) = \bar{\mathbf{z}}(t, \mathbf{y})$ has a unique, globally attracting equilibrium $\phi(\mathbf{y})$ for each value of the slow variables $\mathbf{y}$.

Under this assumption, we can apply Theorem 2 and conclude that, for each $\mathbf{Y}$, it holds that $\mathbf{Z}_{\mathbf{Y}}^N(\infty) \to \delta_{\phi(\mathbf{Y})}$ in probability. At this stage, however, we need a further technical assumption (see also Remark 1):

Assumption 6. $\mathbf{Z}_{\mathbf{Y}}^N(\infty)$ converges to $\delta_{\phi(\mathbf{Y})}$ uniformly in $\mathbf{Y}$, i.e. $\forall \varepsilon > 0$,

$$\lim_{N \to \infty} \mathbb{P}\left\{ \sup_{\mathbf{Y} \in \mathcal{N}^N} \|\mathbf{Z}_{\mathbf{Y}}^N(\infty) - \phi(\mathbf{Y})\| > \varepsilon \right\} = 0.$$

Under these two additional assumptions, it is easy to show that

$$\frac{\widetilde{W}_{0,i}^{\infty}(\mathbf{y})}{N} \xrightarrow[N\to\infty]{} w_{0,i}(\mathbf{y}) \tag{20}$$

uniformly in $\mathbf{y}$. This readily implies that the drift of the QE-reduced process $\widetilde{\mathbf{Y}}^N(t)$, $\widetilde{F}^N(\mathbf{y}) := \sum \nu_i \frac{\widetilde{W}_{0,i}^{\infty}(\mathbf{y})}{N}$ converges uniformly to the drift $G(\mathbf{y}, \phi(\mathbf{y}))$, defining the vector field of the QE-reduced mean field limit, as in equation (12), which is sufficiently regular by hypothesis 3.(a). Hence, the conditions of Theorem 1 are satisfied by the sequence of processes $\widetilde{\mathbf{Y}}^N(t)$, and we can conclude that

Theorem 5. *Under Assumptions 5 and 6 above, with $T < \infty$ fixed and for each $t \le T$, $\mathrm{M}(\mathbb{Q}(\mathbf{X}^N(t)))$ exists and $\mathrm{M}(\mathbb{Q}(\mathbf{X}^N(t))) = \mathbb{Q}(\mathrm{M}(\mathbf{X}^N(t)))$.* □

Remark 1. Assumption 2 requires that the convergence of the sequence of steady state measures of the fast subsystem to their limit point-wise distribution is uniform in the slow state $\mathbf{Y}$. We conjecture this is in fact true without any further requirement on the MPM. A heuristic argument goes as follows: by the functional central limit [12], we know that the fast subsystem will behave like a Gaussian process for N large enough. In particular, the steady state distribution of $\mathbf{Z}_{\mathbf{Y}}^N(\infty)$ will be approximatively Gaussian with mean $\phi(\mathbf{Y})$ and Covariance matrix $C^N(\mathbf{Y}) = \frac{1}{\sqrt{N}} C(\mathbf{Y})$, where $C(\mathbf{Y})$ does not depend on N and it is the steady state solution of the covariance linear noise equations [13]. As such, it will depend continuously on $\mathbf{Y}$. Using similar arguments as in the proofs of Kurtz theorem, we can guarantee that the eigenvalues of $C(\mathbf{Y})$ are uniformly bounded by a constant $\Lambda < \infty$, which implies that we can find a uniform bound in $\mathbf{Y}$ on the spread of the steady state distribution, going to zero as the population size N diverges. A formal proof of Assumption 6 seems to be strictly related to the availability of explicit bounds for the convergence in probability of $\mathbf{Z}_{\mathbf{Y}}^N(\infty)$ to $\delta_{\phi(\mathbf{Y})}$, which is still an open issue, see also [8].

Running Example. The mean field equation for the fast variable z in the self-repressing gene example is linear, so that it is easy to see that it has a unique globally attracting equilibrium for each y. Furthermore, for any N and y, it holds that $\widetilde{W}_{0,1}^{\infty}(y)/N = w_{0,1}(y)$ (cf. the expression of $\widetilde{W}_{0,1}^{\infty}(y)$ computed at the end of last section), hence Assumption 6 is trivial in this case. Therefore Theorem 5 applies: mean field and time scale reduction commute.

5.1   On the Necessity of Assumption 5

Assumption 5, on the other hand, is quite crucial for Theorem 5 to hold. Without it, we cannot say much about the limit behaviour of the sequence of steady state measures of the fast subsystem, a part from the fact that each limit point will be supported in the Birkhoff center of the limit mean field dynamical system [9,3]. If this system has only stable and unstable equilibria as invariant sets (e.g. it satisfies the conditions of [20]), then each limit point of the sequence of steady state measure will be supported in those equilibria, but this is as much as we can say. In particular, *we cannot guarantee the existence of a limit for such a sequence*, hence *the reduced stochastic model may not be amenable of mean field approximation*. However, we can argue that, in case the limit of $\mathbf{Z}_{\mathbf{Y}}^{N}(\infty)$ is defined, then $\mathbb{M}$ and $\mathbb{Q}$ will not generally commute. The reason for this is to be found in the large deviations theory for (population) CTMC [26, Ch. 6], which guarantees that each trajectory of the stochastic system will remain close to all stable equilibrium of the mean field limit a non-negligible fraction of time. Hence, the limit steady state measure, if any, must be a *mixture of point-wise masses* concentrated on (stable) equilibria[1]. On the other hand, the fast subsystem $\bar{\mathbf{z}}$ of the mean-field limit will converge to a single stable equilibrium (assuming no bifurcation event happens in the fast subsystem as $\bar{\mathbf{y}}(t)$ varies, i.e. that Assumption 3.(e) is in force). This implies that the limit for $N \to \infty$ of the rates $\frac{\widetilde{W}_{0,i}^{\infty}(\mathbf{y})}{N}$ will not converge to $w_{0,i}(\mathbf{y})$, which is evaluated on the single equilibrium $\mathbf{z} = \phi(\mathbf{y})$, but rather to a weighted average of the rate function w_i evaluated on all (stable) equilibria.

To render this discussion more concrete, we illustrate this phenomenon by means of a genetic network model of a toggle switch [14]. We have three protein species, whose number is given by variables $\mathbf{X} = (X_1, X_2, X_3)$, living in a volume N, with density $x_j = X_j/N$ (possibly exceeding unit value). The MPM is specified by the following six transitions:

production of X_1 : $\boldsymbol{\nu}_1 = (\ 1, 0, 0)^{\mathsf{T}}$, $W_1(\mathbf{X}) = \alpha_1 N^{\beta_1+1}/(N^{\beta_1} + X_2^{\beta_1})$,
degradation of X_1 : $\boldsymbol{\nu}_2 = (-1, 0, 0)^{\mathsf{T}}$, $W_2(\mathbf{X}) = X_1$,
production of X_2 : $\boldsymbol{\nu}_3 = (0,\ \ 1, 0)^{\mathsf{T}}$, $W_3(\mathbf{X}) = \alpha_2 N^{\beta_2+1}/(N^{\beta_2} + X_1^{\beta_2})$,
degradation of X_2 : $\boldsymbol{\nu}_4 = (0, -1, 0)^{\mathsf{T}}$, $W_4(\mathbf{X}) = X_2$,
production of X_3 : $\boldsymbol{\nu}_5 = (0, 0,\ \ 1)^{\mathsf{T}}$, $W_5(\mathbf{X}) = \varepsilon X_1$,
degradation of X_3 : $\boldsymbol{\nu}_6 = (0, 0, -1)^{\mathsf{T}}$, $W_6(\mathbf{X}) = \varepsilon X_3$.

The proteins '1' and '2' mutually repress each other, and thus properly constitute the toggle switch. Molecule '3' instead, is a slow product of the protein '1', and does not influence the toggle switch. This example is cooked up so that if $\varepsilon \ll 1$ then the variable x_3 and transitions τ_5 and τ_6 are trivially the slow ones. It should still be possible to see the breakdown of the assumptions 5 & 6 in the long time expectation value of the molecule '3'. First we consider the mean field limit

$$\frac{dx_1}{dt} = \frac{\alpha_1}{1 + x_2^{\beta_1}} - x_1, \qquad \frac{dx_2}{dt} = \frac{\alpha_2}{1 + x_1^{\beta_2}} - x_2, \qquad \frac{dx_3}{dt} = \varepsilon(x_1 - x_3).$$

For a symmetric toggle model with parameters $\alpha_1 = \alpha_2 = 10$, $\beta_1 = \beta_2 = 1.4$, the two stable equilibria are $(\bar{x}_1, \bar{x}_2) = (a, b), (b, a)$ where $a = 0.764$, $b = 5.931$. The limiting behavior of x_3 is $x_3 \xrightarrow[\tau \to \infty]{} \bar{x}_1$, where $\bar{x}_1$ is either a or b, depending on whose basin of attraction covers the initial condition of the trajectory. We

---

[1] The role of unstable equilibria is unclear. It is plausible that they will be visited only for a vanishing fraction of time, but we know no proof of this fact.

took the initial conditions that are below the diagonal $x_1 = x_2$. Such initial conditions are attracted to the equilibrium $\bar{x}_1 = b$. The mean field time series x_3 vs t is displayed in figure 1, where the mean field trajectory saturates at b (blue curve).

Next we consider the stochastic dynamics. A representative stochastic time series of X_3/N vs t is shown in figure 1. Its variations are wider than a Gaussian approximation of the probability would imply. Sufficient insights can be gained by looking at the expectations of the form $\langle \mathbf{X} \rangle (t) = \sum_{\mathbf{X}} \mathbf{X} P(\mathbf{X}; t)$. The expectation of molecule '3' satisfies an exact differential equation

$$d \langle X_3 \rangle / dt = \varepsilon \langle X_1 \rangle - \varepsilon \langle X_3 \rangle$$

Making a QE approximation to this equation is equivalent to replacing $\langle X_1 \rangle$ with the equilibrium expectation $\overline{X}_1^\infty$ of the fast ('1+2') subsystem, and $\langle X_3 \rangle (t)$ – with $\widetilde{X}_3(\tau)$, each of which should be expressed in terms of their respective reduced probabilities. Since X_3 is decoupled from X_1 in the full model, $\overline{X}_1(t) = \langle X_1 \rangle (t)$. Moreover, if $\varepsilon \ll 1$, we can also take $\overline{X}_1^\infty \approx \langle X_1 \rangle (t)$, resulting in

$$d\widetilde{X}_3(\tau)/d\tau = \overline{X}_1^\infty - \widetilde{X}_3(\tau).$$

Within this approximation, the solution tends to $\widetilde{X}_3(\tau) \xrightarrow[\tau \to \infty]{} \overline{X}_1^\infty$. Then, comparison of $x_3(t)$, obtained from the mean field limit, and $\widetilde{x}_3(\tau) = \widetilde{X}_3(\tau)/N$, obtained from the stochastic model, provides a good measure of differences between the two approximations. The mean field trajectory, discussed in the previous paragraph, should be compared with the expectation $\widetilde{X}_3(\tau)$, shown as a solid gray line in figure 1. There is a significant difference between the two, suggesting the non-equivalence of reduced models in this particular case. Applying large deviations arguments [26, Ch. 6], one may expect $P(X_3)$ (shown as a density in figure 1) to look like, as $N \to \infty$, a mixture of point masses, concentrated equilibria. Conjecturing that the mass is distributed only on stable equilibria and owing to the symmetry between X_1 and X_2, such weights will be equal to $\frac{1}{2}$, so $\widetilde{x}_3(\infty) = (a + b)/2$. A simulation supports this conjecture, as the curve for $\widetilde{X}_3(\tau)$ is roughly in the middle between the two peaks of the probability density shown in 1.

6   Discussion

In this paper, we discussed in a homogeneous way two approximation techniques for Markov Population Models: the mean-field limit and the quasi-equilibrium reduction in the presence of multiple time-scales. Both approaches are based on a notion of limit: for large population in the former case, and for a diverging separation of time scales in the latter. Our first contribution of this paper is to formalise in a clear way the quasi-equilibrium reduction for MPM, proving also the convergence of the original model to the reduced one in the stochastic setting. The second original ingredient of this work is the investigation of the relationship between QE and mean-field. In particular, we identified sufficient conditions under which the two limits commute. We also argued that the commutation should not hold in general. The situation here is intimately connected with the nature of mean field convergence for steady state distributions.

The take-home message is that care must be exercised when time scale separation techniques are combined with mean field limits. The behaviour of the system that we obtain by first taking the mean field limit and then the QE reduction, the most common way in literature, may not reflect at all the actual behaviour of the original stochastic model. Hence, one has to additionally show that the fast subsystem is well behaved (i.e., it satisfies assumption 5).

We note here that most of the assumptions we introduced hold in almost all practical cases, and are generally easy to verify. The most challenging ones are the separation of time scales (Assumption 2), and those related to the steady state behaviour of ODE models, i.e. Assumptions 5 and 3.(e).

This line of research can be extended in few directions. First of all, the literature on time scale separation for MPM is not as well developed as the literature for ODE models [6]. Many ideas developed in this context can possibly be exported to MPM, especially techniques that automatically identify multiple time scales [19]. Finally, we are investigating how QE reduction propagates to moment closure-based approximations of variance and of higher order moments of the stochastic population process.

References

1. Benäim, M., Le Boudec, J.-Y.: A class of mean field interaction models for computer and communication systems. Perf. Eval. 65(11), 823–838 (2008)
2. Benäim, M., Le Boudec, J.-Y.: On mean field convergence and stationary regime. CoRR, abs/1111.5710 (2011)
3. Benäim, M., Weibull, J.W.: Deterministic approximation of stochastic evolution in games. Econometrica 71(3), 873–903 (2003)
4. Bortolussi, L.: Limit behavior of the hybrid approximation of stochastic process algebras. In: Al-Begain, K., Fiems, D., Knottenbelt, W.J. (eds.) ASMTA 2010. LNCS, vol. 6148, pp. 367–381. Springer, Heidelberg (2010)
5. Bortolussi, L., Hillston, J.: Fluid model checking. In: Koutny, M., Ulidowski, I. (eds.) CONCUR 2012. LNCS, vol. 7454, pp. 333–347. Springer, Heidelberg (2012)
6. Bortolussi, L., Paškauskas, R.: Multiscale reductions of mean field and stochastic models. Technical Report TR-QC-04-2014, QUANTICOL Tech. Rep. (2014)
7. Bortolussi, L., Paškauskas, R.: Mean-Field approximation and Quasi-Equilibrium reduction of Markov Population Models (2014), http://arxiv.org/abs/1405.4200
8. Bortolussi, L., Hayden, R.: Bounds on the deviation of discrete-time Markov chains from their mean-field model. Perf. Eval. 70(10), 736–749 (2013)
9. Bortolussi, L., Hillston, J., Latella, D., Massink, M.: Continuous approximation of collective systems behaviour: a tutorial. Perf. Eval. 70, 317–349 (2013)
10. Bortolussi, L., Lanciani, R.: Model checking markov population models by central limit approximation. In: Joshi, K., Siegle, M., Stoelinga, M., D'Argenio, P.R. (eds.) QEST 2013. LNCS, vol. 8054, pp. 123–138. Springer, Heidelberg (2013)

11. Cao, Y., Gillespie, D.T., Petzold, L.R.: The slow-scale stochastic simulation algorithm. The Journal of Chemical Physics 122(1), 014116 (2005)
12. Ethier, S.N., Kurtz, T.G.: Markov processes: characterization and convergence. series in probability and statistics. Wiley Interscience. Wiley (2005)
13. Gardiner, C.W.: Handbook of stochastic methods, vol. 3. Springer (1985)
14. Gardner, T.S., Cantor, C.R., Collins, J.J.: Construction of a genetic toggle switch in escherichia coli. Nature 403(6767), 339–342 (2000)
15. Gillespie, D.T.: Exact stochastic simulation of coupled chemical reactions. The Journal of Physical Chemistry 81(25), 2340–2361 (1977)
16. Hayden, R., Bradley, J., Clark, A.: Performance specification and evaluation with unified stochastic probes and fluid analysis. IEEE TSE 39(1), 97–118 (2013)
17. Henzinger, T., Jobstmann, B., Wolf, V.: Formalisms for specifying Markovian population models. International Journal of Foundations of Computer Science 22(04), 823–841 (2011)
18. Johnson, K.A., Goody, R.S.: The original Michaelis constant: Translation of the 1913 Michaelis–Menten paper. Biochemistry 50(39), 8264–8269 (2011)
19. Lam, S.H., Goussis, D.A.: The CSP method for simplifying kinetics. International Journal of Chemical Kinetics 26(4), 461–486 (1994)
20. Le Boudec, J.-Y.: The stationary behaviour of fluid limits of reversible processes is concentrated on stationary points. NHM 8(2), 529–540 (2013)
21. Marquez-Lago, T.T., Stelling, J.: Counter-intuitive stochastic behavior of simple gene circuits with negative feedback. Biophysical Journal 98(9), 1742–1750 (2010)
22. Maas, U., Pope, S.B.: Simplifying chemical kinetics: intrinsic low-dimensional manifolds in composition space. Combustion and Flame 88(3), 239–264 (1992)
23. Mastny, E.A., Haseltine, E.L., Rawlings, J.B.: Two classes of quasi-steady-state model reductions for stochastic kinetics. The Journal of Chemical Physics 127(9), 094106 (2007)
24. Rao, C.V., Arkin, A.P.: Stochastic chemical kinetics and the quasi-steady-state assumption: Application to the Gillespie algorithm. Journal of Chemical Physics 118(11), 4999–5010 (2003)
25. Schnakenberg, J.: Network theory of microscopic and macroscopic behavior of master equation systems. Reviews of Modern Physics 48(4), 571–585 (1976)
26. Shwartz, A., Weiss, A.: Large Deviations for Performance Analysis. C&H (1995)
27. Stewart, G.W., Sun, J.: Matrix perturbation theory. Computer Science and Scientific Computing. Academic Press (1990)
28. Verhulst, F.: Methods and applications of singular perturbations: boundary layers and multiple timescale dynamics. Springer (2005)
29. Weinan, E., Liu, D., Vanden-Eijnden, E.: Nested stochastic simulation algorithms for chemical kinetic systems with multiple time scales. Journ. of Comp. Phys. 221(1), 158–180 (2007)

On Performance of Gossip Communication in a Crowd-Sensing Scenario

Marcel C. Guenther and Jeremy T. Bradley

Imperial College London,
180 Queen's Gate,
SW7 2RH, London, United Kingdom
{mcg05,jb}@doc.ic.ac.uk

Abstract. Many applications associated with the smart city experience rely on spatio-temporal data. Specific use-cases include location-dependent real-time traffic, weather and pollution reports. Data is traditionally sampled using stationary sensors, however, in densely populated areas one could envisage crowd-sensing data collection schemes where cars, bikes and pedestrians collect information in transit and transmit it to a service provider through one of either a fast mobile network such as LTE(4G) or by Wifi/Gossip communication. While mobile sensors reduce the need for expensive infrastructure, the downside is that performance characteristics of data coverage and transmission are less reliable and harder to predict. In this paper we present a generic model to investigate the robustness and efficiency of LTE/Gossip hybrid data transmission strategies for crowd-sensing networks that are not amenable to mean-field analysis. To illustrate our model's scalability, we fit it to journey data from the London Cycle Hire scheme.

Keywords: Crowd-sensing, Gossip networks, Spatial modelling, Smart city, Time-inhomogeneous delay-only population CTMC models.

1 Introduction

Smart city research promotes the use of technology to improve the quality of living and to reduce the cost of services in urban areas. Specifically, smart city applications take advantage of data obtained from sensor readings or local social network activity. Applications range from simple services for the urban population, e.g. providing local temperature, traffic or pollution reports, to complex applications that monitor leaks in water systems [1]. As installation and maintenance of sensor and network infrastructure are major cost factors [2], researchers have proposed solutions which make use of mobile sensors to sample data. In particular *crowd-sensing* or *participatory-sensing*, where pedestrians [3], bikes [2] or vehicles [4] are equipped with sensors and radio hardware, are being discussed as cost effective alternatives for data collection in densely populated areas. To reduce the need for network infrastructure, protocols for *ad hoc, opportunistic*

G. Norman and W. Sanders (Eds.): QEST 2014, LNCS 8657, pp. 122–137, 2014.

Gossip networks have been suggested [5,4]. Aside from their cost saving potential, opportunistic networks can act as backups when infrastructure dependent mobile networks are overloaded or down.

While a number of studies [5,6], have addressed non-functional, high-level[1] performance aspects of information spread and collection in Gossip protocols, performance analysis of high-level, large-scale opportunistic networks for data collection is less abundant in the literature. On the other hand many studies on microscopic mobile *ad hoc* network (*MANET*) protocol performance have been published [7,4], however, their analysis is usually limited to simulation or empirical trials. In this paper we introduce a generic probabilistic model that enables system designers to estimate the high-level performance of large-scale crowd-sensing systems, taking into account accurate geographical topology. Opportunistic data gathering [8] is a key component of our model, but like [6] we assume that data producing agents are mobile while Wifi upload points are static. Moreover, we assume that agents can exchange messages in a Gossip fashion akin to [5]. In combining these concepts in a single model, we provide a flexible mobile crowd-sensing performance modelling technique that is suitable for infrastructure planning purposes.

More specifically, the model described in this work analyses the performance of an LTE(4G)/Gossip hybrid transmission protocol for large crowd-sensing systems. The idea is that clients can choose to use LTE, a fast mobile data network, but are encouraged to rely on *ad hoc* Gossip networks in order to reduce the peak time load on the LTE network as well as the overall transmission costs so long as service level agreements are likely to be met. Despite having a specific analysis goal, many of the model features described in Section 2 are kept generic and could easily be adapted to evaluate other performance features, such as spatial crowd-sensing coverage. We are particularly interested in the modelling scenario that arises when the number of network participants is spatially sparse with respect to radio ranges, which can lead to large approximation errors when mean-field assumptions are made. However, since the analysis of such non-linear population models outside a mean-field regime [5] is hard as well as computationally demanding, our model reduces complexity by making use of the findings of Jahnke *et al.* [9] using a linear time-inhomogeneous population CTMC (IPCTMC) model with extra deterministic delays [10] instead. As a consequence we can use a hybrid simulation technique to analyse city-scale models with complex agent movement, rather than having to resort to a less efficient full simulation approach.

The rest of the paper is structured as follows; In Section 2 we introduce a generic model to compare the efficiency of different LTE/Gossip transmission policies in mobile crowd-sensing networks. Section 3 applies the model to journey data from the London cycle hire scheme to show that it scales well. Section 4 summarises our findings and suggests possible future research directions.

---

[1] Models where low-level details such as wireless communication protocol details are abstracted for the sake of efficient macroscopic system analysis.

2 Model Description

Smart city applications, such as traffic and weather apps that provide local information in real time, are usually subject to service level agreements (*SLAs*). An SLA might postulate that 90% of all data must arrive within 10 minutes of being collected. To meet such demands while keeping transmission costs low, data collecting agents should always attempt to use Wifi for uploading data to the service provider and only resort to LTE for untransmitted samples that approach the SLA deadline. To increase the utilisation of Wifi uploads, it is conceivable that a crowd-sensing data transmission protocol would further feature opportunistic Gossip communication, where agents exchange sensor readings whenever they are within each other's radio range. Any *foreign* data is then uploaded at the next available Wifi hotspot. This reduces the Wifi delivery time and increases the chance of conforming to the SLA without resorting to LTE transmission. Evaluating SLA compliance of a combined LTE/Gossip(Wifi) strategy is naturally challenging. Whenever a sample approaches an SLA deadline, i.e. whenever the age of a sample that has not been uploaded via Wifi reaches the SLA deadline, the sampling agent needs to decide whether it should rely on the Gossip network or whether to fall back on LTE transmission. The communication protocol therefore requires a policy which makes this decision based on information such as the number of Gossip contacts the agent has had since sampling the data. Our model offers the opportunity to compare such policies for large crowd-sensing networks. The most important performance aspects of a policy are its chance of SLA violation and its efficiency, relative to a theoretically optimal policy, at which the Gossip network is utilised. The model we are about to present is kept as generic as possible so that it can be applied to various forms of crowd-sensing networks where participants move in a *non-congesting* manner. By non-congesting, we mean modes of transport such as walking or cycling, where faster traffic participants can overtake slower ones at any point and where arrival order at junctions or at traffic lights does not require a queueing model. These assumptions are crucial as they drastically reduce the complexity of the analysis. The model comes in two parts, a movement model (see Section 2.1) and a communication model (see Section 2.3), which uses the notion of a measure agent discussed in Section 2.2.

2.1 Movement Model

Various studies [11,12] have shown that the performance of mobile *ad hoc* networks (MANETs) and their vehicular counterparts (VANETs), is heavily impacted by the underlying choice of movement model. Our movement model assumes that agents are born according to a time-inhomogeneous Poisson process. Each agent moves independently from all other agents and belongs to a specific class that specifies and route that it travels on. Agents travel at constant speed, but are subjected to occasional exponentially distributed time-inhomogeneous stop and start delays. The time-inhomogeneous nature of delays is important as

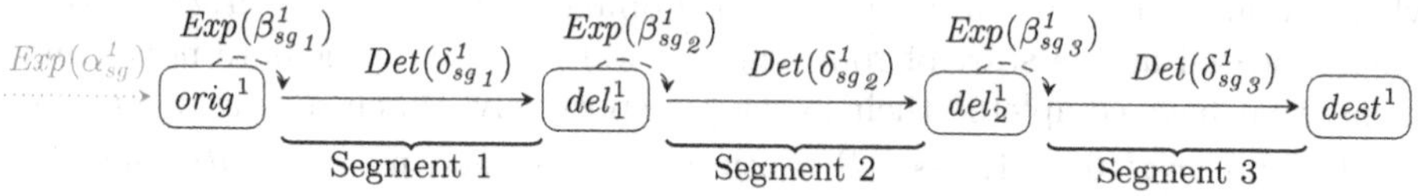

Fig. 1. Movement of agent of class (r_1, sg), with origin $orig^1$, destination $dest^1$ and delay nodes del_*^1 (cf. Eq. (1))

it enables us to ensure that all agents experience the same delay in certain locations, say at red traffic lights. We chose this particular level of abstraction to fit complex inner-city road topologies, which has not been done in comparable studies [5,6] and required expensive discrete event simulation studies in others [11,12]. Results for the *CSM w/pauses* model [11], which our approach was inspired by, suggest that simple stop and go models are decent macroscopic abstractions in sparse, non-congesting traffic scenarios. Furthermore, the microscopic movement analysis of cars in [12] shows that the constant speed assumption is reasonable for many city traffic scenarios.

The states of an individual agent's movement are shown in Figure 1. To represent agent journeys, e.g. a path taken from start to end docking station in a bicycle hire network, we introduce the notion of a *route*. A route is a specific path connecting two locations and consists of *road segments*, *delay* and *non-delay* nodes. On a road segment connecting two nodes, agents move at a constant speed defined by their *speed group*. While nodes generally represent changes in direction and influence an agent's radio range, *delay nodes* further cause an agent to wait for a time-dependent, exponentially distributed time. Some delay nodes, e.g. traffic lights, may cause start, stop dynamics that result in temporary Poisson event rates of $\beta_{sg_*}^r = 0$. When analysing the movement model, we only keep track of the time-evolution of agent populations at delay nodes (see Figure 1 and Eq. (1)), however, when analysing inter-agent communication, we use all nodes to determine exact agent radio range overlaps. A speed group sg for route r defines the Poisson birth rate parameter α_{sg}^r, as well as the speed of agents for all road segments of r and hence the movement delay $\delta_{sg_*}^r$ between any two delay nodes. The pair (r, sg) will be referred to as an *agent class* and $\mathcal{A}$ as the set of all agent classes in the model. Mathematically, we can represent the movement of an agent class, i.e. of a number of *agent instances* of that class, as a time-inhomogeneous [13], delay-only [10], population CTMC model *(ID-PCTMC)*. If birth rates are Poisson and all populations are initially empty, then *Delay-Differential equations (DDEs)* numerically determine the time-evolution of the Poisson distribution rate for the number of agents located at any of the delay nodes at time t. For the agent described in Figure 1 the system of DDEs is

$$
\begin{aligned}
\frac{\delta orig^1(t)}{\delta t} &= \alpha_{sg}^1(t) - \beta_{sg_1}^1(t)\, orig^1(t) \\
\frac{\delta del_1^1(t)}{\delta t} &= \beta_{sg_1}^1(t - \delta_{sg_1}^1)\, orig^1(t - \delta_{sg_1}^1) - \beta_{sg_2}^1(t)\, del_1^1(t) \\
\frac{\delta del_2^1(t)}{\delta t} &= \beta_{sg_2}^1(t - \delta_{sg_2}^1)\, del_1^1(t - \delta_{sg_2}^1) - \beta_{sg_3}^1(t)\, del_2^1(t) \\
\frac{\delta dest^1(t)}{\delta t} &= \beta_{sg_3}^1(t - \delta_{sg_3}^1)\, del_2^1(t - \delta_{sg_3}^1)
\end{aligned}
\tag{1}
$$

where population values are 0 for $t \leq 0$. Populations $orig^1$, del^1_*, $dest^1$ capture agents located at the start of route, agents waiting at intermediate nodes and agents that have completed their journey, respectively. When setting $\alpha^1_{sg}(t) = 0$ for all t and $orig^1(0) = 1$, $dest^1(t)$ captures the passage-time *cumulative density function (CDF)* for a single agent on route r_1. Moreover, both α^1_{sg} and $\beta^1_{sg_*}$ are time-dependent. When solving Eq. (1) we assume that they change according to a deterministic *schedule*. If rate schedules are probabilistic, e.g. due to

---

Algorithm 1. Hybrid simulation for stochastic time-dependent delay rates

---

1 distributions ← new List();
2 **for** *r* *in* numRuns **do**
3   schedule ← sampleTimeDependentRates();
4   distributions.add(ddeSolution(schedule));
5 **return** histogram(distributions);

---

pedestrians that randomly push traffic light buttons, then a hybrid simulation technique as shown in Algorithm 1 is required to analyse the movement and communication model. The hybrid solution simply solves the DDEs representing the movement model *numRuns* times using a randomly sampled rate *schedule* for each repetition. The hybrid analysis becomes more efficient than full simulation as agent birth rates increase.

2.2 Measure Agent, Measure Route

To decide whether to rely on LTE or on Gossip communication as an SLA deadline approaches, a protocol policy requires information regarding the number of other agents the data was communicated to. In combination with knowledge about passage-time distributions to reach the next Wifi spot, this allows agents to make informed decisions in an effort to meet SLA requirements at minimal transmission costs. In this section we introduce the concept of a *measure agent*, which is similar to a *tagged customer* in a Stochastic Petri net. We define the set of measure agents as $\mathcal{M} \subset \mathcal{A} \times \mathbb{R} \times \mathbb{R}$, where $(ac, t, p) \in \mathcal{M}$ is a measure agent whose movement is defined by agent class $ac = (r, sg)$. The policy is then tested for a single data sample taken at time t, $p\%$ down route r. Furthermore, we assume that a measure agent experiences the average delay at every delay node, i.e. for a given delay rate schedule it moves deterministically and its exact radio range is known. That way we avoid non-linear communication dynamics, keep the underlying IPCTMC model linear and all populations independently Poisson distributed. In our case-study analysis (see Section 3.3) we found that the measure agent approach is a good proxy for the actual non-linear model, most likely because the randomness of the delay rate schedule, the agent birth rate and the communication dynamics dominate the effect of random delay node sojourn times. When analysing a model we usually consider a *measure route* rather

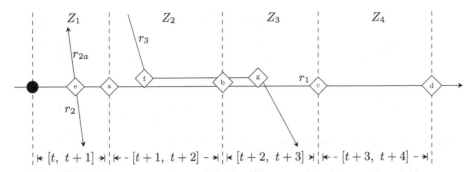

Fig. 2. Zones for a measure agent $ma \in \mathcal{M}$ on r_1 in a model with 4 routes (r_2 is bidirectional). The ma gathers its sample at the black dot and diamonds mark locations of first and last contact points between agents on r_* and ma in a zone.

than a single measure agent. This means we do the analysis in Section 2.4 for all measure agents on the same route and weight each measure agent's contribution to the solution proportional to their speed group's birth rate. This allows us to get a better understanding of how transmission strategies fare across a range of participants.

2.3   Communication Model

Since transmission mode decisions of measure agents depend on the agents it has communicated with before the deadline, we need to extend the model shown in Figure 1 to keep track of unique contacts made with other agents. A unique *first-time contact* occurs when a measure agent communicates with another agent it has not previously communicated with. The number of contacts made during $[t, t + $ SLA deadline$]$ and the information about where contacts were made, form the evidence on which decisions are taken. While it is straightforward to capture this information in a simulation, we have to discretise time into intervals to obtain these measures through DDE analysis. Figure 2 illustrates a discretisation concept that enables us to estimate the Poisson distribution parameter for the total amount of contacts made between a measure agent and agents from other agent classes in a particular time *zone* interval. Zones avoid the need to keep track of precise moments in time at which communication occurs. If our deadline was $t + 10$ minutes, we might for instance have zones $[t, t+1], [t+1, t+2], \ldots, [t+9, t+10]$. Note that due to the time-inhomogeneous nature of delays, the distance covered by a measure agent in each zone varies with the rate schedule. In fact, Figure 2 gives a rather idealised spatial notion of the spatial radio coverage of ma in each time zone, as it can happen that different time zone intervals have overlapping radio ranges. Having defined zone intervals, we can work out the first-time contacts for each zone using the time-inhomogeneous model shown in Figure 3, which extends the movement model from Figure 1. This is done by translating ma's radio range into time-inhomogeneous rates, so that communication events between agents and the

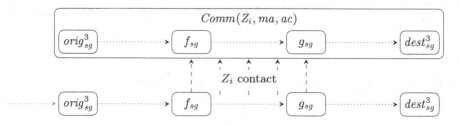

Fig. 3. Contacts made in each zone between measure agent $ma \in \mathcal{M}$ and agents of class $ac = (r_3, sg)$. States f, g correspond to delay nodes on r_3 closest to f, g in Figure 2. Dotted transitions indicate omitted intermediate states. When agents of ac communicate with ma for the first time in zone i, they enter state $Comm(Z_i, ma, ac)$.

measure agent are Poisson distributed with a rate proportional to the amount of time that they were in each other's radio range. When a contact event between an agent of agent class $ac \in \mathcal{A}$ and a measure agent $ma \in \mathcal{M}$ occurs in zone i, the agent continues its journey in state $Comm(Z_i, ma, ac)$, where it can no longer communicate with ma. As for the movement model (see Eq. (1)) linear DDEs can be derived that compute the Poisson zone contact rates between any pair of measure agent $ma \in \mathcal{M}$ and another agent class $ac \in \mathcal{A}$. It is important to bear in mind that all $Comm(Z_i, ma, ac)$ contact populations have mutually independent Poisson distributions. Mathematically this can easily be justified by extending the argument made in [9] for delay-only [10] monomolecular reaction systems. Moreover, since we can solve the DDE system for a single measure agent, agent class pair at a time, the analysis is parallelisable. With respect to scalability, Figure 2 shows that the number of routes we need to consider in our analysis depends on the distance covered by a measure agent from collecting its sample to reaching the SLA deadline. To reduce the size of our case-study model, our policies only consider the contacts made in first 5 minutes of a 10 minutes delivery time deadline. Naturally, this implies that contacts made between $t + 5$ and $t + 10$ are ignored by our protocol, which may not be adequate for other studies.

2.4 Policy Analysis

The multivariate distribution we are interested in has 4 real-valued non-negative random variables (L_L, L_G, G_L, G_G) that sum up to 1 and represent a categorical distribution where $L\_L = \mathbb{P}(\text{Choose LTE} \wedge \text{Gossip missed SLA})$, $L\_G = \mathbb{P}(\text{Choose LTE} \wedge \text{Gossip met SLA})$, $G\_L = \mathbb{P}(\text{Choose Gossip} \wedge \text{Gossip missed SLA})$ and $G\_G = \mathbb{P}(\text{Choose Gossip} \wedge \text{Gossip met SLA})$. Obviously, a good decision strategy must aim to maximise L_L and G_G. G_L measures the proportion of messages for which the SLA was breached by the protocol and L_G indicates the amount of LTE resources that were wasted. Line 4 in Algorithm 1 computes the categorical distribution for a specific measure agent, schedule combination using Algorithm 2 or by simulation. A histogram for the multivariate distribution, which collects all categorical distribution samples from a number of random schedules, is then computed in l. 5. We can consider multiple measure agents on the same route by averaging the distributions of several measure agents returned from l. 5. For a fixed schedule, the categorical distribution can be computed exactly using simulation analysis, which keeps track of where and when agents first communicated with measure agents, of their protocol decision and if the Gossip delivery was timely. With DDEs, however, we cannot capture the precise time at which an exchange took place within a zone. Instead, we compute upper and lower bounds for the distribution, where an upper bound will have the highest possible G_G with the lowest G_L for a given rate schedule and vice versa for the lower bound. The remainder of this section describes how these bounds can be computed for a single measure agent and rate schedule.

First we use DDEs to compute the first-time contact Poisson rate $R_{Z_i,ma,ac}$ (cf. $Comm(Z_i, ma, ac)$) for all zones for all pairs $(ma, ac) \in \mathcal{M} \times \mathcal{A}$, where $ma = (*, t_s, p_s)$ and m_{SLA} the SLA deadline in minutes. For each zone we also keep track of the earliest and the latest possible time $t_{Z_i,e}$, $t_{Z_i,l}$ at which ma and ac can communicate as well as the first and last position $p_{Z_i,f}$, $p_{Z_i,l}$ on ac's route at which communication with ma can occur (cf. diamond locations in Figure 2). We use these to calculate the probability that an agent of class ac will reach a Wifi hotspot in $m_{SLA}-(t_{Z_i,e}-t_s)$ minutes from $p_{Z_i,l}$ and the probability of doing so in $m_{SLA}-(t_{Z_i,l}-t_s)$ minutes from $p_{Z_i,f}$. $D^U_{Z_i,ma,ac}$ and $D^L_{Z_i,ma,ac}$ denote these best and worst case Gossip delivery probabilities for agents of ac met by ma in zone i, respectively. Since all agents move independently, we can express $R_{Z_i,ma} = \sum_{ac} R_{Z_i,ma,ac}$ and $D^*_{Z_i,ma} = \sum_{ac} D^*_{Z_i,ma,ac} \cdot R_{Z_i,ma,ac}/R_{Z_i,ma}$. Furthermore, let $\boldsymbol{R}_{ma} = (R_{Z_1,ma}, \ldots, R_{Z_n,ma})$ be the vector of zone contact rates and $\boldsymbol{D}^*_{ma} = (D^*_{Z_1,ma}, \ldots, D^*_{Z_n,ma})$ be the vector of best (U) and worst case (L) delivery probabilities. Moreover, let $\boldsymbol{H}_{ma} = (H_{Z_1,ma}, \ldots, H_{Z_n,ma})$ be the equivalent of $\boldsymbol{D}_{ma}$, but computed from historical data available for the local area that each zone spans, assuming that the message is exchanged at the end of a zone interval. Naturally, both $\boldsymbol{D}_{ma}$ and $\boldsymbol{H}_{ma}$ depend on the location of Wifi hotspots in our model. Generally, there is no restriction on how many there are on a route, but since we did not make them explicit model features in Section 2.1, we assume that they are always located at the end of a route. Note that for both DDE analysis and simulation analysis, we assume that protocol decisions use position-independent delivery time distributions $H_{Z_i,ma}$ for each zone. Hence protocol delivery time estimates only depend on how many agents of a certain class where met in a zone. While this makes both simulation and DDE analysis dependent on the zone size, it does not matter much in practice if zone intervals are short.

Given $\boldsymbol{R}_{ma}, \boldsymbol{D}^U_{ma}, \boldsymbol{D}^L_{ma}$ and $\boldsymbol{H}_{ma}$ we can evaluate each LTE/Gossip policy for ma using the procedure described in Algorithm 2. We compare the policy decision with the actual delivery probability for all probable combinations of zone contacts (see l. 4). For a valid zone *contacts* combination, we first compute the probability with which the transmission protocol chooses the LTE network $pLTE$ (see l. 6). By valid we mean any combination of zone contacts that are obtained when considering that the number of first-time contacts

---

Algorithm 2. Evaluating the LTE/Gossip strategy performance of an ma for a specific delay rate schedule (see l. 4 in Algorithm 1)

---

1 maxContacts $\leftarrow$ (PoisCDFInv($R_{Z_1,ma}, 0.999$), ..., PoisCDFInv($R_{Z_n,ma}, 0.999$));
2 contacts $\leftarrow$ $(1, 0, 0, \ldots, 0)$;
3 L_L $\leftarrow$ 0, L_G $\leftarrow$ 0, G_L $\leftarrow$ 0, G_G $\leftarrow$ 0;
4 **while** contacts $\neq (0, 0, 0, \ldots, 0)$ **do**
5   pContacts $\leftarrow$ 1, pSLAVio $\leftarrow$ 1;
6   pLTE $\leftarrow$ LTEProbStrat(H_{ma},contacts);
7   **for** $i \leftarrow 1$ *to* n **do**
8   pContacts $\leftarrow$ pContacts * PoisCDF($R_{Z_i,ma}$,contacts(i));
9   pSLAVio $\leftarrow$ pSLAVio * $(1 - D^*_{Z_i,ma})$^contacts(i);
10   L_L $\leftarrow$ L_L + pContacts * pLTE * pSLAVio ;
11   L_G $\leftarrow$ L_G + pContacts * pLTE * (1 - pSLAVio);
12   G_L $\leftarrow$ G_L + pContacts * (1 - pLTE) * pSLAVio;
13   G_G $\leftarrow$ G_G + pContacts * (1 - pLTE) * (1 - pSLAVio);
14   contacts $\leftarrow$ nextValidContactVector(contacts, maxContacts);
15 **return** (L_L,L_G,G_L,G_G);

---

in each zone i is likely to be between 0 and the $99.9th$ percentile of the underlying Poisson distribution. Next we compute the probability *pContacts* of the *contacts* vector and *pSLAVio*, the probability of violating the SLA when relying on Gossip transmission given *contacts*. Since first-time contact distributions as well as passage times in different zones are mutually independent (see Section 2.3), we can obtain the joint probability for *pContacts* (see l. 8) and *pSLAVio* (see l. 9) through multiplication of the respective zone probabilities. Note that *pSLAVio* is simply the probability that all agents that we meet fail to deliver on time. When using D^U_{ma}, Algorithm 2 yields the upper DDE bound for the given rate schedule and the lower bound when D^L_{ma} is used. A simple strategy for computing *pLTE* would be to compute it the same way as *pSLAVio* using H_{ma} instead of D^*_{ma}. However, the strategy can also choose to apply further heuristics, for instance thresholds as in Section 3. Finally in ll. 10–13, the algorithm adds the current *contacts'* contribution to the categorical distribution for the given delay rate schedule and measure agent ma.

3 Case Study: Crowd-Sensing Bikes in London

In this section we fit the model described in Section 2 to journey data from the London Barclays cycle hire scheme. For our analysis we use weekday morning rush hour journey data from May and June 2012. We describe procedures to generate routes and speed groups, show how to create a submodel for specific measure routes and how we implement traffic light and roundabout behaviour. Finally, in Section 3.3 we analyse two policies for two rate parameter setups on 4 measure routes with 4 measure agents each.

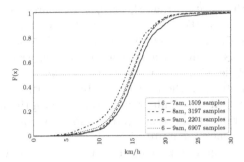

Fig. 4. Distribution of speed estimated for journeys made on routes with no traffic lights

3.1 Route and Speed Group Estimation

Although the publicly available data provides information about journey origin, destination as well as duration, it does not specify the actual route taken by cyclists. To estimate routes, along with roundabout and traffic light locations between any two stations, we extended Routino [14], a routing tool for OpenStreetMap (OSM) [15]. As there are about 200k origin–destination pairs in our data set, this process had to be automated. Using the standard Routino configuration for bikes, we found that 20% of all generated routes were more than twice as long as their aerial distance. To avoid such overlong, unrealistic routes, we always chose the shortest one from a number of alternative routes. The resulting speed distribution for weekday mornings in May, June 2012 for routes with no traffic lights are shown in Figure 4. Clearly, speeds are lower the later we measure, which agrees with observations made in [16]. Further analysis on the data indicated dependencies between route length and speed distribution. The average route length of the 20% slowest journeys, for instance, was shorter than that of the 20% fastest journeys. While this might be due to incorrect route estimates, it could also be down to journeys being affected by start, stop delays. As $speed = distance/time\ cycling = distance/(journey\ time - delay)$, speeds on shorter routes are more likely to be underestimated, since start, stop delays have a larger impact. Of course, there are other traffic related events, such as being slowed down rather than coming to a halt, which have a big impact on the average speed on shorter journeys, but this is beyond the dynamics we can capture in our model. For our analysis in Section 3.3 we use the speed distribution from Figure 4. We decided to use 4 speed groups, the 25, 50, 75, 95 speed quantiles, which are 13, 15, 16.8 and 20.0 km/h for the 7–8am data. Each of these speed groups is assumed to have an equal share of a route's total birth rate.

3.2 Fitting a Network Model

Another challenge in building the model for our case-study was to create a topology that truly reflects the physical road network of the inner city of London. Each measure route is assumed to start at a particular point on an existing route and ends at the location that the fastest measure agent (20 km/h) can reach by the end time of the last zone interval, assuming no delay at delay nodes. Measure routes must end before reaching a docking station, otherwise no transmission decision is required. Having chosen a measure route, we need to find all other routes that lie within its radio range of $30m$ and compute *subroutes*, i.e. radio range overlaps with the measure route. Note that two distinct routes have the same subroute if their radio range overlaps with the measure route are identical. Subroutes range from simple intersections to longer overlaps going in either direction of the measure route.

When looking at a 5 minute zone on a measure route, a 20 km/h fast agent can move up to $1650m$. In central London, a measure route can easily have up to 120 distinct subroutes for a $30m$ radio range. To reduce the number of subroutes, we compute cliques of similar subroutes, which share $\geq 90\%$ of the GPS way-points produced by Routino. For each clique we merge the subroutes into a new subroute, whose way-points that are most similar to all other subroutes in the clique. Moreover, the merged subroute retains total birth rates as well as passage-time samples from the end of the clique to docking stations. Cliques reduce the number of subroutes by up to 60%, hence reducing the size of the agent class set $\mathcal{A}$ drastically, which in turn speeds up the analysis. Next we discard all subroute cliques with less than 3 journeys per hour to save another 50% of subroutes while only losing 5%–10% of all journeys. Finally, we compute the overall birth rates of speed groups for every clique and create discrete passage-time distributions for agents from the end of their subroute to their destination. When computing D, H for the analysis described in Section 2.4, each D is calculated as the convolution of the passage-time required to reach the end of the subroute, either computed exactly in simulation analysis or as upper and lower bounds using DDEs, and the empirical passage-time distribution from the end of the subroute to the docking station according to the speed group quantile. To determine the H distributions, we partition the area around each measure route into $250m \times 250m$ squares and calculate local passage-time H distributions for each partition from training journeys. For a given measure agent and schedule, the zone delivery time distribution is chosen to be the one of the partition that the measure agent spends most time in, during the zone interval.

Traffic lights are clustered by proximity to ensure that phases of co-located signals are synchronised. For our case-study we used 2 phases for each cluster. Phase membership depends on the angle that a subroute approaches the centre of a traffic light cluster. This way, routes from north to south experience a different phase than routes from east to west. In the model with *short delay* rates, time-inhomogeneous rate schedules representing phase changes of traffic light clusters, are sampled from an independent normal distribution with a mean of 25 and a variance of 2 seconds using an initial uniformly-distributed offset with a mean of 10 seconds, while in the *long delay* model it is 50, 4 and 10 seconds respectively. Moreover, in the short delay model the delay experienced at a green traffic light is assumed to be exponentially distributed with rate 2, but for the first 5 seconds after switching from red to green the exponential rate is reduced to 0.2 to capture

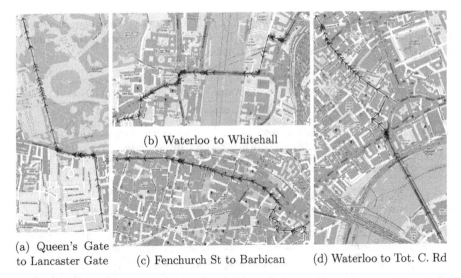

(b) Waterloo to Whitehall

(a) Queen's Gate
to Lancaster Gate   (c) Fenchurch St to Barbican   (d) Waterloo to Tot. C. Rd

Fig. 5. Measure routes. Traffic lights are green and red dots, roundabouts are blue.

acceleration dynamics, while roundabouts always cause an exponential delay at rate 0.2. For the long delay model we halve these rates except for the green light rate and assume that the acceleration time interval extends to up to 10 seconds after a traffic light turns green. Although we exclusively use exponential delays, we could alternatively deploy phase-type delays.

3.3   Analysis

We consider 4 measure routes (see Figure 5) with one measure agent per speed group, for different areas of London. While the *Queen's Gate to Lancaster Gate* and the *Waterloo to Whitehall* routes are moderately used between 7–8am, *Fenchurch St to Barbican* as well as *Waterloo to Tottenham Court Rd* are very busy as commuters head for the city. Measure agents sample data at the beginning of their route.

We compare the performance of two strategies S1 and S2. S1 is a simple threshold policy that decides to use Gossip, whenever the Gossip success probability is believed to be > 95%. S2 uses the same threshold, but assumes a tighter deadline of 9 minutes at the time of making the decision. Hence, S2 is a more conservative strategy and should meet SLA deadlines at least as often as S1. S1 on the other hand, being a more optimistic strategy, should offload more data to the Gossip network. In Figure 6 we show the *Upper* and *Lower* DDE performance bounds for S1 and S2 and the exact measure agent simulation results *Sim* for short delay models. Moreover, Table 1 further shows long delay model results and simulation results for the non-linear model *RndMA*, where measure agents experience probabilistic delays. We used 1000 traffic light schedules for each measure route and 7500 simulation runs for each schedule. The performance measures we study

Table 1. Mean Gossip usage *Efficiency* $= \mathbb{E}[G\_G/(L\_G+G\_G)]$ vs. mean probability of SLA violation $\mathbb{E}[G\_L]$ in % between 7–8am. The *CI-Width* column shows the maximum relative confidence interval width of the DDE bounds.

	S1: Threshold 95%, 10mins					S2: Threshold 95%, 9mins				
	Lower	Sim	RndMA	Upper	CI-Width	Lower	Sim	RndMA	Upper	CI-Width
Queen's Gate to Lancaster Gate (short delay)										
$\mathbb{E}[G\_L]$	0.035	0.031	0.032	0.027	2.099%	0.022	0.019	0.020	0.016	2.349%
Efficiency	21.070%	20.847%	21.103%	20.590%	1.525%	14.469%	14.333%	14.305%	14.099%	1.756%
Queen's Gate to Lancaster Gate (long delay)										
$\mathbb{E}[G\_L]$	0.049	0.044	0.049	0.038	2.015%	0.031	0.028	0.030	0.024	2.302%
Efficiency	25.803%	25.486%	27.901%	25.240%	1.437%	18.507%	18.273%	19.344%	18.039%	1.720%
Waterloo to Whitehall (short delay)										
$\mathbb{E}[G\_L]$	0.104	0.081	0.082	0.065	1.619%	0.034	0.024	0.024	0.019	3.221%
Efficiency	48.487%	47.384%	47.434%	46.429%	1.026%	22.985%	22.206%	21.751%	21.419%	1.925%
Waterloo to Whitehall (long delay)										
$\mathbb{E}[G\_L]$	0.097	0.074	0.075	0.057	2.989%	0.030	0.021	0.022	0.015	3.504%
Efficiency	48.015%	46.680%	46.731%	45.582%	2.002%	22.745%	21.747%	21.549%	20.905%	2.517%
Waterloo to Tottenham Court Rd (short delay)										
$\mathbb{E}[G\_L]$	0.077	0.060	0.061	0.046	0.657%	0.069	0.054	0.054	0.040	0.646%
Efficiency	97.329%	97.218%	97.349%	97.098%	0.044%	95.081%	94.908%	94.610%	94.711%	0.063%
Waterloo to Tottenham Court Rd (long delay)										
$\mathbb{E}[G\_L]$	0.075	0.061	0.061	0.042	0.488%	0.068	0.055	0.055	0.037	0.462%
Efficiency	98.607%	98.569%	98.479%	98.464%	0.028%	96.692%	96.612%	96.314%	96.407%	0.045%
Fenchurch St to Barbican (short delay)										
$\mathbb{E}[G\_L]$	0.168	0.137	0.136	0.100	1.055%	0.147	0.118	0.117	0.084	0.924%
Efficiency	93.414%	93.054%	93.658%	92.868%	0.292%	88.935%	88.405%	89.231%	88.151%	0.451%
Fenchurch St to Barbican (long delay)										
$\mathbb{E}[G\_L]$	0.180	0.147	0.147	0.111	1.959%	0.163	0.132	0.132	0.097	1.664%
Efficiency	95.948%	95.756%	95.984%	95.594%	0.302%	92.774%	92.471%	92.837%	92.221%	0.493%

are the *Efficiency*, which is $G\_G/(L\_G + G\_G)$, i.e. the proportion of messages that an optimal decision policy would have offloaded to the Gossip network and G_L, the SLA violation rate of a policy. The distributions shown in Figure 6 show the joint distribution of the Efficiency and SLA violation probability in %, while Table 1 shows mean values. As expected, S2 always has a better chance of SLA compliance, but never exceeds the efficiency of S1. Interestingly, the upper bound efficiency is lower than the lower bound efficiency for both policies, though reassuringly a look at the raw data shows that G_G is higher for upper bounds in all models. On the other hand G_L is always lower for the upper bound, as is expected. Moreover, the data in the table shows that while the results of the non-linear model are not always bound by the DDE results, the differences in first-order moments are small compared to simulation results for deterministi-cally moving measure agents. Despite being rather simplistic, on average our strategies have at most a 14% chance of violating the SLA and often exhibit a high degree of efficiency. It is likely that more sophisticated strategies, possibly using additional information like weather or weekday dependent distributions to get better estimates for H, can improve upon this further. Furthermore, a look at the raw G_G and L_G data, which was omitted due to space constraints,

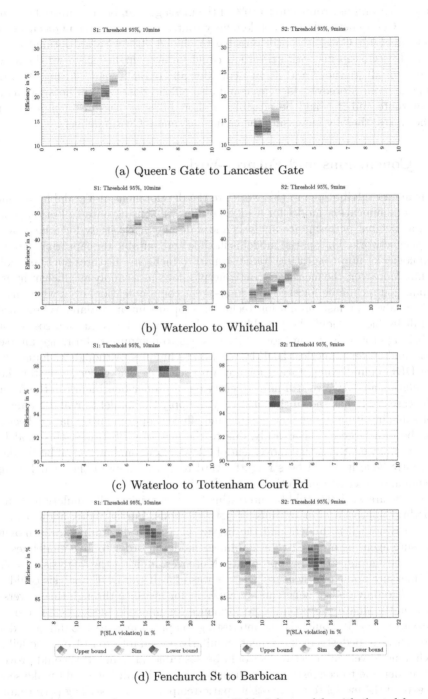

(a) Queen's Gate to Lancaster Gate

(b) Waterloo to Whitehall

(c) Waterloo to Tottenham Court Rd

(d) Fenchurch St to Barbican

Fig. 6. Distributions for measures shown in Table 1 for models with short delays

shows that on busy routes up to 80% of the messages can be transmitted on time via the Gossip network. Even on less busy routes, such as between Queen's Gate and Lancaster Gate, it is still possible to offload 40% with an optimal policy. In reality, Gossip network utilisation could be even higher since we discarded 5–10% of all journeys and do not consider all contacts made during the SLA time interval. Moreover, bicycles could also offload data to Wifi hotspots along their route and it would be possible to allow bikes to pass on foreign data to other bikes, thereby increasing the chance of timely Gossip delivery.

4   Conclusions and Future Work

Our main contributions in this work are the development of a scalable mobility and communication model for a high-level crowd-sensing data-collection protocol; associated performance analysis and a detailed case-study for a large-scale bicycle network. The spatial model is flexible and suitable for capturing detailed topologies of non-congesting traffic scenarios, including features such as traffic lights. Moreover, the measure agent concept allows us to apply an efficient hybrid simulation technique for studying inter-agent communication, even when mean-field analysis or fast simulation techniques [5] are inappropriate. While easily parallelisable, it should be possible to further reduce the evaluation cost of our hybrid approach by using more efficient ways to compute radio range intersections and better numerical DDE integration techniques. Although we found that the DDE bound computation for any rate schedule was generally faster than simulation analysis, we still need to carry out a proper benchmark in the future. The case-study has shown that it is easy to apply the model to real data and to get some intuition about how reliable and efficient opportunistic data collection can be in a large city. Naturally, due to the lack of GPS traces, this model is only a proxy for the actual movement of bikes in London, however, if better data were available it would be straightforward to reuse many of the model fitting techniques described in Section 3 to reflect it.

For future work it would be interesting to benchmark a full simulation against a hybrid approach after optimising both methods. Furthermore, additional research is needed to understand the impact of using a deterministically moving measure agent as opposed to a non-linear model in order to give guidelines for when the simplification is appropriate. Moreover, limiting the analysis to a simple measure point per measure agent is wasteful. If the DDE analysis could be altered to provide last zone contact analysis, for instance through time-reversal, we could evaluate the protocol performance for multiple messages for a single measure agent. It would also be interesting to apply our model to different data sets, such as pedestrian movement. Similarly, an adaptation of our modelling technique to vehicular networks would be desirable, although this could prove a lot harder due to congestion and queueing effects. Finally, it would make sense to extend the model to derive other spatio-temporal crowd-sensing performance metrics such as area coverage of crowd-sensing approaches.

References

1. Difallah, D.E., Cudre-Mauroux, P., McKenna, S.A.: Scalable Anomaly Detection for Smart City Infrastructure Networks. IEEE Internet Computing 17, 39–47 (2013)
2. Campbell, A., Eisenman, S., Lane, N., Miluzzo, E., Peterson, R., Lu, H., Zheng, X., Musolesi, M., Fodor, K., Ahn, G.-S.: The Rise of People-Centric Sensing. IEEE Internet Computing 12(4) (2008)
3. Kanjo, E., Bacon, J., Roberts, D., Landshoff, P.: MobSens: Making Smart Phones Smarter. IEEE Pervasive Computing 8(4) (2009)
4. Lee, U., Zhou, B., Gerla, M., Magistretti, E., Bellavista, P., Corradi, A.: Mobeyes: smart mobs for urban monitoring with a vehicular sensor network. IEEE Wireless Communications 13(5) (2006)
5. Chaintreau, A., Le Boudec, J.-Y., Ristanovic, N.: The age of gossip. In: Proceedings of the Eleventh International Joint Conference on Measurement and Modeling of Computer Systems - SIGMETRICS 2009, p. 109. ACM Press, New York (2009)
6. Feng, C.: Patch-based Hybrid Modelling of Spatially Distributed Systems by Using Stochastic HYPE - ZebraNet as an Example. In: Twelfth International Workshop on Quantitative Aspects of Programming Languages and Systems, QAPL (2014)
7. Mathur, S., Jin, T., Kasturirangan, N., Chandrasekaran, J., Xue, W., Gruteser, M., Trappe, W.: ParkNet: drive-by sensing of road-side parking statistics. In: Proceedings of the 8th International Conference on Mobile Systems, Applications, and Services, MobiSys 2010, p. 123. ACM Press (2010)
8. Bortolussi, L., Galpin, V., Hillston, J.: Hybrid performance modelling of opportunistic networks. Electronic Proceedings in Theoretical Computer Science 85, 106–121 (2012)
9. Jahnke, T., Huisinga, W.: Solving the chemical master equation for monomolecular reaction systems analytically. Journal of Mathematical Biology 54(1), 1–26 (2007)
10. Hayden, R.A.: Mean Field for Performance Models with Deterministically-Timed Transitions. In: 2012 Ninth International Conference on Quantitative Evaluation of Systems (QEST), London, pp. 63–73. IEEE (September 2012)
11. Fiore, M., Härri, J.: The networking shape of vehicular mobility. In: International Symposium on Mobile Ad Hoc Networking and Computing, MobiHoc 2008, p. 261. ACM, ACM Press, New York (2008)
12. Viriyasitavat, W., Bai, F., Tonguz, O.K.: Dynamics of Network Connectivity in Urban Vehicular Networks. IEEE Journal on Selected Areas in Communications 29(3), 515–533 (2011)
13. Guenther, M.C., Bradley, J.T.: Journey Data Based Arrival Forecasting for Bicycle Hire Schemes. In: Dudin, A., De Turck, K. (eds.) ASMTA 2013. LNCS, vol. 7984, pp. 214–231. Springer, Heidelberg (2013)
14. Bishop, A.M.: Routino (2013), http://www.routino.org
15. Haklay, M.M., Weber, P.: OpenStreetMap: User-Generated Street Maps. IEEE Pervasive Computing 7, 12–18 (2008)
16. Borgnat, P., Abry, P., Flandrin, P., Robardet, C., Rouquier, J.-B., Fleury, E.: Shared Bicycles in a City: A Signal Processing and Data Analysis Perspective. Advances in Complex Systems 14, 415–438 (2011)

Probabilistic Model Checking
of DTMC Models of User Activity Patterns

Oana Andrei[1], Muffy Calder[1], Matthew Higgs[1], and Mark Girolami[2]

[1] School of Computing Science, University of Glasgow, G12 8RZ, UK
[2] Department of Statistics, University of Warwick, CV4 7AL, UK

Abstract. Software developers cannot always anticipate how users will actually use their software as it may vary from user to user, and even from use to use for an individual user. In order to address questions raised by system developers and evaluators about software usage, we define new probabilistic models that characterise user behaviour, based on activity patterns inferred from actual logged user traces. We encode these new models in a probabilistic model checker and use probabilistic temporal logics to gain insight into software usage. We motivate and illustrate our approach by application to the logged user traces of an iOS app.

1 Introduction

Software developers cannot always anticipate how users will *actually* use their software, which is sometimes surprising and varies from user to user, and even from use to use, for an individual user. We propose that temporal logic reasoning over formal, probabilistic models of actual logged user traces can aid software developers, evaluators, and users by: *providing insights* into application usage, including differences and similarities between different users, *predicting* user behaviours, and *recommending* future application development.

Our approach is based on systematic and automated logging and reasoning about users of applications. While this paper is focused on mobile applications (apps), much of our work applies to any software system. A logged user trace is a chronological sequence of in-app actions representing how the user explores the app. From logged user traces of a population of users we infer *activity patterns*, represented each by a Discrete-Time Markov Chain (DTMC), and for each user we infer a user *strategy* over the activity patterns. For each user we deduce a meta model based on the set of all activity patterns inferred from the population of users and the user strategy, and we call it the *user metamodel*. We reason about the user metamodel using probabilistic temporal logic properties to express hypotheses about user behaviours and relationships within and between the activity patterns, and to formulate app-specific questions posed by developers and evaluators.

We motivate and illustrate our approach by application to the mobile, multiplayer game Hungry Yoshi [1], which was deployed in 2009 for iPhone devices and has involved thousands of users worldwide. We collaborate with the Hungry Yoshi developers on several mobile apps and we have access to all logged

G. Norman and W. Sanders (Eds.): QEST 2014, LNCS 8657, pp. 138–153, 2014.
© Springer International Publishing Switzerland 2014

user data. We have chosen the Hungry Yoshi app because its functionality is relatively simple, yet adequate to illustrate how formal analysis can inform app evaluation and development.

The main contributions of the paper are:

- a formal and systematic approach to formal user activity analysis in a probabilistic setting,
- inference of user activity patterns represented as DTMCs,
- definition of the DTMC user metamodel and guidelines for inferring user metamodels from logged user data,
- encoding of the user metamodel in the PRISM model checker and temporal logic properties defined over both states and activity patterns as atomic propositions,
- illustration with a case study of a deployed app with thousands of users and analysis results that reveal insights into real-life app usage.

The paper is organised as follows. In the next section we give an overview of the Hungry Yoshi app, which we use to motivate and illustrate our work. We list some example questions that have been posed by the Hungry Yoshi developers and evaluators; while these are specific to the Hungry Yoshi app, they are also indicative questions for any app. In Sect. 3 we give background technical definitions concerning DTMCs and probabilistic temporal logics. In Sect. 4 we define inference of user activity patterns, giving a small example as illustration and some example results for Hungry Yoshi. In Sect. 5 we define the user metamodel, we illustrate it for Hungry Yoshi and we give an encoding for the PRISM model checker. In Sect. 6 we consider how to encode some of the questions posed in Sect. 2.1 in probabilistic temporal logic, and give some results for an example Hungry Yoshi user metamodel. In Sect. 7 we reflect upon the results obtained for Hungry Yoshi and some further issues raised by our approach. In Sect. 8 we review related work and we conclude in Sect. 9.

2 Running Example: Hungry Yoshi

The mobile, multiplayer game Hungry Yoshi [1] is based on picking pieces of fruit and feeding them to creatures called *yoshis*. Players' mobile devices regularly scan the available WiFi access points and display a password-protected network as a *yoshi* and a non-protected network as a *fruit plantation*. Plantations grow particular types of *fruit* (possibly from *seeds*) and yoshis ask players for particular types of fruit. Players score points if they pick the fruit from the correct plantations, store them in a basket, and give them to yoshis as requested. There is further functionality, but here we concentrate on the key user-initiated events, or *button taps*, which are: *see a yoshi*, *see a plantation*, *pick fruit* and *feed a yoshi*. The external environment (as scanned by device), combined with user choice, determines when yoshis and plantations can be observed. The game was instrumented by the developers using the SGLog data logging infrastructure [2], which streams logs of specific user system operations back to servers on

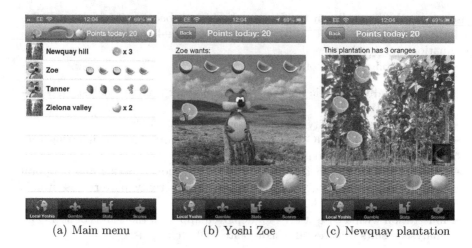

(a) Main menu   (b) Yoshi Zoe   (c) Newquay plantation

Fig. 1. Hungry Yoshi screenshots: two plantations (Newquay hill and Zielona valley) and two yoshis (Zoe and Taner) are observed. The main menu shows the available plantations and yoshis with their respective content and required types of fruit. The current basket contains one orange seed, one apricot and one apple.

the developing site as user traces. The developers specify directly in the source code what method calls or contextual information are to be logged by SGLog. A sample of screenshots from the game is shown in Fig. 1.

2.1 Example Questions from Developers and Evaluators

Key to our formal analysis is suitable hypotheses, or *questions*, about user behaviour. For Hungry Yoshi, we interviewed the developers and evaluators of the game to obtain questions that would provide useful insights for them. Interestingly most of their hypotheses were *app-specific*, and so we focus on these here, and then indicate how each could be generalised. We note that to date, tools available to the developers and evaluators for analysis include only SQL and iPython stats scripts.

1. When a yoshi has been fed n pieces of fruit (which results in extra points when done without interruption for n equal 5), did the user interleave *pick fruit* and *feed a yoshi* n times or did the user perform n *pick fruit* events followed by n *feed a yoshi* events? And afterwards, did he/she continue with that pick-feed strategy or change to another one? Which strategy is more likely in which activity pattern? More generally, when there are several ways to reach a goal state, does the user always take a particular route and is this dependent on the activity pattern?
2. If a user in one activity pattern does not *feed a yoshi* within n button taps, but then changes to another activity pattern, is the user then likely to *feed a yoshi* within m button taps? More generally, which events cause a change of activity pattern, and which events follow that change of activity pattern?

3. What kind of user tries to *pick fruit* 6 times in a row (a basket can only hold 5 pieces of fruit)? More generally, in which activity pattern is a user more likely to perform an inappropriate event?
4. If a user reads the instructions once, then does that user reach a goal state in fewer steps than a user who does not read the instructions at all? (Thus indicating the instructions are of some utility.) More generally, if a user performs a given event, then it is more likely that he/she will perform another given event, within n button taps, than users that have not performed the first event? Is this affected by the activity pattern?

3   Technical Background

We assume familiarity with Discrete-Time Markov Chains, probabilistic logics PCTL and PCTL*, and model checking [3,4]; basic definitions are below.

A *discrete-time Markov chain* (DTMC) is a tuple $\mathcal{D} = (S, \bar{s}, \mathbf{P}, L)$ where: S is a set of states; $\bar{s} \in S$ is the initial state; $\mathbf{P} : S \times S \to [0,1]$ is the transition probability function (or matrix) such that for all states $s \in S$ we have $\sum_{s' \in S} \mathbf{P}(s, s') = 1$; and $L : S \to 2^{AP}$ is a labelling function associating to each state s in S a set of valid atomic propositions from a set AP. A *path* (or execution) of a DTMC is a non-empty sequence $s_0 s_1 \ldots$ where $s_i \in S$ and $\mathbf{P}(s_i, s_{i+1}) > 0$ for all $i \geq 0$. A path can be finite or infinite. Let $Path^{\mathcal{D}}(s)$ denote the set of all infinite paths of $\mathcal{D}$ starting in state s.

Probabilistic Computation Tree Logic (PCTL) [3] and its extension PCTL* allow one to express a probability measure of the satisfaction of a temporal property. Their syntax is the following:

$$
\begin{array}{rl}
\textit{State formulae} & \Phi ::= true \mid a \mid \neg\Phi \mid \Phi \wedge \Phi \mid \mathbf{P}_{\bowtie p}[\Psi] \\
\textit{PCTL Path formulae} & \Psi ::= \mathsf{X}\,\Phi \mid \Phi\,\mathsf{U}^{\leq n}\,\Phi \\
\textit{PCTL* Path formulae} & \Psi ::= \Phi \mid \Psi \wedge \Psi \mid \neg\Psi \mid \mathsf{X}\,\Psi \mid \Psi\,\mathsf{U}^{\leq n}\,\Psi
\end{array}
$$

where a ranges over a set of atomic propositions AP, $\bowtie\, \in \{\leq, <, \geq, >\}, p \in [0,1]$, and $n \in \mathbb{N} \cup \{\infty\}$.

A state s in a DTMC $\mathcal{D}$ satisfies an atomic proposition a if $a \in L(s)$. A state s satisfies a state formula $\mathbf{P}_{\bowtie p}[\Psi]$, written $s \models \mathbf{P}_{\bowtie p}[\Psi]$, if the probability of taking a path starting from s and satisfying Ψ meets the bound $\bowtie p$, i.e., $\mathrm{Pr}_s\{\omega \in Path^{\mathcal{D}}(s) \mid \omega \models \Psi\} \bowtie p$, where Pr_s is the probability measure defined over paths from state s. The path formula $\mathsf{X}\,\Phi$ is true on a path starting with s if Φ is satisfied in the state following s; $\Phi_1\,\mathsf{U}^{\leq n}\,\Phi_2$ is true on a path if Φ_2 holds in the state at some time step $i \leq n$ and at all preceding states Φ_1 holds. This is a minimal set of operators, the propositional operators *false*, disjunction and implication can be derived using basic logical equivalences and a common derived path operators is the *eventually* operator F where $\mathsf{F}^{\leq n}\,\Phi \equiv true\,\mathsf{U}^{\leq n}\,\Phi$. If $n = \infty$ then superscripts omitted. We assume the following two additional notations. Let φ denote the state formulae from the propositional logic fragment of PCTL, i.e., $\varphi ::= true \mid a \mid \neg\varphi \mid \varphi \wedge \varphi$, where $a \in AP$. Let $\mathcal{D}_{|\varphi}$ denote the DTMC obtained from $\mathcal{D}$ by restricting the set of states to those satisfying φ.

Many of the properties we will examine require PCTL*, because we want to examine sequences of events: this requires multiple occurrences of a bounded until operator. This is not fully implemented in the current version of PRISM (only a single bounded U is permitted[1]) and so we combine probabilities obtained from PRISM *filtered properties* to achieve the same result. Filtered probabilities check for properties that hold *from sets of states* satisfying given propositions. For a DTMC $\mathcal{D}$, we define the filtered probability of taking all paths that start from any state satisfying φ and satisfy (PCTL) ψ by:

$$Prob^{\mathcal{D}}_{filter(\varphi)}(\psi) \overset{def}{=} filter_{s \in \mathcal{D}, s \models \varphi} \text{Pr}_s\{\omega \in Path^{\mathcal{D}}(s) \mid \omega \models \psi\}$$

where *filter* is an operator on the probabilities of ψ for all the states satisfying φ. In the examples illustrated in this paper we always use *state* as the filter operator since φ uniquely identifies a state.

4   Inferring User Activity Patterns

The role of inference is to construct a representation of the data that is amenable to checking probabilistic temporal logic properties. Developers want to be able to select a user and explore that user's model. While this could be achieved by constructing an independent DTMC for each user, there is much to be gained from sharing information between users. One way to do this is to construct a set of *user classes* based on attribute information, and to learn a DTMC for each class. This is the approach taken in [5] for users interacting with web applications, and is a natural way to aggregate information over users and to condition user-models on attribute values. One issue with this approach is that it assumes within-class use to be homogeneous. For example, all users in the same city using the same browser are modelled using the same DTMC.

In this work we take a different approach to inference. We have found the common representations of context - such as location, operating system, or time of day - to be poor predictors of mobile application use. For this reason we construct user models based on the log information alone, without any ad-hoc specification of user classes. By *letting the data speak for itself*, we hope to uncover interesting activity patterns and meaningful representations of users.

4.1   Statistical Model and Inference

We extend the standard DTMC model by introducing a *strategy* for each user over activity patterns. More formally, we assume there exists a finite K number of activity patterns, each modelled by a DTMC denoted $\alpha_k = (S, \iota^{init}, \mathbf{P}_k, L)$, for $k = 1, \dots, K$. Note only the transition probability $\mathbf{P}_k$ varies over the set of DTMCs, all the other parameters are the same. For some enumeration of users $m = 1, \dots, M$, we represent a user's strategy by a vector θ_m such that $\theta_m(k)$ denotes the probability that user m transitions according to α_k.

---

[1] Because currently the LTL-to-automata translator that PRISM uses does not support bounded properties.

Statistical Model. The data for each user is assumed to be generated in the following way. We assume all users to be independent and all DTMCs to be available to all users at all points in time. A user chooses an initial state according to ι^{init}. When in state $s \in S$, user m selects the kth DTMC with probability $\theta_m(k)$. If the user chooses the kth DTMC, then they transition from state s to $s' \in S$ with probability $\mathbf{P}_k(s, s')$. This simple description specifies all the probabilistic dependencies required to compute the likelihood of the data given the parameters of the model. While it is possible to extend the model so θ is state-dependent, this will require us to either lose the distributed representation of the user population, or to increase the number of parameters in a way that leads to a high combinatorial degree of complexity.

Inference. Inference is performed by maximising the log-likelihood of the data over the parameters of the model. This cannot be done analytically and we use a numerical method: the expectation-maximisation (EM) algorithm of [6]. For $K > 1$, the log-likelihood has multiple maxima and we restart the algorithm multiple times and select the output parameters with the highest log-likelihood over all runs. For the data considered here, restarting the algorithm 1000 times was sufficient to reproduce the same output parameters.

4.2 Example Activity Patterns from Hungry Yoshi

In Fig. 2 we give the activity patterns inferred from a dataset of user traces for 164 users randomly selected from the user population, for $K = 2$. A more detailed overview is given in the work-in-progress paper [7]. For brevity, we do not include the exact values of $\mathbf{P}_1$ and $\mathbf{P}_2$, but thicker arcs correspond to transition probabilities greater than 0.1, thinner ones to transition probabilities in $[0.01, 0.1]$, and dashed ones to transition probabilities smaller than 10^{-12}. Intuitively, we can see that given the game is essentially about seeing yoshis and feeding them, α_1 looks like a better way for playing the game. For example in α_2 it is quite rare to reach *feed* from *seeY* and *seeP*, and also rare to move from *seeP* to *pick*. Hungry Yoshi is a simple app with only two distinctive activity patterns, in a more complex setting we might not be able to have any intuition about the activity patterns.

5   User Metamodel

We define the formal model of the behaviour of a user m with respect to the population of users, which we call the *user metamodel* (UMM). The UMM for user m is a DTMC obtained by "flattening" the transition model over states *and* strategies. The resulting DTMC describes how the user transitions between composite states of the form (s, k) where s is an observable state and k indicates the activity pattern at that time. The UMM can be defined formally in the following way.

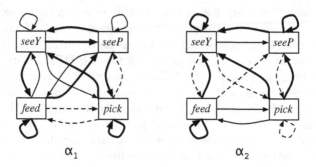

Fig. 2. Two user activity patterns α_1 and α_2 inferred from Hungry Yoshi usage

Definition 5.1 (User Metamodel). *Given K activity patterns $\alpha_1, \ldots, \alpha_K$ and θ_m the strategy of user m for choosing activity patterns, the user metamodel for m is a DTMC $\mathcal{M} = (S_\mathcal{M}, \iota_\mathcal{M}^{init}, \mathbf{P}_\mathcal{M}, \mathcal{L}_\mathcal{M})$ where:*

- $S_\mathcal{M} = S \times \{1, \ldots, K\}$,
- $\iota_\mathcal{M}^{init}(s, k) = \theta_m(k) \cdot \iota^{init}(s)$,
- $\mathbf{P}_\mathcal{M}((s, k), (s', k')) = \theta_m(k') \cdot \mathbf{P}_{k'}(s, s')$,
- $\mathcal{L}_\mathcal{M}(s, k) = \mathcal{L}(s) \cup \{\alpha = k\}$.

We label each state (s, k) with the atomic proposition $\alpha = k$ to denote that the state belongs to the activity pattern α_k.

5.1   Example UMM from Hungry Yoshi

An intuitive graphical description of the UMM for the Hungry Yoshi game for $K = 2$ is illustrated in Fig. 3. For example, $\theta_m(1)$ is the probability that user m continues with activity pattern α_1, i.e. takes a transition between states in α_1. The probability that the user changes the activity pattern and makes a transition according to α_2 is proportional to $\theta_m(2)$. Figure 3 is not a direct representation of the transition probability matrix of the UMM DTMC, but it illustrates how that matrix is derived from the matrices of the individual user activity patterns. Note that the activity patterns have the same sets of states. For instance, in the Hungry Yoshi example, consider we are in state *seeY* with α_1; we can move to state *feed* following the same pattern α_1 with the probability $\theta_m(1) \cdot P_1(seeY, feed)$, *or* we can change the activity pattern and move to state *feed* following α_2 with the probability $\theta_m(2) \cdot P_2(seeY, feed)$.

5.2   Encoding a UMM in PRISM

We use the probabilistic model checker PRISM [8]. We assume some familiarity with the modelling language (based on the language of reactive modules), which includes global variables, modules with local variables, labelled-commands corresponding to transitions and multiway synchronisation of modules. Below we

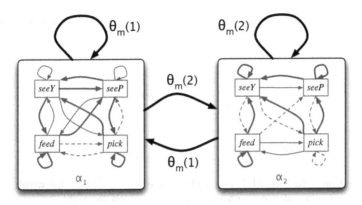

Fig. 3. An intuitive view of computing the transition probability matrix of the user metamodel for the Hungry Yoshi app

illustrate the PRISM encoding of the UMM for user m, where K is the number of activity patterns, n is the number of states in each activity pattern α_k.

> **module** *UserMetamodel_m*
> $s : [0 .. n]$ **init** 0;
> $k : [0 .. K]$ **init** 0;
>
> $[] \ (s = 0) \longrightarrow \theta_m(1) * \iota^{init}(1) : (s' = 1) \& (k' = 1) + \ldots +$
> $\qquad \theta_m(K) * \iota^{init}(n) : (s' = n) \& (k' = K);$
> $[] \ (s = 1) \longrightarrow \theta_m(1) * \mathbf{P}_1(1, 1) : (s' = 1) \& (k' = 1) + \ldots +$
> $\qquad \theta_m(K) * \mathbf{P}_K(1, n) : (s' = n) \& (k' = K);$
>
> $\vdots$
>
> $[] \ (s = n) \longrightarrow \theta_m(1) * \mathbf{P}_1(n, 1) : (s' = 1) \& (k' = 1) + \ldots +$
> $\qquad \theta_m(K) * \mathbf{P}_K(n, n) : (s' = n) \& (k' = K);$
> **endmodule**

The representation is straightforward, consisting of one module with $(n + 1)$ commands for all n states of any activity pattern and for one initial state. The initial state $(s = 0, k = 0)$ is a dummy that encodes the global initial distribution ι^{init} for the user activity patterns. All activity patterns have the same set of states and we enumerate them from 1 to n; we can label them conveniently with atomic propositions. For instance, in a Hungry Yoshi UMM the states $(0, k)$ to $(4, k)$ are labelled by the atomic proposition *init*, *seeY*, *feed*, *seeP*, *pick* respectively. For each state $(s, \cdot)$, with $s > 0$, we have a command defining all possible $n \cdot K$ probabilistic transitions. $\mathbf{P}_k(i, j)$ is the transition probability from state i to state j in α_k, and $\theta_m(k)$ is the probability of user m to choose the activity patterns α_k, for all $i, j \in \{1, \ldots, n\}$, $k \in \{1, \ldots, K\}$. If the probability of an update is null, then the corresponding transition does not take place.

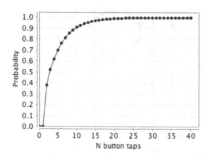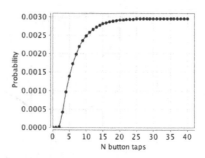

Fig. 4. Question 1: the probability of feeding a yoshi for the first time within N button taps for the activity pattern α_1 on the left and for α_2 on the right

6   Analysing the Hungry Yoshi UMM

In this section, we give some example analysis for a UMM. Namely, we encode and evaluate quantitatively several example questions from Sect. 2.1 for the UMM with the user strategy for transitioning between activity patterns defined by $\theta = (0.7, 0.3)$. The PRISM models and property files are freely available[2].

Recall that to score highly, a user must feed one or more yoshis (the appropriate fruit) often. An informal inspection of α_1 and α_2 indicates that α_2 is a less effective strategy for playing the game, since paths from *seeP* and *seeY* to *feed* are unlikely. Now, by formal inspection of the UMM (encoded in PRISM), we can investigate this hypothesis more rigorously. We consider properties that are parametrised by a number of button taps (e.g. N, $N1$, $N2$) and by activity pattern (e.g. α_1, α_2), so we use the PRISM experiment facility that allows us to evaluate and then plot graphically results for a range of formulae.

Question 1. How many button taps N does it take to feed a yoshi for the first time? We encode this by the probabilistic until formula:

$$p_1(i) = Prob^{\mathcal{M}}_{init}((\neg feed)\, \mathsf{U}^{\leq N}\, ((\alpha = i) \wedge feed))$$

and equivalently in PRISM:P=?[(!"feed") U<=N (alpha=i)&("feed")]..

For activity pattern α_1, Figure 4 shows that within 2 button taps the probability increases rapidly, and after 5 button taps the probability is more than 70%. Contrast this with the results for α_2: the probability increases rapidly after 3 button taps but soon it reaches the upper bound of 0.003. Comparing the two results, α_1 is clearly more effective.

Now we consider more complex questions concerning sequences of feeding and picking; recall that a basket can hold at most 5 fruits and extra points are gained by feeding a yoshi its required 5 fruits without any other interruption. In Question 2 we consider feeding a full basket to a yoshi, without any interruptions;

---

[2] Available from http://dcs.gla.ac.uk/~oandrei/yoshi

in Question 3 we consider picking a full basket, without being interrupted by a *feed*, followed by feeding the full basket to a yoshi, which is again defined by five consecutive *feeds*, without any interruptions. Note that when considering feeding the full basket to a yoshi, we exclude all interruptions, i.e. any interleavings with *pick*, *seeY*, and *seeP*.

Question 2. What is the probability of feeding the same yoshi a full basket? We calculate the probability of reaching the state *feed* within N button taps and then visiting it (with the same activity pattern $i \in \{1, 2\}$) for another four times without visiting any other state:

$$p_2(i) = Prob_{init}^{\mathcal{M}}(\mathsf{F}^{\leq N}(\alpha = i \wedge feed)) \cdot (Prob_{feed}^{\mathcal{M}_{|\alpha=i}}(\mathsf{X}\, feed))^4$$

We calculate this probability in PRISM using the property:

```
P=?[F<=N((alpha=i)&"feed")] *
   pow(filter(state,P=?[X(alpha=i&"feed")],(alpha=i&"feed")),4)
```

The results are shown in Fig. 5 for both activity patterns and a range of number of button taps. While the results for α_1 (converging to 0.018) are higher than for α_2 (effectively 0); they are both small. There could be several causes for this. For example, players are only made aware of the possibility of extra points at the end of the instructions pages, or available fruit depends on the external environment. If designers/evaluators want this investigated further, then we would require to record and extract more detail from the logs, for example to log numbers of available WiFi access points and scrolls through instruction pages.

Question 3. What is the probability of filling up a basket of fruit without feeding a yoshi, and only after the basket is full feeding the same yoshi the whole basket? We calculate the probability of reaching the state *feed* only after visiting the state *pick* five times (without feeding) and then visiting the state *feed* four more times without visiting any other state, for each activity pattern $i \in \{1, 2\}$:

$$p_3(i) = Prob_{init}^{\mathcal{M}}[(\neg pick)\, \mathsf{U}^{\leq N}((\alpha = i) \wedge pick)] \cdot$$
$$(Prob_{pick}^{\mathcal{M}_{|\alpha=i}}[\mathsf{X}((\neg feed \wedge \neg pick)\, \mathsf{U}\, (pick))])^4 \cdot$$
$$Prob_{pick}^{\mathcal{M}_{|\alpha=i}}[(\neg feed)\, \mathsf{U}\, feed] \cdot (Prob_{feed}^{\mathcal{M}_{|\alpha=i}}[\mathsf{X}\, feed])^4$$

The corresponding PRISM property is:

```
P=?[!("pick") U<=N ((alpha=i)&"pick")] * pow(filter(state,
   P=?[X ((alpha=i)&(!"feed")&(!"pick") U ((alpha=i)&"pick"))],
   ((alpha=i)&"pick")),4) * filter(state,
   P=?[(alpha=i)&(!"feed")U((alpha=i)&"feed")],((alpha)=i&"pick")) *
   pow(filter(state,P=?[X((alpha=i)&"feed")],((alpha=i)&"feed")),4)
```

The results are presented in Fig. 6. Again, while the probabilities are low (presumably for the reasons outlined above for Question 2) the user that picks

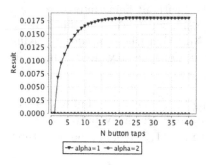
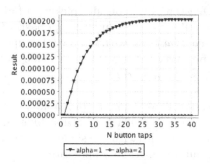
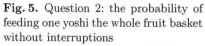

Fig. 5. Question 2: the probability of feeding one yoshi the whole fruit basket without interruptions

Fig. 6. Question 3: the probability of picking five pieces of fruit and then feeding one yoshi the whole basket

a full basket and feeds it to a yoshi by following activity pattern α_1 does it with around 0.00019 probability within 20 steps into the game, whereas if they follow α_2 from the beginning, they almost never empty the basket. So again, α_1 proves to be more effective.

Now we turn our attention to a question that involves a *change* of activity pattern, i.e. a change in the playing strategy.

Question 4. What is the probability of starting with an activity pattern and not feeding a yoshi within N button taps, then changing to the other activity pattern and eventually first feeding a yoshi within N_2 button taps? We compute this probability as follows, where $\mathcal{L}_0 = \{feed, pick, seeY, seeP\}$:

$$p_4(i) = \sum_{\ell \in \mathcal{L}_0} Prob^{\mathcal{M}}_{init}((\neg(\alpha = i) \wedge \neg feed)\, \mathsf{U}^{\leq N}((\alpha = i) \wedge \ell)) \cdot$$
$$Prob^{\mathcal{M}_{|\alpha=i}}_{\ell}((\neg feed)\, \mathsf{U}^{\leq N_2} feed)$$

The corresponding PRISM property is:

```
P=?[(!(alpha=i)&!("feed")) U<=N (alpha=i&"feed")] *
  filter(state,P=?[(alpha=i)&!("feed") U<=N2 (alpha=i&"feed")],
  alpha=i&"feed") + P=?[(!(alpha=i)&!("feed")) U<=N (alpha=i&"pick")] *
  filter(state,P=?[(alpha=i)&!("feed") U<=N2 (alpha=i&"feed")],
  alpha=i&"pick") + P=?[(!(alpha=i)&!("feed")) U<=N (alpha=i&"seeY")] *
  filter(state,P=?[(alpha=i)&!("feed") U<=N2 (alpha=i&"feed")],
  alpha=i&"seeY") + P=?[(!(alpha=i)&!("feed")) U<=N (alpha=i&"seeP")] *
  filter(state,P=?[(alpha=i&!"feed")U<=N2(alpha=i&"feed")],alpha=i&"seeP")
```

Figure 7 shows the results for switching from activity patterns α_1 to α_2 and vice-versa respectively for less than 10 button taps to feed a yoshi after switching the activity pattern, while Figure 8 shows the same but for an unbounded number of button taps (to feed a yoshi). We can see that success is much more likely by switching from α_2 to α_1, than switching from α_1 to α_2, and a user needs about 4-5 button taps to switch from α_2 to α_1 to maximise their score. This latter

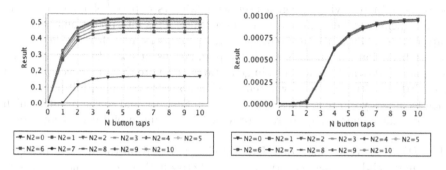

Fig. 7. Question 4 for $N_2 \leq 10$ and $i = 1$ on the left and $i = 2$ on the right

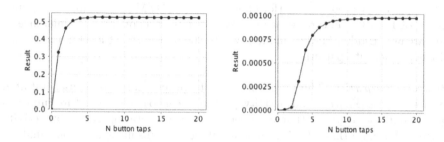

Fig. 8. Question 4 for $N_2 = \infty$ and $i = 1$ on the left and $i = 2$ on the right

result is not surprising, considering that users might first inspect the game, which would involve visiting the 4 states.

All analyses were performed on a standard laptop. Note that for brevity, the mobile app analysed here, and its formal model, are relatively small in size; more complex applications will yield more meaningful activity patterns and complex logic properties that can be analysed on the metamodels. While state-space explosion of the UMM could be an issue, it is important to note that the state-space does not depend on the number of users, but on the granularity of the states (logged in-app actions) we distinguish.

7   Discussion

We reflect upon the results obtained for the Hungry Yoshi example and further issues raised by our approach.

Hungry Yoshi Usage. Our analysis has revealed some insight into how users have actually played the game: α_1 corresponds to a more successful game playing strategy than α_2 and a user is much more likely to be effective if they change from α_2 to α_1 (rather than vice-versa), thus we conclude that α_1 is *expert* behaviour and α_2 is *ineffective* behaviour. (Note that users can, and do, switch between both behaviours, e.g. a user who exhibits expert behaviour can still exhibit ineffective behaviour at some later time.) This interpretation of activity patterns can inform a future redesign that helps users move from ineffective to expert behaviour, or induces explicitly populations of users to follow selected computation paths to reach certain goal states. We note that the developers had very little intuition about how often, or if, users were picking a full basket and then feeding a yoshi (e.g. Questions 3 and 4 in Sect. 6), and so the results, which indicate this scenario is quite rare, provided a new and useful insight for them.

Why DTMCs? Our choice of DTMC models is based on the work of of [9] in modelling web-browsing activity, usage of Microsoft Word commands, and telephone usage across populations of individuals. Girolami et al. used probabilistic convex combinations of DTMCs and demonstrated empirically that such model was superior in predictive performance to single DTMCs and mixture (pointmass) of DTMCs. Future work involves developing algorithms for inference of Hierarchical Hidden Markov models, where the first abstract level in the hierarchy is the activity patterns.

Temporal Properties. The properties refer to propositions about user-initiated events (e.g. *seeY*, *feed*) and activity patterns (e.g. α_1, α_2). A future improvement would be a syntax that parametrises the temporal operators by activity pattern. We note that PCTL properties alone were insufficient for our analysis and we have made extensive use of *filtered* properties. We also note that for some properties we have used PRISM rewards, e.g. to compare scores between activity patterns, but these are omitted in this short paper.

Reasoning about Users. Model checking is performed on the UMM resulting from the augmentation of the set of K activity patterns with a strategy θ_m. It is simple to select a user by selecting a θ_m and to analyse the resulting UMM. Metrics on the set $\{\theta_m \mid m = 1, \ldots, M\}$ will be used in future work to characterise how the results of the analysis change depending on the value of one θ_m, in the hope that results of the analysis for one user can be generalised to users close by (under the given metric).

Formulating Hypotheses: Domain Specific and Generic. We have considered domain specific hypotheses presented by developers and evaluators, but could a formal approach help with hypothesis generation? For example, we could frame questions using the specifications patterns for probabilistic quality properties as defined in [10] (probabilistic response, probabilistic precedence, etc.). Referring to our questions in Sect. 2.1, we recognise in the first item the *probabilistic precedence* pattern, in the second one the *probabilistic response* pattern, and in the last two the *probabilistic constrained response* pattern. However, these patterns refer only to the top level structure, whereas all our properties consist of multiple levels of embedded patterns. Perhaps more complex patterns are required for our domain? The patterns of [10] were abstracted from avionic, defence, and automotive systems, which are typical reactive systems; does the mobile app domain, or domains with strong user interaction exhibit different requirements? We remark also that analysis of activity patterns is just one dimension to consider: there are many others that are relevant to tailoring software to users, for example software variability and configuration, and user engagement. These are all topics of further work.

Choosing K Activity Patterns. What is the most appropriate value for K, can we guide its choice? While we could use model selection or non-parametric methods to infer it, there might be domain-based reasons for fixing K. For example, we can start with an estimate value of K and then compare analysis activity patterns: if properties for two different activity patterns give very close quantitative results then we only need a smaller K.

What to Log? This is a key question and depends upon the propositions we examine in our properties, as well as the overheads of logging (e.g. on system performance, battery, etc.) and ethical considerations (e.g. what have users agreed). Formal analysis will often lead to new instrumentation requirements, which in turn will stimulate new analysis. For example, our analysis of Hungry Yoshi has indicated a need for logged traces to include more information about current context, e.g. the observable access points (yoshis).

8 Related Work

Our work is a contribution to the new *software analytics* described in [11], focusing on local methods and models, and user perspectives. It is also resonates with their prediction that by 2020 there will be more use of analytics for mobile apps and games. Recent work in analysis of user behaviours in systems, especially XBox games, is focused on understanding how features are used and how to drive users to use desirable features. For example, [12] investigates video game skills development for over 3 million users based on analysis of users' TrueSkill rating [13]. Their statistical analysis is based on a single, abstract "skill score", whereas our approach is based on reasoning about computation paths relating to in-app events and temporal property analysis of activity patterns. Our approach can be considered a form of run-time quantitative verification (by probabilistic model checking) as advocated by Calinescu et al. in [14]. Whereas they consider functional behaviour of service-based systems (e.g. power management) and software evolution triggered by a violation of correctness criteria because software does not meet the specification, or environment change, we address evolution based on behaviours users *actually* exhibit and how these behaviours relate to system requirements, which may include subtle aspects such as user goals and quality of experience. Perhaps of more relevance is the work on off-line runtime verification of logs in [15] that estimates the probability of a temporal property being satisfied by program executions (e.g. user traces). Their approach and results could help us determine how logging sampling in-app actions and app configuration affects analysis of user behaviour. The work of [16] employing

Hidden Markov Chains models (HMMs) is related to our approach, however our focus on capturing behavioural characteristics that are shared across a population forces us to consider a model whose distributed representation cannot be captured by HMMs. Finally we note the very recent work of [5] on a similar approach and comment the major differences in Sect. 4. In addition they analyse REST architectures (each log entry corresponds to a web page access), whereas the mobile apps we are analysing are not RESTful, we can include more fine grained and contextual data in the logged user data.

9   Conclusions and Future Work

We have outlined our contribution to software analytics for user interactive systems: a novel approach to probabilistic modelling and reasoning over actual user behaviours, based on systematic and automated logging and reasoning about users. Logged user traces are computation paths from which we infer *activity patterns*, represented each by a DTMC. A user meta model is deduced for each user, which represents users as mixtures over DTMCs. We encode the user metamodels in the probabilistic model checker PRISM and reason about the metamodel using probabilistic temporal logic properties to express hypotheses about user behaviours and relationships within and between the activity patterns.

We motivated and illustrated our approach by application to the Hungry Yoshi mobile iPhone game, which has involved several thousands of users worldwide. We showed how to encode some example questions posed by developers and evaluators in a probabilistic temporal logic, and obtained quantitative results for an example user metamodel. After considering our formal analysis of two activity patterns, we conclude the two activity patterns distinguish *expert* behaviour from *ineffective* behaviour and represent different strategies about how to play the game. While in this example the individual activity pattern DTMCs are small in number and size, in more complex settings it will be impossible to gain insight into behaviours informally, and in particular to insights into relationships between the activity patterns, so automated formal analysis of the user metamodels will be essential.

In this paper we have focused on defining the appropriate statistical and formal models, their encoding, and reasoning using model checking. We have not explored here the types of insights we can gain into user behaviours from our approach, nor how we can employ these in system redesign and future system design, especially for specific subpopulations of users. Further, in this short paper, we have not considered the role of prediction from analysis and the possibilities afforded by longitudinal analysis. For example, how do the activity patterns and properties compare between users in 2009 and users in 2013? This is ongoing work within the *A Population Approach to Ubicomp System Design* project, where we are working with system developers on the practical application of our formal analysis in the design and redesign of several new apps. We are also investigating metrics of user engagement, tool support, and integration of this work with statistical and visualisation tools.

Acknowledgments. This research is supported by EPSRC Programme Grant *A Population Approach to Ubicomp System Design* (EP/J007617/1). The authors thank all members of the project, and Gethin Norman for fruitful discussions.

References

1. McMillan, D., Morrison, A., Brown, O., Hall, M., Chalmers, M.: Further into the Wild: Running Worldwide Trials of Mobile Systems. In: Floréen, P., Krüger, A., Spasojevic, M. (eds.) Pervasive 2010. LNCS, vol. 6030, pp. 210–227. Springer, Heidelberg (2010)
2. Hall, M., Bell, M., Morrison, A., Reeves, S., Sherwood, S., Chalmers, M.: Adapting ubicomp software and its evaluation. In: Proc. of EICS 2009, pp. 143–148. ACM, New York (2009)
3. Baier, C., Katoen, J.P.: Principles of Model Checking. The MIT Press (2008)
4. Kwiatkowska, M., Norman, G., Parker, D.: Stochastic Model Checking. In: Bernardo, M., Hillston, J. (eds.) SFM 2007. LNCS, vol. 4486, pp. 220–270. Springer, Heidelberg (2007)
5. Ghezzi, C., Pezzè, M., Sama, M., Tamburrelli, G.: Mining Behavior Models from User-Intensive Web Applications. In: Jalote, P., Briand, L.C., van der Hoek, A. (eds.) Proc. of ICSE 2014, pp. 277–287. ACM (2014)
6. Dempster, A.P., Laird, N.M., Rubin, D.B.: Maximum Likelihood from Incomplete Data via the EM Algorithm. Journal of the Royal Statistical Society. Series B (Methodological) 39(1), 1–38 (1977)
7. Higgs, M., Morrison, A., Girolami, M., Chalmers, M.: Analysing User Behaviour Through Dynamic Population Models. In: Proc. of CHI 2013, Extended Abstracts on Human Factors in Computing Systems, CHI EA 2013, pp. 271–276. ACM (2013)
8. Kwiatkowska, M., Norman, G., Parker, D.: PRISM 4.0: Verification of Probabilistic Real-Time Systems. In: Gopalakrishnan, G., Qadeer, S. (eds.) CAV 2011. LNCS, vol. 6806, pp. 585–591. Springer, Heidelberg (2011)
9. Girolami, M., Kaban, A.: Simplicial Mixtures of Markov Chains: Distributed Modelling of Dynamic User Profiles. In: Thrun, S., Saul, L., Schölkopf, B. (eds.) Advances in Neural Information Processing Systems 16. MIT Press, Cambridge (2004)
10. Grunske, L.: Specification patterns for probabilistic quality properties. In: Schäfer, W., Dwyer, M.B., Gruhn, V. (eds.) Proc. of ICSE 2008, pp. 31–40. ACM (2008)
11. Menzies, T., Zimmermann, T.: Software Analytics: So What? IEEE Software 30(4), 31–37 (2013)
12. Huang, J., Zimmermann, T., Nagappan, N., Harrison, C., Phillips, B.: Mastering the art of war: how patterns of gameplay influence skill in Halo. In: Mackay, W.E., Brewster, S.A., Bødker, S. (eds.) Proc. of CHI 2013, pp. 695–704. ACM (2013)
13. Herbrich, R., Minka, T., Graepel, T.: Trueskill^{TM}: A Bayesian skill rating system. In: Proc. of NIPS 2006, pp. 569–576 (2006)
14. Calinescu, R., Ghezzi, C., Kwiatkowska, M.Z., Mirandola, R.: Self-adaptive software needs quantitative verification at runtime. Commun. ACM 55(9), 69–77 (2012)
15. Stoller, S.D., Bartocci, E., Seyster, J., Grosu, R., Havelund, K., Smolka, S.A., Zadok, E.: Runtime Verification with State Estimation. In: Khurshid, S., Sen, K. (eds.) RV 2011. LNCS, vol. 7186, pp. 193–207. Springer, Heidelberg (2012)
16. Bartocci, E., Grosu, R., Karmarkar, A., Smolka, S.A., Stoller, S.D., Zadok, E., Seyster, J.: Adaptive Runtime Verification. In: Qadeer, S., Tasiran, S. (eds.) RV 2012. LNCS, vol. 7687, pp. 168–182. Springer, Heidelberg (2013)

Performance Comparison
of IEEE 802.11 DCF and EDCA
for Beaconing in Vehicular Networks

Geert Heijenk, Martijn van Eenennaam, and Anne Remke

Design and Analysis of Communication Systems,
University of Twente, The Netherlands
{geert.heijenk,e.m.vaneenennaam,a.k.i.remke}@utwente.nl

Abstract. For use in vehicular networks, IEEE 802.11p has been standardized as the underlying wireless system. The 802.11 standard distinguishes two main methods of operation with respect to channel access, the Distributed Coordination Function (DCF) and the Enhanced Distributed Channel Access (EDCA), where the latter is the mandated method to be used in vehicular networks (in 802.11p). We present validated analytical models for both DCF and EDCA, and compare both methods in the context of beaconing. We will show that, surprisingly, DCF outperforms EDCA under assumptions that are realistic for beaconing in vehicular networks.

1 Introduction

Vehicular networks are expected to enable increased traffic safety and efficiency, and reduced environmental impact of road traffic. Many traffic safety and efficiency applications rely on vehicles wirelessly broadcasting, e.g., their position and speed, to vehicles in their surroundings. The rate at which this beaconing occurs, may vary from once per few seconds until up to 25 times per second, depending on the application.

For use in vehicular networks, a dedicated variant of the IEEE 802.11 family of wireless communication protocols [1] has been specified. The most important adaptations made in this IEEE 802.11p standard are the increased symbol time and reduced data rate, in order to deal with the vehicular environment, and the possibility to communicate with other stations in an ad-hoc manner, without pre-establishing an association. The typical communication range for IEEE 802.11p ranges between a few hundred to thousand meters.

In a vehicular environment communication is taking place directly between vehicles, and not via an access point. Hence, it is important to perform medium access control in a fully distributed way. In the traditional IEEE 802.11 standard, access to the medium is governed using the so-called Distributed Coordination Function (DCF), which applies a form of Carrier Sense Multiple Access with Collision Avoidance (CSMA/CA). The DCF uses carier sensing to avoid interference and collisions between different nodes. The start of a new transmission after a busy period is randomized by decrementing a backoff counter (bc). To

G. Norman and W. Sanders (Eds.): QEST 2014, LNCS 8657, pp. 154–169, 2014.

allow for prioritized access for traffic flows, the so-called EDCA has been defined, which treats packets from different Access Categorys (ACs) differently. To make this prioritization possible, the rules for decrementing the bc have been slightly changed. IEEE 802.11p prescribes the use of EDCA in vehicular networks.

Beaconing in vehicular networks is fundamentally different from communication in traditional wireless LANs, where most traffic consists of series of unicast messages to or from an access point. In case of beaconing, nodes broadcast a single packet periodically. Broadcast implies that transmitted packets are not acknowledged by the receiver. As a consequence, packets are never retransmitted, and nodes do not adapt their load on the network, based on the success or failure of previous transmissions. Because nodes are typically dispersed of a large geographical area, and the use of Request-To-Send Clear-To-Send (RTS/CTS) is not possible in a broadcast environment, vehicular networks suffer a lot from hidden terminal problems. Beaconing in vehicular networks is characterized by a large number of nodes, each with a relatively low load. Vehicles may receive beacons from many other vehicles in their surroundings, which poses a scalability challenge to such networks.

This paper evaluates the scalability of beaconing in vehicular networks using the IEEE 802.11p DCF and EDCA. More specifically, we investigate the impact of the modified bc decrement rules on the probability of successful transmission and throughput. We present analytical performance models for the DCF and EDCA. We focus on the bc decrement behaviour, hence, we do not model the effect of hidden terminals. Solving the analytical models, and comparing numerical results reveals remarkable differences in the system performance for both access methods, which we will explain.

In [2], we already introduced the DCF model described in Sec. 3. The contributions of this paper are as follows. (1) We present an analytical model for beaconing in vehicular networks using EDCA. (2) We compare the beaconing performance of EDCA and DCF, and (3) we present a detailed analysis of the effect of the bc decrement rules on the system performance.

In the following, we first describe DCF and EDCA and their differences, in Sec. 2. Models of these mechanisms are described and analysed in Sec. 3. In Sec. 4, we compare and explain numerical results for both mechanisms. For ease of understanding, related work on modeling IEEE 802.11(p) is only reviewed in Sec. 5. Finally, we give conclusions and future work in Sec. 6.

2   Operation of IEEE 802.11DCF and EDCA

2.1   Distributed Coordination Function (DCF)

The mandatory Distributed Coordination Function (DCF) specifies basic rules for medium access and contention resolution in all IEEE 802.11 stations [1]. It coordinates transmission attempts by multiple stations contending for access to the same wireless channel, and prevents collisions. To reduce the probability of collision, the DCF specifies the use of CSMA/CA. The Carrier Sense (CS) part prevents a node from transmitting when an other node is already transmitting.

The medium is idle when the detected signal level is below the carrier sense threshold. In this case, a node may proceed to access the channel. If the channel is busy, the access attempt is deferred until the medium turns idle again.

The Collision Avoidance (CA) part prevents collisions where they are most likely to occur: just after a transmission by an other node. As described above, CS mandates to defer access until the channel turns idle again. If multiple contending nodes are waiting for the channel to become idle, they are somehow synchronised and could cause a collision if they were to transmit immediately. To alleviate this, CA mandates use of a so-called backoff. When the medium turns idle, a node does not immediately begin transmission, but will wait a mandatory gap, called Interframe Space (IFS), and some random extra time by means of a backoff counter (bc). The bc is randomly drawn from the a range $[0, \mathtt{CWmin}]$, the contention window. More precisely, the DCF performs the following operations once the network layer submits a packet to the MAC's transmission queue, assuming the station starts in the idle state:

1. Upon reception of a packet in the transmission queue, the MAC performs CS.
   (a) If the channel is **idle** for at least one IFS, the transmission may commence immediately.
   (b) If the channel is **busy**, the node enters contention, which is divided into a countdown and freeze state.
2. The node draws a bc from the contention window, according to the discrete uniform distribution $[0,\mathtt{CWmin}]$.
3. When the channel turns **idle**, the node waits one IFS.
4. After every idle timeslot σ, the bc is decremented.
5. If the channel turns **busy** during Countdown, bc is frozen and the process continues from step 3.
6. When bc reaches 0, the node transmits the frame. This may also happen if a node chooses 0 as bc in step 2.

The operation of the DCF is illustrated in Fig. 1, where the actions relating to decrementing or transmitting are illustrated on top of the blocks. In this example, node A finds the medium busy and chooses a bc of 1. Another node B has choosen $bc = 0$ during the same medium busy period, and transmits immediately after the IFS. Node A then has to wait for an IFS, and an empty slot σ to decrement its bc to 0, after which it can transmit. These two transmissions follow each other without intermediate σ. However, all nodes that have a non-zero bc freeze until an empty slot is observed on the medium. This consecutive series of transmissions without intermediate empty slots is called a *streak*.

2.2   Extended Interframe Spacing and Post-Backoff

The Extended Interframe Space (EIFS) is used to respond to a frame which failed its CRC check. This implies that the station was also not able to determine the nature of the (badly) received packet, i.e., whether it was a broadcast or

Fig. 1. DCF operation (empty slot needed for counter decrement)

Fig. 2. EDCA operation (no empty slot needed for counter decrement)

unicast transmission. To prevent this station from interfering with an ongoing transaction, it has to wait until the intended recipient had an opportunity to return an acknowledgement frame, which is done after leaving the medium idle for a Short IFS (SIFS). So the EIFS extends the DIFS with a SIFS and the time to transmit a full acknowledgement frame. Nodes that notice a collision on the medium will also refrain from accessing the medium for this prolonged time.

Post-backoff (PBO) prevents unfair advantage of nodes that have just finished a transmission and still have packets to send. Since other nodes in the system are still counting down their bc values, such a node may find the medium idle, and can immediately transmit without performing backoff. The mechanism of PBO prevents the starvation of other nodes, which could be caused by a node with a very high traffic load. PBO is similar to the backoff prior to performing a transmission: (i) After completion of a transmission, the node draws a bc from the contention window. (ii) After an IFS during which the channel remains idle, for every timeslot σ in which the channel remains idle, the bc is decremented. (iii) If the channel turns busy the bc is frozen and the process may only continue after the channel has turned idle, and remained idle for at least an IFS. (iv) When bc reaches 0, the node transmits the frame if there was a frame to transmit. If the transmission queue is empty, the node remains idle.

2.3   Enhanced Distributed Channel Access (EDCA)

To enable service differentiation for packets of different flows, Enhanced Distributed Channel Access has been introduced in IEEE 802.11. In EDCA, packets are classified based on an Access Category (AC), and each AC has its own transmission queue within a station. The service differentiation is defined among traffic from different stations, but also among traffic from different ACs within a single node.

Instead of the standard IFS used in the DCF (DIFS), EDCA uses a different Arbitration Interframe Space (AIFS) for each AC. Stations with a longer Arbitration Interframe Space (AIFS) have less chance of accessing the medium than stations with a shorter AIFS. Another way of differentiation is using different

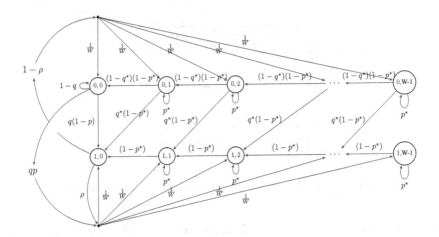

Fig. 3. Markov chain of the DCF model

contention window sizes for different ACs. This applies both to the initial contention window (CWmin) and to the maximum contention window (CWmax). The latter is however not used in broadcast transmissions. Stations with a smaller CWmin value can on average acquire the medium faster than stations with a larger value. The Transmission Opportunity (TXOP) value is defined for use with unicast transmissions. A station with a non-zero TXOP is allowed to send frames in rapid succession for a certain duration, separated by a Short IFS (SIFS) instead of a DIFS or AIFS.

Another difference between DCF and EDCA, which is less widely understood, is the way the backoff counter is decremented. Whereas DCF decrements the bc after an empty slot σ and is allowed to transmit at the moment the bc reaches 0, EDCA decrements the bc at slot boundaries. In this case, a station decrements the bc at the beginning of a timeslot immediately following the IFS (either AIFS or EIFS), irrespective of the channel status in that slot [3]. However, it is not allowed to decrement the bc *and* start transmission simultaneously at the same slot boundary. At a slot boundary, a station shall perform only one of the following actions [1]: start a transmission, decrement the bc, invoke the backoff procedures for an internal collision (a virtual collision with a packet from another AC in the same station), or do nothing. Hence, an EDCA node has to wait for the next slot boundary to transmit, once its bc reaches zero.

Fig. 2 illustrates the operation of EDCA for the same scenario as discussed for DCF in Fig. 1. Note the difference when bc decrement takes place. Station A can decrement its bc at the slot boundary following the first IFS. Immediately after the second IFS, after B's transmission, A can start transmission. This provides EDCA a bc decrement advantage over DCF [3]. However, as we will see this decrement advantage is not always beneficial for the system performance.

3   Modelling and Analysis

3.1   General Model Structure and Assumptions

Along the lines of [4], we model a single station as an $M/G/1$ queueing station with successive beacon arrivals according to a Poisson process, i.e., exponentially distributed inter-arrival times with mean $1/\lambda_g$. A station is assumed to have and infinite queue. This is not a strong simplification, as it turns out that queueing is very limited. The server of the $M/G/1$ queueing station has a general service time distribution, where the service time of a packet includes the time the station contents for medium access before a packet is transmitted. A packet is considered serviced regardless of the success or failure of the transmission, i.e., also if a collision takes place. The mean of this service time $\mathbb{E}[S]$ is derived from an embedded Discrete Time Markov Chain (DTMC), which models the behaviour of the DCF or EDCA in detail. The DTMC model in turn, needs the utilization ρ ($\rho = \lambda_g \mathbb{E}[S]$) from the $M/G/1$ queue.

In the DTMC, which models a single station, time is discretised into generic slots. A slot is either idle or busy; a busy slot is either successful or a collision. When deriving the probability of packet arrivals in a generic slot, the exact duration of a slot, e.g., the deterministic time to transmit a beacon, is taken into account. Using the DTMC, we can determine the probability τ that the station is transmitting in a generic slot. It is assumed that n stations are sharing the medium, and are able to receive each other. No hidden terminals are assumed. Using a mean-field approximation technique [5], a station is assumed to experience the average behaviour for each of the $n-1$ other stations, and will assume that each of them is also accessing the medium in a generic slot with probability τ. This way, we can obtain results for the overall model using a fixed point iteration.

3.2   DCF Model

We have presented and analysed the DTMC model of DCF used in this paper in [2]. For comparison purposes, we recall the model here, without deriving all variables and solving the steady-state equations. Instead, we refer to the original paper.

Fig. 3 shows the DTMC model for a DCF node that is only broadcasting. The state space S of the DTMC consists of a finite set of states $\mathcal{S} = \{s_{j,k}|j \in \{0,1\} \wedge k \in \{0,\ldots,W-1\}\}$, where $j = 0$ holds for a node that is currently not accessing the medium (it is either in PBO or idle) and $j = 1$ means that the node is contending for medium access (BO), or actually transmitting. Parameter k denotes the current bc value, when (1) a station takes a packet from the queue and starts its medium access attempt and finds the medium busy, or (2) when a station starts PBO. Each bc value between 0 and $W-1$ is chosen with probability $1/W$, where $W - 1$ equals the initial contention window, CWmin.

When the bc reaches zero ($s_{1,0}$), the node transmits the current frame. After transmission, with probability ρ the station finds another packet in its queue and performs a new BO for medium access. With probability $1 - \rho$ the queue is empty and the node will enter PBO. While in BO or in PBO, the bc is decremented for every idle slot. If a transmission by an other node is overheard (with probability $p^\star$) the bc is frozen. Countdown resumes when the channel turns idle again, with probability $1 - p^\star$.

During PBO, with probability $q^\star$ a frame enters the transmission queue. The bc countdown will continue, in order to access the medium. This is modelled by the "diagonal transitions" in Fig. 3. When a node reaches $s_{0,0}$, which represents an idle node, it receives a packet in its transmission queue with probability q or remains idle with probability $1 - q$. A node perceives the channel busy with probability p, hence, will perform a BO with probability qp, or a direct transmission with $q(1 - p)$, if it perceives the channel idle. Both the direct transmissions and those mediated by BO will transition into $s_{1,0}$: the transmission state. The probability that the DTMC is in this state, i.e., that a station is transmitting in a generic slot, is denoted as τ, which is used in the mean field approximation.

3.3   EDCA Model

As the behaviour of a station operating according to EDCA is different from a station using DCF operation, also the DTMC model for EDCA is different. Since we assume that all beaconing stations are using the same Access Category (AC), we do not model the service differentiation features of the EDCA, although we do take the modified IFS and CWmin into account. The most important difference sbetween the DCF and EDCA model come from the modified bc decrement rules. Fig. 4 shows the DTMC model for EDCA. Because an EDCA station always decrements at slot boundaries, bc countdown occurs in generic slots, irrespective of the medium condition. Thus, the bc is not frozen with probability $p^\star$, as in DCF, but is decremented at every generic slot. This behaviour was already taken into account in the original Bianchi model [4] where DCF bc freezing was modelled incorrectly. While in the DCF model, the probability of a packet arrival during PBO ("the diagonal transition") was denoted as $q^\star$, the probability of an arrival during a freezing period or streak, in the EDCA model, it is equal to q, the probability of an arrival in a generic slot. Because a generic slot can be empty, contain a successful transmission, or a collision, an arrival occurs in one of these slots with different probability depending on the type of slot. The mix of these types depends on the probability of occurence of such slots. Therefore, the probability q that a packet arrival occurs in a generic slot, is given by a weighted Poisson arrival process with parameter λ_g:

$$q = 1 - \left((1 - p_b^\star)e^{-\lambda_g T_e} + p_s^\star e^{-\lambda_g T_s} + (p_b^\star - p_s^\star)e^{-\lambda_g T_c} \right), \qquad (1)$$

where $p_s^\star$ is the probability that the node under consideration observes a slot containing a successful transmission from one of the $n - 1$ others. This means that out of these other nodes, one does transmit and $n - 2$ do not transmit:

$$p_s^\star = (n - 1)\tau(1 - \tau)^{n-2}. \qquad (2)$$

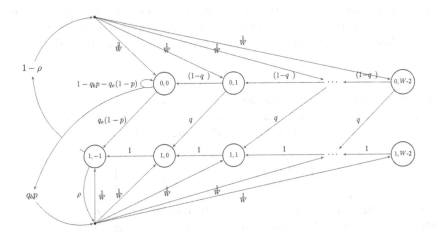

Fig. 4. Markov chain of the EDCA model

The probability of observing a busy slot is obtained as:

$$p_b^* = 1 - (1 - \tau)^{n-1}. \tag{3}$$

Finally, T_e, T_s, and T_c denote the duration of an empty, successful, or collision slot, respectively. A successful slot lasts the time for transmission of the entire packet, including preamble and headers, plus the AIFS period following the transmission. Similarly, a collision slot lasts the time for packet transmission plus the EIFS period following the transmission.

To account for the fact that an EDCA station may not decrement its bc to 0 and transmit in the same slot, an extra state $(s_{1,-1})$ is added to the model, which represents a station transmitting in a generic slot. Upon completion of a transmission, a station goes into BO with probability ρ and into PBO with probability $1 - \rho$. In both cases, the randomly chosen bc can be decremented directly after the IFS following the transmission. This means that if a bc of $W - 1$ is chosen, it is already decremented before the next generic slot. As a result, states $s_{j,W-1}$ are not present in the EDCA model. If the station choses $bc = 0$ while going into BO, it immediately transmits the frame in the next slot, and returns to state $s_{1,-1}$. If the station choses $bc = 0$ while going into PBO, it will go to the same idle state $(s_{0,0})$ as when it choses $bc = 1$. This is why the probability of going from transmission to idle is 2 times higher than the probability of going into another PBO state ($2/W$ versus $1/W$).

When a station is idle ($s_{0,0}$), what happens during the next generic slot depends on the type of of the current slot. The current generic slot is a busy slot with probability, p, that at least one of the other $n - 1$ stations is transmitting:

$$p = 1 - (1 - \tau)^{n-1}. \tag{4}$$

With probability $q_e(1-p)$, the current slot is idle and a packet arrival occurs within the slot, so that the station can transmit in the next slot, i.e., go to $s_{1,-1}$. Here, q_e denotes the probability of a packet arrival during an empty slot;

$$q_e = 1 - e^{-\lambda_g T_e}. \tag{5}$$

With probability $q_b p$, the current slot is busy, and a packet arrival occurs within the slot, so that the station will draw a new bc and go into BO. Similarly, q_b denotes the probability of a packet arrival during a busy slot;

$$q_b = 1 - \left(\frac{p_s}{p_b} e^{-\lambda_g T_s} + \left(1 - \frac{p_s}{p_b} \right) e^{-\lambda_g T_c} \right). \tag{6}$$

If no packet arrival occurs while in $s_{0,0}$, with probability $1 - q_b p - q_e(1-p)$, the station will remain in the same state. (6) is again a weighted Poisson arrival process, where p_b denotes the probability that a slot is busy, i.e., at least one of the n nodes is transmitting,

$$p_b = 1 - (1 - \tau)^n. \tag{7}$$

Furthermore, p_s denotes the probability that a generic slot contains a successful transmission, i.e., one of the n stations transmits, and the other $n-1$ do not transmit,

$$p_s = n\tau(1 - \tau)^{n-1}. \tag{8}$$

3.4   Steady State Distribution of the EDCA Model

Let $b_{0,k}$, $b_{1,k}$, and $b_{1,-1}$ denote the stationary probability of being in states $s_{0,k}$, $s_{1,k}$, and $s_{1,-1}$ for $k \in \{0, \ldots, W-2\}$. By working recursively from right to left (see Fig. 4), the following expressions for the steady state probabilities can be derived. Complete derivations can be found in [6].

The steady-state probability for a node in PBO is given by:

$$b_{0,k} = \frac{1-\rho}{W} b_{1,-1} \frac{1 - (1-q)^{W-k-1}}{q}, \quad \text{for } k = 1, \ldots, W-2. \tag{9}$$

A node is idle with steady-state probability:

$$b_{0,0} = \frac{(1-\rho)}{W(q_b p + q_e(1-p))} b_{1,-1} \left(1 + \frac{1 - (1-q)^{W-1}}{q} \right). \tag{10}$$

The steady-state probability for a node in BO is given by:

$$b_{1,k} = \frac{b_{1,-1}}{W} \left((W-k-1) \left(\rho + \frac{q_b p(1-\rho)}{W(q_b p + q_e(1-p))} \left(1 + \frac{1-(1-q)^{W-1}}{q} \right) \right) \right.$$
$$\left. + (1-\rho) \left((W-k-1) - \frac{1-(1-q)^{W-k-1}}{q} \right) \right), \quad \text{for } k = 0, \ldots, W-2. \tag{11}$$

The probability that a node transmits in a generic slot, τ, equals the probability of being in the state in which transmission is performed, which is obtained by normalisation:

$$\tau = b_{1,-1} = \left(1 + \frac{(W-1)}{2} + \left(\frac{1-\rho}{W(q_b p + q_e(1-p))}\right.\right.$$
$$\left.\left.\left(1 - \frac{1-(1-q)^{W-1}}{q}\right)\right)\left(1 + \frac{q_b p(W-1)}{2}\right)\right)^{-1}. \tag{12}$$

3.5   Service Time

We can now derive an expression for the service time of the EDCA. This is the sum of the time it takes to transmit a frame (including the IFS), and the time spent in contention. Recall that whether or not to perform contention depends on the state of the channel upon arrival of a packet. The probability that a slot is observed busy is expressed as p, see (4). However, arrivals can happen at random moments in time (and not only on slot boundaries), so following the PASTA property, these Poisson arrivals see time averages. In this line, we need to find the observed real-time channel utilisation (μ) by multiplying the probability of encountering a busy slot with the duration of such a slot, and divide by the duration of a generic slot:

$$\mu = \frac{p\mathbb{E}[T_b]}{\mathbb{E}[T]}, \tag{13}$$

where the average duration of a generic slot is:

$$\mathbb{E}[T] = (1 - p_b)T_e + p_s T_s + (p_b - p_s)T_c, \tag{14}$$

and the average duration of a busy slot:

$$\mathbb{E}[T_b] = \frac{p_s}{p_b}T_s + \left(1 - \frac{p_s}{p_b}\right)T_c. \tag{15}$$

The expected service time $\mathbb{E}[S]$ is obtained as follows. Transmission of a message, including the IFS, has a duration of $\mathbb{E}[T_b]$. A station observes the medium busy and has to perform BO with probability μ. In this case, the event which caused it to back off has a mean remaining duration of $\frac{\mathbb{E}[T_b]}{2}$, after which BO starts. On average, the station has to count down $\frac{(W-1)}{2}$ empty slots of average slot length $\mathbb{E}[T]$. The expected service time of the EDCA, including contention, then becomes:

$$\mathbb{E}[S] = \mathbb{E}[T_b] + \mu\left(\frac{\mathbb{E}[T_b]}{2} + \frac{(W-1)}{2}\mathbb{E}[T]\right). \tag{16}$$

Then, Little's Law can be used to obtain $\rho = \lambda_g \mathbb{E}[S]$.

4   Performance Comparison

We will now use the models presented in Sec. 3 to compare the performance of the DCF and EDCA. The analytical results from the DTMC models are obtained by solving the system of equations using a fixed-point iteration approach in Matlab. The iterations terminate once $\tau - \tau_{new} < \varepsilon$, with $\varepsilon = 1 \cdot 10^{-6}$.

The parameters values used in the experiments are as follows. The number of nodes is varied from $n = 1, \ldots, 200$, to analyze the scalability of DCF and EDCA. The generation rate λ_g is either kept constant at 10 beacons per second, or varied between 0 and 25 beacons per second. The data rate of beacons is assumed to be 3 Mbps (for highly robust beacon broadcasting), and a beacon is assumed to have 3200 bits of data. For EDCA, Access Class 0 is assumed, leading to a duration of a successful packet of $T_s = 1.336 \cdot 10^{-3}$s, and, because of the longer EIFS, a duration of a collision of $T_c = 1.480 \cdot 10^{-3}$s. For the DCF, these durations are slightly shorter, because of the use of a DIFS instead of AIFS: $T_s = 1.224 \cdot 10^{-3}$s; $T_c = 1.368 \cdot 10^{-3}$s. For both EDCA and DCF, the duration of an empty backoff slot is $T_e = 16 \cdot 10^{-6}$s. Finally, W=16, i.e., the (initial) contention window, CWmin, is 15 for both DCF and EDCA (AC0).

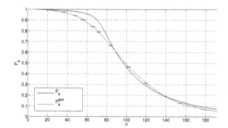

Fig. 5. Success probability EDCA, analysis and simulation

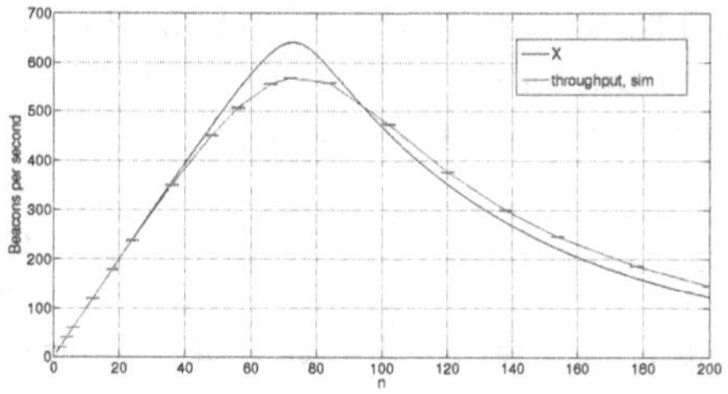

Fig. 6. Throughput EDCA, analysis and simulation

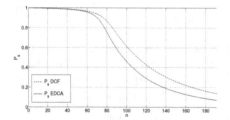

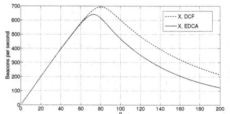

Fig. 7. Success probability, DCF vs. EDCA

Fig. 8. Throughput, DCF vs. EDCA

We compare DCF and EDCA with respect to the success probability and throughput. The success probability of a transmitted beacon is the probabability that none of the other stations transmits in the same slot and given by $P_s = (1 - \tau)^{n-1}$. The throughput, $X = \frac{p_s}{\mathbb{E}[T]}$, is defined as the mean number of successful beacon transmissions per second, can be found by multiplying the expected number of slots per second, $\frac{1}{\mathbb{E}[T]}$, with the probability that these slots contain successful transmission, p_s.

Both models have been validated against simulation experiments performed using OMNeT++ and MiXiM with extensions to simulate vehicular networking [6]. The validation results for the DCF model have been published in [2] and the validation for the EDCF model is presented in Fig. 5 and Fig. 6. As is the case for the DCF model, the EDCA model also retains inaccuracies in the semi-saturated areas for n between 60 and 80.

Fig. 7 shows the beacon success probability, P_s, for both DCF and EDCA. The general trend is that with increasing number of nodes, n, P_s first decreases slowly, and after a critical point drops sharply. DCF is able to achieve a significantly larger P_s than EDCA with increasing n. For DCF, the sharp decrease in P_s occurs also at a larger n, and for $n > 150$, it is nearly double the P_s for EDCA. Note that these values of n correspond to a saturated network. In general, one would like to avoid these highly congestion situations. However, beaconing in vehicular networks will have to deal with a wide range of vehicle densities, and from time to time, the network will experience situations with high overload, e.g., around a highway junction with (road traffic) congestion a vehicle may find hundreds of cars within transmission distance. It is therefore of paramount importance to have a reasonably smooth performance degradation with increasing load.

The difference between DCF and EDCA is even more visible in Fig. 8, which shows the throughput, X, for both mechanisms. The DCF is able to achieve a significantly larger throughput and its saturation point occurs at larger n.

The performance difference between the two mechanisms can partially be attributed to the smaller IFS of the DCF, which leaves more channel resources for effective use. However, the dominant factor determining the better performance of the DCF is the difference in bc decrement rules. The explanation of the performance difference between DCF and EDCA, given below, has been confirmed by other model variables (streak length and collision multiplicity) and by simulation experiments in [6].

As the medium gets more heavily loaded, more and more of the generic slots on the medium are busy slots. Since the EDCA does not need empty slots to decrement its bc, stations typically do a transmission in a randomly chosen slot within CWmin slots after the beacon generation. With the parameter setting used in the experiments, beacons are mostly transmitted before the next beacon arrives. As a result, the EDCA tends to spread its transmissions randomly and evenly over all generic slots, and the system behaviour resembles slotted Aloha behaviour at high load.

For the DCF, the behaviour at high load is different. Since the DCF is freezing its bc in case of busy slots, it needs empty slots to move from backoff to transmission. As a result, transmissions are done in streaks, where a streaks is an empty slot followed by a number of busy slots (See Fig. 1). During the empty

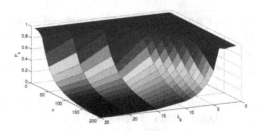

Fig. 9. Influence of λ_g and n on P_s for the DCF

slot, all stations in backoff will decrement their counter. A significant fraction of those (on average $1/\texttt{CWmin}$) will decrement to 0, and start transmission in the first slot of a streak. Only a small subset of the stations is allowed to transmit in the second slot of a streak. Those are the stations that were idle and generated a new packet for transmission during the first slot, and have chosen 0 as the bc value. Furthermore, stations that transmitted in the first slot of the streak, found another packet in their queue, and have chosen 0 as bc value are also allowed in the second slot. Similarly, even fewer stations are allowed to transmit in the third slot of a streak, if any. As a result, in case of DCF, we can observe streaks on a highly loaded medium, where on average many stations do a transmission in the first slot of a streak, and relatively few do a transmission in successive slots of the streak. Therefore, the first slot will most often yield a collision, whereas subsequent slots have a much higher probability of success. This uneven distribution of transmissions over time increases the probability that in an average slot exactly one station transmits, and hence the success probability.

We now explore the joint effect of increasing the beacon generation rate, λ_g, and the number of stations, n, on the success probability, P_s, for DCF (Fig. 9) and EDCA (Fig. 10). We can observe that also if either λ_g or n is high, the DCF gives a somewhat higher beacon success probability than the EDCA. However, as can also be observed, for a large range of parameter values, the beaconing performance of DCF and EDCA is very poor. It can be concluded that the scalability of the IEEE 802.11 multiple access mechanisms towards high rate beaconing and high node density is limited. Adapting the beaconing rate and/or the number of stations in range (by reducing transmission power) as the medium becomes heavily loaded, as for instance described in [7], is essential.

5   Related Work

IEEE 802.11 standards have been widely studied over the past fifteen years. Bianchi introduced a foundational model of the IEEE 802.11 DFC [4], [8], [9] which focuses on saturation throughput for the Basic Access mechanism and RTS/CTS in the DCF under ideal channel conditions for the 802.11b Medium Access Control (MAC). This sparked a whole family of models, each adding protocol features or extracting different metrics. For example, [10] models the

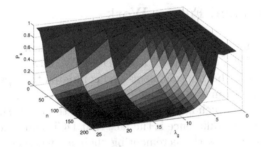

Fig. 10. Influence of λ_g and n on P_s for the EDCA

IEEE 802.11e EDCA based on [8], [9] and [11] and adds priority differentiation with respect to contention window size and a finite retransmission limit.

In [12] the impact of the CA feature of the IEEE 802.11 DCF is evaluated using a 2-dimensional Markov Chain, focusing on the state the MAC is in. The IEEE 802.11 MAC is modelled as a gated system with no buffer; arrivals after the beginning of the current contention period will not be served until the next contention period. [13] adds bc freezing behaviour to Bianchi's model by using a separate DTMC to model the channel state. [14] explicitly models the queue size in the third dimension of the Markov chain. As a consequence, all 802.11 DCF system characteristics can be obtained directly from the Markov model.

[15] extends the EDCA model by adding AIFS differentiation and modifying for use in the whole saturation range, from non-saturated to fully saturated. In addition, post-backoff behaviour is added, which finally ends up in an idle state. Whether or not to exhibit saturation behaviour is governed by a probability ρ that the queue contains another packet after transmission completes.

Yang *et al.* [16] identify that the EDCA brings difficulties and complexities to the per-slot based Markov chain modelling techniques widely used for analysis of the DCF. One problem is that, due to AIFS differentiation, it is no longer possible to accurately define a common time scale across all nodes. This common time scale is a fundamental property of many Markov-chain based models. To cope with these problems, [16] uses a channel access cycle-based modelling approach and adapt this for use in non-saturated conditions.

As opposed to IEEE 802.11 as used in WLAN situations where unicast dominates, the large scale use of broadcast as envisioned in vehicular networks has received little attention in the early modelling work. Chen *et al.* [17] analyse IEEE 802.11 broadcast performance using a one-dimensional Markov chain, modelling the bc decrementing behaviour of the DCF under saturation conditions. Ma and Chen [18] provide a model for broadcast in VANETs, including the presence of hidden terminals. Their model always performs backoff, exhibiting saturation behaviour without bc blocking. In [19], Vinel *et al.* address the trade-off between generation rate and network performance using deterministic arrivals, assuming backoff prior to transmission–also saturated behaviour. To the best of our knowledge, there is no existing work that analytically models and compares the performance of the DCF and EDCA for beaconing in vehicular networks.

6   Conclusions and Future Work

In this article, we have identified differences in the backoff counter (bc) decrement rules for the IEEE 802.11 DCF and EDCA. We have described analytical performance models for both mechanisms and compared their performance. The surprising result of our analysis is that the original DCF exhibits better scalability than the EDCA, which is mandated for use in the IEEE 802.11p standard. The explanation for this difference is the fact that DCF nodes need empty timeslots in order to proceed with decrementing their bc, whereas EDCA nodes do not. As a result, DCF transmissions occur in streaks, where towards the end of the streak very few nodes are contending for the medium, yielding a relatively high success probability. In EDCA, transmission attempts are spread out evenly, leading to a relatively low success probability if many nodes are contending. It can be concluded that in vehicular networks based on IEEE 802.11p, the service differentiation probabilities of the EDCA come at the cost of a reduced beaconing performance.

In future work, the interaction of hidden terminals with the described behaviour of both mechanisms needs to be taken studied. It is also important to derive other performance metrics, e.g., related to delay and the freshness of beacons from the presented models. Finally, the presented models can also be used to make simulation models at the traffic (safety) application level more accurate, yet computationally efficient. Incorporating approaches for congestion control by means of reducing transmit power or beaconing rate in our models is another important topic for future work.

Acknowledgment. The authors would like to thank Boudewijn Haverkort and Georgios Karagiannis for their contributions to discussions regarding this work. This work was supported by the Dutch NL Agency/HTAS (High Tech Automotive Systems) Project Connect&Drive, Project no. HTASD08002.

References

1. IEEE Standard for Information technology–Telecommunications and information exchange between systems Local and metropolitan area networks–Specific requirements Part 11: Wireless LAN Medium Access Control (MAC) and Physical Layer (PHY) Specifications, IEEE Std 802.11-2012 (Revision of IEEE Std 802.11-2007), pp. 1–2793 (2012)
2. van Eenennaam, E.M., Remke, A., Heijenk, G.J.: An analytical model for beaconing in vanets. In: Vehicular Networking Conference, pp. 9–16 (2012)
3. Xiong, L., Mao, G.: An Analysis of the Coexistence of IEEE 802.11 DCF and IEEE 802.11e EDCA. In: Wireless Communications and Networking Conference, pp. 2264–2269 (2007)
4. Bianchi, G., Fratta, L., Oliveri, M.: Performance evaluation and enhancement of the CSMA/CA MAC protocol for 802.11 wireless LANs. In: 7th IEEE Int. Symp. on Personal, Indoor and Mobile Radio Communications, pp. 392–396 (1996)

5. Benaïm, M., Le Boudec, J.-Y.: A class of mean field interaction models for computer and communication systems. In: Int. Symp. on Modeling and Optimization in Mobile, Ad Hoc, and Wireless Networks and Workshops, pp. 589–590 (2008)
6. van Eenennaam, E.: Scalable beaconing for cooperative adaptive cruise control. Ph.D. dissertation, University of Twente (November 2013)
7. Tielert, T., Jiang, D., Hartenstein, H., Delgrossi, L.: Joint power/rate congestion control optimizing packet reception in vehicle safety communications. In: ACM Int. Workshop on VehiculAr InterNETworking, Systems, and Applications (2013)
8. Bianchi, G.: IEEE 802.11-saturation throughput analysis. IEEE Communications Letters 2(12), 318–320 (1998)
9. Bianchi, G.: Performance analysis of the IEEE 802.11 Distributed Coordination Function. IEEE Journal on Selected Areas in Communications 18(3), 535–547 (2000)
10. Xiao, Y.: Performance analysis of IEEE 802.11e EDCF under saturation condition. In: IEEE Int. Conf. on Communications, vol. 1, pp. 170–174 (2004)
11. Ziouva, E., Antonakopoulos, T.: CSMA/CA performance under high traffic conditions: throughput and delay analysis. Computer Communications 25(3), 313–321 (2002)
12. Ho, T.-S., Chen, K.-C.: Performance analysis of IEEE 802.11 CSMA/CA medium access control protocol. In: 7th IEEE Int. Symp. on Personal, Indoor and Mobile Radio Communications, vol. 2, pp. 407–411 (1996)
13. Felemban, E., Ekici, E.: Single Hop IEEE 802.11 DCF Analysis Revisited: Accurate Modeling of Channel Access Delay and Throughput for Saturated and Unsaturated Traffic Cases. IEEE Transactions on Wireless Communications 10(10), 3256–3266 (2011)
14. Liu, R.P., Sutton, G., Collings, I.: A New Queueing Model for QoS Analysis of IEEE 802.11 DCF with Finite Buffer and Load. IEEE Transactions on Wireless Communications 9(8), 2664–2675 (2010)
15. Engelstad, P.E., Østerbø, O.N.: Non-saturation and saturation analysis of IEEE 802.11e EDCA with starvation prediction. In: ACM Int. Symp. on Modeling, Analysis and Simulation of Wireless and Mobile Systems, pp. 224–233 (2005)
16. Yang, X., Liu, R.P., Hedley, M.: A channel access cycle based model for IEEE 802.11e EDCA in unsaturated traffic conditions. In: IEEE Wireless Communications and Networking Conference, pp. 1496–1501
17. Chen, X., Refai, H., Ma, X.: Saturation Performance of IEEE 802.11 Broadcast Scheme in Ad Hoc Wireless LANs. In: IEEE Vehicular Technology Conference, pp. 1897–1901 (1901)
18. Ma, X., Chen, X.: Delay and broadcast reception rates of highway safety applications in vehicular ad hoc networks. In: Mobile Networking for Vehicular Environments, pp. 85–90 (2007)
19. Vinel, A., Vishnevsky, V., Koucheryavy, Y.: A simple analytical model for the periodic broadcasting in vehicular ad-hoc networks. In: IEEE GLOBECOM Workshops, pp. 1–5 (2008)

A New GreatSPN GUI for GSPN Editing and CSL^{TA} Model Checking

Elvio Gilberto Amparore

Università di Torino, Dipartimento di Informatica, Italy
amparore@di.unito.it

Abstract. This tool demonstration paper describes a new Graphical User Interface for the interactive modeling and verification of GSPN systems with the stochastic logic CSL^{TA}. The GUI provides a modern and fully-featured environment designed around a complete modeling workflow: The user designs a GSPN model, a DTA (automaton describing properties for the CSL^{TA} logic), and can simulate the GSPN behavior and the model checking process with an interactive simulation (a sort of "joint token game"). The tool then supports CSL^{TA} model-checking and the computation of classical performance indices and qualitative properties. The aim is to provide a state-of-the-art integrated environment for the quantitative and qualitative analysis of GSPNs with the support of GreatSPN solvers and of the MC4CSL^{TA} model checker.

Keywords: CSL^{TA} stochastic logic, GreatSPN, GUI, GSPN editor.

1 Objectives and Contributions

Many concurrent formalisms, like Petri nets or Timed Automata, have a *graphical* representation, that requires the availability of graphical editors to be able to gain the full advantages. A graphic schema is usually simple to specify and understand. Editors for these formalisms have flourished in the past, providing a significant help in the adoption of these performance engineering techniques.

This paper introduces a new graphical editor for Generalized Stochastic Petri Nets (GSPNs) and Deterministic Timed Automata (DTAs), integrated in the GreatSPN framework. While there are a certain amount of editors for GSPNs, this is (to the best of our knowledge) the first graphical editor for CSL^{TA} DTAs. The editor presents a complete workflow for modeling and verification, that allows to edit models, simulate their behaviors, inspect their structural properties, test them with numerical solvers and visualize the computed results. The application is designed both to be an easy-to-use interface for the MC4CSL^{TA} [2] model checker, and to be integrated in the GreatSPN pipeline [3].

2 Graphical Modeling with the New GreatSPN GUI

Figure 1 shows the main application window, taken while editing a GSPN model. The editor is designed around the idea of *multi-page* projects. Each project

G. Norman and W. Sanders (Eds.): QEST 2014, LNCS 8657, pp. 170–173, 2014.

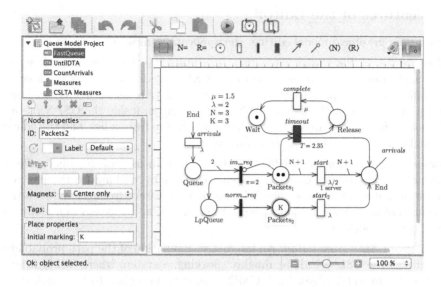

Fig. 1. The interface of the application while editing a GSPN model

correspond to a file, and is made by several pages, listed in the upper-left panel. In the current version of the editor, pages can be GSPN models, DTA models or table of measures. New model formalisms or composition of models can be added easily to the editor by specifying a new type of page. The property panel on the lower-left shows the editable properties of the selected objects. The central canvas shows the content of the selected page, that is in this case a GSPN.

Drawing GSPN Models. GSPNs are drawn with the usual graphical notation. Transitions may be immediate (thin black bars), exponential (white rectangles) or general (black rectangles). Names, arc multiplicities, transition delays, weights and priorities are drawn as small movable labels near the corresponding GSPN elements. Arcs may be "broken", like the one from place End to transition *arrivals*, so that only the endpoints of the arrow are shown. The editing process supports all the common operations of interactive editors, like undo/redo of every actions, cut/copy/paste of objects, drag selection of objects with the mouse, single and multiple editing of selected objects, etc. Name labels for elements (places, transitions, constants, etc) can be optionally substituted with a LATEX string, allowing for more readable models that express better their meanings.

Drawing CSL^{TA} DTAs. A DTA is a control automaton of the CSL^{TA} stochastic logic that measures the behavior of a GSPN. The CSL^{TA} stochastic logic works by measuring stochastic GSPN behaviors using a control automaton, a DTA. A DTA is an automaton that reads the language of GSPN paths, and separates accepted and rejected paths. The formal semantics of the DTA can be found in [9] (single clock), and in [6] (with multiple clocks). The new GreatSPN editor can also draw DTA models using their graphical formalism.

3   Features of the Editor

Other than graphical model editing, the application implements these features:

Visualization of Place and Transition Invariants: the interface may show interactively the structural invariants of a GSPN model. Figure 2(A) shows a P-semiflow (a set of places where the weighted sum of tokens remains constant) drawn in the GUI, with the flow multiplicities written inside the GSPN places. The user selects the semi-flow that wants to visualize from a list, and the editor highlights the involved places and transitions.

Interactive Simulation of GSPN Models: also known as "token game", allows the modeler to observe how the marking changes by clicking on the enabled transitions, and the editor responds by moving the tokens from the input to the output places. The reached marking is then shown, and the user can continue firing new transitions.

Interactive CSL^{TA} Simulation: An interactive simulation of the path probability operator of the CSL^{TA} logic is, roughly speaking, a system where GSPN firings are checked by DTA edges. Each GSPN transition firing has to be matched by a corresponding DTA edge, otherwise the path is rejected. Figure 2(B) shows the GUI window for the joint simulation of a GSPN and a DTA. The user may advance the simulation by clicking the enabled GSPN transitions or DTA edges.

Definition of Batches of Measures: The user can define batches of measures in multiple languages (CTL, CSL, CSL^{TA}, basic performance indexes), and the tool implements the invocation of the supported solvers and the interactive visualization of the results.

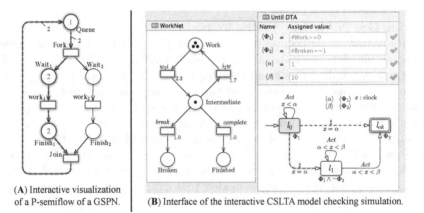

(A) Interactive visualization of a P-semiflow of a GSPN.

(B) Interface of the interactive CSLTA model checking simulation.

Fig. 2. Some features of the new GreatSPN editor

The application is cross platform and runs on Windows, MacOSX and Linux. Numerical solvers are all written in C/C++, and are invoked in pipeline as external processes. The new GreatSPN interface is freely available at http://www.di. unito.it/~greatspn/index.html, in the "New Java GUI" section.

4 Related Work and Conclusions

While we believe that this new tool provides a solid base for quantitative/qualitative analysis of models in the GSPN formalism, there are other GUIs that provide similar features. The Moebius tool [8] has similar aims, integrating multiple formalisms, multiple solvers, and providing a complete analysis workflow, from design to verification. However, the main formalism is different (SAN network) and there is no support for DTAs. The tool Snoopy [10] also provides a unified editor for GSPN models, with support for hierarchical composition and multiple solvers, but does not support DTAs. The editor Coloane [7] supports both Petri net and Timed Automata, but (at the time of writing) is not focused on stochastic models. Other GSPN editors exist, with different approaches to the modeling workflow. The DTA editing and the CSL^{TA} stochastic logic support is, to the best of our knowledge, a unique feature of the new GreatSPN GUI.

We plan to extend the tool in various directions. First of all, the Petri net formalism can be augmented to support various extensions, such as colors, fluid models, compositional formalisms like Kronecker partitioning [5], and others. Similarly, DTAs can be extended to cover statistical control automata, like Linear Hybrid Automata [4]. Work is ongoing to integrate the tool RGMEDD, which supports CTL model checking and counterexample generation[1].

References

1. Amparore, E.G., Beccuti, M., Donatelli, S.: (Stochastic) model checking in greatSPN. In: Ciardo, G., Kindler, E. (eds.) PETRI NETS 2014. LNCS, vol. 8489, pp. 354–363. Springer, Heidelberg (2014)
2. Amparore, E.G., Donatelli, S.: MC4CSL^{TA}: an efficient model checking tool for CSL^{TA}. In: International Conference on Quantitative Evaluation of Systems, pp. 153–154 (2010)
3. Babar, J., Beccuti, M., Donatelli, S., Miner, A.: GreatSPN Enhanced with Decision Diagram Data Structures. In: Lilius, J., Penczek, W. (eds.) PETRI NETS 2010. LNCS, vol. 6128, pp. 308–317. Springer, Heidelberg (2010)
4. Ballarini, P., Djafri, H., Duflot, M., Haddad, S., Pekergin, N.: HASL: An expressive language for statistical verification of stochastic models. In: Proceedings of VALUETOOLS 2011, Cachan, France, pp. 306–315 (May 2011)
5. Bause, F., Buchholz, P., Kemper, P.: Hierarchically combined queueing petri nets. In: Proc. 11th Int. Conf. on Analysis and Optimization of Systems, Discrete Event Systems, Sophie-Antipolis, France, pp. 176–182 (1994)
6. Chen, T., Han, T., Katoen, J.P., Mereacre, A.: Quantitative Model Checking of Continuous-Time Markov Chains Against Timed Automata Specifications. In: Symposium on Logic in Computer Science, pp. 309–318 (2009)
7. Coloane webpage, https://coloane.lip6.fr/
8. Courtney, T., Daly, D., Derisavi, S., Gaonkar, S., Griffith, M., Lam, V., Sanders, W.: The Mobius modeling environment: recent developments. In: International Conference on Quantitative Evaluation of Systems (QEST), pp. 328–329 (2004)
9. Donatelli, S., Haddad, S., Sproston, J.: Model checking timed and stochastic properties with CSL^{TA}. IEEE Trans. Softw. Eng. 35(2), 224–240 (2009)
10. Heiner, M., Herajy, M., Liu, F., Rohr, C., Schwarick, M.: Snoopy – A unifying petri net tool. In: Haddad, S., Pomello, L. (eds.) PETRI NETS 2012. LNCS, vol. 7347, pp. 398–407. Springer, Heidelberg (2012)

The Octave Queueing Package

Moreno Marzolla

Dipartimento di Informatica–Scienza e Ingegneria (DISI), Università di Bologna,
Mura Anteo Zamboni 7, I-40127 Bologna, Italy
moreno.marzolla@unibo.it

Abstract. Queueing Networks are a widely used performance modeling tool that has been successfully applied to evaluate many kind of systems. In this paper we describe the `queueing` package, a collection of numerical solution algorithms for Queueing Networks and Markov chains written in GNU Octave (an interpreted language for numerical computations). The `queueing` package allows users to compute steady-state performance measures for product-form and some types of non product-form Queueing Networks. Additionally, the package provides functions to analyze single station queueing systems and Markov chains. Therefore, the `queueing` package can be used for reliability analysis, capacity planning and general systems modeling and evaluation.

1 Introduction

Queueing Networks (QNs) are a powerful modeling notation that can be used for capacity planning, bottleneck analysis and performance evaluation of many kinds of systems. In its basic form, a QN consists of K service centers, each containing one or more servers sharing a common queue. Requests circulate through the system, joining the queues from which they are extracted according to a queueing policy, e.g. First-Come First-Served (FCFS), to receive service from one of the associated servers. After service completion, a request may join another queue or, for open queueing networks, leave the system. Many extensions to this basic model have been proposed in the literature (e.g., networks with multiple request classes, jobs with priorities, passive queues, finite capacity regions and so on).

Despite the vast literature on numerical solution techniques for QN models (see [1] and references therein), there is a shortage of software packages for QN analysis[1]. To this aim, we developed the `queueing` package for GNU Octave [4], an interpreted language for numerical computations. The `queueing` package implements numerical algorithms for stationary analysis of product-form queueing models; open, closed and mixed networks with different classes of customers are supported. Furthermore, the package allows transient and steady-state analysis of discrete and continuous-time Markov chains (e.g., state occupancy probabilities, first passage times, mean time to absorption).

---

[1] A list of tools is available at http://web2.uwindsor.ca/math/hlynka/qsoft.html, although many links are broken as packages disappear.

G. Norman and W. Sanders (Eds.): QEST 2014, LNCS 8657, pp. 174–177, 2014.

2   Design Principles

queueing is a collection of Octave functions for computing various transient and steady-state performance measures of queueing models and Markov chains. The Octave interactive environment provides the glue which allows complex models to be built and evaluated programmatically. This can be useful, e.g., to do parametric evaluation of complex models, or to perform ad-hoc analysis not already covered by one of the functions provided. While this allows the greater degree of flexibility, it imposes a steep learning curve.

The following usage scenarios for queueing can be identified: (*i*) **Incremental model development:** the queueing package and GNU Octave can be used for rapid prototyping and iterative refinement of QN models. (*ii*) **Modeling environment:** large and complex performance studies can be done quickly, since models involving repetitive or embedded structure can be easily defined. (*iii*) **Queueing Network research:** new algorithms can be programmed and tested against existing ones. The Octave language is well suited for implementing numerical algorithms which operate on arrays or matrices; QN algorithms fall in this category. (*iv*) **Reference implementations:** as observed in [2], some large research communities (e.g., linear algebra and parallel computing) have a long history of sharing implementations of standard algorithms. The queueing package aims at providing reference implementations of core QN algorithms. (*v*) **Teaching:** queueing is being used in some Universities to teach performance modeling courses. Since the package implements many textbook QN algorithms, students can immediately put those algorithms at work to solve practical problems, encouraging "learning by doing".

Special care has been put to make queueing a useful tool for research, education of practical use. The documentation of each function can be accessed using the help() Octave command (e.g., help(ctmc) prints the usage documentation of the ctmc() function, that computes the transient or stationary probability of a continuous-time Markov chain). Usage demos are available as well, and can be accessed using the demo() command, e.g., demo("ctmc") displays and executes all demo blocks for the ctmc() function.

One important issue of numerical software is to make sure that the computed results can be relied upon. Most of the functions included in the queueing package embed unit tests as specially-formatted comments inside the source code. These tests can be executed automatically to check the results against known values. When reference results are not available, cross-validation may be possible by executing two different functions on the same model and comparing the results. For example, the same closed network can be analyzed by Mean Value Analysis (MVA), or using the convolution algorithm. Finally, results can be compared with those produced by different tools.

3   Usage Example

The model depicted in Figure 1, taken from [3], shows a three-tier enterprise system with $K = 6$ service centers. The first tier contains the *Web server* (node

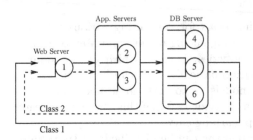

App. Servers   DB Server

Web Server

	Demands	
# Name	Class 1	Class 2
1 Web Server	12	2
2 App. Server 1	14	20
3 App. Server 2	23	14
4 DB Server 1	20	90
5 DB Server 2	80	30
6 DB Server 3	31	33

Fig. 1. Three-tier enterprise system model from [3]

1), which is responsible for generating Web pages and transmitting them to clients. The application logic is implemented by nodes 2 and 3, and the storage tier is made of nodes 4–6. The system is subject to two workload classes, both represented as closed populations of N_1 and N_2 requests, respectively. Let $D_{c,k}$ denote the service demand of class c requests at center k. We use the parameter values given in [3] and reported on Figure 1.

We set the total number of requests $N = N_1 + N_2$ to 100, and we study how different population mixes (N_1, N_2) affect the system throughput and response time. Let $\beta_1 \in (0, 1)$ denote the fraction of class 1 requests: $N_1 = \beta_1 N$, $N_2 = (1 - \beta_1)N$. The following Octave code defines the model for $\beta_1 = 0.1$:

```
N = 100; beta1 = 0.1;
S = [12 14 23 20 80 31; 2 20 14 90 30 33 ];
V = ones(size(S));
pop = [fix(beta1*N) N-fix(beta1*N)];
[U R Q X] = qncmmva(pop, S, V);
```

The qncmmva(pop, S, V) function uses the multiclass MVA algorithm to compute per-class utilizations $U_{c,k}$, response times $R_{c,k}$, mean queue lengths $Q_{c,k}$ and throughputs $X_{c,k}$ at each service center k, given a population vector pop, mean service times S and visit ratios V. Since we are given the service demands $D_{c,k} = S_{c,k}V_{c,k}$, but function qncmmva() requires separate service times and visit ratios, we set the service times equal to the demands, and all visit ratios equal to one. Overall class and system throughputs and response times can be computed as [5]:

```
X1 = X(1,1) / V(1,1);   X2 = X(2,1) / V(2,1);
XX = X1 + X2;            # system throughput
R1 = dot(R(1,:), V(1,:));   R2 = dot(R(2,:), V(2,:));
RR = N / XX;            # system resp. time
```

For $\beta_1 = 0.1$ we get X1 = 0.0044219, X2 = 0.010128, XX = 0.014550, R1 = 2261.5, R2 = 8885.9, RR = 6872.7. We can iterate the computations above for various values of β_1 to obtain the results shown in Figure 2.

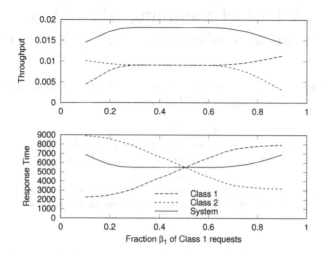

Fig. 2. Throughput and Response Times as a function of the population mix β

4 Conclusions

In this paper we presented the `queueing` package for GNU Octave. The `queueing` package is available from `http://octave.sourceforge.net/` and can be freely used under the terms of the GNU General Public License (GPL) version 3.

References

1. Bolch, G., Greiner, S., de Meer, H., Trivedi, K.: Queueing Networks and Markov Chains: Modeling and Performance Evaluation with Computer Science Applications. Wiley (1998)
2. Casale, G., Gribaudo, M., Serazzi, G.: Tools for performance evaluation of computer systems: Historical evolution and perspectives. In: Hummel, K.A., Hlavacs, H., Gansterer, W. (eds.) PERFORM 2010 (Haring Festschrift). LNCS, vol. 6821, pp. 24–37. Springer, Heidelberg (2011)
3. Casale, G., Serazzi, G.: Quantitative system evaluation with Java modeling tools. In: Proc. Second Joint WOSP/SIPEW International Conference on Performance Engineering, ICPE 2011, pp. 449–454. ACM, New York (2011)
4. Eaton, J.W.: GNU Octave Manual. Network Theory Limited (2002)
5. Lazowska, E.D., Zahorjan, J., Graham, G.S., Sevcik, K.C.: Quantitative System Performance: Computer System Analysis Using Queueing Network Models. Prentice Hall (1984)

A Perfect Sampling Algorithm
of Random Walks with Forbidden Arcs[*]

Stéphane Durand[1,4], Bruno Gaujal[4,2,3], Florence Perronnin[2,3,4],
and Jean-Marc Vincent[2,3,4]

[1] ENS of Lyon, France
[2] Univ. Grenoble Alpes, LIG, F-38000 Grenoble, France
[3] CNRS, LIG, F-38000 Grenoble, France
[4] Inria, France
stephane.durand@ens-lyon.fr, bruno.gaujal@inria.fr,
{florence.perronnin,jean-marc.vincent}@imag.fr

Abstract. In this paper we show how to construct an algorithm to sample the stationary distribution of a random walk over $\{1, \ldots, N\}^d$ with forbidden arcs. This algorithm combines the rejection method and coupling from the past of a set of trajectories of the Markov chain that generalizes the classical sandwich approach. We also provide a complexity analysis of this approach in several cases showing a coupling time in $O(N^2 d \log d)$ when no arc is forbidden and an experimental study of its performance.

Keywords: Perfect simulation, Markov chain, random walks.

1 Introduction

Random walks are well studied in the literature on Markov chains (see for example [Aldous and Fill, 2002]). Several fundamental questions in Markov chains such as *hitting time* (the first time a state is reached) and the *mixing time* (the time it takes for the measure of a chain to be close to its stationary measure) are of particular interest for random walks.

In this paper, we consider the problem of sampling from the stationary distribution of a random walk over a multidimensional grid. To the best of our knowledge, this problem has not been studied before, neither at the theoretical nor at the algorithmic level.

This problem can be approached using Monte-Carlo simulation, which only converges asymptotically. Instead here, we use a *perfect sampling* approach that provides a perfect sample in finite time. The efficiency of Monte-Carlo simulation is given by the mixing time of the chain. However, for perfect sampling, the time complexity is given by the *coupling time* of the chain (the duration until two coupled chains, starting from any two states, coalesce). To our knowledge, coupling times (or coalescence times) of random walks have not been studied

---

[*] This work was partially supported by ANR Marmote project ANR-12-MONU-0019.

G. Norman and W. Sanders (Eds.): QEST 2014, LNCS 8657, pp. 178–193, 2014.
© Springer International Publishing Switzerland 2014

before and must be evaluated to estimate the number of steps needed by a perfect sampler.

Furthermore, two additional difficulties must be solved to design an effective perfect sampling algorithm that can handle large state spaces.

The first difficulty is the design of a constructive definition of the Markov chain of the form $X_{n+1} = \phi(X_n, U_{n+1})$ (where U_{n+1} is a uniform random variable on $[0,1]$). Using this construction by starting in all possible positions in the grid defines a *grand coupling* of the random walk, using the terminology of [Levin et al., 2008]. We propose one solution in Section 3.1 where the discrete time random walk is transformed into a continuous time Markov chain, for which a grand coupling can be constructed. The bias introduced by this transformation is removed using a rejection method.

The second difficulty comes from the fact that the random walk (discrete or continuous) with forbidden arcs is not monotone so that classical perfect sampling techniques based on the simulation of extreme points [Propp and Wilson, 1996] are not valid here. Furthermore, specific techniques for non-monotone chains based on upper and lower bounds of all trajectories (often called the *sandwich* method [Kendall and Møller, 2000]) are not possible here either, because such bounds will not coalesce in most cases. This second problem is solved by introducing a more sophisticated version of the bounds/split paradigm of [Bušić et al., 2008], presented in Section 3.3. We construct a tight superset of all trajectories of the random walk starting from all possible states. This superset is made of a set of isolated points and one interval (given by its two extreme points). While the number of isolated points increases during the simulation, we show that this increase remains moderate (logarithmic in the size of the state space) and does not hamper the applicability of the algorithm even for large state spaces, at least when the number of forbidden arcs remains reasonably small (see the complexity Section 4).

Finally, we are able to analyse the time and space complexity of our Algorithm. Its time complexity (equal to the coalescence time of the chain, up to a factor 4) is shown to be logarithmic in the size of the state space without forbidden arcs. Experimental results show that this complexity remains of the same order when random forbidden arcs are added. We also make sure that the space complexity of the algorithm remains linear in the number of forbidden arcs.

Here is the structure of the paper. Section 2 shows several domains where random walks in high dimensions are useful objects. Section 3 presents the construction of the sampling algorithm based on two original ideas: first we explain how rejection can be used in this context, then we introduce the generalization of sandwich simulation using intervals and isolated points. In Section 4 we analyze the time and space complexity of our algorithm and we report experimental studies in Section 5. A long version of this paper with detailed proofs and some extensions is available on-line as a research report [Durand et al., 2014].

2   Random Walks in $\mathbb{Z}^d$

2.1   Definitions and Notations

Let us consider a random walk over a finite grid in dimension d, $S \stackrel{\text{def}}{=} \{1, \ldots, N\}^d$, where both the *span* N of the grid and its *dimension* d are large.

The set S is equipped with the componentwise order: For any $\mathbf{x}, \mathbf{y} \in S$, $\mathbf{x} \preccurlyeq \mathbf{y}$ if $x_i \leqslant y_i$ for all $i = 1, .., d$. For this order, the smallest point is $\mathbf{1} = (1, 1, \ldots, 1)$ and the largest point is $\mathbf{N} = (N, ..., N)$. The *interval* $[\mathbf{a}, \mathbf{b}]$ denotes the set of points larger than $\mathbf{a}$ and smaller than $\mathbf{b}$: $[\mathbf{a}, \mathbf{b}] \stackrel{\text{def}}{=} \{\mathbf{x} \in S, \text{ s.t. } \mathbf{a} \preccurlyeq \mathbf{x} \preccurlyeq \mathbf{b}\}$. Let $\mathbf{e}_i \stackrel{\text{def}}{=} (0, ..., 1, ..., 0)$ with 1 in the i-th position. From position $\mathbf{x}$, the walker can move uniformly to $\mathbf{x} \pm \mathbf{e}_i$ unless this is a forbidden arc or takes him out of the grid. An example in dimension 2 is given in Figure 1.

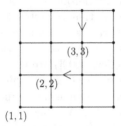

Fig. 1. Random walk over $\{1, 2, 3, 4\}^2$, with two forbidden arcs. The forbidden arcs are displayed with arrows like one-way roads on road-maps. Here a move to the right from point $(2, 2)$ is forbidden as well as a move up from point $(3, 3)$.

The position of the walker after n steps is denoted by $X(n)$ and forms a discrete time Markov chain with state space S. If the state space is strongly connected (we assume this is true in the rest of the paper), this random walk has a unique stationary measure ν. This paper is devoted to the construction of an efficient algorithm that provides a point in S distributed according to the stationary measure of $X(n)$.

It should be clear that as soon as the random walk contains one forbidden arc, it is not reversible in general and computing its stationary measure ν becomes difficult. To our knowledge, the best way to compute ν is then to solve the balance equations $\nu = \nu P$. The complexity of this approach is in $O(N^{3d})$ in time and $O(N^{2d})$ in space, unusable even for reasonable values of N and d. Here, we develop an algorithm that samples according to ν in logarithmic time and space.

2.2   Potential Applications

Random walks have be used in several domains, ranging from statistical physics to models of parallel systems.

Percolation over a Finite Graph. Random walks with forbidden arcs are well studied in the literature. Percolation theory has been studied intensively by mathematicians and physicists since the early seminal work in [Broadbent and Hammersley, 1957].

In general, in percolation theory, the forbidden edges have random positions (arcs in both directions on the edge are forbidden): each edge is forbidden with a given probability p. Note that in that case, the random walk is a reversible Markov chain. Even though it is not reversible, percolation in directed graphs has also been studied, see for example [Schwartz et al., 2002].

Here, we study the case where the state space is finite so the classical question of escape to infinity is not relevant. Instead we analyze the effect of the forbidden arcs on the coupling time of the chain, and therefore on the difficulty to sample from the stationary distribution.

Interacting Particles. The random walk in dimension $d = DM$ can also be used to model the movements of M particles moving in a space of dimension D, with asynchronous movements of the particles. The random walk then consists of the concatenation of the coordinates of all particles. In that context, the forbidden arcs correspond to the fact that certain types of movements of some particles in space, are not allowed due to the current positions of the other particles.

The example of two particles moving on the discrete line (dimension 1) that cannot occupy the same location can be seen as a Markov chain in dimension 2, with forbidden arcs preventing the walker to reach the diagonal (see Figure 2). Of course, in this particular example the two particles cannot cross each other so the Markov chain is not irreducible. However, as soon as the particles evolve in dimension D higher than 1, this problem disappears. In the remainder of the paper we will assume that the Markov chain is irreducible. The forbidden states on diagonal hyperplanes can simply be removed from the state space to ensure irreducibility.

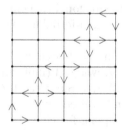

Fig. 2. Two particles moving a discrete finite line as a random walk in dimension 2. Forbidden arcs prevent the walker (the two particles) to reach the diagonal (occupy the same position).

Our analysis (see Section 4) will show that the sampling time of the corresponding Markov chain grows in the number of particles in an acceptable way ($O(M \log M)$) in the case with no forbidden arcs. While this may be unacceptable when studying a system comprising a very large number of particles, we believe this approach is suitable for the study of a moderate number of particles whose interactions are very complex.

Stochastic Automata Networks. Stochastic Automata Networks (SANs) have been shown to be a powerful approach to model parallel systems with a small degree of synchronization [Plateau and Stewart, 1997]. In the simple case where each automaton is a birth and death process and synchronizations correspond to blocking transitions, SANs are random walks with forbidden arcs as studied here, where forbidden arcs correspond to synchronization constraints.

As an illustration, let us consider the famous dining philosophers. This is a random walk where the dimension d is the number of philosophers and the span is $\{0, 1\}$ (states of one philosopher). The arcs from $(a_1, \ldots, a_{i-1}, 0, a_{i+1}, \ldots, a_d)$ to $(a_1, \ldots, a_{i-1}, 1, a_{i+1}, \ldots, a_d)$ are forbidden if a_{i-1} or a_{i+1} is equal to one.

3   Perfect Sampling Algorithm

This section is devoted to the construction of a perfect sampling algorithm of a random walk $X(n)$ over $\mathcal{S}$ where certain arcs are forbidden.

This construction is done in several steps and the final algorithm is provided in pseudo-code (Algorithm 1).

3.1   Grand Coupling and Rejection

The random walk over a grid with forbidden arcs is an irreducible, finite, discrete time Markov chain over a finite state space $\mathcal{S}$ denoted by $X(n)_{n \in \mathbb{N}}$, with transition matrix P. By definition, for any position $\mathbf{a}$ and any non-forbidden direction $\mathbf{m} = \pm \mathbf{e}_i$, $P_{\mathbf{a}, \mathbf{a}+\mathbf{m}} = \frac{1}{q_\mathbf{a}}$ where $q_\mathbf{a}$ is the number of possible moves from $\mathbf{a}$. Otherwise, $P_{\mathbf{a}, \mathbf{a}+\mathbf{m}} = 0$.

From $X(n)_{n \in \mathbb{N}}$, one can construct a continuous time Markov chain $Y(t)_{t \in \mathbb{R}}$ over the same state space in the following way. The generator Q of Y is obtained by multiplying each line $\mathbf{a}$ in P by $q_\mathbf{a}$ and defining the diagonal element $Q_{\mathbf{a}, \mathbf{a}}$ as $Q_{\mathbf{aa}} = -\sum_\mathbf{b} q_\mathbf{a} P_{\mathbf{ab}}$. Therefore, the rates from $\mathbf{a}$ to any non-forbidden $\mathbf{a} + \mathbf{m}$ are all equal to one: $Q_{\mathbf{a}, \mathbf{a}+\mathbf{m}} = 1$.

From $Y(t)$, it is possible to extract a new discrete time Markov chain, $Y(n)_{n \in \mathbb{N}}$ by uniformization. Its transition matrix is $Id + \Lambda^{-1} Q$, where Λ (uniformization constant) is any positive real number larger than all $q_\mathbf{a}$'s. Since the total rate out of any state in Y is bounded by $2d$, it can be uniformized by $\Lambda = 2d$.

While it can be difficult to construct a grand coupling for chain X, such a construction is easy and natural for the chain Y since the rates are all equal. To couple the walks starting from all states, just pick one move uniformly among the $2d$ possibilities and make every walk take this move. The walks where the move is forbidden stay still.

This yields the following constructive definition of the chain $Y(n)$ in the form $Y(n+1) = \phi(Y(n), \mathbf{m}_{n+1})$ with the following definition for the function ϕ.

Definition 1 (of constructive function ϕ). *In any state $\mathbf{y} \in \mathcal{S}$, and for any direction $\mathbf{m}$ among the $2d$ possibilities $\pm\mathbf{e}_1, \ldots, \pm\mathbf{e}_d$ (chosen uniformly) then*

$$\phi(\mathbf{y}, \mathbf{m}) = \begin{cases} \mathbf{y} + \mathbf{m} & \text{if the move is valid} \\ \mathbf{y} & \text{otherwise.} \end{cases}$$

This coupling makes the chain Y more attractive for perfect sampling. Unfortunately, the chain Y and the initial chain X do not have the same stationary probability distribution. However it is possible to construct a procedure that generates a stationary sample of the initial chain X, from a stationary sample of the chain Y. This procedure (given in Algorithm 1) is based on a rejection method.

Let us consider that the construction of $Y(n)$ is given by

$$Y(n+1) = \phi(Y(n), \mathbf{m}_{n+1}),$$

where ϕ is the deterministic function given above and $\mathbf{m}_{n+1}$ a uniform random variable among $\{\pm\mathbf{e}_1, \ldots, \pm\mathbf{e}_d\}$. If Y can be sampled efficiently (using an algorithm PSA_Y that will be described later) then X can also be sampled efficiently provided rejection is unlikely.

Algorithm 1. Sampling algorithm of $X(n)$ (PSA_X)

    Data: Algorithm PSA_Y sampling Y perfectly; a random move $\mathbf{m}_1$.
    Result: A state sampled from the stationary distribution of X
1 **begin**
2   **repeat**
3   Sample $Y(0)$ with stationary distribution (using Algo. PSA_Y)
4   Simulate one step: $Y(1) := \phi(Y(0), \mathbf{m}_1)$;
5   **until** $Y(1) \neq Y(0)$ *//reject if $Y(1) = Y(0)$*;
6   **return** $Y(0)$

Theorem 1. *If the algorithm PSA_Y samples Y under its stationary distribution, then the rejection Algorithm PSA_X gives a sample distributed according to the stationary distribution of $X(n)$.*

Proof. Let us call μ_i the stationary probability of state i for the uniformized (as well as for the continuous time) chain Y and ν_i the stationary probability of state i for original Markov chain $X(n)$. By construction, the original Markov chain $X(n)$ is the embedded chain at jump times of the continuous time version $Y(t)$. It is well known [Brémaud, 1998] that $q_i\mu_i \propto \nu_i$ where q_i is the rate out of state i. Let us compute the probability that the output of the rejection algorithm is i. Let T be the first time when the sample $Y(0)_T$ is not rejected:

$$\mathbb{P}(Y(0)_T = i) = \sum_{t=1}^{\infty} \mathbb{P}(T = t, Y(0)_t = i) = \sum_{t=1}^{\infty} R^{t-1}\mu_i \frac{q_i}{\Lambda},$$

where the probability of rejection R of the t-th sample does not depend on t and is equal to $R = \sum_j \mu_j \frac{\Lambda - q_j}{\Lambda}$. This implies that $\mathbb{P}(Y(0)_T = i) = \frac{1}{1-R}\mu_i \frac{q_i}{\Lambda}$.

Therefore, $\mathbb{P}(Y(0)_T = i) \propto \mu_i q_i$, so that it has to be equal to ν_i.   □

3.2   Coupling from the Past for $Y(n)$

The algorithm to sample $Y(n)$ is based on a procedure $\mathtt{simulate}(E, \mathbf{m})$ whose arguments are E a set of states, and one move $\mathbf{m}$ in $\{\pm\mathbf{e}_1, \ldots, \pm\mathbf{e}_d\}$. $\mathtt{simulate}$ outputs a set of states F ($=\mathtt{simulate}(E, \mathbf{m})$) such that $\phi(E, \mathbf{m}) \subset F$.

Once the elementary procedure $\mathtt{simulate}$ is known, the perfect sampling can be achieved, using coupling from the past. This is done in Algorithm 2. It was shown in [Propp and Wilson, 1996] that if this algorithm (called PSA_Y) terminates, then its output is distributed according to the stationary distribution of $Y(n)$.

Lemma 1 ([Propp and Wilson, 1996]). *Under the foregoing assumptions, the output of Algorithm PSA_Y is distributed according to the stationary distribution of $Y(n)$, if the algorithm terminates.*

Algorithm 2. Sampling algorithm of $Y(n)$ (PSA_Y)

    Data: Function $\mathtt{simulate}$
    Result: A state F sampled from the stationary distribution of Y
1 **begin**
2     $t := 1$;
3   generate event $\mathbf{m}_0$;
4   **repeat**
5        $E := \mathcal{S}$;
6   **for** $i = t - 1$ **downto** 0 **do**
7           $E := \mathtt{simulate}(E, \mathbf{m}_{-i})$;
8        $t := 2t$;
9   generate events $\mathbf{m}_{-t+1}, \ldots, \mathbf{m}_{-t/2}$;
10   **until** *coalescence (E is reduced to a unique state)*;
11   **return** E;

A naive implementation of the function $\mathtt{simulate}(E, \mathbf{m})$ consists in computing the function $\phi(x, \mathbf{m})$ for all $x \in E$. However, its time and space complexity is in $O(|E|)$. Since the size of E is N^d at the start, when $E = \mathcal{S}$, this will be unacceptable, so that compact ways to construct the procedure $\mathtt{simulate}$ are needed if we want the perfect sampling algorithm to be effective.

3.3   Non-monotonicity and Intervals

If there are no forbidden arcs, then the Markov chain Y is monotone in the following sense. Using the previous coupling, $Y_1(n) \preccurlyeq Y_2(n)$ implies $Y_1(n+1) \preccurlyeq Y_2(n+1)$.

The random walk with forbidden arcs is not monotone as shown by the following example. Consider two coupled random walks Y^1 and Y^2, with starting states $Y^1(0) = \mathbf{y}$ and $Y^2(0) = \mathbf{y} - \mathbf{e}_1$ respectively. Consider the case where the arc from $\mathbf{y}$ to $\mathbf{y} + \mathbf{e}_2$ is forbidden, then if the next random move happens to be $\mathbf{e}_2$, then $Y^1(1) = \mathbf{y}$ and $Y^2(1) = \mathbf{y} - \mathbf{e}_1 + \mathbf{e}_2$ respectively. So that $Y^2(0) \preccurlyeq Y^1(0)$ but $Y^1(1)$ and $Y^2(1)$ are not comparable.

When the Markov chain is not monotone, one classical way to sample its stationary distribution is to simulate upper and lower bounds of all the trajectories of the chain instead of its extreme states. Such upper and lower bounds form *intervals* of $\mathcal{S}$.

In [Bušić et al., 2008], the authors show how such intervals can be built iteratively considering a non-monotone Markov chain defined by $Y(n + 1) = \phi(Y(n), U(n + 1))$ over a state space equipped with a complete lattice structure.

Given an interval $E = [\mathbf{a}, \mathbf{b}]$, then the next interval can be defined as

$$\left[\inf_{\mathbf{z} \in [\mathbf{a},\mathbf{b}]} \phi(\mathbf{z}, \mathbf{m}) , \sup_{\mathbf{z} \in [\mathbf{a},\mathbf{b}]} \phi(\mathbf{z}, \mathbf{m}) \right].$$

This approach does not work here because intervals may never collapse. One such case is presented in Figure 3.

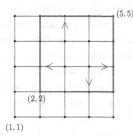

Fig. 3. No contraction of upper and lower bounds, in the presence of forbidden arcs

In this case, starting from the extreme points $(1, 1)$ and $(5, 5)$, and under any sequence of moves, the lower bound will never go over $(2, 2)$ and the upper bound will never depart from $(5, 5)$.

One way to solve this problem is to replace intervals with more complex data structures that still capture all trajectories but remain compact and coalesce with probability one. Here, the structure we choose is composed by one interval and a set of isolated points.

Suppose the set E is made of one interval $I = [\mathbf{x}_1, \mathbf{x}_2]$ and a set of isolated points $\mathbf{y}_1, ..., \mathbf{y}_k$. Then, the execution of $\texttt{simulate}(E, \mathbf{m})$ transforms E in the following manner (with $\mathbf{m} \in \{\pm\mathbf{e}_1, \ldots, \pm\mathbf{e}_d\}$):

1. The new interval $I' = [\mathbf{x}'_1, \mathbf{x}'_2]$ is such that $\mathbf{x}'_1 = (\mathbf{x}_1 + \mathbf{m}) \wedge \mathbf{N} \vee \mathbf{1}$, $\mathbf{x}'_2 = (\mathbf{x}_2 + \mathbf{m}) \wedge \mathbf{N} \vee \mathbf{1}$ (forbidden arcs are not taken into account)[1].

---

[1] The symbol $\wedge$ (resp. $\vee$) stands for point-wise minimum (resp. maximum).

2. For all the isolated points, ϕ is applied (forbidden arcs are taken into account): $\forall 1 \leqslant j \leqslant k$, $\mathbf{y}'_j = \phi(\mathbf{y}_j, \mathbf{m})$
3. Finally, new isolated points are created: If move $\mathbf{e}_i$ is forbidden in a point $\mathbf{y} \in I - I'$, then $\mathbf{y}$ becomes a new isolated point.

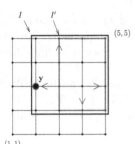

Fig. 4. Executing `simulate` over a set E composed of one interval I under move $\mathbf{e}_1$ creates a new interval I' and one isolated point $\mathbf{y}$

This construction is illustrated in Figure 4, where one isolated point is created by moving the interval to the right.

Theorem 2. *Algorithm 2 terminates with probability one and outputs a sample of the chain $Y(t)$ under its stationary distribution.*

Proof. Using the foregoing notations, consider $E' = \texttt{simulate}(E, \mathbf{m})$, where $E = I \cup \{\mathbf{y}_1, ..., \mathbf{y}_k\}$. For all isolated points, $\mathbf{y}'_j = \phi(\mathbf{y}_j, \mathbf{m}) \in E'$ by construction. As for the points $\mathbf{x} \in I$, two cases can occur. If the move from $\mathbf{x}$ to $\mathbf{x} + \mathbf{e}_i$ is possible, then $\phi(\mathbf{x}, \mathbf{m}) = (\mathbf{x} + \mathbf{m}) \wedge \mathbf{N} \vee \mathbf{1} \in I'$ by monotonicity. Else, if the move from $\mathbf{x}$ to $\mathbf{x} + \mathbf{e}_i$ is forbidden, then $\phi(\mathbf{x}, \mathbf{m}) = \mathbf{x}$ and if $\mathbf{x} \notin I'$, it becomes a new isolated point. In all cases, $\phi(E, \mathbf{m}) \subset E'$ and from Lemma 1, Algorithm 2 outputs a sample of the chain $Y(t)$ under its stationary distribution whenever it terminates. To prove termination, let us consider the evolution over time of the set E. It comprises one interval and a set of points. The evolution of the interval does not depend on the forbidden arcs. Under the sequence of moves: $\mathbf{e}_1, ..., \mathbf{e}_1, \mathbf{e}_2 ..., \mathbf{e}_2, ..., \mathbf{e}_d, ..., \mathbf{e}_d$ (where each move is repeated N times) the initial interval $I = [\mathbf{1}, \mathbf{N}]$ collapses in point $\mathbf{N}$. Using the Borel-Cantelli Lemma, this proves that the interval will collapse with probability one into a single point. From this collapse time on, the set E will comprise a set of isolated points that will coalesce with probability one by irreducibility of the chain. (see [Durand et al., 2014] for the details). This part of the proof is illustrated by Figure 5. □

A detailed implementation of procedure `simulate` is given in Algorithm 3. It is based on a dual point of view. The main loop in Algorithm 3 iterates through the list of forbidden arcs instead of iterating through the list of isolated points in E as suggested by the previous description. Using hash tables and a representation of the interval using relative coordinates instead of absolute ones, its time complexity is independent of the number of isolated points and its space complexity is linear in Nk (k is the number of forbidden arcs). See Lemmas 4 and 5.

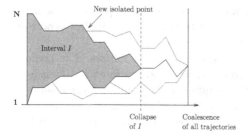

Fig. 5. Illustration of the two phases of coalescence using `simulate`. First, new isolated points are created until the interval collapses. After the collapse of the interval, the points merge up to global coalescence.

4 Coalescence and Complexity Analysis

We will use the $O(..)$ notation with several variables with the following meaning: $x(N, k, d) = O(f(N, k, d))$ means

$$\exists \beta \in \mathbb{R}, \ \forall N, k, d \in \mathbb{N}, \ d \leq N \Rightarrow x(N, k, d) \leq \beta f(N, k, d).$$

4.1 Coalescence Time without Forbidden Arcs

The goal of this part is to bound the execution time of our algorithms. We first analyze the coalescence time without forbidden arcs (that also corresponds to the collapse time of the interval in E), then the complexity of the related part of the algorithms before giving both space and time complexity in the general case.

We will call C the coalescence time in a grid of dimension 1. This is the number of iterations of the function ϕ needed for all the Markov chains starting from all the points in $[1, N]$ to coalesce in a single point.

Although the following result concerns a simple random walk on the line, we have not been able to find a similar asymptotic formula in the literature. The best bound we could find is based on hitting times and involves the spectral gap. It is of the order $O\left(1 - \frac{1}{N^3}\right)^T$ (see [Aldous and Fill, 2002]). This is not good enough to assert the complexity of our algorithms, given below.

Lemma 2 (Coalescence in dimension 1)
For any $T > 0$, $\mathbb{P}(C > T) \leqslant \cos^T\left(\frac{\pi}{N+1}\right)\left(1 + O(\frac{1}{N^2})\right)$.

Proof. The proof is technical and is based on the spectral decomposition of the transition matrix. □

The case of the random walk in dimension d follows. Let C_d be the coalescence time in dimension d. The expectation of C_d can be estimated, based on Lemma 2. This also provides the time complexity of Algorithm 2.

Lemma 3 (Coalescence in dimension d). *Let us consider a random walk in a grid with no forbidden arcs. Let C_d be its coalescence time.*
(i) The sampling time τ_Y of Algorithm PSA_Y (number of calls to `simulate`) is smaller than $4C_d$.
(ii) The expected coalescence time satisfies $\mathbb{E}[C_d] = O\left(N^2 d \log d\right)$.

Proof. (i) The fact that $\tau_Y \leqslant 4C_d$ follows from the doubling period trick used in Algorithm 2. Indeed, $\tau_Y = 1 + 2 + \cdots + 2^c$ where 2^c is the smallest power of 2 larger than C_d. This implies that $\tau_Y \leqslant 4C_d$.

(ii) The order of magnitude of $\mathbb{E}[C_d]$ can be computed as follows. The coalescence in dimension d exactly occurs when coalescence occurs in each dimension, each of them being independent once the number of steps in each dimension is given. Each dimension is chosen with probability $1/d$ at each step. Therefore, $\mathbb{E}[C_d] = d\mathbb{E}[\max(C^{(1)}, \ldots, C^{(d)})]$ where $C^{(i)}$ denotes the coalescence time in the i-th dimension. By independence,

$$\mathbb{P}(\max(C^{(1)}, \ldots, C^{(d)}) > T) = 1 - \mathbb{P}(\max(C^{(1)}, \ldots, C^{(d)}) \leqslant T)$$
$$= 1 - (1 - \mathbb{P}(C^{(1)} > T))^d,$$

$$\mathbb{E}[C_d] = d\sum_{T=0}^{\infty} \mathbb{P}(\max(C^{(1)}, \ldots, C^{(d)}) > T) = d\sum_{T=0}^{\infty}\left(1 - (1 - \mathbb{P}(C^{(1)} > T))^d\right).$$

Using Lemma 2, $\mathbb{E}[C_d] \leqslant d\sum_{T=0}^{\infty}(1 - (1 - h^T)^d) + \varepsilon$, where $h = \cos(\frac{\pi}{N+1})$ and ε is negligible with respect to the first term. From this point, one can show using classical calculus that

$$\mathbb{E}[C_d] \leqslant d + \frac{d\log d}{-\log h} + \varepsilon.$$

Since $\log h \approx \frac{-\pi^2}{N^2}$, we get $\mathbb{E}[C_d] = O\left(N^2 d\log d\right)$. □

4.2 Number of Rejections, without Forbidden Arcs

Let us now focus on the sampling algorithm for X, PSA_X. The reject probability is the probability that the next move cannot be taken by the walker who is in a stationary state (of Y). In all states, this probability is bounded by $\frac{1}{d}$ (the bound is reached if the walker is in a corner of the grid). Therefore, the expected number of rejections is always bounded by d. However this bound is very loose. By using the fact that the stationary measure of Y is uniform over all states, then the probability p_R that the next move is blocked can be computed by considering each dimension independently:

$$p_R = \sum_{i=1}^{2d}\frac{1}{2d}\mathbb{P}(\text{move blocked} \mid \text{move } i \text{ chosen }) = \sum_{i=1}^{2d}\frac{1}{2d}\frac{1}{N} = \frac{1}{N}.$$

Therefore, the number of runs R up to the first non-rejection is geometrically distributed and $\mathbb{E}[R] = \frac{N}{N-1}$.

This allows us to state our main result in the case where no arc is forbidden. Actually, as numerical experiments suggest, we believe this complexity is true as long as the random walk has a large connectivity degree.

Theorem 3. *With no forbidden arcs, the expected sampling time $\mathbb{E}[\tau_X]$ for Algorithm PSA_X (Algorithm 1) is such that $\mathbb{E}[\tau_X] = O(N^2 d\log d)$.*

Proof. The sampling time for Algorithm PSA_X is $\tau_X = \sum_{i=1}^{R} \tau_Y(i)$, where $\tau_Y(i)$ is the sampling time of the i-th run of Algorithm PSA_Y. It satisfies the following assumptions: The sampling times $\tau_Y(i)$ form an iid sequence of positive random variables with finite expectations, and the number of runs, R, is a stopping time of the filtration generated by the sampling algorithm. Using Wald's equation, the expected sampling time is $\mathbb{E}[\tau_X] = \mathbb{E}[R]\,\mathbb{E}[\tau_Y] = O(N^2 d \log d)$, by Lemma 3. □

4.3   Complexity with Forbidden Arcs

For the general case, the coalescence time is harder to estimate analytically because it does not only depend on the number of forbidden arcs but also on their actual configuration. For example, using only $2d$ forbidden arcs, it is possible to cut the state space into several disconnected components, in which case the coalescence time of the algorithm will be infinite.

Even if it is not possible to give tight bounds for the coalescence time of Algorithm 2, one can still use Lemma 3 to bound the memory space used by the algorithm. Also note that the time complexity (number of calls to simulate up to coalescence of the interval) is the same as without forbidden arcs.

Let $|E|$ denote the number of isolated points composing the sets E used by simulate$(E, \cdot)$ in Algorithm 2. This size can be bounded as a function of the number k of forbidden arcs.

Lemma 4. *If forbidden arcs are chosen randomly, uniformly among all arcs in the grid and if k is the expected number of forbidden arcs, then the maximal size $|E|$ of the set E, is bounded in expectation:* $\mathbb{E}[\,|E|\,] \le \frac{kN}{\pi^2} + O(kd + \frac{kN}{d})$.

Proof. The proof is based of a combinatorial argument to bound the number of isolated points created at each execution of simulate. □

The inner loop in Algorithms 1 and 2 uses the procedure simulate that is therefore the main component in the time and space complexity. A detailed version of the procedure simulate is given in Algorithm 3.

The main idea is to replace the iteration through all isolated points (see Sec. 3.3, item 2) by an iteration through all the forbidden arcs (Algorithm 3 line 2).

We use two hash tables $\mathcal{H}$ and $\mathcal{A}$ to store, access and merge isolated points in constant time: $\mathcal{H}$ for the forbidden arcs and $\mathcal{A}$ for each border. They are treated as sets in the algorithm.

Finally, we use relative coordinates so that we do not have to access every element at each move, only those encountering the forbidden arcs. More precisely, we have a reference point $\mathbf{r}$ moving freely on $\mathbb{Z}^d$ (in fact on $[-N, 2N]^d$), and all the other points are coded by their coordinates relative to it (in $[-N, N]^d$). This allows us to virtually move the isolated points as blocks, without recalculating every hash of coordinates.

In the details, the loop in line 2 searches for obstacles (*i.e.* forbidden moves and border) and tests if they block an isolated point (lines 3 and 4) or if they separate an element from the interval (lines 6 and 7). Line 6 implies that k *exits*

Algorithm 3. Procedure `simulate` (detailed)

Data:

    1. Forbidden arcs, coded by their head point sorted in $2d$ sets $\mathcal{K}_{0 \leq i < 2d}$ by their direction.

    2. An initially empty waiting list $\mathcal{W}$

    3. The move $\mathbf{m} = +\mathbf{e}_i$ *(the case* $\mathbf{m} = -\mathbf{e}_i$ *is obtained by replacing N by 1 and* $\mathbf{b}$ *by* $\mathbf{a}$ *in lines 1-17)*

    4. The interval, $I = [\mathbf{a}, \mathbf{b}]$ (initialized at $t = 1$ to $\mathbf{a} = \mathbf{1}$ and $\mathbf{b} = \mathbf{N}$)

    5. A reference point $\mathbf{r}$ (initialized at $t = 1$ to $\mathbf{r} = \mathbf{1}$)

    6. A hash table $\mathcal{H}$ coding the relative positions of the isolated points

    7. Arrays $\mathcal{A}_{1 \leq i \leq d}$, storing the relative positions of the isolated points on the border in each dimension, modulo N.

```
1  begin
2  │  foreach k ∈ Km do
3  │  │     if k ∈ H(k − r) then
4  │  │     │   remove k from H and from A, and add to W    //blocked isolated
5  │  │     └                                                //points
6  │  │     if kᵢ = aᵢ then
7  │  │     └   insert k in W    //creation of a new isolated point if k exits [a, b]
8  │  foreach s ∈ Aᵢ[N − rᵢ] do
9  │  │     remove s from H and from A, and add to W    //isolated points blocked
10 │  └                                                  //by the border
11 │  if bᵢ = (N − rᵢ) then
12 │  └   b ← b − m                                      //compensate for the update of r
13 │  r ← r + m
14 │  foreach x ∈ W do
15 │  │     x ← x − m
16 │  │     if xᵢ ≠ bᵢ then
17 │  │     └   insert x in H and A    //only keep isolated points outside interval
   └
```

$[\mathbf{a}, \mathbf{b}]$. More precisely, the head of the forbidden arc is in the interval before the move and not after.

The loop in line 8 concerns the isolated points that hit the border. Because of the relative notation, we can obtain them by checking the set of indexes $(i, N - r_i)$ containing every element with a relative i^{th} coordinate of $N - r[i]$, hence an absolute one equal to N, at the border. We do the same for the interval.

In line 13 the reference point is moved along $\mathbf{m}$. The last loop (lines 14 to 17) inserts each extracted point in the hash tables after having taken into account the global move, if it has not merged with the interval.

Lemma 5. *Let k be the number of forbidden arcs in the grid. Each execution of* `simulate`$(E, \mathbf{m})$ *takes time $O(d + k)$ and is independent of the number of isolated points in E. Its space complexity is $O(dN + \mathbb{E}[|E|])$, hence $O(dN + kN)$ in the random case.*

Proof. The extraction (line 2) is done by a search of the forbidden arcs and the border aligned against the move. The former contains an average (depending on the direction chosen) of $\frac{k}{2d}$ forbidden arcs. It will make $(1 + \frac{k}{d})$ tests and add at most $1 + \frac{k}{d}$ elements to the waiting list. Each test can be done in time $O(d)$, and hash tables access times in $O(d)$. Hence the time taken by this step is in $O(d + k)$.

Updates and reinsertions are similar to the first step : at most $O(1 + \frac{k}{d})$ elements with operations costing $O(d)$ in time. The third step hence takes a time $O(|\mathcal{W}|) = O(d + k)$. Overall, the global time cost is $O(d + k)$.

As for the space complexity, one hash table with at most $|E|$ element and d arrays of size N need total space of $O(|E| + d\,N)$ in memory. The only other variable with a non fixed size is the waiting list; its size is bounded by $\max \mathcal{K}_d + 1 < k + 1$, which is below $|E|$. □

5   Numerical Experiments

We have implemented Algorithm 1 in C++ (700 lines of code) to validate the approach and test the actual sampling time under several values of the parameters N (the span of the state space), d (the dimension of the state space) and k (the number of forbidden arcs).

To assert the performance of the algorithm with forbidden arcs, for each fixed value of k, we generated 20 random grids, choosing the forbidden arcs uniformly. Then, we ran PSA_X 50 times over each grid. The 95 % confidence intervals are reported in all the following figures. They all remain very small, confirming a small variance in the performance of the algorithm. Finally, in most experiments, the state space is greater than 10^{20} (a number of states that is challenging for current hardware). However each sample was obtained within a few seconds (with $N \geqslant 100$, $d \geqslant 10$) on a desktop PC (2.3 GHz Intel Core I7, with 16 Gb of memory).

We have shown in Sec. 4.1 that the coalescence time of Algorithm 1 is in $\mathcal{O}(N^2 d \log d)$ in the absence of forbidden arcs. We show in Fig. 6(a) that the presence of forbidden arcs does not increase coalescence time, which suggests that Theorem 3 generalizes to (strongly connected) random graphs with forbidden arcs. Actually, further experiments, reported in the long version of this paper [Durand et al., 2014], show that the forbidden arcs do not affect the coalescence time as long as less than 20% of the arcs are forbidden. When more arcs are forbidden, then the coalescence time variance increases and coalescence depends heavily on the position of the forbidden arcs. This observation empirically justifies the computation of the coalescence time without forbidden arcs as one measure of complexity in Sec. 4 as long as the proportion of forbidden arcs is small. This is further corroborated by another measurement: the number of isolated points *at the collapse time* of the interval averaged 1.2, and very rarely exceeded 2. This means that when the interval collapses, all the isolated points that have been created have also merged in most cases. The second phase of coalescence is therefore very short. It is null when all the points have already merged (majority of the cases) and remains short when a few points are left.

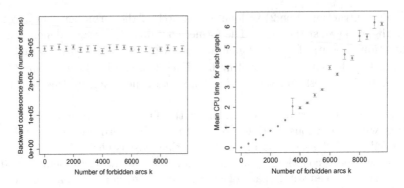

Fig.6(a). Coupling time independent of number of forbidden arcs, k.

Fig.6(b). CPU time for one sample grows linearly with k.

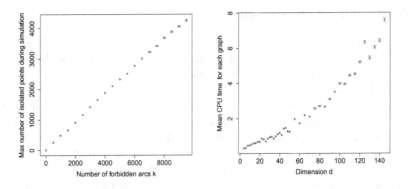

Fig.6(c). Number of isolated points grows linearly with k.

Fig.6(d). CPU time for one sample is a convex function of d.

The coalescence time is not the only factor that affects the complexity of Algorithm 1. We have executed the program on a standard PC. Figure 6(b) shows that the actual sampling time (CPU time, measured with `ctime` library) grows almost linearly with the number k of forbidden arcs, as predicted by our complexity analysis. Indeed, as detailed in Lemma 5, the procedure `simulate` iterates over a set K_m of forbidden arcs in a given direction m (line 2 of Algorithm 3). On average there are $\frac{k}{2d}$ forbidden arcs in each direction, which explains the linear complexity in k. The super-linearity trend for small values of k may come from hidden negligible factors in the complexity.

Figure 6(c) reports the number of isolated points created during the execution as a function of k. As predicted by Lemma 4, this number also grows linearly with k, and has a very small variance over the executions.

Finally, Figure 6(d) reports the execution time (in seconds) to obtain one stationary sample over a grid of size 100^d, with 1000 forbidden arcs, as a function of d. This performance metric is important with application to interacting particles in mind, because $d/3$ represents the number of particles in the 3-dimensional space that are simulated. Our complexity analysis gives an asymptotic complexity bounded by $O(d^2 \log d)$ while the measured complexity seems to have a super-linear but sub-quadratic behavior.

6 Conclusion

This paper presents an algorithm to sample the stationary distribution of a uniform random walk in a grid of size N^d with forbidden arcs. Its time and space complexity are logarithmic in the size of the grid when forbidden arcs are uniformly distributed. Therefore, this algorithm enables the exact sampling of large grids arising in applications such as particle interactions and stochastic automata networks with many pieces. This algorithm can be easily extended to general non-uniform random walks on the grid. However, the time and space complexity in the general case remains a challenging problem.

References

[Aldous and Fill, 2002] Aldous, D., Fill, J.: Reversible Markov Chains and Random Walks on Graphs, University of California, Berkeley (2002)

[Brémaud, 1998] Brémaud, P.: Markov chains Gibbs fields, Monte Carlo simulation, and queues. Springer (1998)

[Broadbent and Hammersley, 1957] Broadbent, S., Hammersley, J.: Percolation processes i. crystals and mazes. Proceedings of the Cambridge Philosophical Society 53, 629–641 (1957)

[Bušić et al., 2008] Bušić, A., Gaujal, B., Vincent, J.-M.: Perfect simulation and non-monotone Markovian systems. In: 3rd Int. Conf. Valuetools 2008, Athens (2008)

[Durand et al., 2014] Durand, S., Perronnin, F., Gaujal, B., Vincent, J.-M.: Perfect sampling algorithm of random walks with forbidden arcs using rejection and a hybrid sandwich technique. Technical Report 8504, Inria (2014)

[Kendall and Møller, 2000] Kendall, W.S., Møller, J.: Perfect simulation using dominating processes on ordered spaces, with application to locally stable point processes. Advances in Applied Probability 32(3), 844–865 (2000)

[Levin et al., 2008] Levin, D., Peres, Y., Wilmer, E.: Markov Chains and Mixing Times. AMS, American Mathematical Society (2008)

[Plateau and Stewart, 1997] Plateau, B., Stewart, W.J.: Stochastic automata networks. In: Computational Probability, pp. 113–152. Kluwer (1997)

[Propp and Wilson, 1996] Propp, J.G., Wilson, D.B.: Exact sampling with coupled Markov chains and applications to statistical mechanics. Rand. Struct. Alg. 9(1-2), 223–252 (1996)

[Schwartz et al., 2002] Schwartz, N., Cohen, R., Ben-Avraham, D., Barabási, A.-L., Havlin, S.: Percolation in directed scale-free networks. Phys. Rev. E, 66 (2002)

Modelling Replication in NoSQL Datastores

Rasha Osman[1] and Pietro Piazzolla[2]

[1] Department of Computing, Imperial College London
London SW7 2AZ, UK
rosman@imperial.ac.uk
[2] Dip. di Elettronica e Informazione, Politecnico di Milano,
via Ponzio 34/5, 20133 Milano, Italy
piazzolla@elet.polimi.it

Abstract. Distributed NoSQL datastores have been developed to cater for the usage scenarios of Web 2.0 applications. These systems provide high availability through the replication of data across different machines and data centers. The performance characteristics of NoSQL datastores are determined by the degree of data replication and the consistency guarantees required by the application. This paper presents a novel performance study of the Cassandra NoSQL datastore deployed on the Amazon EC2 cloud platform. We show that a queueing Petri net model can scale to represent the characteristics of read workloads for different replication strategies and cluster sizes. We benchmark one Cassandra node and predict response times and throughput for these configurations. We study the relationship between cluster size and consistency guarantees on cluster performance and identify the effect that node capacity and configuration has on the overall performance of the cluster.

1 Introduction

Scalable NoSQL datastores have been developed for simple read/write operations that mainly characterize distributed Web 2.0 applications [20]. They have been designed for horizontal scalability and high availability and thus are more suitable for elastic cloud deployment than traditional databases. High availably is achieved through the replication of data across different machines and different data centers. Replication of data is asynchronous, in which updates/writes are written to replicas in the background allowing for faster response times, albeit with a weaker guarantee of data consistency [4] in comparison to traditional databases. Current web applications and services require low response times, with high latency having a direct impact on revenue for large service providers [13,14,19]. The trade-off between response time and consistency in NoSQL data stores means that providers need to be able to strike a balance between these two factors.

To improve system deployment and user satisfaction performance predication techniques are needed to estimate NoSQL datastore response times given a consistency level guarantee. Performance prediction techniques would give service

G. Norman and W. Sanders (Eds.): QEST 2014, LNCS 8657, pp. 194–209, 2014.

providers the ability to test the resilience, scalability and latency of a configuration for different workloads and hardware environments. With pay for use
cloud hosting, providers would be able to deploy the correct amount of servers
for their configurations to meet service level agreements.

Performance modelling has mostly concentrated on traditional databases [18]
with queueing networks being the main modelling formalism. Other forms of
quorum systems have been modelled in [16,21]. In this paper, we present a
novel performance modelling study of the Cassandra [12] NoSQL datastore using
queueing Petri nets (QPNs) [2]. Queueing Petri nets have been applied in the
performance evaluation of distributed systems [10,9,11], grid environments [15]
and relational database concurrency control [17,6]. This work contributes a simple QPN model that is parameterized by benchmarking one Cassandra node. We
show that the model can scale to represent the performance of read workloads
for different replication strategies and cluster sizes and predict response times
and throughput for these configurations. The relationship between cluster size
and consistency guarantees is identified for scaled and non-scaled workloads. In
addition, we identify the effect of node capacity on the overall performance of
the cluster.

The rest of this paper is organized as follows. Section 2 gives an introduction
to replication in Cassandra. Section 3 presents the specification of the QPN
model for a Cassandra cluster. Experimental results and analysis are in Section
4. Finally, Section 5 concludes the paper and provides directions for future work.

2   Replication in Cassandra

Apache Cassandra is a distributed extensible column store [20,4] originally developed at Facebook [12] to provide high scalability and availability with no single
point of failure. To guarantee even distribution of data across the cluster, the
key value space is mapped onto a ring. The ring is divided into ranges, in which
each node is assigned one or more random ranges on the ring. A node can only
store keys that fall within its assigned range and thus the node becomes the *coordinator* node for these key values. Cassandra supports multi-master replication
with configurable quorum style read and write consistency [20,4]. Multi-master
consistency means that a client can contact any node, irrespective of whether it
stores/replicates the requested key.

For a cluster of N nodes, the *replication factor* (RF) is the number of times a
key is replicated across the cluster. Each coordinator node is in charge of storing
its data locally and replicating it to $RF - 1$ nodes. Hence, a node will store
$RF/N\%$ of the key ranges. The *consistency level* (CL) is the number of replicas
in which the read/write operation must succeed before returning a success to
the client. The effect of the consistency level is different for reads and writes [7].
For the purpose of this paper, we will explain replication for read operations.

Each node in a Cassandra cluster has local information about the key ranges
stored at other nodes. For read requests, a coordinator node will forward requests
to CL of the RF replicas of a key for requests not stored locally and to $CL - 1$

replicas for locally stored requests. The coordinator forwards the key requests to the replica nodes that are responding the fastest.

Figure 1 shows a cluster with six nodes, a replication factor of 3 and a consistency level of 2. A client contacts node 5 with a read request for key A, which is replicated on nodes 1, 2 & 3. Therefore, node 5 will act as the coordinator node for the client and send a read request for key A to 2 of the 3 replicas, i.e., the configured consistency level. The client's request will be blocked until *both* of the contacted replicas reply with their locally stored values for key A. Node 5 will compare the returned values and relay the most recent value of key A to the client.

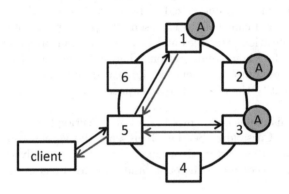

Fig. 1. Example of a read operation on a 6 node Cassandra cluster with replication factor 3 and consistency level 2

3   QPN Model of a Cassandra Node

To model a node in a Cassandra cluster, we need to represent (1) the scheduling of requests processed by a node, (2) the synchronization and blocking of requests at the coordinator node that need to be fulfilled by remote nodes and (3) the routing of replica requests based on the replication factor and consistency level. To model scheduling and processing of requests, a queueing station is the most suitable formalism. For synchronization between nodes and blocking of requests Petri nets are more suitable. Our modelling approach is based on queueing Petri nets [2], an extension of coloured stochastic Petri nets, which provide a natural formalism in combining scheduling and synchronization in one model.

In queueing Petri nets, queueing places consist of two components: the *queue*, and the *depository*. Tokens enter the queueing place through the firing of input transitions, as in other Petri nets; however, as the entry place is a queue, they are placed in the queue according to the scheduling strategy of the queue's server. When a token has been serviced it is placed in the depository where it can be used in further transitions. *Timed queueing* places have variable scheduling strategies and service distributions; while *immediate queueing* places impose a scheduling discipline on arriving tokens without a service delay. Due to space limitations, we refer the reader to [2] for a more detailed description of QPNs.

Figure 2 represents the QPN model of a Cassandra cluster showing two nodes. The bordered area highlights the parts of the model representing one node. We refer to an arbitrary node in the cluster by $node_i$. As all nodes are similar in the cluster, we will describe our QPN model by referring to $node_i$. The token colours used in the model are summarized in Table 1.

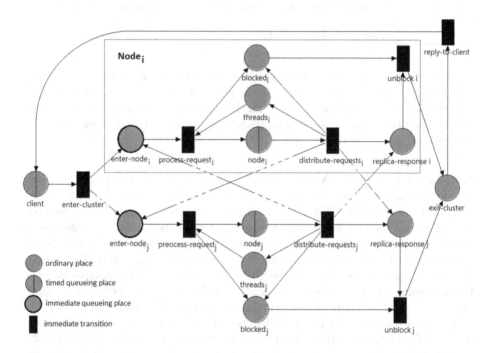

Fig. 2. A queueing Petri Net model of a Cassandra cluster

In the QPN model of Figure 2, the clients are represented by a timed queueing place with an infinite-server queue and an exponentially distributed think time. The *client* place is initialized with the number of client read requests sent to the cluster. Read requests are represented by *read* tokens. The *enter-cluster* immediate transition distributes the incoming requests randomly between the different nodes in the cluster by depositing *read* tokens in the $enter\text{-}node_i$ immediate queueing place of each node. This ensures that all nodes are equal in receiving requests, similar to Cassandra's multi-master architecture.

The default configuration of Cassandra restricts the number of concurrent read operations to 32 [7]. This is modelled using the ordinary place, $threads_i$, which is initialized to 32 tokens. Incoming requests queue in FIFO order in the $enter\text{-}node_i$ place and the immediate transition $process\text{-}requests_i$ fires, providing

Table 1. Token Colours

Token Color	Description	Place
T	thread token (initialized to 32)	$threads_i$
local-read	local incoming read request	$node_i$
read	remote read request	enter-$node_i$ & $node_i$
	blocked reads	$blocked_i$
	client incoming and returning requests	client & exit-cluster
$read_i$	forwarded read request from $node_i$	enter-$node_j$ & $node_j$, where $j \neq i$
R	replication responses between nodes	replica-$response_i$

access to the processor when there is at least one token in the $threads_i$ place. The immediate transition $process\text{-}requests_i$ determines the proportion of local and remote requests entering $node_i$ based on the replication factor. This is discussed below. The processing at $node_i$ is represented by $node_i$, a timed queueing place, representing an $-/M/1 - PS$ queue.

Non-Local Requests. A non-local or remote request is a request for a key that is not stored at $node_i$. The probability of a non-local request entering $node_i$ is equal to $1 - RF/N$. When the immediate transition $process\text{-}requests_i$ fires producing a non-local request, it removes a $read$ token from the enter-$node_i$ place and deposits a $read$ token in the $node_i$ timed queueing place. For simplicity, we assume that the mean service time of a remote read request is similar to that of a local read request. However, we note that a remote request does not use thread tokens, thus not affecting a node's ability to accept local reads. After processing, a remote request is blocked awaiting replies from replica nodes.

To synchronize replica requests between the coordinator $node_i$ and replica nodes, $node_i$ blocks remote read requests awaiting replies by firing the immediate transition $distribute\text{-}requests_i$. This is executed by placing a $read$ token in the $blocked_i$ ordinary place and a token of color $read_i$ in the enter-$node_j$ place of $node_j$, where $j \neq i$ & the total number of contacted nodes is equal to CL. The choice of $node_j$ depends on the consistency level and replication factor (discussed in Section 3.1). The color $read_i$ distinguishes forwarded replica requests by node index so that the reply can be returned back to the correct coordinator.

Any $node_j$ receiving a token $read_i$ in the enter-$node_j$ place will process it similarly to a local-request (see below). When a $read_i$ token leaves $node_j$ the immediate transition $distribute\text{-}requests_j$ will return a thread token to the $threads_j$ place and notify $node_i$ of the arrival of the request by depositing a token in the corresponding replica-$response_i$ place of $node_i$. This will enable the $unblock_i$ immediate transition, which will fire when the number of tokens in the replica-$response_i$ place is equal to the consistency level, i.e., the required nodes have responded. One $read$ token (representing the blocked read request) is removed from the *blocked$_i$* place and CL tokens are removed from the *replica-response$_i$* place and one *read* token is deposited in the *exit-cluster* place. The read request is then transferred back to the client.

Local Requests. A local request is a request for a key that is stored at node$_i$. The probability of a local request entering node$_i$ is equal to RF/N. For a local request, when the immediate transition *process-requests$_i$* is enabled, it will remove a *read* token from the *enter-node$_i$* place and a thread token from the *threads$_i$* place and deposit a *local-read* token in the *node$_i$* timed queueing place. In addition, a *read* token is deposited in the *blocked$_i$* place so that the functionality of the release of a blocked read request is the same for both local and remote requests when $CL > 1$.

After processing the local request, the *distribute-requests$_i$* transition is enabled and the thread token is returned and *one* token is deposited in the *replica-response$_i$* place, representing the completion of one request. If $CL > 1$, then the *distribute-requests$_i$* transition will deposit a *read$_i$* token in the *enter-node$_j$* place of $CL-1$ replica nodes. The contacted replica nodes will process the *read$_i$* tokens in the same manner described in the previous section for non-local requests.

The *unblock$_i$* transition will fire when the number of tokens in the *replica-response$_i$* place is equal to the consistency level, i.e. all the required nodes have responded (this includes the token deposited by node$_i$). In the case when $CL = 1$, the *unblock$_i$* transition will be immediately enabled. Similar to non-local requests, one *read* token (representing the blocked read request) is removed from *blocked$_i$* and CL number of tokens is removed from the *replica-response$_i$* place and one *read* token is deposited in the *exit-cluster* place. Finally, the read request is transferred back to the client.

3.1 Forwarding to Replicas

To model replica forwarding, we assume that the coordinator contacts the replica nodes randomly. The probability of a replica node$_j$ receiving a request from node$_i$ is modelled as the probability of the firing of the arc connecting the *distribute-request$_i$* immediate transition with the *enter-node$_j$* place of node$_j$. We describe this below.

Figure 3 illustrates a token ring for an N node cluster. Assume that a node's index represents its position on the virtual range ring and hence the key space is divided into N ranges. A node will replicate its assigned keys by moving clockwise on the ring.[1] Therefore, when $RF > 1$, node$_i$ replicates its key range on the subsequent $RF - 1$ nodes on the ring, i.e. nodes of indices:

$$S = [(i \bmod N) + 1, ((i + RF - 2) \bmod N) + 1] \ . \tag{1}$$

In addition, node$_i$ is a replica node for the preceding $RF - 1$ nodes, i.e., it stores a replica for the key values of nodes:

---

[1] This is the default method used by Cassandra1.2.

$$P = [((i - RF) \bmod N) + 1, ((i - 2) \bmod N) + 1] .\tag{2}$$

When $node_i$ receives a local key request and $CL > 1$, it randomly chooses a key value, k, from the local key range:

$$L = P \cup \{i\} = [((i - RF) \bmod N) + 1, i] .\tag{3}$$

$Node_i$ will identify the $RF - 1$ replica nodes of k and randomly choose $CL - 1$ nodes in which to forward a replica key request. The number of random node combinations is $\binom{RF}{CL-1}$, thus forming a collection of sets, F_l, where $l = 1, \ldots, \binom{RF}{CL-1}$, and for any $node_j \in F_l$, $j \in L \,\&\, j \neq i$. Each set in F_l represents a set of connections/arcs between the $distribute\text{-}request_i$ immediate transition of $node_i$ and the $enter\text{-}node_j$ place of all nodes in F_l. When the $distribute\text{-}request_i$ transition is enabled it will randomly fire one set of connections/arcs and $CL - 1$ downstream nodes will receive tokens simultaneously. Algorithm 1 details the method for identifying the connections between the $distribute\text{-}request_i$ immediate transition and the $enter\text{-}node_j$ places. Figure 4 illustrates an example of the different combinations for the incidence function of the $distribute\text{-}request_i$ immediate transition when $RF = 4$ and $CL = 3$. In general, if $CL = RF$, then $node_i$ will send the request to *all* replica nodes and await all their replies. If $CL = 2$, it will choose one node randomly.

Request for keys not stored on $node_i$ will be key ranges assigned to the remaining nodes on the ring. Assume the set of all N nodes in the ring is A, then the remaining nodes on the ring are:

$$O = A \setminus L = A \setminus [((i - RF) \bmod N) + 1, i] .\tag{4}$$

When $node_i$ receives a remote key request, it randomly chooses a key value from the nonlocal key range O and identifies its RF replica nodes, then randomly chooses CL nodes to forward a replica request. Similar to local requests, the number of random node combinations is $\binom{RF}{CL}$, which represents the number of connection/arc sets between the $distribute\text{-}request_i$ immediate transition of $node_i$ and the $enter\text{-}node_j$ places, where $j \in O \,\&\, j \neq i$. Algorithm 2 describes the method for identifying the nonlocal replica forwarding connections between the $distribute\text{-}request_i$ immediate transition and the $enter\text{-}node_j$ places. In general, if $CL = RF$, then $node_i$ will send the request to *all* replica nodes and await all their replies. If $CL = 1$, it will choose one node randomly.

4   Experimental Results

4.1   Experimental Setup

For our experiments, we use Cassandra version 1.2, Datastax distribution [7]. The loading and benchmarking of the Cassandra cluster was conducted using the Cassandra-stress tool [3], a Java-based stress testing utility included with the Cassandra Datastax distribution. The nodes of the Cassandra cluster used

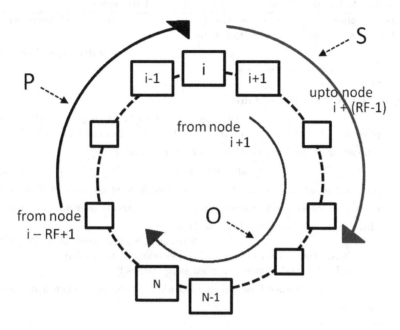

Fig. 3. A token ring of a cluster of N nodes. S represents the replica nodes for $node_i$'s assigned key range. P represents the $RF - 1$ predecessors to $node_i$ and O are the coordinator nodes for key ranges not stored on $node_i$.

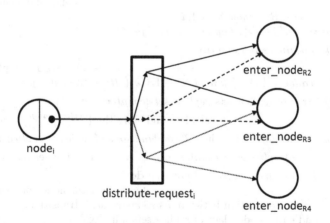

Fig. 4. The combination of connections for the *distribute-request$_i$* transition when $RF = 4$ and $CL = 3$ on processing a local read request at $node_i$. For each firing one connection set is chosen randomly.

Algorithm 1. Replica Request Forwarding: Local Requests

Require: N: cluster size, RF: replication factor, CL: consistency level where $RF > 1$
     & $1 < CL \leq RF$ & i: index of current node

                                              ▷ define local key range

1:   $key\_begin = ((i - RF) \bmod N) + 1)$
2:   $key\_end = i$
3:   **for** $k = key\_begin \rightarrow key\_end$ **do**

                          ▷ calculate $RF - 1$ replica indices, excluding node$_i$

4:     $replica\_range = \{k\} \cup [(k \bmod N) + 1, ((k + RF - 2) \bmod N) + 1] \setminus \{i\}$
5:     Define array of replica sets $replica\_sets[counter, nodes[\,]]$

                    ▷ calculate and store the $\binom{replica\_range}{CL-1}$ combinations

6:     $replica\_sets[counter][\,] =$ the $CL - 1$ *combinations* of nodes of $replica\_range$
7:     **for** $j = 1 \rightarrow |replica\_sets.counter|$ **do**   ▷ iterate over all combinations
8:       **for** $l = 1 \rightarrow |replica\_sets[j].nodes[\,]|$ **do**

                          ▷ iterate over the nodes of combination j

9:         Create connection between $distribute\text{-}request_i$ transition
10:        and entry_node place of replica_sets[j].nodes[l]

                 ▷ connections are enabled when a $local\_read$ token leaves node$_i$

11:      **end for**
12:    **end for**
13: **end for**

Algorithm 2. Replica Request Forwarding: NonLocal Request

Require: N: cluster size, RF: replication factor, CL: consistency level where $RF > 1$
     & $1 \leq CL \leq RF$, A: set of the indices of the N nodes & i: index of current node

                                        ▷ define nonlocal key range

1:   Let $O = A \setminus [(i - RF) \bmod N) + 1, i\,]$
2:   $key\_begin = min\{j \in O\}$; $key\_end = max\{j \in O\}$
3:   **for** $k = key\_begin \rightarrow key\_end$ **do**

                                 ▷ calculate RF replica indices

4:     $replica\_range = \{k\} \cup [(k \bmod N) + 1, ((k + RF - 2) \bmod N) + 1]$
5:     Define array of replica sets $replica\_sets[counter, nodes[\,]]$

                       ▷ calculate and store the $\binom{replica\_range}{CL}$ combinations

6:     $replica\_sets[counter][\,] =$ the CL *combinations* of nodes of $replica\_range$
7:     **for** $j = 1 \rightarrow |replica\_sets.counter|$ **do**   ▷ iterate over all combinations
8:       **for** $l = 1 \rightarrow |replica\_sets[j].nodes[\,]|$ **do**

                          ▷ iterate over the nodes of combination j

9:         Create connection between $distribute\text{-}request_i$ transition
10:        and entry_node place of replica_sets[j].nodes[l]

                    ▷ connections are enabled when a $read$ token leaves node$_i$

11:      **end for**
12:      **end for**
13: **end for**

---

in the experiments were hosted on the Amazon EC2 cloud environment [1]. More specifically, each machine hosting a Cassandra node was an instance of the m1.xlarge[1] AMI type. Since we were not investigating the impact of multi-core processing we disabled one of the virtual cores of the instance. Hence, each node behaved as a single core CPU with all the available cache capacity at its disposal.[2]

Cassandra 1.2 was set up using the default configuration, except for the *read_repair_chance* value that was set to 0 in order to avoid possible interferences by the random automatic updating of data across the cluster. The stress tool used the default configurations, which included the default keyspace options. To control the number of requests entering the cluster we modified the default number of threads of the stress tool to represent different request rates. The client machine hosting the Cassandra-stress tool was external to the cluster (similar to Figure 1) and hosted on the Amazon EC2 cloud environment using the same configurations as the Cassandra nodes, except with both cores active to avoid becoming the bottleneck in the system. The stress tool randomly sent requests to each of the nodes on the cluster.

All the experiments discussed in the following Section focused on read replication. For each experiment, the Cassandra cluster had a specific configuration in terms of number of nodes, replication factor and consistency level. This was controlled by the Cassandra-stress tool, in which each experiment specified the replication factor and the consistency level (see [3] for details). During each test two million keys were written to the cluster, and then separate runs would read these keys while varying the number of concurrent threads used by the stress tool for each run.

The QPN models in this paper were developed and solved using QPME2.0 [8]. QPME2.0 (Queueing Petri net Modeling Environment) is an open source performance modelling tool based on the QPN modelling formalism. QPME2.0 is composed of two components, a QPN editor (QPE) and a simulator for QPNs (SimQPN). All our simulation runs used the *method of non-overlapping batch means* (with the default settings) to estimate the steady state mean token residence times with 95% confidence intervals.

To parameterize the queueing Petri net models, we setup a Cassandra cluster with one node and ran the Cassandra-stress tool with concurrent threads ranging from 1 to 60 threads. The number of concurrent threads run by the stress tool corresponds to the number of jobs entering the system. We noticed that the throughput of the node stabilized at 5 and more threads, with an average throughput of 4.37 requests per ms. We used the mean request response time for five threads to approximate the response time for one thread, and thus the service rate of one read request. We have found the service rate of a read request to be 4.6 requests/ms. To calculate the client (thread) think time, we used the

---

[1] Specification of a m1.xlarge instance (as of January 2014): 4 ECU, 2 cores, 7.6 GB RAM, 8GB HD.

[2] The cores of physical processors used by Amazon for the m1.xlarge instance share the L1 and L2 caches.

utilization law. We measured the mean utilization for the client when running the stress tool, in addition to the mean throughput arriving at the client. From that we calculated the mean client think time as 0.074 ms.

4.2   Results

As described in the previous Section, the Cassandra-stress tool was deployed on a client node and a set of experiments on different sized Cassandra clusters was conducted. Each experiment had n nodes, RF replication factor and CL consistency level, where $n = 2$ to 4, $RF = 1$ to n and $CL = 1$ to RF. We refer to each configuration using a triple: (a-b-c), where a represents the number of nodes, b represents the replication factor and c represents the consistency level. This convention will be used throughout to discuss our results. Figure 5 shows the measured and predicted mean response times and mean throughput for a 4 node cluster for different RF and CL configurations. The results for clusters of size 2 and 3 exhibit similar behaviour as that of a 4 node cluster and have been omitted due to space restrictions. Measured and predicted mean response times for 1, 2, 3 and 4 node clusters for 20, 40 and 60 threads are shown in Table 2.

Response Time and Throughput. From Figure 5 and Table 2, when the cluster size is fixed, in this case 4 nodes, the read request response time increases linearly as the workload intensity increases. This linear increase within a fixed RF and CL is due to the increase in job queueing time as jobs wait for free processing threads at each node. This effect is evident in the mean throughput of the cluster, which increases and then reaches its maximum at about 40 jobs/threads for the 4 node cluster. Increasing the RF and CL within a fixed cluster size, decreases the performance of the system. This is due to increased inter-node traffic due to increased replica request forwarding between nodes, thus leading to a sharper linear increase as shown in Figure 5.

The QPN model accurately predicts the linear increase in response times for all cluster sizes and configurations, as it accurately reflects the blocking, queueing and processing of the requests. However, the model generally underestimates the response time as the model does not represent the effect of network delay and the increased synchronization processing between the nodes which will increase response times and lower throughput. The QPN model gives excellent predictions for the 2 node configurations. For the 3 node and 4 node configurations, the predictions have an average error ranging from 2% to 20% for the majority of configurations for both mean response times and throughput.

However, for the (4-3-1) & (4-4-2) configurations the mean prediction error for response time and throughput are 25% and 30% respectively. For (3-3-1) configuration the mean error for response time and throughput is 30% & 28%, for (3-3-2) it is 35% & 35% and for (4-4-1) it is 40% & 60%. When conducting the experiments we noticed that as the replication factor approached the cluster size, the stability of the cluster and its performance was affected. This can be due to the high synchronization between the nodes, increasing the inter-node traffic as the consistency level increased. In addition, as the clusters are hosted on virtual machines, it is probable that the increased incoming requests to each node affected the ability of the virtual processors. Moreover, the instability of the 4 node cluster was evident with high variability in comparison to the 2 and 3 node clusters. The accuracy of the model was not affected by this instability at higher consistency levels. The increased inter-node traffic increased request queueing time for processing threads; which becomes more dominant in overall response times in comparison to other factors. Previous studies have shown that CPU intensive workloads on the Amazon EC2 environment have unstable response

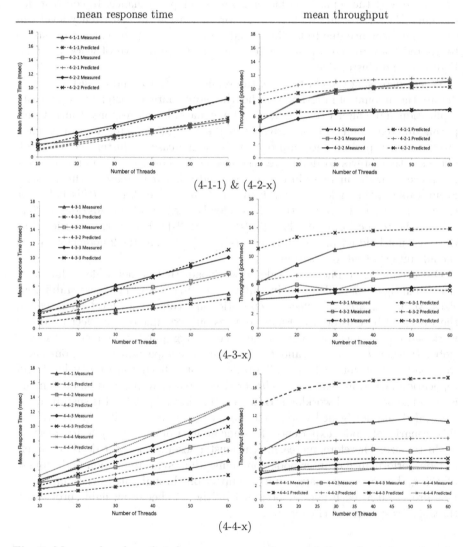

Fig. 5. Measured and predicted mean response times and throughput for a 4 node cluster for different configurations

times with up to 25% variance [5]. Within this range, the model predictions are reliable for capacity planning, giving the upper bounds on performance and throughput of the clusters based on benchmarking the performance of one node. This is clear from the absolute difference between measured and predicted values presented in Table 2.

Replication and Consistency. Higher replication levels increase the likelihood of faster request responses. This is closely tied to the consistency level. From the results (and intuitively), given the same consistency level and cluster size, increasing the replication factor gives better performance. In contrast, for the same cluster size and replication factor, increasing the consistency level will worsen performance due to the increased waiting time for the extra requests to be routed back to the original node. Both hold irrespective of workload (e.g. Figure 5) and cluster size.

When increasing the cluster size, the replication factor and consistency level dictate the amount of inter-node traffic, therefore a larger cluster size with more replication and a higher consistency level may mean longer response times. Comparing different configurations, given that the replication factor and consistency level are fixed, performance improves with the increase in cluster size when job queueing time starts to dominate response time. Thus, we see improved performance with the increase in cluster size and a linear increase in throughput. From Table 2, this holds for workloads of 30 and more threads (this holds for the predicted results irrespective of workload), e.g., at 40 threads/jobs (4-2-1) exhibits better performance than (3-2-1) and (2-2-1). However, we notice that this does not hold for (3-2-x) in comparison to (2-2-x), as the 2 node cluster has the advantage of full replication.

Our original experiments considered workloads that do not scale with cluster size. To mitigate the differences in request queueing between clusters of different sizes, we investigated the effect of replication and consistency level when scaling the workload by cluster size. Figure 6 gives one example of the effect of scaled workload on the mean response times and throughput for number of threads (jobs) t: when $t = n \times 10$ and $t = n \times 20$, in comparison to the non-scaled workload $t = 40$ for (n-1-1) configurations. For a fixed consistency level and replication factor, increasing the cluster size increases response time in comparison to the non-scaled workload. However, the throughput level is sustained for the scaled and non-scaled workloads and increases linearly with cluster size. In this scenario, all nodes experience the same number of incoming jobs irrespective of cluster size. Therefore, increasing the cluster size increases the detrimental effect on performance of inter-node communication, synchronization and request blocking. These aspects have minimal impact on throughput, as the number of jobs processed at each node is bounded by the number of available processing threads, irrespective of cluster size.

Table 2. Measured and predicted mean response times for different configurations and cluster sizes for 20, 40, & 60 thread workloads

threads	CL	1 node		2 nodes				3 nodes						4 nodes							
	RF	1		1		2		1		2		3		1		2		3		4	
		Meas.	Pred.	Meas.	Pred.	Meas.	Pred.	Meas.	Pred.	Meas.	Pred.	Meas.	Pred.	Meas.	Pred.	Meas.	Pred.	Meas.	Pred.	Meas.	Pred.
20	1	4.72	4.25	3.25	3.35	2.19	2.21	2.44	2.59	2.10	2.06	2.24	1.53	2.38	2.05	2.40	1.81	2.25	1.50	2.04	1.19
	2					4.90	4.33			4.66	3.47	4.32	2.94			3.52	2.93	3.30	2.65	3.16	2.37
	3											5.56	4.37					4.58	3.72	4.28	3.45
	4																			5.39	4.51
40	1	8.93	8.38	5.58	6.61	5.16	4.38	5.11	5.00	5.22	3.99	4.71	2.97	3.87	3.88	3.89	3.44	3.38	2.86	3.60	2.27
	2					8.04	8.68			9.16	6.94	9.23	5.86			5.95	5.64	5.89	5.10	5.49	4.53
	3											10.47	8.74					7.48	7.31	7.41	8.33
	4																			9.05	8.84
60	1	13.01	12.58	8.40	9.87	6.86	6.56	7.57	7.42	6.68	5.94	6.19	4.42	5.41	5.71	5.36	5.07	4.99	4.24	5.34	3.35
	2					11.75	13.02			12.64	10.79	11.77	8.78			8.44	8.48	7.92	7.67	8.09	6.70
	3											15.83	13.10					10.14	11.22	11.13	9.95
	4																			13.06	13.17

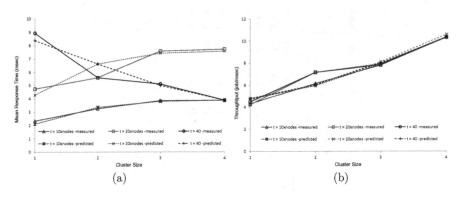

Fig. 6. Measured and predicted mean (a) respone times and (b) throughput for (n-1-1) configurations comparing scaled and non-scaled workloads

5   Conclusions and Future Work

This paper presents a generic QPN model of a Cassandra multi-master cluster under different read workload intensities and consistency configurations. Given the performance measures of one Cassandra node the model was able to predict response times and throughput for different configurations for Cassandra clusters running on the Amazon EC2 cloud platform. This work detailed the relationship between cluster size, replication factor and consistency level on response time and throughput for scaled and non-scaled workloads. Moreover, our experiments identified the effect that node capacity and configuration has on the overall performance of the cluster. The model presented can be applied to other NoSQL datastores that share a similar replication model with Cassandra.

The results presented are based on random access to keys; skewed access can be modelled by modifying Algorithms 1 & 2 through setting the connection probabilities of different node set combinations. We plan to extend the QPN model to model write and mixed workloads and explicitly represent network delays. Furthermore, as the QPN simulation time increases with cluster size, we will investigate deriving analytical formulas based on the cluster size, replication factor, consistency level and node capacity to determine the trade-off between different configurations. This would require further experimentation with larger clusters and increased runs to mitigate the effect of virtualization.

Acknowledgments. The authors would like to thank the anonymous reviewers for their comments that helped improve the quality of this paper. This work has been partially supported by the *AWS in Education* research grant from Amazon and by the *ForgeSDK* project sponsored by Reply S.R.L.

References

1. Amazon Web Services - Elastic Cloud Computing, http://aws.amazon.com
2. Bause, F.: Queueing Petri Nets–A Formalism for the Combined Qualitative and Quantitative Analysis of Systems. In: Fifth Intl Workshop Petri Nets and Performance Models (1993)
3. Cassandra-stress benchmarking tool, http://www.datastax.com/documentation/cassandra/1.2/cassandra/tools/toolsCStress_t.html
4. Cattell, R.: Scalable SQL and NoSQL Data Stores. SIGMOD Rec. 39(4), 12–27 (2011)
5. Cerotti, D., Gribaudo, M., Piazzolla, P., Serazzi, G.: End-to-End Performance of Multi-core Systems in Cloud Environments. In: Balsamo, M.S., Knottenbelt, W.J., Marin, A. (eds.) EPEW 2013. LNCS, vol. 8168, pp. 221–235. Springer, Heidelberg (2013)
6. Coulden, D., Osman, R., Knottenbelt, W.J.: Performance Modelling of Database Contention using Queueing Petri Nets. In: 4th ACM/SPEC International Conference on Performance Engineeering (2013)
7. Datastax Documentation Apache Cassandra 1.2 (2014), http://www.datastax.com/docs
8. Kounev, S., Spinner, S., Meier, P.: QPME 2.0 - A Tool for Stochastic Modeling and Analysis Using Queueing Petri Nets. In: Sachs, K., Petrov, I., Guerrero, P. (eds.) Buchmann Festschrift. LNCS, vol. 6462, pp. 293–311. Springer, Heidelberg (2010)
9. Kounev, S.: Performance Modeling and Evaluation of Distributed Component-based Systems using Queueing Petri Nets. IEEE Trans. Software Engineering 32(7), 486–502 (2006)
10. Kounev, S., Buchmann, A.: Performance Modelling of Distributed e-business Applications using Queuing Petri Nets. In: Proc. IEEE International Symposium on Performance Analysis of Systems and Software, pp. 143–155 (2003)
11. Kounev, S., et al.: Automated Simulation-Based Capacity Planning for Enterprise Data Fabrics. In: Proc. 4th SIMUTOOLS 2011, pp. 27–36 (2011)
12. Lakshman, A., Malik, P.: Cassandra: a Decentralized Structured Storage System. SIGOPS Operating Systems Review 44(2), 35–40 (2010)
13. Linden, G.: Make Data Useful (2006), http://www.gduchamp.com/media/StanfordDataMining.2006-11-28.pdf
14. Linden, G., Mayer, M.: Presented at Web 2.0 (2006), http://glinden.blogspot.com/2006/11/marissa-mayer-at-web-20.htm
15. Nou, R., Kounev, S., Julia, F., Torres, J.: Autonomic QoS Control in Enterprise Grid Environments using Online Simulation. Journal of Systems and Software 82(3), 486–502 (2009)
16. Omari, T., Derisavi, S., Franks, G., Woodside, M.: Performance Modeling of a Quorum Pattern in Layered Service Systems. In: QEST 2007 (2007)
17. Osman, R., Coulden, D., Knottenbelt, W.J.: Performance Modelling of Concurrency Control Schemes for Relational Databases. In: Dudin, A., De Turck, K. (eds.) ASMTA 2013. LNCS, vol. 7984, pp. 337–351. Springer, Heidelberg (2013)
18. Osman, R., Knottenbelt, W.J.: Database System Performance Evaluation Models: A Survey. Performance Evaluation 69(10), 471–493 (2012)
19. Schurman, E., Brutlag, J.: Performance Related Changes and their User Impact. Presented at Velocity Web Performance and Operations Conference (2009)
20. Stonebraker, M., Cattell, R.: 10 Rules for Scalable Performance in *Simple Operation* Datastores. Commun. ACM 54(6), 72–80 (2011)
21. Tadj, L., Rikli, N.E.: Matrix Analytic Solution to a Quorum Queueing System. Mathematical and Computer Modelling 32(3-4), 481–491 (2000)

On Queues with General Service Demands and Constant Service Capacity

Herwig Bruneel, Wouter Rogiest*, Joris Walraevens, and Sabine Wittevrongel

Department of Telecommunications and Information Processing
Ghent University (UGent), St.-Pietersnieuwstraat 41, B-9000 Gent, Belgium
{hb,wrogiest,jw,sw}@telin.UGent.be

Abstract. This paper analyzes a single-server discrete-time queueing model with general independent arrivals, where the service process of the server is characterized in two steps. Specifically, the model assumes that (i) each customer represents a random, arbitrarily distributed, amount of work for the server, the *service demand*, and (ii) the server disposes of a fixed number of work units that can be executed per slot, the *service capacity*.

For this non-classical queueing model, we obtain explicit closed-form results for the probability generating functions (pgf's) of the unfinished work in the system and the queueing delay of an arbitrary customer. The pgf of the number of customers is derived explicitly in case of either geometrically distributed service demands, and/or for a geometric arrival distribution. The analysis is complemented by several numerical examples.

Keywords: queueing, discrete-time, independent arrivals, general service requirements, fixed service capacity.

1 Introduction

A queueing phenomenon occurs when some kind of customers require some kind of service from a given service facility, usually with finite resources. In particular, queues arise when the customers arrive randomly at the service facility and/or the time required to serve the customers is nondeterministic. Classical queueing models characterize the random nature of arrivals and services by modelling the interarrival times (or, alternatively, the number of arrivals in given time intervals) and the service times as random variables [11,2]. In particular, this means that the *service time* is considered as the fundamental quantity that characterizes the speed of the service process: it expresses the time the service facility needs to process exactly one customer. In reality, however, the service time of a customer is the synthesis of two different aspects of the service process: the *service requirement* or *service demand* of the customer and the *service capacity* of the service facility.

In this paper, we introduce the term *work unit* to model both aspects separately. On the one hand, the service demand of each customer is expressed as

---

* Corresponding author.

G. Norman and W. Sanders (Eds.): QEST 2014, LNCS 8657, pp. 210–225, 2014.

an integer number of work units. On the other hand, the service capacity of the service facility is expressed as an integer number of work units that the service facility is able to execute in one time unit ("time slot"). More specifically, we consider a model in which the service demands of the consecutive customers are random and arbitrarily distributed, and where the service capacity of the system is constant from time slot to time slot. Customers demanding variable amounts of work enter the system according to a general independent arrival process (numbers of arrivals are independent from slot to slot) and are served in First-Come-First-Served (FCFS) order, i.e., during each slot the service facility executes as many work units as possible with respect to the service of the customers present in the system, in accordance with their order of arrival. The service of a next customer is only started when the previous customer has received complete service.

Some literature exists on similar models in which the number of work units that can be performed per slot is variable. E.g., in [10,9,7,1,12], a discrete-time queueing model is examined with deterministic "service times" of 1 slot each and a constant number (say, m) of servers, of which a variable number is available from slot to slot. In terms of our present model, this comes down to assuming that the service demand of each customer represents exactly one work unit and the service-capacity distribution has *finite support*. For this model, the number of customers in the system (equivalent to the amount of work, in this case) is examined for $m = 1$ in [10] (at arbitrary slot boundaries) and in [9] (at service-completion times), and for arbitrary $m \geq 1$ in [7] and [1]. For the latter (most general) model, the delay analysis was performed in [12]. Although these papers consider nondeterministic service-capacity distributions, they are not really more general than the present study: the analysis in all these papers relies heavily on the deterministic (single-slot) nature of the service times (a severe restriction that the present model does not have). In fact, the current paper is more related to [3], in which a model with general service-demand distribution and *geometrically distributed* service capacities was examined. The geometric (and, hence, memoryless) nature of the service-capacity distribution simplified the analysis considerably in that case, but is undoubtedly less realistic than the deterministic counterpart considered in the current paper. Nevertheless, as we shall see, the fixed service capacity also allows for a nearly completely analytic solution of the model.

Although this paper is quite generic and does not focus on a specific application, the model we study is relevant in a wide variety of application areas. The fixed service capacity could model, for instance, the available bandwidth on a communication link [13], the printing capacity of a high-speed printer printing a fixed number of pages per time unit, the capacity of immigration desks, the capacity of a border checkpoint [8], the number of nurses or operating room time slots [4] in a hospital, or the number of dockers in a port or a shipyard [14], the number of workers in a building company, the production capacity of a plant [5], etc.

The structure of the paper is as follows. Section 2 introduces the mathematical model. Section 3 presents the analysis of the (steady-state) *unfinished work* in the system. In section 4, we derive the pgf of the (steady-state) *delay* of an arbitrary customer from the pgf of the unfinished work. In section 5 we show that, in general, the derivation of the pgf of the (steady-state) *system content* is much harder. Therefore, we confine ourselves to the special case where the service-demand distribution is geometric in section 6. Section 7 presents the analysis for the special case of a geometric arrival process. In section 8, we discuss our main results and provide a number of instructive numerical examples.

2   Mathematical Model

We consider a discrete-time queueing system with infinite waiting room and a service facility (henceforth also referred to as the "server" of the system) which can deliver a fixed amount of service per time unit. As in all discrete-time models, the time axis is divided into fixed-length intervals referred to as *(time) slots*. New customers may enter the system at any given (continuous) point on the time axis, but services are synchronized to (i.e., can only start and end at) slot boundaries. Specifically, we assume that the service of a customer can start no earlier than during the slot following his arrival slot, i.e., we adopt a *late-arrival-with-delayed-access* convention with respect to the sequence of events at slot boundaries.

The arrival process of new customers in the system is characterized by means of a sequence of independent and identically distributed (i.i.d.) nonnegative discrete random variables with common probability mass function (pmf) $a(n)$ and common probability generating function (pgf) $A(z)$ respectively. More specifically, $a(n) \triangleq \text{Prob}[\, n \text{ customer arrivals in one slot}\,]$, with $n \geq 0$. The corresponding pgf and mean number of customer arrivals per slot are given by

$$A(z) \triangleq \sum_{n=0}^{\infty} a(n)\, z^n \;, \qquad \lambda \triangleq A'(1) \;.$$

The latter is referred to as *(mean) arrival rate* in the following.

The service process of the customers is described in two steps. First, we characterize the *demand* that customers place upon the resources of the system, by attaching to each customer a corresponding *service requirement* or *service demand*, which indicates the number of *work units* required to give complete service to the customer at hand. The service demands of consecutive customers arriving at the system are modelled as a sequence of i.i.d. positive discrete random variables with common pmf $s(n)$ and common pgf $S(z)$ respectively. More specifically, $s(n) \triangleq \text{Prob}[\, \text{service demand equals } n \text{ work units}\,]$, with $n \geq 1$. The corresponding pgf and mean service demand of the customers are given by

$$S(z) \triangleq \sum_{n=1}^{\infty} s(n)\, z^n \;, \qquad \frac{1}{\sigma} \triangleq S'(1) \;.$$

Next, we describe the *resources* of the server, by attaching to each time slot a corresponding *service capacity*, which indicates the number of work units that the server is capable of delivering in that slot. In this paper, we assume that the service capacity is fixed from slot to slot and given by r work units per slot. Note that the term *service time*, used in traditional queueing models to indicate the total time the service facility needs to serve one customer, is not used in this paper. Instead, it is replaced by its two components, the *service demand* and the *service capacity*, which are usually left implicit in classical models.

The operation of the queueing system is as follows. Customers arrive in the system according to an independent arrival process, characterized by the pgf $A(z)$, and take place in the queue in their order of arrival. The amount of service required by each customer (expressed in work units) is given by their corresponding service demand, described by the pgf $S(z)$. The server serves customers from the queue sequentially in FCFS order, spending no more than r work units in each slot. If, at the start of a slot, the remaining service demand of the customer in service is less than r work units, the server completes the service of this customer and immediately (during the same slot) starts the service of the next customer in the queue (if any) – using the remaining part of its service capacity – or (otherwise) becomes idle. If a customer whose service is initiated in a slot requires more work units than the server has left in that slot, the service of that customer continues in the next slot.

3   Unfinished Work

Investigating the time evolution of the total amount of work in the system, we first define a number of relevant random variables. Specifically, let a_k indicate the number of customers entering the system during slot k (with known pgf $A(z)$), let $s_{k,i}$ denote the service demand (with pgf $S(z)$) of the i-th of these customers and let u_k (with pgf $U_k(z)$) represent the unfinished work, i.e., the total number of work units "present in" the system, at the beginning of slot k. Then, the following recursive system equation can be established:

$$u_{k+1} = \sum_{i=1}^{a_k} s_{k,i} + (u_k - r)^+ , \qquad (1)$$

where $(x)^+$ denotes the quantity $\max(0, x)$. In terms of pgf's, this equation is

$$U_{k+1}(z) = E\left[z^{\sum_{i=1}^{a_k} s_{k,i}}\right] \cdot E\left[z^{(u_k - r)^+}\right] , \qquad (2)$$

where $E[x]$ denotes the expected value of the random variable x. This recursive equation is well-known in discrete-time queueing analysis, and was analyzed in [2] (analysis of the GI-D-c model). There, a steady-state solution was obtained under the condition that the system is stable. Here, we also assume that the system is stable, with the mean number of work units entering the system per slot, given by $A'(1)S'(1) = \lambda/\sigma$, strictly smaller than the per-slot service capacity r, i.e., if and only if

$$\frac{\lambda}{\sigma} < r . \tag{3}$$

Moving to steady state, we let the time parameter k go to infinity. Then, both functions $U_k(z)$ and $U_{k+1}(z)$ in (2) converge to a common limit function $U(z)$, which denotes the pgf of the unfinished work at the beginning of an arbitrary slot in steady state. As shown in [2], following expression for $U(z)$ can be derived,

$$U(z) = \left(r - \frac{\lambda}{\sigma}\right) \frac{(z-1)A(S(z))}{z^r - A(S(z))} \prod_{j=1}^{r-1} \frac{z - z_j}{1 - z_j} , \tag{4}$$

where the quantities $\{z_j \mid 1 \le j \le r - 1\}$ are the $r - 1$ complex zeroes of $z^r - A(S(z))$ strictly inside the complex unit disk. As soon as the zeroes $\{z_j \mid 1 \le j \le r-1\}$ have been determined (numerically), equation (4) represents a known function of z. Closed-form expressions for the moments of the unfinished work can be easily obtained from this by computing derivatives of $U(z)$ at $z = 1$. Further, note that in some special cases, the computation of the zeroes *inside* the unit disk can be circumvented, and just one zero *outside* the unit disk has to be determined instead. An example of this approach is treated in section 7.

4   Customer Delay

In this section, we analyze the delay customers incur in the system. Specifically, let C denote an arbitrary ("tagged") customer entering the system in steady state, and let J denote the arrival slot of C. We define the *(discrete) delay d* of C as the total number of (full) slots between the arrival instant of C in the system and the departure time of C from the system, i.e., d indicates the number of slots between the end of slot J and the end of the slot during which the last work unit of the service demand of C is being executed. Owing to the FCFS queueing discipline, the delay d of C is equal to the time needed to execute the unfinished work present in the system just after slot J, but to be performed before or during the service of C. In this section, we first compute the pgf of this amount of work. Next, from this, we derive the pgf of d.

Let $\tilde{u}$ denote the unfinished work at the beginning of slot J and f the number of customers entering the system during slot J but to be served before C. Then, the total amount of work to be performed before or during the service of C, still "present in" the system just after slot J, is given by

$$v = (\tilde{u} - r)^+ + \sum_{i=1}^{f} \tilde{s}_i + s_C , \tag{5}$$

where $\tilde{s}_i$ refers to the service demand of the i-th customer entering the system during slot J, but to be served before C, and s_C represents the service demand of C.

It is well-known from previous papers (e.g. [12]) that the random variable f has as pgf

$$F(z) \triangleq E[z^f] = \frac{A(z) - 1}{\lambda(z - 1)} \, . \tag{6}$$

On the other hand, the independent nature of the arrival process implies that the probability distribution of $\tilde{u}$, i.e., the unfinished work at the beginning of the *arrival slot* of C, is identical to the probability distribution of the unfinished work at the beginning of an *arbitrary slot* in the steady state. This implies that the pgf of $\tilde{u}$ is equal to $U(z)$ (see equation (4)). For the same reason, the random variables f and $\tilde{u}$ are mutually independent. Putting all these elements together, we conclude that the pgf of v can be obtained as

$$V(z) \triangleq E[z^v] = E\left[z^{(\tilde{u} - \tilde{r})^+}\right] \cdot E\left[z^{\sum_{i=1}^{f} \tilde{s}_i}\right] \cdot E[z^{sC}] = \frac{U(z)}{A(S(z))} \cdot F(S(z)) \cdot S(z) \, ,$$

where, in the last step, we have used the steady-state version of equation (2), i.e., equation (2) for $k \to \infty$.

Using (4) and (6), we can derive from the above result the following explicit expression for $V(z)$:

$$V(z) = \left(r - \frac{\lambda}{\sigma}\right) \frac{(z - 1)S(z)[A(S(z)) - 1]}{\lambda[S(z) - 1][z^r - A(S(z))]} \prod_{j=1}^{r-1} \frac{z - z_j}{1 - z_j} \, . \tag{7}$$

As explained before, the delay d of customer C is given by the number of slots the service facility needs to perform v work units. Owing to the constant service capacity of r work units per slot in the system at hand, we can therefore simply express d as

$$d = \left\lceil \frac{v}{r} \right\rceil \, , \tag{8}$$

where the so-called ceiling of x, denoted $\lceil x \rceil$, is equal to the smallest integer larger than or equal to the (real) quantity x. Deriving the pgf $D(z)$ of d from the known pgf $V(z)$ of v thus amounts to determining the pgf of the quotient (d) of a division by an integer (r) from the known pgf of the dividend (v). From [2] it therefore follows that an explicit expression for $D(z^r)$ can be obtained as follows:

$$D(z^r) = \frac{1 - z^r}{r} \sum_{j=0}^{r-1} \frac{V(\theta^j z)}{1 - \theta^j z} \, , \tag{9}$$

where, as before, $\theta \triangleq \exp(\frac{i2\pi}{r})$ is the complex r-th root of 1. Although not very transparent, equations (9) and (7) provide explicit results for the pgf of the customer delay. In particular, by computing derivatives at $z = 1$, explicit closed-form expressions can be derived from (9) and (7) for the moments of the customer delay. Although such expressions contain various complex numbers, such as the zeroes z_j and the powers of θ, the results they deliver are all real-valued.

5   System Content

The *system content* is defined as the number of customers present in the system. In this paper, we denote the system content at the beginning of the k-th slot as b_k; the steady-state system content is denoted as b. We note that the *mean* system content can always be derived from the *mean* delay, by applying (the discrete-time version of) Little's result [6]:

$$E[b] = \lambda E[d] .$$

In this section, however, we try to find an expression for the whole *pgf* of the system content from the results obtained above, i.e., either for the unfinished work or for the delay.

First, we examine the connection between system content and unfinished work. The following relationships can be established between the system content b_k and the unfinished work u_k, defined in section 3:

$$u_k = 0 \quad , \quad \text{if} \quad b_k = 0 ,$$

$$u_k = \hat{s}_1 + s_2 + \ldots + s_{b_k} \quad , \quad \text{if} \quad b_k > 0 , \tag{10}$$

where $\hat{s}_1$ indicates the remaining service demand of the customer in service at the beginning of slot k and $s_2, \ldots, s_{b_k}$ denote the (full) service demands of the other $b_k - 1$ customers in the system at the beginning of slot k. It is not straightforward to derive from the above equations a relationship between the pgf's of the unfinished work and the system content. There are two main reasons for this. First, it is not obvious how to find the distribution (or pgf) of the random variable $\hat{s}_1$: the classical results from renewal theory [15,2] on the distribution of the residual lifetime in a sequence of i.i.d. random variables are not applicable here, as the remaining service demand of an ongoing service does not simply decrease by one unit per slot in the system under study. Second, the random variables $\hat{s}_1$ and b_k, appearing in equation (10), are not necessarily independent, as both are connected to the elapsed service demand of the customer in service. In the next section, however, we shall see that these obstacles do not exist if the service demands have a geometric distribution, and we shall be able to determine the pgf of the system content completely in that case.

Next, we try to relate the system content to the customer delay. In order to do so, we exploit the FCFS nature of the service discipline. This implies that the customers left behind in the system by a departing customer are exactly those customers that have entered the system during the sojourn time of the departing customer in the system. As before, let C indicate an arbitrary tagged customer visiting the system in the steady state, with J the arrival slot of C, b the system content at the beginning of slot J, and d the delay of C. Furthermore, let b_a and b_d denote the system content as seen by C just before its arrival and just after its departure respectively. It is well-known [11] that the quantities b_a (with pgf $B_a(z)$) and b_d (with pgf $B_d(z)$) have the same (steady-state) distributions. In terms of pgf's this translates into

$$B_a(z) = B_d(z) \ . \tag{11}$$

It is also easily seen that $b_a = b + f$, which implies

$$B_a(z) = B(z)F(z) \ , \tag{12}$$

where $F(z)$ is given by (6). Finally, the FCFS nature of the service process entails

$$b_d = g + \sum_{k=1}^{d} \tilde{a}_k \ , \tag{13}$$

where g denotes the number of customers arriving during slot J but to be served *after* C, and $\tilde{a}_k$ indicates the number of arrivals during the k-th slot of C's delay. Note that the random variable g can be expressed as

$$g = \tilde{a} - f - 1 \ , \tag{14}$$

where $\tilde{a}$ is the total number of customers arriving during slot J. The problem with equation (13) is that the quantities g and d appearing in it are not necessarily independent, as both are connected to f, as can be clearly concluded from equations (14), (5) and (8). It is clear, however, that g and d are independent if f and g are independent. In order to examine when this is the case, let us compute the joint pgf of f and g:

$$P(x, y) \triangleq E\left[x^f y^g\right] = \sum_{n=1}^{\infty} \mathrm{Prob}[\tilde{a} = n] \, E\left[x^f y^g | \tilde{a} = n\right] \ . \tag{15}$$

As known, for instance from [12], $\mathrm{Prob}[\tilde{a} = n]$ is given by

$$\mathrm{Prob}[\tilde{a} = n] = \frac{na(n)}{\lambda} \ , \tag{16}$$

whereas the expectation in the right hand side of equation (15) follows from the fact that customer C is an arbitrary customer among the customers entering the system during slot J:

$$E\left[x^f y^g | \tilde{a} = n\right] = \frac{1}{n} \sum_{i=0}^{n-1} x^i y^{n-1-i} \ . \tag{17}$$

Combining (15)-(17) then leads to

$$P(x, y) = \frac{1}{\lambda} \sum_{n=1}^{\infty} a(n) \frac{x^n - y^n}{x - y} = \frac{A(x) - A(y)}{\lambda(x - y)} \ . \tag{18}$$

In accordance with (6) (for f), the marginal pgf's of f and g are given by

$$F(z) = P(z, 1) = \frac{A(z) - 1}{\lambda(z - 1)} \ , \quad G(z) = P(1, z) = \frac{A(z) - 1}{\lambda(z - 1)} = F(z) \ ,$$

which shows that f and g are identically distributed, which is intuitively acceptable for reasons of symmetry. The point is, however, that, in general, f and g are not independent unless $P(x, y) = F(x)F(y)$, for all values of x and y. This condition is only fulfilled for specific choices of the arrival process, e.g. for geometric arrivals, as we shall discuss in section 7. If f and g are independent, then (13) implies

$$B_d(z) = G(z)D(A(z)) = F(z)D(A(z)) \ . \tag{19}$$

In these circumstances, it follows from (11), (12) and (19) that the pgf $B(z)$ of the system content can indeed be derived from the pgf $D(z)$ of the customer delay through the simple equation

$$B(z) = D(A(z)) \ . \tag{20}$$

6   Geometric Service Demands

So far, we have made no specific assumptions as to the precise nature of the service-demand distribution, i.e., the pgf $S(z)$ was arbitrary. In this section, we explore the special case where the service demands are *geometrically distributed* with mean value $1/\sigma$, such that

$$s(n) = \sigma(1-\sigma)^{n-1} \ , \ \ n \geq 1 \ , \qquad S(z) = \frac{\sigma z}{1-(1-\sigma)z} \ . \tag{21}$$

The pgf of the unfinished work is obtained by using (21) in equation (4). Likewise, the pgf of the customer delay is determined by equations (9) and (7), using the same substitution. In both cases, the assumption of geometric service demands does not simplify the analysis or the results very much. In the analysis of the system content, however, the assumption of geometric service demands brings about very substantial simplifications. The main reason for this lies in the memoryless nature of the geometric distribution [15,16], which implies that, in (10), the distribution of the remaining service demand ($\hat{s}_1$) is identical to the distribution of a full service demand (such as $s_2, \ldots, s_{b_k}$). Also, the distribution of $\hat{s}_1$ is not influenced by the value of b_k in this case: although b_k may be correlated with the received amount of service of the customer in service (at the beginning of slot k), this does not affect in any way the distribution of the remaining service demand. As a consequence, equations (10) are equivalent to

$$u_k = \sum_{i=1}^{b_k} s_i \ ,$$

where the s_i's are i.i.d. random variables with geometric distribution (with mean $1/\sigma$), which, in addition, are independent of the random variable b_k. It then simply follows that the (steady-state) pgf of u_k is given by

$$U(z) = B\left(\frac{\sigma z}{1-(1-\sigma)z}\right) \triangleq B(\hat{z}) \ . \tag{22}$$

The relationship $\hat{z} = \sigma z / (1 - (1 - \sigma) z)$ introduced in (22) can be easily inverted so that equation (22) leads to $B(\hat{z}) = U(\hat{z}/(\sigma + (1 - \sigma)\hat{z}))$. Replacing $\hat{z}$ by z again and using equation (4), we then get the following explicit expression for the pgf of the system content:

$$B(z) = \frac{(r\sigma - \lambda)(z - 1)A(z)}{z^r - [\sigma + (1 - \sigma)z]^r A(z)} \prod_{j=1}^{r-1} \frac{z - \hat{z}_j}{1 - \hat{z}_j} , \tag{23}$$

where the quantity $\hat{z}_j$ is defined as

$$\hat{z}_j \triangleq \frac{\sigma z_j}{1 - (1 - \sigma) z_j} . \tag{24}$$

It is interesting to observe that, in this special case of geometric service demands, the pgf of the system content can also be derived directly, without requiring the pgf of the unfinished work as an intermediate result. In order to show this, we first compute the pgf of the number of *customers* (rather than *work units*) that can be completed in the k-th slot. As the service facility disposes of exactly r work units in slot k and σ corresponds to the probability that slot k is the last slot of some customer's service demand (owing to the geometric distribution of the service demands), the number of customers (c_k) that can complete their service in slot k has a binomial distribution with parameters r and σ, i.e.,

$$\mathrm{Prob}[c_k = n] = \frac{r!}{n!(r - n)!} \sigma^n (1 - \sigma)^{r-n} , \quad 1 \le n \le r ,$$

with corresponding pgf $C(z) = (1 - \sigma + \sigma z)^r$.

The system content can then be analyzed by means of an equivalent (classical) discrete-time queueing model with deterministic service times equal to 1 slot, a general independent arrival process, characterized by pgf $A(z)$, and r servers, each of which is subject to server interruptions, such that the number of available servers per slot is characterized by pgf $C(z)$. The result can be retrieved from e.g. [1] as

$$B(z) = \frac{[C'(1) - A'(1)](z - 1)A(z)}{z^r - z^r C(1/z)A(z)} \prod_{j=1}^{r-1} \frac{z - \hat{z}_j}{1 - \hat{z}_j}$$

$$\tag{25}$$

$$= \frac{(r\sigma - \lambda)(z - 1)A(z)}{z^r - [\sigma + (1 - \sigma)z]^r A(z)} \prod_{j=1}^{r-1} \frac{z - \hat{z}_j}{1 - \hat{z}_j} ,$$

where the quantities $\{\hat{z}_j \mid 1 \le j \le r - 1\}$ indicate the $r - 1$ complex zeroes of the denominator $z^r - [\sigma + (1 - \sigma)z]^r A(z)$ strictly inside the complex unit disk.

It is clear that the expressions in (23) and (25) are identical, provided we can prove that the quantities $\{\hat{z}_j \mid 1 \le j \le r - 1\}$, defined by (24), are zeroes of $z^r - [\sigma + (1 - \sigma)z]^r A(z)$, as soon as the quantities $\{z_j \mid 1 \le j \le r - 1\}$ are zeroes of $z^r - A(S(z))$. This is indeed the case. First, we invert (24), in order to express z_j in terms of $\hat{z}_j$. Now, as z_j is a zero of $z^r - A(S(z))$, it follows that

$$z_j^r = A\left(\frac{\sigma z_j}{1 - (1 - \sigma)z_j}\right) ,$$

or, using (24),

$$\left(\frac{\hat{z}_j}{\sigma + (1 - \sigma)\hat{z}_j}\right)^r = A(\hat{z}_j) ,$$

which shows that $\hat{z}_j$ is a zero of $z^r - [\sigma + (1 - \sigma)z]^r A(z)$.

7 Geometric Arrivals and Geometric Service Demands

The analysis in this paper did not assume any specific form for the distribution of the number of arrivals per slot. In this section, we treat the special case of *geometric arrivals*, defined by $A(z) = 1/(1 + \lambda - \lambda z)$. It turns out that considerable simplifications of the results are possible, especially under the assumption of a geometric distribution for the service demands. First, we observe that for geometric arrivals the joint pgf $P(x, y)$, defined in (15), factorizes as $P(x, y) = F(x)F(y)$, implying that the simple relationship (20) is valid. Indeed, in view of the expression for $A(z)$,

$$P(x, y) = \frac{A(x) - A(y)}{\lambda(x - y)} = \frac{1}{1 + \lambda - \lambda x} \cdot \frac{1}{1 + \lambda - \lambda y} .$$

Now, introducing $A(z) = 1/(1 + \lambda - \lambda z)$ and $S(z) = \sigma z/(1 - (1 - \sigma)z)$ in equation (4), we obtain the following result for the pgf of the unfinished work:

$$U(z) = \left(r - \frac{\lambda}{\sigma}\right) \frac{(z - 1)[1 - (1 - \sigma)z]}{[z^r(1 + \lambda) - 1][1 - (1 - \sigma)z] - \lambda\sigma z^{r+1}} \prod_{j=1}^{r-1} \frac{z - z_j}{1 - z_j} , \quad (26)$$

where the z_j's are the zeroes of the denominator of (26) – which is now a simple polynomial of degree $r + 1$ – *inside* the closed complex unit disk. By cancelling out common factors in the numerator and the denominator of (26), we therefore obtain

$$U(z) = \frac{1 - (1 - \sigma)z}{\sigma} \cdot \frac{1 - z_u}{z - z_u} , \quad (27)$$

where z_u is the only zero of $[z^r(1 + \lambda) - 1][1 - (1 - \sigma)z] - \lambda\sigma z^{r+1}$ *outside* the complex unit disk. It simply follows that the pmf of the unfinished work can be expressed as

$$u(0) = \frac{1}{\sigma}\left(1 - \frac{1}{z_u}\right) ,$$

$$ \quad (28)$$

$$u(i) = \frac{1 + (\sigma - 1)z_u}{\sigma}\left(1 - \frac{1}{z_u}\right)\left(\frac{1}{z_u}\right)^i , \quad i \geq 1 ,$$

while the mean unfinished work is given by

$$E[u] = \frac{z_u}{z_u - 1} - \frac{1}{\sigma} .$$

(29)

Along the same lines, it is easily seen that the pgf $V(z)$, given in general by (7), reduces to

$$V(z) = \frac{(1 - z_u)z}{z - z_u} ,$$

such that the pmf of the random variable v is simply given by

$$v(n) = \left(1 - \frac{1}{z_u}\right)\left(\frac{1}{z_u}\right)^{n-1} , \quad n \geq 1 ,$$

i.e., the random variable v is geometrically distributed with parameter $1/z_u$ in this case. The delay distribution can be derived from this, departing from equation (8):

$$d(k) \triangleq \text{Prob}[d = k] = \sum_{j=0}^{r-1} v((k-1)r + 1 + j) = \left(1 - \frac{1}{z_u}\right)\left(\frac{1}{z_u}\right)^{(k-1)r} \sum_{j=0}^{r-1}\left(\frac{1}{z_u}\right)^j .$$

Resulting, the pmf $d(k)$ is simply given by

$$d(k) = \left(1 - \frac{1}{z_u^r}\right)\left(\frac{1}{z_u^r}\right)^{k-1} , \quad k \geq 1 ,$$

(30)

and the pgf $D(z)$ reads

$$D(z) = \sum_{k=1}^{\infty} d(k)z^k = \frac{(1 - z_u^r)z}{z - z_u^r} .$$

(31)

The distribution of the delay turns out to be (positive) geometric with parameter $1/z_u^r = 1/A(S(z_u))$ and mean value

$$E[d] = \frac{z_u^r}{z_u^r - 1} .$$

(32)

Finally, the pgf of the system content, resulting from equation (20) is given by

$$B(z) = D\left(\frac{1}{1 + \lambda - \lambda z}\right) = \frac{1 - z_u^r}{1 - z_u^r(1 + \lambda - \lambda z)} ,$$

(33)

which can be easily inverted into

$$b(j) = \frac{z_u^r - 1}{z_u^r(1 + \lambda) - 1}\left(\frac{\lambda z_u^r}{(1 + \lambda)z_u^r - 1}\right)^j , \quad j \geq 0 .$$

(34)

In other words, the system content is geometrically distributed with parameter

$$\frac{\lambda z_u^{\ r}}{(1+\lambda)z_u^{\ r} - 1} = \frac{1 - (1-\sigma)z_u}{\sigma z_u} = 1/S(z_u)$$

in this case. The mean system content is given by

$$E[b] = \frac{\lambda z_u^{\ r}}{z_u^{\ r} - 1} \ . \tag{35}$$

We note that our results are in accordance with Little's result. Finally, note that introducing $A(z) = 1/(1 + \lambda - \lambda z)$ in (25) equally allows to derive expression (33), as required.

8   Discussion of Results

8.1   General Discussion

In this paper, we have obtained a large number of explicit expressions for the main performance measures of a single-server discrete-time queueing model with FCFS queueing discipline and a fixed service capacity per slot. More specifically, we have derived generic results for the pgf's of the unfinished work in the system and the customer delay. The pgf of the system content could not be obtained explicitly in general. However, for geometric service demands and/or geometric arrivals, we also succeeded in deriving a closed-form expression for this function.

Finally, we have shown that in case a geometric service-demand distribution is combined with geometric arrivals, we are able to obtain explicit closed-form results for the pgf, the pmf and the expected value of the unfinished work, the delay and the system content, in terms of the parameters of the model on the one hand and just one single zero (z_u) of a complex function outside the complex unit disk.

In section 7, we have observed that the relevant probability distributions have *geometric tails*, i.e., the tail distributions of the unfinished work, the customer delay and the system content all take the form

$$\text{Prob}[x > n] = C\alpha_x^{\ n} \ , \quad n > 0 \ , \tag{36}$$

where the random variable x may be either u, d or b. The decay rate α_x, in general, turns out to be given by

$$\alpha_u = 1/z_u \ , \quad \alpha_d \ = 1/z_u^{\ r} = 1/A(S(z_u)) \ , \quad \alpha_b = 1/S(z_u) \ , \tag{37}$$

where z_u is the only root of the equation $z^r = A(S(z))$ outside the complex unit disk. Although these results were obtained exclusively for the combination of geometric service demands with geometric arrivals, we are tempted to believe that, at least for sufficiently large values of n, the tail probabilities $\text{Prob}[x > n]$ can be derived with formulas similar to (36) and (37), where z_u denotes the *dominant* pole of the pgf of u, in much more general circumstances. This, however, is subject to further research.

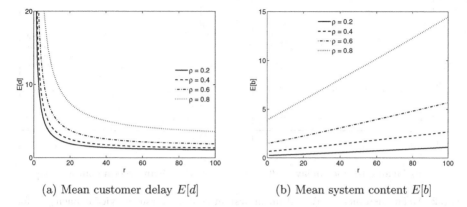

(a) Mean customer delay $E[d]$ (b) Mean system content $E[b]$

Fig. 1. Mean customer delay and mean system content versus service capacity r, for fixed mean service demand $1/\sigma = 20$ and various values of the load $\rho = \lambda/r\sigma$, keeping the ratio λ/r fixed

8.2   Numerical Examples

We now discuss a number of numerical examples. Specifically, we illustrate the explicit results obtained in section 7 for geometric arrivals and geometric service demands. In this case, the system is completely determined by just three parameters: the arrival rate λ, the mean service demand $1/\sigma$ and the fixed service capacity r. The global load of the system is given by

$$\rho = \lambda/r\sigma \tag{38}$$

and the stability condition (3) requires $\rho < 1$.

In figures 1a and 1b, we show the mean customer delay $E[d]$ and the mean system content $E[b]$ as functions of the service capacity r, while keeping the ratio λ/r fixed, i.e., as the value of the service capacity r on the horizontal axis increases, so does the mean arrival rate λ. The mean service requirement of the customers is kept fixed at $1/\sigma = 20$ work units, and various values of the global load ρ are considered. Figure 1a shows that the mean delay of the customers reduces considerably when the service capacity r is increased from its lowest value $r = 1$ to about $r = 20 = 1/\sigma$; for yet larger values of r, the mean delay basically becomes independent of r (and, hence, λ) and converges to a relatively small value. An intuitive explanation of this behavior could be the fact that as soon as the service capacity per slot is (much) larger than the average service demand of the customers, multiple customers can be handled in one slot and delays are inherently limited, whereas for low values of the service capacity, the service of one customer already requires multiple slots and therefore also the delays get higher. Figure 1b shows a completely different picture for the mean system content, which appears to increase *more or less linearly* with the service capacity r (and the arrival rate λ).

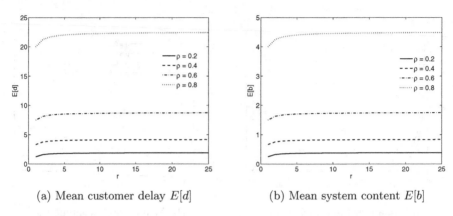

(a) Mean customer delay $E[d]$ (b) Mean system content $E[b]$

Fig. 2. Mean customer delay and mean system content versus service capacity r, for fixed mean arrival rate $\lambda = 0.2$ and various values of the load $\rho = \lambda/r\sigma$, keeping the ratio $1/\sigma$ versus r fixed

In figures 2a and 2b, we have again plotted the mean customer delay $E[d]$ and the mean system content $E[b]$ as functions of the service capacity r. In this case, however, we keep the ratio of the average service demand per customer $(1/\sigma)$ versus the service capacity per slot (r) fixed, i.e., the mean number of slots required to serve one customer basically remains unchanged, but as the value of the service capacity r on the horizontal axis increases, so does the mean service requirement $1/\sigma$. The mean arrival rate is kept fixed at $\lambda = 0.2$ customers per slot, and again various values of the global load ρ are considered. It is easily seen that each value of ρ corresponds to a given value of the mean "service time" of the customers in the classical sense of the word: $\rho = 0.2/0.4/0.6/0.8$ is equivalent to a mean service time of $1/2/3/4$ slots, respectively. We note that, in this case, both figures 2a and 2b are very similar, which can be explained by the simple observation that λ is fixed and the mean system content is therefore proportional with the mean delay by virtue of Little's result: $E[b] = \lambda E[d] = 0.2E[d]$. Both figures show that, in these circumstances, the effect of varying the service capacity r (and the mean service demand $1/\sigma$ accordingly) on mean delay and mean system content is extremely limited: only for very low values of r, the mean delay and the mean system content are somewhat lower than for all other values of r. A possible conclusion could be that classical queueing models, in which the notion of "service time" is used instead of its two more basic components "service demand" and "service capacity", are basically adequate in case the value of r (and hence also the value of the mean service demand $1/\sigma$) is not too small, at least within the limits of the examples considered in this section (i.e., geometric arrivals and geometric service demands).

Acknowledgement. Part of this research has been funded by the Interuniversity Attraction Poles Programme initiated by the Belgian Science Policy Office. The second author is a postdoctoral fellow with the Research Foundation, Flanders (FWO-Vlaanderen).

References

1. Bruneel, H.: A general model for the behaviour of infinite buffers with periodic service opportunities. European Journal of Operational Research 16(1), 98–106 (1984)
2. Bruneel, H., Kim, B.G.: Discrete-time models for communication systems including ATM. Kluwer Academic, Boston (1993)
3. Bruneel, H., Walraevens, J., Claeys, D., Wittevrongel, S.: Analysis of a discrete-time queue with geometrically distributed service capacities. In: Al-Begain, K., Fiems, D., Vincent, J.-M. (eds.) ASMTA 2012. LNCS, vol. 7314, pp. 121–135. Springer, Heidelberg (2012)
4. Creemers, S., Belien, J., Lambrecht, M.: The optimal allocation of server time slots over different classes of patients. European Journal of Operational Research 219(3), 508–521 (2012)
5. Dong, M., Hou, F.: Modelling and implementation of manufacturing direct labour allocation: a case study in semiconductor production operations. International Journal of Production Research 50(4), 1029–1044 (2012)
6. Fiems, D., Bruneel, H.: A note on the discretization of Little's result. Operations Research Letters 30, 17–18 (2002)
7. Georganas, N.D.: Buffer behavior with Poisson arrivals and bulk geometric service. IEEE Transactions on Communications 24, 938–940 (1976)
8. Haughton, M., Isotupa, K.P.S.: Flow control in capacity-constrained queueing systems with non-stationary arrivals. Journal of the Operational Research Society 64(2), 283–292 (2013)
9. Heines, T.S.: Buffer behavior in computer communication systems. IEEE Transactions on Computers 28, 573–576 (1979)
10. Hsu, J.: Buffer behavior with Poisson arrival and geometric output process. IEEE Transactions on Communications 22, 1940–1941 (1974)
11. Kleinrock, L.: Queueing systems, part I. Wiley, New York (1975)
12. Laevens, K., Bruneel, H.: Delay analysis for discrete-time queueing systems with multiple randomly interrupted servers. European Journal of Operations Research 85, 161–177 (1995)
13. Ling, X., Hu, M., Long, J., Ding, J., Shi, Q.: Traffic resource allocation for complex networks. Chinese Physics B 22(1) (2013)
14. Liu, Z., Chua, D., Yeoh, K.: Aggregate production planning for shipbuilding with variation-inventory trade-offs. International Journal of Production Research 49(20), 6249–6272 (2011)
15. Mitrani, I.: Modelling of Computer and Communication Systems. Cambridge University Press, Cambridge (1987)
16. Papoulis, A., Pillai, S.U.: Probability, random variables, and stochastic processes, 4th edn. Mc Graw-Hill, New York (2002)

Simulation Debugging and Visualization in the Möbius Modeling Framework

Craig Buchanan and Ken Keefe

Information Trust Institute
staff@mobius.illinois.edu

Abstract. Large and complex models can be difficult to analyze using static analysis results from current tools, including the Möbius modeling framework, which provides a powerful, formalism-independent, discrete-event simulator that outputs static results such as execution traces. The Möbius Simulation Debugger and Visualization (MSDV) feature adds user interaction to running simulations to provide a more transparent view into the dynamics of the models under consideration. This paper discusses the details of the design and implementation of the feature in the Möbius modeling environment. Also, a case study is presented to demonstrate the new abilities provided by the feature.

Keywords: discrete-event simulation, simulation visualization, model debugging, multi-formalism modeling.

1 Introduction

Because of its high flexibility and relative simplicity, discrete-event simulation remains a popular technique for complex analysis in many technical disciplines, as it is used in applications that range from availability assessments in computer science [1], to environmental impact assessments [2], to disease propagation assessments [3]. Despite its powerful benefits, acquisition of appropriate parameters and design of correct models of systems can be quite complicated because of the multitude of uncertainties inherent to the complex systems under study. Currently employed discrete-event simulation tools, such as Möbius [4], Simul8 [5], and Vensim [6], require complete models coupled with complete simulation runs to return any useful results; tweaking of model and simulation parameters can become time-consuming and error-prone, as human operators must complete each modeling workflow from beginning to end. We address that problem by introducing the Möbius Simulation Debugger and Visualization (MSDV) feature, an extension of the discrete-event simulator available in the Möbius modeling framework [7][8], which adds user interaction and visibility to running simulations.

The goal of the MSDV feature is to provide the analyst with a highly transparent view of the running simulation, rather than simply provide results at the end of the simulation. While other discrete event simulators, such as OMNeT++ [9], SAS/OR [10], and AnyLogic [11], also offer visualizations of running

G. Norman and W. Sanders (Eds.): QEST 2014, LNCS 8657, pp. 226–240, 2014.

discrete-event simulations, the MSDV feature extends this functionality by providing the analyst with full leverage to modify and pause any running discrete-event simulation in the Möbius tool. The transparency resulting from both the visualization and model state modification functionalities can aid analysts in designing correct, complete models of the complex systems under consideration. The additional functionality effectively increases the ease, speed, and reliability of the model validation and verification phases of the overall simulation analysis.

In section 2, we present an overview of the Möbius discrete-event simulation. In section 3, we detail the functionality provided by the MSDV feature. In section 4, we examine how it is implemented in the Möbius modeling framework. In section 5, we consider a case study that reveals the utility of the new features. Finally, in section 6, we provide concluding remarks.

2   Möbius Discrete-Event Simulator Overview

Each solver in the Möbius modeling framework, including the Möbius discrete-event simulator, executes modeling-formalism-independent solution techniques by decoupling the solution technique used from the specific modeling formalism of the model under consideration. That powerful feature makes it possible to solve a large subset of modeling formalisms, as well as to easily combine submodels created in different modeling formalisms within this subset. To accomplish such independence in the solution technique, the Möbius modeling framework utilizes the Abstract Functional Interface (AFI), a general modeling formalism that leverages the two overarching modeling characteristics shared by many modeling formalisms: the model state and the transition system [12].

In the AFI, a **state variable** is a basic modeling element that represents the state of a component within the model [13]. For example, when a queue is being modeled, a state variable can represent the number of items currently in the queue. Then, the full model state can be represented as the set of all state variables' values.

Also, in the AFI, an **action** is a basic modeling element that changes the model state [13]. Each action is associated with a timing distribution (e.g., exponential or Weibull) that determines when it will fire, thus changing the model state. Each action is also associated with a Boolean "enabled" status to determine whether it is currently eligible to fire. That status is determined by certain specified conditions of the model state. For example, in the queue model mentioned in the previous paragraph, an action can represent the popping of an item from the front of the queue. The action could be defined as exponential with a rate of 1.0, where it is only "enabled" when there is at least one item in the queue. Therefore, the full transition system of the model can be represented as the set of all actions in the model.

In addition to representing the model, the Möbius discrete-event simulator also employs a **future event list** to determine the specific sequence of **events** in the given simulation batch [7]. In the list, each event couples an action with a deterministic simulation time at which it will fire. The list contains one event item per "enabled" action.

3   Features

In order to achieve the desired transparency and usability, we first had to consider the additional interface features needed for our simulation tool. We needed a way to access the model state during the simulation, a way to modify the model state during the simulation, a way to pause the progress of the running simulation, and a way for users to interact with the simulation. For the first requirement, we implemented the functionality of model state analysis. For the second, we implemented a reliable way to effectively modify the model state. For the third, we implemented a way to apply explicitly defined breakpoints to the running simulation and a way to implicitly step through the simulation. Finally, for the last requirement, we implemented a graphical user interface for the model that provides access to all those new features. The new features are discussed in the following sections.

3.1   Model State Analysis

In the Möbius modeling framework, the model state is composed of both the culmination of the values of the state variables [13] and the contents of the future event list [7]. The contents of the state variables are stored in a contiguous memory block, and the contents of the future event list can be accessed in a straightforward manner. The values must be serialized into a message and sent to the Möbius visualization front-end over the communication layer, as discussed in Section 4.2.

3.2   Model State Modification

Model state modification is more complicated than model state analysis in the Möbius modeling framework. Its added complexity is a result of the dependencies between the elements of the model state. For example, modifying the value of a single state variable could result in a change to the "enabling" status of an action, thus affecting the contents of the future event list. To address those dependencies, we use the built-in dependency mechanisms of the Möbius modeling framework.

Those dependency mechanisms, as presented in [7], operate by associating state variables with actions by declaring the state variables to be either `enabling` or `affecting` with respect to the actions. If a state variable is marked as `enabling` to an action, then modifying that given state variable would require that given action to reevaluate its enabled status. If an action is marked as `affecting` a state variable, then when that action fires, the state variable value may be altered by the firing event.

For example, consider the simple AFI model derived from [7], pictured in Figure 1. The model shows the `enabled` and `affected` relationships between the state variables and actions. As can be seen, an `enabling` relationship exists between state variable P2 and action A2, since the enabled status of A2 depends on the value of P2. Also, an `affected` relationship exists between the action A2 and the two state variables P2 and P3, since the firing of A2 could result in a

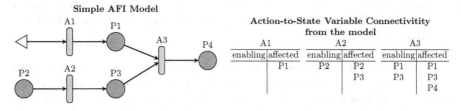

Simple AFI Model

Action-to-State Variable Connectivitity from the model

A1		A2		A3	
enabling	affected	enabling	affected	enabling	affected
	P1		P2	P1	P1
			P3	P3	P3
					P4

Derived State Variable-to-Action Connectivity

P1		P2		P3		P4	
enabling	affected	enabling	affected	enabling	affected	enabling	affected
A3	A1	A2	A2	A3	A2		A3
	A3				A3		

Fig. 1. Connectivity List Example

change in the values of both state variables. Use of those relationships simplifies the modification of the model state, as only **affected** state variables must be reevaluated.

Since the model state modifications available through this feature only include modifying state variables and the firing times of the events in the future events list, the only model state relationships we need to examine are the **enabling** relationships.

3.3 Breakpoints

To access or modify the model state at a given simulation time t, the user must have a way to pause a running simulation at that given simulation time t. One method of pausing the simulation is through user-defined simulation breakpoints. Those breakpoints allow the user to explicitly define conditions in which the simulation should pause. There are three types of breakpoints: simulation time breakpoints, action breakpoints, and state variable breakpoints. Each type returns a **Boolean** value, allowing the user to easily create combinations of the three types using the Boolean logical operators **AND**, **OR**, and **NOT**.

Simulation Time Breakpoints. A simulation time breakpoint allows the user to pause the simulation at a certain simulation time t. For example, if the user wants to run the simulation until simulation time $t = 5.5$, the simulation would pause at simulation time $t = 5.5$, allowing the user to access and modify the resulting model state at this simulation time.

Action Breakpoint. An action breakpoint allows the user to pause the simulation at a certain action event of a specific action. Currently, available action events include **OnFired**, **OnStatusToEnabled**, and **OnStatusToDisabled**. Respectively, the breakpoints are triggers at the simulation times immediately following firing of actions, switching of an action from disabled status to enabled status, and switching of an action from enabled status to disabled status. This breakpoint can be used if the user wants to run a simulation until a certain event has fired. This functionality can be useful during examination of actions that rarely fire.

State Variable Breakpoint. A state variable breakpoint allows a user to pause the simulation when certain conditions concerning state variable values have been met. Specifically, arithmetic combinations of state variable values and literal values are compared using standard comparison operators: $\{<, >, =\}$. For example, if a user wants to pause a simulation when a certain state variable value is greater than another state variable value by 7.5 or more, the user can specify the breakpoint $sv1 > sv2 + 7.5$. This functionality is useful during the examination of the quantitative relationships between state variables.

3.4 Simulation Stepping

In addition to breakpoints, the MSDV feature provides simulation stepping as another means of pausing a running simulation. Whereas breakpoints are explicitly defined by users, simulation stepping is an implicitly defined operation that runs the simulation until the next action fires. Since the model state of a discrete-event simulation does not change until an action is fired, simulation stepping gives users a way to easily examine all of the successive model states of a running simulation in chronological order. The examination can occur from any given paused state, including the initial model state, a state reached through the use of breakpoints, or a state reached through previous stepping. This functionality is useful for examining the fine-grained details of the operation of a running simulation from a given simulation time t.

3.5 Model State Visualization

To effectively access the functionality of the previously discussed features, the user requires a powerful and intuitive interface to control and view the running simulation. After considering potential designs for this interface, we decided that the most useful interface would be one already familiar to the user. Thus, we implemented the user interface to mimic the specific modeling formalism with which the user had specified the model. For example, if the initial model is defined as a stochastic activity network (SAN) model [14], then the user interface should display a SAN-like presentation of the model. Specifically, the user interface will display the model state as a combination of SAN elements identical to the SAN elements of the original model. Thus, the visualization interface becomes an effortless way to bring the user's model to life, rather than a complicated and unfamiliar tool that the user must painstakingly learn.

Although this design decision simplifies the use of the tool, it would be impractical to create a different user interface for every different modeling formalism, not only because of the large number of existing formalisms, but also because of the constant introduction of new modeling formalisms. To address the issue, we leveraged the underlying Abstract Functional Interface (AFI) of the Möbius

Fig. 2. Möbius Layer Interaction

modeling framework [12]. The model-level Möbius AFI is a modeling formalism that is the base of all other modeling formalisms in Möbius. Since all of the specific modeling formalisms are forms of their parent AFI modeling formalism, each can be represented as an AFI model. Therefore, we started by implementing the user interface in AFI. We then continued to develop user interfaces for specific modeling formalisms. The idea is that if the user interface for a specific modeling formalism has not yet been implemented (e.g., a newly developed modeling formalism is being used), then the MSDV tool will default to the AFI visualization and user interface. Although the general AFI visualization and user interface will not be as familiar to a user as a modeling-formalism-specific version would have, it still provides the same power as formalism-specific visualizations and editing interfaces in MSDV. Currently, the MSDV feature supports the AFI, SAN, ADVISE, and Rep/Join modeling formalisms.

4   Implementation

The implementation of the MSDV tool in the Möbius modeling framework relies on its integration into the currently existing discrete-event simulator, which is composed of three different layers [7], pictured in Figure 2. 1) The back-end Möbius simulation processes, implemented in C++, execute the actual simulation, leveraging the power and speed available when running natively on the host machine. 2) The communication layer provides the medium for the communication between the back-end Möbius simulation processes and the front-end visualization interface. 3) The front-end visualization interface, implemented in Java, allows users to control and receive feedback from the back-end Möbius simulation processes. The implementation of the MSDV tool with respect to those three simulator layers is discussed in the following sections.

4.1   Back-End Möbius Simulation

The back-end Möbius simulation processes are responsible for executing the actual simulation of the model. This layer must be modified to allow model state analysis, model state modification, and simulation pausing through the use of breakpoints and stepping functionality.

Model State Analysis. So that we could add the functionality of model state analysis, we needed for the back-end Möbius simulation processes to have the ability to forward the model state to the front-end visualization interface to be interpreted by the end user. We decided to meet this requirement by making it possible to serialize the model state at a particular simulation time t into a message that can be forwarded through the communication layer. Therefore, we designed the MSDV back-end to serialize the state variable data and future event list into a `char` array message, as discussed in Section 4.2. The `char` array is forwarded to the front-end visualization interface to be interpreted and displayed.

Model State Modification. Because of the dependencies between elements of the model state, model state modification, as discussed in Section 3.2, is not as trivial as model state analysis. To simplify its implementation, the user is restricted to modifying only one state variable primitive value, or only one firing time of an event in the future event list, at once. That does not mean that the user could not modify multiple model state elements at one paused point in the simulation. It simply means that the entire model state must adjust to a single modification before the user can specify another change. The restriction simplifies model state modification since the elements of the model state that rely on `enabling` relationships with the modified model state element only need to adjust to a single change in the model state at a time. `Affecting` relationships do not need to be considered in this context since they only affect the model state when an action is directly modified, a feature that is not available through the MSDV feature.

To modify the model state, the MSDV back-end receives a model state modification message from the front-end visualization interface. It specifies a state variable primitive and its new value. The MSDV back-end updates the specified state variable with the new value, and reevaluates the status of each of the actions with which the state variable shares an `enabling` relationship. Consequently, if the status of an action switches from enabled to disabled, then the associated event in the future event list is removed from the list. Similarly, if the status of an action switches from disabled to enabled, then the timing distribution of the action is sampled, and it is added to the future event list. If the status of the action does not change, then it does not modify its associated event, or lack thereof, in the future event list.

Note that when an action's associated event is added to the future event list, its timing distribution is sampled. Consequently, if the modification of a state variable results in the removal of the action's event from the future event list, then even if the state variable is modified back to its original value, the overall model state is unlikely to return to the same state. Since the timing of events is based on the statistically random distribution of the actions, the event will be added back to the future event list with a different associated time. However, the firing time for events in the event list may also be changed through MSDV, so the old firing time can be restored if desired.

In addition to modifying state variables, the MSDV tool also allows users to directly modify the firing times of the events in the future event list in the simulation time interval $t \geq$ `currentSimulationTime`. Since those event times are independent of the rest of the model state, no further consideration must be paid by the MSDV back-end.

Although model state modification is a helpful feature in analysis of running simulations, it is important to note that any modifications to a running simulation could result in statistical differences to runs without modification. Thus, simulation batches utilizing model state modification should not be considered the final results of a system model analysis. Rather, they should be used to help the analyst determine more appropriate parameters and model designs for a complete model that better describes the complex system under consideration.

Breakpoints and Stepping. As described in Sections 3.3 and 3.4, the simulation-pausing capabilities are provided through breakpoints and simulation stepping. The back-end MSDV contributes to this capability by determining the point at which to stop, and by waiting for further instructions from the front-end visualization interface. The evaluation of both explicit and implicit (stepping) breakpoints occurs in the back-end, rather than the front-end, to eliminate the need to forward the entire model state to the front-end after the firing of each event. Thus, the simulation can proceed at near-optimal solution speed until a breakpoint is hit. Both breakpoint and simulation-stepping messages are discussed in Section 4.2.

4.2   Communication Layer

The communication layer of the Möbius simulator is responsible for providing the medium between the back-end Möbius simulation processes and the front-end visualization interface. This layer operates by forwarding TCP/IP messages between the Unix domain sockets of each of these end layers. Each of these message is represented as a raw byte string, and is parsed by the receiving end layer. The several message types available in the MSDV feature are discussed throughout this section.

Model State Message. The model state message contains a serialized representation of the entire current model state to be forwarded from the back-end Möbius simulation processes to the front-end visualization interface. This message contains both the values of all of the state variables and the contents of the future event list.

Modify State Variable Message. The modify state variable message allows the user to modify specified state variable primitive values. This message, which is forwarded from the front-end visualization interface to the back-end simulation processes, contains the unique index of the state variable under consideration, the memory offset of the primitive value in the contiguous memory representing the entire state variable value, the type of the primitive value to be modified, and the new desired value of the state variable primitive value.

Modify Future Event List Message The modify future event list message allows the user to modify the firing time of an event in the future event list to a simulation time in the interval $t \geq$ `currentSimulationTime`. This message contains the index of the event in the future event list, and the new desired time at which the event will fire.

Breakpoint Message. The breakpoint message allows the user to specify explicit breakpoint conditions for a running simulation. This message, which is forwarded from the front-end visualization interface to the back-end Möbius simulation processes, is specified as shown in the UML diagram in Figure 3. As described in Section 3.3, breakpoints are logical combinations of simulation time breakpoints, action breakpoints, and state variable breakpoints. Breakpoint messages can be imagined as serialized versions of a tree data structure containing these breakpoint conditions. For example, consider the tree-like representation of a breakpoint message in Figure 4. This breakpoint message representation specifies that the simulation will pause when the simulation time reaches $t = 4.25$, the condition that the value of `sv1` $< 7 +$ `sv2` after the simulation reaches $t = 3.1$ is `false`, or `action[1]` fires. Serialization of this tree-like representation results in the message listed in Table 1. Such messages are then parsed and interpreted by the back-end Möbius simulation process to determine when to pause the running simulation.

Step Message. The step message allows a user to continue a simulation until immediately after the next event in the future event list fires. This simple message type, which is forwarded from the front-end visualization interface to the back-end Möbius simulation processes, contains no additional parameters. Although this message could be represented explicitly as a breakpoint message combining all action fire events with the `OR` logical operator, this implicit message type is simpler to use and requires less communication overhead to accomplish this frequently useful operation.

Table 1. Breakpoint Message Example

Message	Byte String			
3 BPs	0x00 0x00 0x00 0x03			
off[1]	0x00 0x00 0x00 0x09			
off[2]	0x00 0x00 0x00 0x3a			
ST 4.25	0x06	0x40 0x11 0x00 0x00 0x00 0x00 0x00 0x00		
NOT	0x03	0x00		
AND	0x04	0x00	0x00 0x00 0x00 0x09	
ST 3.1	0x06	0x40 0x08 0xcc 0xcc 0xcc 0xcc 0xcc 0xcd		
<	0x08	0x02	0x00 0x00 0x00 0x09	
sv[2][5]	0x0c	0x01	0x00 0x00 0x00 0x02	0x00 0x00 0x00 0x05
+	0x0a	0x00	0x00 0x00 0x00 0x09	
7	0x0b	0x40 0x1c 0x00 0x00 0x00 0x00 0x00 0x00		
sv[1][4]	0x0c	0x03	0x00 0x00 0x00 0x01	0x00 0x00 0x00 0x04
a[3].fire	0x07	0x00	0x00 0x00 0x00 0x02	

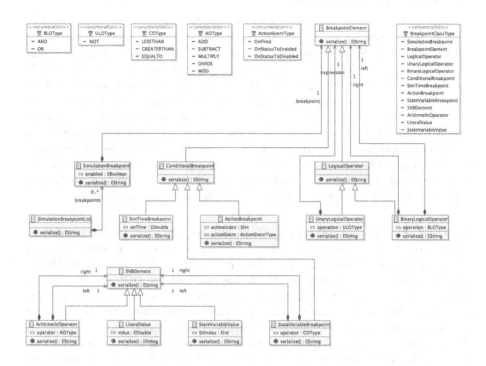

Fig. 3. Breakpoint Message Protocol UML

```
; ( BreakpointList  OR
    ( SimTimeBreakpoint 4.25 )
    ( UnaryOperator NOT
      ( BinaryOperator AND
        ( SimTimeBreakpoint 3.1 )
        ( StateVariableBreakpoint <
          ( StateVariableValue short 1 )
          ( ArithmeticOperator +
            ( LiteralValue 7 )
            ( StateVariableValue double 2 )
          )
        )
      )
    )
  )
  ( ActionBreakpoint OnFired 1 )
; )
```

Fig. 4. Breakpoint Message Protocol OCL

4.3 Front-End Visualization

As discussed in Section 3.5, the front-end visualization interface is designed to replicate the interface of the specific modeling formalism used to create the original model under consideration. For example, if the original model under consideration was created using the SAN modeling formalism, then the MSDV visualization interface should display the running simulation through the SAN model elements of that original model. If the specific modeling formalism has not yet been implemented in the MSDV tool, then the front-end visualization interface should default to the visualization of its parent modeling formalism in the Möbius modeling framework: AFI. In this section, we first discuss the front-end visualization interface of the AFI modeling formalism. Then, we discuss the implementation of the specific modeling formalism SAN to demonstrate how to apply our design paradigm to additional specific modeling formalisms.

Abstract Functional Interface. The AFI, as described in Section 2, is a network of state variables and actions that are connected by `enabling` and `affecting` relationships. To simplify the representation of elements in the MSDV tool, we define them through the UML diagram and the OCL specification in Figures 5 and 6, respectively. In this representation, the visible elements include the `DefaultStateVariable`, the `CustomStateVariable`, the `TimedAction`, the `InstantaneousAction`, and the `Arc` elements. The `StateVariable` and `Action` classes have been split into more specific elements to assist in the specific modeling formalism implementation process. The `Arc` elements represent the `enabling` and `affecting` relationships between the `source` and `target` elements.

To provide useful visual feedback to the user, MSDV can expand an AFI representation to an AFI debugging representation while retaining all information from the parent AFI representation. In a debugging representation, 1) `StateVariable` classes contain the `avgMark`, `minMark`, and `maxMark` fields, which respectively represent the average, minimum, and maximum mark values of the state variable primitive values in the simulation time interval $t = [0, \text{currentSimulationTime}]$; 2) the `AFIModel` class contains the `simTime` field, which represents the current simulation time, t; and 3) `Action` classes contain the `isEnabled`, `timeToFire`, `isNextToFire`, `wasLastToFire`, and `numTimesFired` fields. `isEnabled` indicates whether the action is enabled; `timeToFire` is the simulation time t of the associated event in the future event list. `isNextToFire` indicates whether the associated event is the first element in the future event list. `wasLastToFire` indicates whether the associated event was the last to fire. Finally, `numTimesFired` is the number of times the action fired in the time interval $t = [0, \text{currentSimulationTime}]$. In the corresponding visualization interface, all of those fields are `readonly`, except for the `mark` field of the `StateVariable` classes (for modifying state variable values) and the `timeToFire` field of the `Action` classes (for modifying the event firing times of the future event list).

Stochastic Activity Network. The SAN visualization interface extends the functionality of the AFI visualization interface to represent the data in the

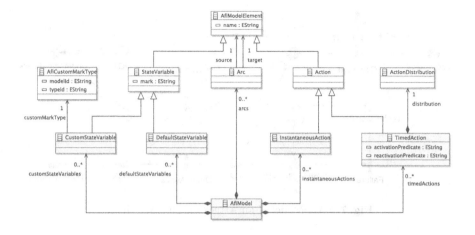

Fig. 5. AFI UML

context Arc **inv**: source . oclIsTypeOf (StateVariable) <>
          target . oclIsTypeOf (StateVariable)
context Arc **inv**: source . oclIsTypeOf (Action) <>
          target . oclIsTypeOf (Action)
context DefaultStateVariable **inv** : CanParseAsShort (mark)
context CustomStateVariable **inv** :
          CanParseAs<customMarkType . toType ( ) >(mark)

Fig. 6. AFI OCL

form of the SAN elements of which the original model is composed. Specifically, the SAN visualization interface is composed of `Place`, `ExtendedPlace`, `TimedActivity`, `InstantaneousActivity`, `Arc`, `InputGate`, and `OutputGate` elements, the first five of which are directly inherited from the AFI `DefaultStateVariable`, `CustomStateVariable`, `TimedAction`, `InstantaneousAction`, and `Arc` elements, respectively. Although the last two SAN elements, `InputGate` and `OutputGate`, affect the model during simulation, they themselves are static elements that do not change during the simulation. Therefore, they can be displayed to the user as static elements, allowing the SAN visualization interface to rely on the AFI parent methods to perform the majority of the necessary functionality.

In addition to representing the model state data as text, the SAN visualization interface also includes the option to display each SAN model element as a visual representation of its current contents. For example, the size of a `Place` visual element is associated with the number of tokens currently contained by the associated `Place` element, increasing as it gains more tokens. Also, each `Activity` visual element is highlighted with a different color if it was the last `Activity` to fire or will be the next `Activity` to fire.

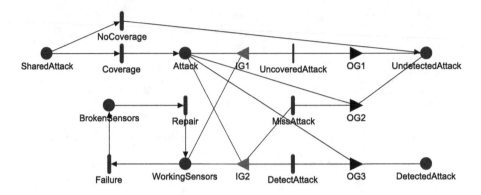

Fig. 7. SAN Model of an Attack on an AMI Meter

5   Case Study of an Attack on an AMI Network

To examine the utility of the additional functionality provided by the MSDV
tool, we explored a SAN representation of an attack on an Advanced Metering
Infrastructure (AMI) smart meter (Figure 7) [15]-[18]. In this model, each intrusion detection sensor (IDS) in the system is represented as a token residing
in either the WorkingSensors or BrokenSensors place, depending on whether
the given IDS is currently operational. Tokens alternate between the two states
through the Failure and Repair activities, which have exponential rates of 0.05
and 0.95, respectively. The SharedAttack place is initialized with a single token
that represents an attack on the AMI network. The token causes a race condition between the NoCoverage and Coverage activities, which indicate whether
the attack occured outside or inside the coverage area of the IDS network. If
the NoCoverage activity fires before the Coverage activity, then the attack token moves directly to the UndetectedAttack place, indicating that the attack
was not detected by the IDS network. If the Coverage activity fires before the
NoCoverage activity, then the attack token moves to the Attack place to be
evaluated by the IDS network. If the attack type is not recognizable to the IDS
network, then the attack token moves directly through the UncoveredAttack
instantaneous activity to the UndetectedAttack place. Otherwise, a race condition between the MissAttack and DetectAttack activities occurs; each of those
activities has an exponential distribution whose rate relies on the number of tokens currently in the WorkingSensors place. If the MissAttack activity occurs
first, signifying that the IDSs in the coverage area are currently broken, then
the attack token moves to the UndetectedAttack place. If the DetectAttack
activity occurs first, then the attack token moves to the DetectedAttack place,
indicating that the attack was detected by at least one of the operational IDSs.

First, consider the case in which the rate of the NoCoverage activity is significantly higher than the rate of the Coverage activity. In that case, the attack
token will immediately move to the UndetectedAttack place in most simulation
batches, leaving the rest of the model unexplored. One way to move the token to

```
( BinaryOperator AND
    ( ActionBreakpoint  OnFired  5)   ; MissAttack->index=5
    ( SimTimeBreakpoint  3.0)
)
```

Fig. 8. Case Study Breakpoint Example

the `Attack` place without changing the model, is to modify the event associated with the `Coverage` activity so that it occurs before the event associated with the `NoCoverage` activity in the future event list. That will force the token to move from the `SharedAttack` place to the `Attack` place to allow the user to examine the rest of the model.

Also, consider the case in which the modeler wants to quickly examine the effects of different numbers of sensors without building a complete model for each test. The modeler can send model state modification messages to the running simulation to update the `Attack` place to contain multiple attacks, and to update the `WorkingSensors` and `BrokenSensors` places to contain different numbers of sensors. The user can then step through the simulation, updating the number of sensors as desired.

For another example, imagine that the modeler would like to examine the numbers of working and broken sensors when the `MissAttack` activity fires after a certain threshold time $t = 3.0$. The modeler would simply define a breakpoint composed of those two conditions, as shown in Figure 8. After the simulation has run, it will continue until the breakpoint condition occurs, allowing the user to examine the entire model state at that simulation time.

6   Conclusion

In this paper, we discuss the benefits of including debugging and visualization capabilities within discrete-event simulations, instead of simply applying them to the original model or final results. We also describe the specific implementation of those capabilities that we enabled by expanding the Möbius discrete-event simulator with the MSDV tool. After introducing the new tool, we describe its usefulness in the context of attack detection in AMI networks.

Acknowledgments. The authors would like to thank Jenny Applequist for her editorial comments. This material is based on research sponsored by the U.S. Department of Homeland Security, under agreement number HSHQDC-13-C-B0014. The U.S. Government is authorized to reproduce and distribute reprints for Governmental purposes notwithstanding any copyright notation thereon.

References

1. Jazouli, T., Sandborn, P., Kashani-Pour, A.: A Direct Method for Determining Design and Support Parameters to Meet an Availability Requirement. International Journal of Performability Engineering 10(2), Paper 09, 211–225 (2014)

2. Andersson, J.: Environmental Impact Assessment using Production Flow Simulation. Research series from Chalmers University of Technology, Department of Product and Production Development: report, ISSN 1652-9243; nr 85

3. Viana, J., Brailsford, S.C., Harindra, V., Harper, P.R.: Combining Discrete-event Simulation and System Dynamics in a Healthcare Setting: A Composite Model for Chlamydia Infection. European Journal of Operational Research (March 2014)

4. Clark, G., Courtney, T., Daly, D., Deavours, D., Derisavi, S., Doyle, J.M., Sanders, W.H., Webster, P.: The Möbius Modeling Tool. In: Proceedings of the 9th International Workshop on Petri Nets and Performance Models, Aachen, Germany, September 11-14, pp. 241–250 (2001)

5. Simul8, http://www.simul8.com/

6. Vensim, http://vensim.com/

7. Williamson, A.L.: Discrete Event Simulation in the Möbius Modeling Framework. Master's Thesis, University of Illinois at Urbana-Champaign (1998)

8. Kuratti, A.: Improved Techniques for Parallel Discrete Event Simulation. Ph.D. Thesis, University of Illinois at Urbana-Champaign (1997)

9. OMNeT++, http://www.omnetpp.org/

10. SAS/OR, http://support.sas.com/rnd/app/or/qsim.html

11. AnyLogic, http://www.anylogic.com/

12. Derisavi, S., Kemper, P., Sanders, W.H., Courtney, T.: The Möbius State-Level Abstract Functional Interface. In: Field, T., Harrison, P.G., Bradley, J., Harder, U. (eds.) TOOLS 2002. LNCS, vol. 2324, pp. 31–50. Springer, Heidelberg (2002)

13. Deavours, D.: Formal Specification of the Möbius Modeling Framework. Doctoral Dissertation, University of Illinois at Urbana-Champaign (2001)

14. Sanders, W.H., Meyer, J.F.: Stochastic Activity Networks: Formal Definitions and Concepts. In: Brinksma, E., Hermanns, H., Katoen, J.-P. (eds.) FMPA 2000. LNCS, vol. 2090, pp. 315–343. Springer, Heidelberg (2001)

15. Berthier, R., Jetcheva, J., Mashima, D., Huh, J., Grochocki, D., Bobba, R., Cárdenas, A., Sanders, W.: Reconciling Security Protection and Monitoring Requirements in Advanced Metering Infrastructures. In: Proceedings of the IEEE International Conference on Smart Grid Communications (SmartGridComm), Vancouver, Canada, October 21-24 (2013)

16. Berthier, R., Sanders, W.H., Khurana, H.: Intrusion Detection for Advanced Metering Infrastructures: Requirements and Architectural Directions. In: Proceedings of the 1st IEEE International Conference on Smart Grid Communications (SmartGridComm), Gaithersburg, Maryland, October 4-6, pp. 350–355 (2010)

17. Cárdenas, A., Berthier, R., Bobba, R., Huh, J., Jetcheva, J., Grochocki, D., Sanders, W.H.: A Framework for Evaluating Intrusion Detection Architectures in Advanced Metering Infrastructures. IEEE Transactions on Smart Grid

18. Grochocki, D., Huh, J., Berthier, R., Bobba, R., Cárdenas, A., Jetcheva, J., Sanders, W.H.: AMI Threats, Intrusion Detection Requirements and Deployment Recommendations. In: Proceedings of the 3rd IEEE International Conference on Smart Grid Communications (SmartGridComm), Tainan City, Taiwan, November 5-8, pp. 395–400 (2012)

Scalar: A Distributed Scalability Analysis Framework

Thomas Heyman, Davy Preuveneers, and Wouter Joosen

iMinds-DistriNet, KU Leuven
3001 Leuven, Belgium
first.last@cs.kuleuven.be

Abstract. Analyzing the scalability and quality of service of large scale distributed systems, such as cloud based services, requires a systematic benchmarking framework that is at least as scalable to sufficiently stress the system under test. This paper summarizes Scalar, our distributed, extensible load testing tool that can generate high request volumes using multiple coordinated nodes. It has support for communication and synchronization between user threads, and built-in node monitoring to detect resource bottlenecks in the benchmark framework deployment itself. Furthermore, it offers highly scalable results analysis that exploits data locality and characterizes the overall system scalability in terms of the Universal Scalability Law.

1 Introduction and Problem Statement

Over the last decade, both the scale of online systems and the degree to which we depend on them has increased tremendously. This makes software qualities such as availability, scalability and performance essential. However, as the scale of a system increases in number of users and complexity, assessing its actual capacity and future scalability potential becomes even harder. The problem is twofold: We need to simulate ever more complex work flows while generating large enough loads to sufficiently stress the system under test.

Workflows become more complex due to the user fulfilling more actions or following more involved business processes. They often also depend on the collaboration of multiple users, which requires inter-user communication and synchronisation in the load generation and benchmarking platform. Similarly, complex workflows might require out of band data processing and a high volume data storage capacity. As the computational overhead increases, care must be taken that the load generator itself does not become the bottleneck. This makes increasing workflow complexity and generating sufficient loads a compound problem.

To solve the problem, the ideal scalability analysis tool would realise the following requirements. First, it needs to explicitly *support multiple concurrent usage scenarios*, and provide statistical breakdowns per scenario. Some distinct usage scenarios are not independent, and users that execute one scenario depend on the actions performed by users in another scenario. Therefore, second, the tool should explicitly *support inter user communication and data exchange*. Third, as

G. Norman and W. Sanders (Eds.): QEST 2014, LNCS 8657, pp. 241–244, 2014.

the complexity of interdependent usage scenarios and the number of simulated users increase, the tools should also *support synchronisation.*

Scalability and performance are two crucial qualities for our ideal scalability analysis tool. Clearly, in order to analyze large scale systems, scalability analysis tools should be highly scalable themselves. This includes both horizontal scalability (i.e., deploying more instances in parallel), as well as vertical scalability (i.e., extensibility by means of plug-ins). As load tests of the envisioned distributed setups easily involve hundreds of thousands of requests per minute, tools should support intelligent results processing that takes data locality into account. When scaling up, care must be taken that the load generation itself is performant enough to not become the bottleneck. To facilitate this, we would need at least a warning mechanism when the tool cannot handle the required load, a way to offload computationally intensive tasks, and a way to find how far the tool can scale on the underlying hardware.

Many load testing tools exist, ranging from load tests embedded in integrated development environments (such as Microsoft Visual Studio) to web testing frameworks with support for distribution (such as The Grinder [1] and Apache JMeter [2]). However, many are lacking inter-user communication and synchronisation facilities, built-in analytics and bottleneck detection, or both. For instance, JMeter has no inter machine communication facility, except for passing static data in configuration files. And although it is fully extensible by means of plug-ins, there is no default support for scalability analysis (e.g., by means of applying the Universal Scalability Law [3]). Similarly, while The Grinder has distributed agents that collate the data and send it back to the coordinator, it does not offer default built-in support for scalability analysis. In the next section, we document Scalar, a highly scalable distributed load generation and benchmarking platform that is developed specifically to support these features.

2   Scalar Architecture

Scalar (`https://distrinet.cs.kuleuven.be/software/scalar/`) is a fully distributed system implemented in Java, and consists of multiple individual, collaborating Scalar instances. Scalar instances automatically discover others, and perform master election. The master coordinates the start of an experiment (i.e., a scalability analysis), which consists of a number of individual runs (i.e., single load tests). A run consists of a lower load warm-up phase, followed by a gradual ramp up to full load, the peak load phase during which statistics are collected, a ramp down phase, and finally another lower load cool down phase. The master collates the results and publishes a scalability report consisting of a quantification of the relative throughput of the system under test in function of user load, as characterized by the Universal Scalability Law, and a statistical breakdown of request residence times and their results.

Representative user behaviour against which the system is to be tested, is encoded in one or more specific user and request types. The abstract User class represents individual simulated users that follow a business flow which encodes the anticipated way in which the system will be used. All scalability analysis results are relative towards that behaviour. Inter-user communication is implemented by means of the blackboard architectural pattern: There is one central data repository, implemented by the DataProvider abstraction, which allows user objects to store and retrieve arbitrary objects. The interface of a DataProvider is similar to that of a map. This abstraction allows for many flexible data provider implementations to be used interchangeably. The default data provider, HazelCastProvider, leverages the underlying HazelCast distributed in-memory database [4], which allows inter-machine communication.

Synchronization is also built on top of the data provider abstraction. A data provider offers both lock(key) and unlock(key) operations, which allows synchronisation of both Scalar instances and user objects on specific key values; the HazelcastProvider leverages the underlying distributed Hazelcast locking mechanisms. As the overall Scalar functionality (including master election, instance discovery, experimental synchronization and results exchange) is built on top of this abstraction, fine tuning the Scalar cluster behavior can be achieved by selecting a correct underlying data provider implementation.

The overall functionality of the Scalar platform can be modified and extended by means of plug-ins. A plug-in is notified of different system events via callback methods: When it is loaded and destroyed, and when the different load testing phases (i.e., warm-up, ramp up, peak load, ramp down, and cool down) take place. This allows plug-ins to perform platform wide initialisation tasks, such as populating the data provider with certain transactions to be executed, configuring the server under test, etc. Similarly, plug-ins can clean up the platform state in between different runs. Plug-ins can also be used to inspect requests—every plug-in receives a call-back for every request that has been executed. This allows plug-ins to perform real-time request analysis and reporting. Plug-ins can use the underlying data providers to store results.

Scalar comes with a number of domain independent plug-ins, such as monitoring the underlying platform resources and visualising results in real time on a web-based dashboard. The most important plug-in for large scale analyses is the ExperimentalResultsPublisher, which handles distributed processing of request data and quantifies the scalability of the system under test in two dimensions. First, it calculates statistics per request type, and provides an overview of the distribution of request type residence times. That allows experimenters to calculate the residence time density function, which provides answers to questions such as "How many requests were handled within 10ms?". Second, the plug-in computes the relative capacity of the system under test for various user loads, and fits the relative capacity data to the Universal Scalability Law. That allows experimenters to extrapolate how many users the system under test would be able to handle under different circumstances. It additionally allows pinpointing of the optimal load point, and provides a precise characterisation of the coherency and serial fraction parameters of that system, as per [3]. Scalar exploits data locality by making every node responsible for calculating aggregate statistics on raw data, and only exchanging these aggregate values with the master node.

3   Discussion and Conclusion

We have presented Scalar, a distributed platform for scalability analysis of large distributed systems. The platform is developed specifically to support complex workflows that involve both intra- and inter-machine communication and synchronisation, and is fully extensible by means of plug-ins. Built-in functionality includes monitoring of the underlying load generating platform, support for data aggregation and analysis by means of the Universal Scalability Law, real-time visualisation via a web based dashboard, and time synchronisation over NTP.

Scalar inherits the scalability of the underlying DataProvider system. In the case of the HazelcastProvider, the underlying system is explicitly designed to scale up to clusters of hundreds of nodes. However, in specialized contexts (e.g., a real-time or embedded domain), it is fairly straightforward to plug in a different communication and synchronisation layer, as the dependency on Hazelcast is not hard coded. Similarly, the distributed statistics aggregation enables longer, high volume experiments involving many Scalar instances.

Scalability analysis is rife with pitfalls. The most common one is that the bottleneck is not the system under test, but the load generation process itself. In order to avoid this, Scalar comes with a number of built-in protection features. First, the tool contains various domain independent test user implementations that can be used to perform a scalability analysis of a Scalar deployment itself, to detect problems early. Second, Scalar will automatically generate warnings when scheduled requests exceed the inter-request waiting time (i.e., the 'think time') by more than 5%. Experience shows that that is a good indicator for detecting bottlenecks internal to the load generation process. Third, the tool comes with built-in resource monitoring of the underlying platform.

Scalar has already been applied successfully to a number of in-house projects, as well as commercial systems. We conclude that it is capable of characterizing both the scalability and quality of service of complex, distributed services. Future work involves automating the instantiation of Scalar for very large cloud-based deployments. That would allow us to achieve scalability analysis as a service.

Acknowledgment. This research is partially funded by the Research Fund KU Leuven.

References

1. Aston, P.: The Grinder, http://htmlunit.sourceforge.net/ (accessed March 6, 2014)
2. The Apache Software Foundation: Apache JMeter, http://jmeter.apache.org/ (accessed February 17, 2014)
3. Gunther, N.J.: Guerrilla capacity planning - a tactical approach to planning for highly scalable applications and services. Springer (2007)
4. Hazelcast, Inc.: The Hazelcast Open Source In-Memory Data Grid, http://www.hazelcast.org/ (accessed March 6, 2014)

Non-intrusive Scalable Memory Access Tracer

Nobuyuki Ohba, Seiji Munetoh, Atsuya Okazaki, and Yasunao Katayama

IBM Research - Tokyo, Kawasaki 2120032, Japan
{ooba,munetoh,a2ya,yasunaok}@jp.ibm.com

Abstract. Memory access tracing is one of the widely used methods to evaluate, analyze, and optimize hardware and software designs. We are developing a non-intrusive, scalable, full-address-range memory tracer. The tracer hardware board is compliant with the JEDEC DDR3 DIMM form factor, and fits in a DIMM slot. It is so compact that we can populate up to 16 tracer boards in a 4-CPU server chassis, and record the commands and addresses of all the memory accesses. Each board drives four SSDs to record the memory access addresses without a break until the SSDs are full. For example, we can make a trace of a full SPECjbb 2005 run, which lasts 26 minutes and generates over 11TB trace data. In addition to recording memory accesses, it collects various types of statistical data, such as a large number of segmented read/write statistics and DRAM bank utilization rates, and displays them on the control dashboard in real time.

Keywords: memory trace, memory system, performance measurement.

1   Introduction

Memory access tracing has been widely used to analyze, evaluate, and optimize memory systems [1]. Memory tracing by snooping DIMM signals was studied by Ban et al. [2]. Many recent servers have two or more CPUs in a chassis, and each CPU has multiple memory channels. It is important to capture all of the memory accesses generated by all of the CPUs for the comprehensive analysis of a memory system. In addition, the tracer should be non-intrusive and have no affect on the target system.

Recently, memory systems using non-volatile memory devices have appeared [3–5]. Non-volatile memory devices have advantages, such as lower per-bit costs, space requirements, and power consumption, but they have different characteristics than DRAM. Some of them need to be erased before they can be written. Some have lower endurance than DRAM, and need wear-leveling. Some take more time to write data. Therefore, it is important to understand how the memory system is accessed to make best use of the advantages of non-volatile memory. More precisely, the access address, frequency, timing, and spatial/time locality are essential information in designing memory systems. Although software simulators that imitate CPUs and memory systems can be used for evaluations, they are too slow to run practical benchmark programs that run longer than a few

G. Norman and W. Sanders (Eds.): QEST 2014, LNCS 8657, pp. 245–248, 2014.

minutes on real machines. Trace data obtained by running real applications and OSes on real servers are crucial to obtaining deep insight into memory systems.

We are developing a non-intrusive, scalable, full-address-range memory tracer for comprehensive analyses of memory systems. This is one of the tools of a memory system analyzer we are currently developing, as shown in Fig. 1. This paper presents the hardware and software tools of the tracer.

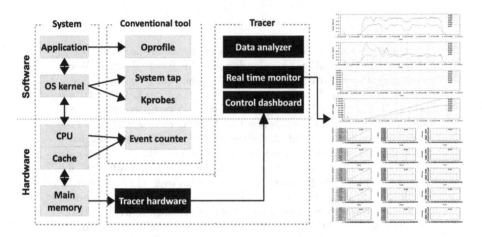

Fig. 1. Conceptual view of the memory system analyzer

2   Hardware

To implement a non-intrusive, scalable, full-address-range memory tracer, we focused on the ease of hardware installation and high link bandwidth from the tracer board to external storage. The tracer hardware uses the JEDEC DDR3 DIMM form factor, and acts like a standard 16GB 1333MHz DDR3 memory module. It consists of a base board and a power module, as shown in the left side of Fig. 2. The base board has a Xilinx FPGA, 18 8-Gb DDR3 DRAMs, and auxiliary components, such as crystal oscillators and an FPGA configuration PROM. The power module, which is piggybacked on the base board, generates all of the power lines for the FPGA and auxiliary components from the standard 1.5-V power supply available at the DIMM edge connector.

The FPGA on the board probes the DRAM command and address signals, adds timestamps at the DDR clock resolution, and directly stores the trace data to four SSDs via a custom-made flexible SATA cable. To sustain the SSD's maximum write speed, the trace data are always recorded sequentially in the SSDs, which are configured using sector-level striping (RAID0). To minimize the file management overhead, the set of trace data from each run is stored in an individual partition managed by a GUID Partition Table (GPT) [6].

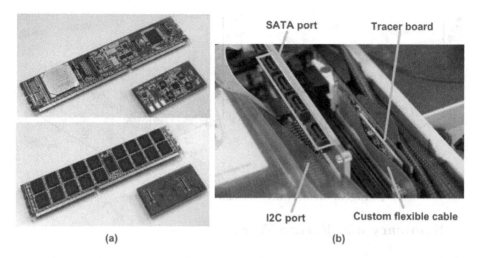

Fig. 2. (a) Front and back views of the base board and power module, (b) Tracer board with custom flexible cable

The board has an I2C port, with which we control the tracing operations, such as starting or stopping traces. The I2C port is also used to collect real-time statistical data. We can install up to 16 tracer boards in a 4-CPU server chassis, and simultaneously start and stop the tracing operations by using broadcast commands on the I2C bus. The maximum skew of the start and stop times among 16 tracer boards is 20 x DDR clocks.

3   Software Tools

The tracer software tools are Control dashboard, Real time monitor, and Data analyzer, as shown in Fig. 1. Control Dashboard is an HTML5 based GUI, with which the user can start/stop the trace, monitor the status of the system (e.g., temperature and voltage), and manage the trace data stored in the SSDs. Real time monitoring allows the user to see various statistical data, such as a large number of segmented read/write statistics and the number of active DRAM banks, in real time. Data analyzer is an off-line tool, which reads the trace data stored in the SSDs and analyzes it. Fig. 3 is an example of the off-line analysis results obtained by running SPECjbb [7] warehouse 8 on a 2-CPU CentOS-6.4 server with 112GB of memory. The figure shows the numbers of read and write accesses per DRAM page on the first memory channel of the first CPU during the 63 second run. In the figure there are several sharp peaks, which indicate that the memory accesses have a great deal of time and spatial locality, and therefore we have to track the wear of the memory cells if non-volatile memory is used for the main memory.

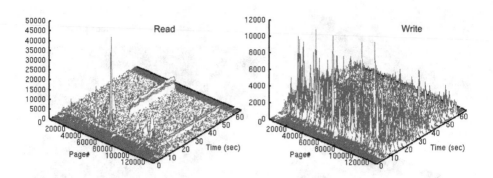

Fig. 3. Memory access counts per DRAM page

4   Summary and Future Work

We devised a memory tracer that allows us to capture full memory access traces from real applications running on a real OS. We are collecting traces in various situations and analyzing them for the design of future memory systems. The integration of the tracer with conventional tools is continuing.

Acknowledgments. We would like to express our special thanks to Tadayuki Okada, Makoto Ono, Joe Jakubowski, Dan Colglazier, Maurice Bland, and Chris Floyd for their support in various phases of the project.

References

1. Uhlig, R.A., Mudge, T.N.: Trace-Driven Memory Simulation: A Survey. Computing Surveys (CSUR) 29(2) (1997)
2. Ban, Y., Chen, M., Ruan, Y., Lin, L., Fan, J., Yuan, Q., Song, B., Xin, J.: HMTT: A Platform Independent Full-System Memory Trace Monitoring System. In: ACM SIGMETRICS International Conference on Measurement and Modeling of Computer Systems, pp. 229–240 (2008)
3. Wu, M., Zwaenepoel, W.: eNVy: a non-volatile, main memory storage system. In: ASPLOS VI, pp. 86–97 (1994)
4. Mogul, J.C., Argollo, E., Shah, M., Faraboschi, P.: Operating System Support for NVM+DRAM Hybrid Main Memory. In: 12th Conference on Hot Topics in Operating Systems, HotOS 2009 (2009)
5. Ramos, L., Gorbatov, E., Bianchini, R.: Page Placement in Hybrid Memory Systems. In: International Conference on Supercomputing, ICS 2011, pp. 85–95 (2011)
6. Intel white paper: GUID Partition Table (GPT), http://download.intel.com/support/motherboards/server/sb/gpt_white_paper_1_0.pdf
7. Standard Performance Evaluation Corporation: SPECjbb 2005 (2005), http://www.spec.org/jbb2005

Probabilistic Programming Process Algebra

Anastasis Georgoulas[1], Jane Hillston[1],
Dimitrios Milios[1], and Guido Sanguinetti[1,2]

[1] School of Informatics, University of Edinburgh, UK
[2] SynthSys — Synthetic and Systems Biology, University of Edinburgh, UK

Abstract. Formal modelling languages such as process algebras are widespread and effective tools in computational modelling. However, handling data and uncertainty in a statistically meaningful way is an open problem in formal modelling, severely hampering the usefulness of these elegant tools in many real world applications. Here we introduce ProPPA, a process algebra which incorporates uncertainty in the model description, allowing the use of Machine Learning techniques to incorporate observational information in the modelling. We define the semantics of the language by introducing a quantitative generalisation of Constraint Markov Chains. We present results from a prototype implementation of the language, demonstrating its usefulness in performing inference in a non-trivial example.

1 Introduction

Stochastic process algebras are an established tool for modelling and analysing the behaviour of dynamical systems, combining theoretical elegance with a range of attractive and practically useful features — compositionality, formal interpretation of models and the ability to verify their behaviour using model-checking, to list a few. The starting point for process algebras, as for many other formal modelling methods, is a full specification of the system being modelled, both in terms of interaction structure and of parameters quantifying the (infinitesimal) dynamics of the system. It is increasingly clear, however, that such complete knowledge is seldom achieved, and unattainable in a number of important application domains. This problem is particularly acute in the biomedical field, where parameters such as reaction rates are estimated from noisy measurements which are often highly sensitive to the experimental conditions. How to quantify and propagate this uncertainty in formal models is an open problem of fundamental importance in any scientific application.

Constraining models from empirical observations is a very large and active research field in machine learning and signal processing; however, these fields usually work directly with low-level mathematical descriptions of the system, which negate some of the major advantages of formal languages. Within the process algebra community, attempts to address this problem have mostly used greedy optimisation methods — the Evolving Process Algebra framework ([17,18]), for instance, uses evolutionary computation algorithms to fit the parameters and

G. Norman and W. Sanders (Eds.): QEST 2014, LNCS 8657, pp. 249–264, 2014.

structure of a process algebra model. While this and similar approaches (further discussed in Section 7) are valuable contributions, they do present practical and conceptual limitations. First and foremost, the optimization methods used lack a statistical framework, and hence cannot quantify the uncertainty associated with their predictions. In general, optimisation methods return a single optimal value of the parameters: this implies an assumption that measurements are sufficient to *completely remove* uncertainty, clearly an untenable assumption. Secondly, the learning and the modelling take place on separate levels, since the modelling language itself does not really include the uncertainty about the system. Thus, the model does not reflect our understanding, and inference is done in an *ad hoc* manner, independently from the modelling. While this orthogonality may appear attractive, it introduces a degree of conceptual dissonance between what we try to capture and the way we represent it.

In this paper we aim to address these problems by introducing a formal modelling language which directly incorporates observations (and the associated uncertainty) and can leverage cutting-edge statistical machine learning tools to perform inference and quantify uncertainty. We are inspired by recent progress in probabilistic programming languages, which aim to perform *inference by programming*; however, current probabilistic programming languages are all relatively low-level. We introduce ProPPA, a Probabilistic Programming Process Algebra; to our knowledge, this is the first time the probabilistic programming paradigm is extended to a higher-level, formal system description language like a process algebra. Note that the ability to perform inference from data qualitatively distinguishes our approach from general stochastic modelling methodologies such as stochastic process algebras, which simply incorporate uncertainty in model evolution through the use of random variables to determine rates.

ProPPA is based on the stochastic process algebra Bio-PEPA [8], and inherits many of its qualities. We show how to include uncertainty in the definition of the language (Section 3), and propose an appropriate semantic model for uncertain models (Sections 4 and 5). We adopt a modular approach to construct our language, so that the core language is capable of adopting different machine learning methodologies to perform inference from possibly very different types of data. We demonstrate the power of this approach by performing inference in a nontrivial example in Section 6.

2   Background

This section gives some information on the language on which ProPPA is based, the frameworks and mathematical objects used for the definition of its semantics, and the field of probabilistic programming from which we draw inspiration.

2.1   Process Algebras and Bio-PEPA

Process algebras are a family of languages first used to model concurrent systems, by specifying the system's components and the actions that these may perform.

The formal nature of the languages allow one to reason about the behaviour of the modelled system, such as verifying that undesirable states (configurations of the system) are avoided or that simple properties hold. The sub-family of stochastic process algebras (e.g. PEPA [15], IMC [14], EMPA [1]) extend this framework by introducing time into the system and assuming that the time for a transition to occur is an exponentially distributed random variable. The parameter of the distribution is called the rate of the transition, and, when multiple transitions are possible, the probability of choosing a particular transition is proportional to its rate. This is formalized through the concept of a Continuous Time Markov Chain (CTMC), a mathematical description of the possible states of the system and the transitions between them.

The description of a system in a process algebra can be used to implicitly generate its state space, in the form of a labelled transition system (LTS). A LTS is a graph whose nodes are the system's states and whose edges are the possible transitions between states, labelled with some information (e.g. what reaction causes the transition or at what rate the transition occurs). Analysing the LTS can give important insights into the behaviour of the system, such as whether a state is reachable under certain conditions or within a specified time frame. Verifying whether a system description satisfies such properties is the subject of model checking algorithms and tools, with the properties to be checked often being expressed in a temporal logic, such as CSL [2] or CTL [9].

Bio-PEPA [8] is a stochastic process algebra based on PEPA but designed for the modelling of biological processes. In Bio-PEPA, system components (termed *species*) are defined through their behaviour, that is, how they interact with each other, reflecting a reagent-centric modelling style. The definition of a species takes the form

$$A = (\alpha_1, k_1)op_1 + \cdots + (\alpha_n, k_n)op_n \text{ where } op_i = \downarrow, \uparrow, \oplus, \ominus \text{ or } \odot$$

which means that species A takes part in reaction α_i with stoichiometry k_i. The different options for op_i correspond to different roles of A in α_i: reactant, product, catalyst, inhibitor or generic modifier, respectively. These definitions are composed using the choice operator $(+)$ to describe species that can take part in multiple reactions.

Each reaction has an associated rate law, which can be specified either as a formula or using a predefined law (such as mass-action or Hill kinetics). Parameters can be defined and used, for example, in kinetic laws or as initial concentrations, but their values must be specified and are considered fixed. The language results in a modular or compositional approach, wherein the behaviour of the system emerges as a direct consequence of the behaviour of the species (without the need to, for instance, explicitly write out ODEs or chemical equations for reactions, as they can be automatically computed). This means modifications to the model can be performed by changes to the "local" species definitions.

Formally, a Bio-PEPA system is defined as a tuple $\langle \mathcal{V}, \mathcal{N}, \mathcal{K}, \mathcal{F}_R, Comp, P \rangle$, where $\mathcal{V}$ is the set of compartments (locations) in the system; $\mathcal{N}$ is a set of quantities associated with each species, such as its maximum concentration; $\mathcal{K}$ is the set of parameters; $\mathcal{F}_R$ is the set of rate laws; $Comp$ is the set of sequential components (species definitions); and P is the model component, which describes how the various species cooperate with each other as well as their initial concentrations. An example of a model component comprising three species A_i with initial quantities l_i is

$$A_1[l_1] \bowtie_* A_2[l_2] \bowtie_* A_3[l_3]$$

2.2   FuTS

The FuTS (state-to-function transition system) framework [20] is a way of describing semantics of process algebras. In a FuTS, the possible transitions from a state s can be represented collectively as $s \xrightarrow{\alpha} f$, where f is called the _continuation_ and is a function over states. The value of $f(s')$ then gives some information (such as the rate) about the transition $s \xrightarrow{\alpha} s'$. Depending on the codomain of the continuation functions, FuTS can represent different kinds of behaviour and their associated information, such as non-determinism (continuations take boolean values to denote possible next states), discrete time systems (values in $[0, 1]$ give transition probabilities) or continuous time systems (values in $\mathbb{R}$ to denote transition rates). The notation $[s_1 \mapsto v_1, s_2 \mapsto v_2, \ldots, s_n \mapsto v_n]$ is shorthand for a function f such that $f(s_i) = v_i$, $i = 1 \ldots n$ and $f(s)$ takes the zero value of its codomain (0 for real values, _false_ for boolean etc.) for all states besides the specified s_i. As ProPPA represents uncertainty using probability distributions, the FuTS style, which already makes use of functions, seems a natural fit for expressing its semantics.

2.3   Constraint Markov Chains

Constraint Markov Chains (CMCs, [6]) are a generalisation of Discrete Time Markov Chains in which the probability of transitioning from a given state to another does not have a fixed value. Instead, the CMC specifies a constraint that the values of the various transition probabilities must obey or, equivalently, a set of acceptable values for them. Formally, a Constraint Markov Chain is a tuple $\langle S, o, A, V, \phi \rangle$, where:

- S is the set of states, of cardinality k.
- $o \in S$ is the initial state.
- A is a set of atomic propositions.
- $V : S \to 2^{2^A}$ gives a set of acceptable labellings for each state.
- $\phi : S \times [0, 1]^k \to \{0, 1\}$ is the _constraint function_.

The constraint function indicates whether a given set of transition probabilities satisfies the constraints: $\phi(i, \boldsymbol{r}) = 1$ if and only if $\boldsymbol{r} = (r_1, r_2, \ldots, r_k)$ is an acceptable vector of transition rates from state i.

As explained in [24], there are two ways of interpreting the uncertainty in the transition probabilities, which give rise to different behaviours when simulating a CMC. One way is to assume that the transition probabilities can change during the run of the system, so that each time we visit a state we must choose new values for them. This is referred to as Markov Decision Process (MDP) semantics. Alternatively, under the Uncertain Markov Chain (UMC) semantics, we assume that each probability has a constant (but unknown) value. In this case, the values are fixed before the simulation and maintained throughout.

2.4 Probabilistic Programming

Probabilistic programming is a framework for reasoning about uncertain processes in a statistically consistent manner. In a probabilistic program, uncertain aspects of the system, such as unknown parameters, are treated as random variables and can be assigned probability distributions that express this uncertainty. Additionally, one can specify observations of the system, from which information about the unobserved aspects can be gleaned. In other words, the program specifies a probability distribution, which can be viewed in two ways: one can sample from it, essentially simulating the system; or, if one has additional information about the system, one can condition the distribution on this data, inferring an updated distribution over the unknown variables that takes into account this new knowledge. Probabilistic programming offers an elegant approach for treating uncertain systems in these two ways, automating the process to a degree and eliminating the need for bespoke inference solutions, as the inference algorithm can be configured and executed automatically based on the structure of the program.

Previous work has focused mainly on integrating the paradigm into traditional programming languages, giving rise to frameworks like Church [13], IBAL [21] and Infer.NET [19]. These languages, however, describe systems at a low level: one must explicitly specify all the statistical dependences between the different variables, yielding potentially large descriptions which are difficult to manage. This limits the range of systems that can be modelled, with continuous-time dynamical systems being particularly hard or even impossible to deal with. We therefore advocate combining the principle of probabilistic programming with a formal language like a process algebra, for a flexible, high-level framework in which to model and analyse complex systems with uncertain aspects.

3   A Probabilistic Programming Process Algebra

The ProPPA syntax is based on Bio-PEPA, with the addition of two key features that introduce aspects of probabilistic programming. The first concerns the representation of uncertainty in the system. We should note that we are only considering uncertainty in the kinetics, and assume that we fully know what reactions each species can take part in. In Bio-PEPA, parameters can be used in the definition of kinetic rate functions, but their values must be fixed. With this in mind, we allow uncertain parameters, whose values are given as probability distributions rather than concrete numbers. The second feature is a way of incorporating information about the behaviour of the system into the model. These will be the *observations*, which may take the form of actual partial observations of the state of the system (as a time series) or, more generally, could be any observed function of the specific trajectory of the system (specified through a temporal logic formula, for instance).

On a formal level, we will need to modify the definition of a Bio-PEPA system (given previously in Section 2.1), mainly by reconsidering the role of the set of parameters $\mathcal{K}$. We extend the system definition in two ways, corresponding to the two features described above. Firstly, since the uncertain quantities are represented as parameters in the model, we extend the definition of a parameter to include a distribution rather than a concrete value. These are the *prior distributions* or *priors* over parameters, which express our belief about a parameter's values before seeing any data. The set of parameters $\mathcal{K}$ is then partitioned into two subsets: $\mathcal{K}_c$ comprises the concrete parameters, while $\mathcal{K}_u$ contains the uncertain ones, along with the priors associated with them. We write $(k \sim \mu) \in \mathcal{K}_u$ if the parameter k is drawn *a priori* from the distribution μ. Importantly, the functional rates $\mathcal{F}_R$ can refer to any parameter in $\mathcal{K}_u$ as well as those in $\mathcal{K}_c$; in this sense, a functional rate can represent a family of functions.

Secondly, we add a new component $\mathcal{O}$ representing the observations. These impose restrictions on the acceptable parameter values and modify our belief about their distribution, as described in more detail in Section 5.4. Extending the syntax to accommodate these features is straightforward. A ProPPA system is therefore a tuple $\langle \mathcal{V}, \mathcal{N}, \mathcal{K}_c, \mathcal{K}_u, \mathcal{F}_R, \mathcal{O}, Comp, P \rangle$, with the other components retaining the meaning they have in Bio-PEPA. Following the terminology of [11], we will also write a system as $\langle \mathcal{T}, P \rangle$ where $\mathcal{T} = \langle \mathcal{V}, \mathcal{N}, \mathcal{K}_c, \mathcal{K}_u, \mathcal{F}_R, \mathcal{O}, Comp \rangle$ is called the context.

3.1   A Rumour Spreading Example

As an example of a ProPPA model, we will consider a population CTMC model of rumour spreading over a network [10]; this consists of three types of agents (Figure 1). A spreader (S) is someone who has already encountered the rumour and is actively trying to spread it. When an ignorant (I) meets a spreader, the ignorant also becomes a spreader. When two spreaders meet, one of them stops spreading and becomes a blocker (R), reflecting the idea that only new rumours are worth spreading. A blocker can then convert spreaders into other blockers. The dynamics of the system can exhibit qualitatively different behaviours depending on the parameter values: in particular, two possible steady state regimes exist, where all agents are in blocker state, or where a blocker and an ignorant population coexist.

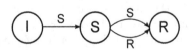

Fig. 1. State transitions of a rumour-spreading agent. The arrow labels indicate the type of agent that must be encountered for the corresponding transition to take place.

```
1   k_s = Uniform(0,1);
2   k_r = Uniform(0,1);
3
4   kineticLawOf spread : k_s * I * S;
5   kineticLawOf stop1 : k_r * S * S;
6   kineticLawOf stop2 : k_r * S * R;
7
8   I = (spread,1) ↓ ;
9   S = (spread,1) ↑ + (stop1,1) ↓ + (stop2,1) ↓ ;
10  R = (stop1,1) ↑ + (stop2,1) ↑ ;
11
12  I[10] ⋈ S[5] ⋈ R[0]
13
14  observe('trace')
15  infer('ABC')
```

Fig. 2. ProPPA model of the rumour spreading example

Figure 2 shows the description of the system in ProPPA for an initial population of 15 agents. The behaviour of the different agents is described in lines 8-10: line 8 says that the count of ignorants is decreased by 1 when the **spread** interaction occurs, and the other types of agents are similarly defined by the changes to their count through the various interactions. Line 12 shows the initial population of each kind of agent; the cooperation ⋈ means that the agents synchronise on all shared reactions. We assume that these interactions happen at rates that obey mass-action kinetics, i.e. they are proportional to the count of the agents involved. We also assume that the rate constants for the spreader-spreader and spreader-blocker interactions are the same (k_r), while the spreader-ignorant interaction has a rate constant k_s; this is shown in lines 4-6. The definition of k_s and k_r as uniformly distributed (lines 1-2) reflects our prior belief that, without seeing any data, they are not biased towards any value in their domain, which in this case we chose to be $[0,1]$. We will use this model as a running example for the rest of this paper; lines 14-15 are discussed in Section 5.4.

4   Probabilistic Constraint Markov Chains

As mentioned earlier, Bio-PEPA models can be mapped to CTMCs. In a CTMC, every transition between states has a concrete rate, making this interpretation unsuitable for a language with uncertainty, such as ProPPA. We must therefore use a different object to define the semantics of our language, one that is more suited to describing uncertain models, such as a CMC.

The way CMCs were first proposed (shown in Section 2.3) presents two limitations. Firstly, they have been defined only for discrete-time systems, whereas we are interested in modelling in continuous time. Their definition can be adapted to the continuous-time domain through simple alterations which are presented below. Secondly, while a CMC defines the set of possible values for a rate, it gives no information on the relative likelihood of those values. We would like to move from a purely non-deterministic to a probabilistic setting, where, instead of a binary decision, we have quantitative information about our belief in the plausibility of a value.

Based on the original definition, we can define a Probabilistic Constraint Markov Chain as a tuple $\langle S, o, A, V, \phi \rangle$, where:

- S is the set of states, of cardinality k.
- $o \in S$ is the initial state.
- A is a set of atomic propositions.
- $V : S \rightarrow 2^{2^A}$ gives a set of acceptable labellings for each state.
- $\phi : S \times [0, \infty)^k \rightarrow [0, \infty)$ is the *constraint function*.

The changes concern the constraint function and address the limitations described. The constraints are now on rates rather than transition probabilities, reflecting the shift to continuous time. Additionally, ϕ now describes a probability density function, therefore it takes values in $\mathbb{R}_+$ instead of $\{0, 1\}$, with the additional restriction that $\int_0^\infty \cdots \int_0^\infty \phi(i, \boldsymbol{r}) d\boldsymbol{r} = 1$ for every $i \in \{1, k\}$. The resulting object is richer, and the additional information it can capture means we can use Probabilistic CMCs to define the semantics of ProPPA models, as explained in the next section.

5   ProPPA Semantics

We now describe the semantics of a ProPPA model, eventually mapping to a Probabilistic CMC. The semantics of Bio-PEPA are given in terms of two relations. The *capability relation* describes what transitions may occur between states, without giving any quantitative information about the rates — in other words, it gives the structure of the transition system. The *stochastic relation* uses that information, as well as the definition of the kinetic functions, to provide the rates of the transitions, thus completing the labelling of the transition system.

We have kept this two-step approach, and have in fact found the separation of the two steps to be useful for our extension. Since there is no uncertainty in the qualitative behaviour of the species, the capability relation remains unchanged by our additions. We present both the existing capability and the new stochastic relation in a uniform way using the FuTS framework.

5.1   Capability Relation

To consider the capability relation as a FuTS, we use boolean-valued continuations: $P \overset{l}{\rightarrowtail}_c f$ means that P can transition to those states Q for which $f(Q)$ is *true*. For simple terms, there is at most one reachable state for any reaction, as described by the rules below, where op is one of the modifier operators $\oplus, \ominus$ and $\odot$, and N is the maximum count for the species S:

$$\text{PrefixReac } (a,k) \downarrow S(l) \xrightarrow{(a,[S:\downarrow(l,k)])}_c [S(l-k) \mapsto \text{true}] \quad k \leq l \leq N$$

$$\text{PrefixProd } (a,k) \uparrow S(l) \xrightarrow{(a,[S:\uparrow(l,k)])}_c [S(l+k) \mapsto \text{true}] \quad 0 \leq l \leq N-k$$

$$\text{PrefixMod } (a,k) \text{ op } S(l) \xrightarrow{(a,[S:op(l,k)])}_c [S(l) \mapsto \text{true}] \quad \begin{cases} k < l \leq N \text{ if } op = \oplus \\ k \leq l \leq N \text{ otherwise} \end{cases}$$

The label (α, w) records two kinds of information: α is the reaction name, while w is a list of *roles*, indicating how a species takes part in a reaction, with what stoichiometry and at what initial count. The rules for the choice and cooperation operators are simple:

$$\text{Choice1} \frac{P_1 \xrightarrow{(a,w)}_c f}{P_1 + P_2 \xrightarrow{(a,w)}_c f} \qquad \text{Choice2} \frac{P_2 \xrightarrow{(a,w)}_c f}{P_1 + P_2 \xrightarrow{(a,w)}_c f}$$

$$\text{Coop1} \frac{P_1 \xrightarrow{(a,w)}_c f_1 \quad a \notin \mathcal{L}}{P_1 \bowtie_{\mathcal{L}} P_2 \xrightarrow{(a,w)}_c g_1} \qquad \text{Coop2} \frac{P_2 \xrightarrow{(a,w)}_c f_2 \quad a \notin \mathcal{L}}{P_1 \bowtie_{\mathcal{L}} P_2 \xrightarrow{(a,w)}_c g_2}$$

$$\text{Coop3} \frac{P_1 \xrightarrow{(a,w_1)}_c f_1 \quad P_2 \xrightarrow{(a,w_2)}_c f_2 \quad a \in \mathcal{L}}{P_1 \bowtie_{\mathcal{L}} P_2 \xrightarrow{(a,w_1::w_2)}_c g}$$

where :: indicates concatenation of lists and we make the usual distinction according to whether the reaction in question is shared between the cooperating components. The functions g_1, g_2 and g are as follows:

$$g_1(Q) = \begin{cases} f_1(Q_1) & \text{if } Q = Q_1 \bowtie_{\mathcal{L}} P_2 \\ false & \text{otherwise} \end{cases} \qquad g_2(Q) = \begin{cases} f_2(Q_2) & \text{if } Q = P_1 \bowtie_{\mathcal{L}} Q_2 \\ false & \text{otherwise} \end{cases}$$

$$g_1(Q) = \begin{cases} f_1(Q_1) \wedge f_2(Q_2) & \text{if } Q = Q_1 \bowtie_{\mathcal{L}} Q_2 \\ false & \text{otherwise} \end{cases}$$

Finally, for completeness, we have a rule for named species definition:

$$\text{Constant} \frac{P \xrightarrow{(a,w)}_c f \quad Q \stackrel{def}{=} P}{Q \xrightarrow{(a,w[P \to Q])}_c f}$$

where $w[P \to Q]$ means renaming all instances of P in w to Q.

For example, in the rumour-spreading network (Figure 2), we have that $S(2) \xrightarrow{(stop2,[S:\downarrow(2,1)])}_c [S(1) \mapsto \text{true}]$ and $R(1) \xrightarrow{(stop2,[R:\uparrow(1,1)])}_c [R(2) \mapsto \text{true}]$ from the prefix rules. By applying the cooperation rules, we can infer that

$$I(0) \bowtie_* S(2) \bowtie_* R(1) \xrightarrow{(stop2,w)}_c \left[I(0) \bowtie_* S(1) \bowtie_* R(2) \mapsto \text{true} \right]$$

where $w = [S :\downarrow (2,1), R :\uparrow (1,1)]$.

5.2 Stochastic Relation

The capability relation is defined, as above, between species. The stochastic relation, in contrast, is between whole systems and gives information on the rate of transitioning from one system to another. In Bio-PEPA, the reaction rates depend on the model's parameters, and this dependence is also true in ProPPA. This means that any uncertainty about the values of the parameters must lead to uncertainty about the rates. The ProPPA stochastic relation, therefore, gives a distribution over possible rates instead of a single value. If all the parameters on which a particular rate depends are concrete, this will simply be a Dirac δ distribution, i.e. one where all the probability mass is assigned to a single value.

When building the stochastic relation, we use both the capability relation, which indicates if a state is reachable, as well as the system's context $\mathcal{T}$ (see Section 3), which holds information related to the reaction rates. The context provides two pieces of information we require: the kinetic law for the reaction, and the distribution of the uncertain parameters involved in that law. As the rate normally depends on the concentrations of the species involved, we also need this information, which is contained in the roles w, in the label of the capability relation. The stochastic relation is defined by the rule:

$$\frac{P \xrightarrow{\ (a,w)\ }_c g}{\langle \mathcal{T}, P \rangle \xrightarrow{\ a\ }_s h_{g,w,\mathcal{T}}}$$

As described above, the function h maps systems to distributions of rates:

$$h_{g,w,\mathcal{T}}(\langle \mathcal{T}', s' \rangle) = \begin{cases} \delta(0) & \text{if } \mathcal{T}' \neq \mathcal{T} \\ \delta(0) & \text{if } g(s') = \textit{false} \\ \mu & \text{otherwise} \end{cases}$$

According to this, transitions to systems with a different context or where the species is not reachable can only occur at zero rate, i.e. never. We now show how to derive the distribution over rates (μ above) in the non-trivial cases.

Assume the rate Y of a reaction depends on a parameter Θ and let $Y = T(\Theta)$ express this dependence. We know, from the context, that Θ is distributed according to a probability density function (pdf) f_Θ. Y, being a transformation of Θ, will also follow a distribution, whose pdf we denote f_Y. If the function T is strictly monotonic, f_Y can be obtained through a simple change of variable:

$$f_Y(y) = \left| \frac{dT^{-1}(Y)}{dY} \right|_{Y=y} f_\Theta(T^{-1}(y)) \tag{1}$$

where $T^{-1}(y)$ is the (unique) value of the parameter Θ for which the rate is y.

To see why this is valid, let us first consider the case where T is strictly increasing, which implies that the inverse function T^{-1} is well-defined and is also increasing. The cumulative distribution function of Y is then:

$$F_Y(y) = P(Y \leq y) = P(T(\Theta) \leq y) = P(\Theta \leq T^{-1}(y)) = F_\Theta(T^{-1}(y))$$

and the corresponding pdf is:

$$f_Y(y) = \left.\frac{dF_Y(y)}{dY}\right|_{Y=y} = \left.\frac{dF_\Theta\left(T^{-1}(Y)\right)}{dY}\right|_{Y=y} = f_\Theta\left(T^{-1}(y)\right)\left.\frac{dT^{-1}(Y)}{dY}\right|_{Y=y}$$

Considering the case where T is decreasing leads to the general result (1).

Once the stochastic relation has been constructed, it is then easy to build a Probabilistic CMC that captures its behaviour by appropriately defining the constraint function. To do so, we need the probability density of a vector of rates, which can be obtained as the product of the densities for each individual rate. The latter are given directly by the stochastic relation. Note that there can be more than one reaction that leads to the same transition in the state space; the total transition rate is then the sum of the rates of the individual reactions, and the pdf of the sum of random variables is the convolution of their individual pdfs. For a system with k states, then, the constraint function ϕ over the rates r_{ij} of transitioning from state i to state j is given by:

$$\phi(i, (r_{i1}, r_{i2}, \ldots, r_{ik})) = \prod_{j=1}^{k} f_j(r_{ij})$$

where $\langle T, i \rangle \xrightarrow{\alpha}_s f_\alpha$ and $f_j = \bigotimes_\alpha f_\alpha(\langle T, j \rangle)$ is the convolution described above.

5.3 Concretization

We now describe how the choice of UMC or MDP interpretation (discussed in Section 2.3) affects the behaviour of the system. Essentially, in the UMC setting we replace the prior distribution over each parameter with a Dirac δ distribution, to reflect the fact that a value is chosen only once and remains fixed thereafter. First we note that the concretization procedure only affects the context of a system, so the corresponding transition relation will have the form $T \rightarrowtail g_X(T)$, where T is a context (as before) and X can be UMC or MDP. The corresponding continuation functions for each scenario are defined as:

$$g_{\text{UMC}}(\langle \mathcal{V}, \mathcal{N}, \mathcal{K}_c, \mathcal{K}_u, \mathcal{F}_R, \mathcal{O}, Comp \rangle)(T') =$$

$$\begin{cases} \prod_i f_i(v_i) & \text{if } T' = \langle \mathcal{V}, \mathcal{N}, \mathcal{K}_c, \{k_i \sim \delta(v_i)\}, \mathcal{F}_R, \mathcal{O}, Comp \rangle \\ 0 & \text{otherwise} \end{cases}$$

and

$$g_{\text{MDP}}(T)(T') = \begin{cases} 1 & \text{if } T' = T \\ 0 & \text{otherwise} \end{cases}$$

5.4   Observations and Inference

We have shown (Section 5.2) how the prior beliefs about the parameters induce a distribution over rates, but we have so far assumed that no observations are present. Observations represent new information which can affect our belief about how likely different values are. Inference can then be thought of as a transformation of the context, which takes the observations into account and updates the distributions of the uncertain parameters accordingly. From a probabilistic point of view, this corresponds to conditioning the prior distribution $P(\theta)$ on the observed data D, obtaining the *posterior distribution* $P(\theta \mid D)$. The relation between these quantities is given by Bayes' Theorem:

$$P(\theta \mid D) \propto P(\theta)P(D \mid \theta)$$

where $P(D \mid \theta)$ is the likelihood of the data, a measure of how likely we are to see these observations for a particular assignment of values to the parameters.

This view does not give any information on *how* to perform inference, however. Indeed, it is generally not possible to calculate the likelihood or the posterior analytically, so approximations must be used. The specification of the language allows for some freedom in the inference implementation; this approach creates a modular framework that can employ different algorithms, some examples of which are given in the next section. The choice may depend in part on the type of observations available, and here we focus on two possible ways of specifying those: a time-series of measurements of the species in the system, or a set of temporal logic formulae, which describe properties of the system's behaviour. A statement of the form infer(*algorithm*) in the model sets the desired inference algorithm.

In line 14 of Figure 2, the observe statement refers to an external file which holds our observations of the system. Line 15 specifies what inference algorithm will be used.

6   Inference

In this section, we illustrate the capabilities of the language by showing some ways we can perform inference on the model of Figure 2.

6.1   Inference from Time-Series Observations

We first assume that our observations are measurements of the species at different times during a run of the system. We can use an Approximate Bayesian Computation (ABC) [25] algorithm that takes into account both the prior beliefs and the data to obtain the posterior distribution over parameter values. ABC provides an efficient way of exploring the parameter space and keeping those values which better fit the data, and is particularly useful in cases where the likelihood is unknown or intractable to calculate. The algorithm returns an approximation to the posterior distribution as a (multi)set of parameter samples.

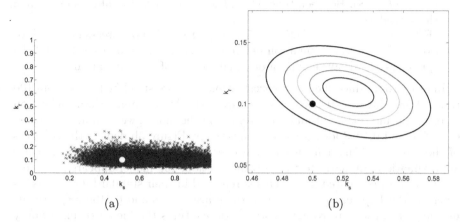

(a)   (b)

Fig. 3. Results from the two inference methods: (a) accepted samples after running ABC; (b) contours of the Laplace approximation of the posterior when using formulae as inputs (the outer ring corresponds to the 98th quantile). The dots show the true values.

For this experiment, we simulated the system with $k_s = 0.5$, $k_r = 0.1$, and used ten points from the resulting trajectory as the input time-series. We gathered 100,000 parameter samples using ABC in approximately 20 seconds, plotted in Figure 3a. We can see the distribution of k_r is centred on its true value. It is also narrower than that of k_s, indicating that the behaviour of the system is not as sensitive to the latter. Simulating the system for values of k_s in $[0.3, 1.0]$ shows that its behaviour does not change much in this range, validating our results. Even with this wide variance, however, the posterior differs from the prior in that it assigns little or no probability to values of k_s under 0.3, which produce significantly different behaviour to the one in the input observations.

6.2   Inference from Specification

We now examine the possibility of inference from specification in the form of logical constraints, rather than quantitative observations in the form of time-series. In many cases detailed quantitative information is not readily available, whereas we have an idea of how the model should behave with respect to certain properties. We follow the framework of Bortolussi & Sanguinetti [4], in which observations are satisfaction values of formulae specified in a suitable temporal logic (e.g. MiTL). This method yields an estimate for the parameter value that optimises the posterior distribution (Section 5.4); we can then use this to approximate the whole posterior with a Gaussian distribution by employing the *Laplace approximation*.

We have considered the following three properties, expressed in MiTL:

- $\mathbf{G}^{[3,5]}(I > 0)$: There are still ignorants at all time points between time 3 and 5, i.e. the rumour has not reached everyone in the population.

- $\mathbf{G}^{[0.5,1]}(R \geq S)$: Between time 0.5 and 1, the blockers are always more than the spreaders.
- $(\mathbf{F}^{[0,1]}(S \geq 7.5)) \wedge (\mathbb{G}^{[1,2]}(S \leq 3.75))$: The proportion of spreaders reaches or exceeds 50% of the population before time 1, and between time 1 and 2 the spreaders are always less than or equal to 25% of the population.

In order to produce the input specification data, we simulated the system 100 times after fixing $k_s = 0.5$, $k_r = 0.1$ as previously. After running the optimisation algorithm using this input and the priors from the model, we obtained estimates of 0.5236 and 0.1098 for k_s and k_r respectively. Figure 3b depicts the contours of the posterior Gaussian distribution over the parameters, given via Laplace approximation. Similarly to the previous section, the model appears to be less sensitive to k_s compared to k_r. This is reflected in their standard deviations, as given by the Laplace approximation; these are 0.0288 and 0.0199 respectively, meaning that the approximate distribution captures the increased uncertainty regarding k_s.

7   Related Work

Some previous work has explored parameter estimation methods for formal models. The Evolving Process Algebra [17] framework uses genetic algorithms to find parametrizations of models written in the PEPA language, such that the behaviour of the model matches an observed time-series. It has been further applied to optimising the structure of Bio-PEPA models, going beyond parameter search [18]. The work in [23] deals with parameter estimation for the Calculus of Wrapped Components, employing a different search algorithm. The problem has also been considered in the case of BIOCHAM in [7], where the space of possible parameters is searched exhaustively to find values giving a good fit.

The somewhat related problem of reasoning about the behaviour of incompletely specified process algebra models, rather than inferring their parameters, is dealt with in [3]. Brim *et al.* [5] propose a method of approximating quantitative model checking results over an entire parameter space as an alternative to parameter estimation. Although not directly relevant to inference, the work in [16] is another example of combining machine learning with formal modelling, presenting a Bayesian approach to statistical model checking.

With regards to probabilistic programming, as mentioned previously, languages like Church [13], IBAL [21] and Infer.NET [19] are not particularly suited to modelling complex dynamical systems. A language for describing continuous-time systems has been proposed in [22], but still lacks the formal features of a process algebra. An initial attempt to apply probabilistic programming to continuous-time models of biological systems also used a lower-level description language [12].

8   Conclusions

We have presented a process algebra that incorporates elements of probabilistic programming, the first such attempt at combining the two fields. This approach integrates uncertainty and observations into the system description, allowing us to model systems for which we have incomplete knowledge and giving us access to techniques from machine learning for inferring the unknown parameters.

The new features, while affecting the syntax of the language only minimally, significantly extend its expressivity. Additionally, our system is modular and flexible in the choice of inference algorithm. The application of two such algorithms on an example gives promising results for the effectiveness of our approach.

An interesting question for future work is whether the observations can be further integrated into the semantics of the language — for instance, whether the specification of a system can let us reject some transitions when building the underlying transition system. We are also interested in examining other benefits afforded by the use of a formal language, such as the description of equivalences and how they can be adapted to this probabilistic programming-like setting. Furthermore, we plan to explore more inference algorithms and test our framework on larger systems.

Acknowledgements. This work was supported by Microsoft Research through its PhD Scholarship Programme, the EU FET-Proactive programme through QUANTICOL grant 600708, the SysMIC project through BBSRC grant BB/I014713/1 and the European Research Council through grant MLCS306999. The authors thank Luca Cardelli and Vashti Galpin for their comments and suggestions.

References

1. Aldini, A., Bernardo, M., Corradini, F.: A process algebraic approach to software architecture design. Springer (2010)
2. Aziz, A., Sanwal, K., Singhal, V., Brayton, R.: Verifying continuous time Markov chains. In: Alur, R., Henzinger, T.A. (eds.) CAV 1996. LNCS, vol. 1102, pp. 269–276. Springer, Heidelberg (1996)
3. Baldan, P., Bracciali, A., Brodo, L., Bruni, R.: Deducing Interactions in Partially Unspecified Biological Systems. In: Anai, H., Horimoto, K., Kutsia, T. (eds.) AB 2007. LNCS, vol. 4545, pp. 262–276. Springer, Heidelberg (2007)
4. Bortolussi, L., Sanguinetti, G.: Learning and designing stochastic processes from logical constraints. In: Joshi, K., Siegle, M., Stoelinga, M., D'Argenio, P.R. (eds.) QEST 2013. LNCS, vol. 8054, pp. 89–105. Springer, Heidelberg (2013)
5. Brim, L., Češka, M., Dražan, S., Šafránek, D.: Exploring parameter space of stochastic biochemical systems using quantitative model checking. In: Sharygina, N., Veith, H. (eds.) CAV 2013. LNCS, vol. 8044, pp. 107–123. Springer, Heidelberg (2013)
6. Caillaud, B., Delahaye, B., Larsen, K.G., Legay, A., Pedersen, M.L., Wsowski, A.: Constraint Markov Chains. Theor. Comp. Science 412(34), 4373–4404 (2011)
7. Calzone, L., Chabrier-Rivier, N., Fages, F., Soliman, S.: Machine Learning Biochemical Networks from Temporal Logic Properties. In: Priami, C., Plotkin, G. (eds.) Trans. on Comput. Syst. Biol. VI. LNCS (LNBI), vol. 4220, pp. 68–94. Springer, Heidelberg (2006)

8. Ciocchetta, F., Hillston, J.: Bio-PEPA: A framework for the modelling and analysis of biological systems. Theor. Comp. Science 410(33-34), 3065–3084 (2009)
9. Clarke, E.M., Emerson, E.A., Sistla, A.P.: Automatic verification of finite-state concurrent systems using temporal logic specifications. ACM Trans. Program. Lang. Syst. 8(2), 244–263 (1986)
10. Daley, D.J., Kendall, D.G.: Epidemics and Rumours. Nature 204(4963) (1964)
11. Galpin, V.: Equivalences for a biological process algebra. Theor. Comp. Science 412(43), 6058–6082 (2011)
12. Georgoulas, A., Hillston, J., Sanguinetti, G.: ABC–Fun: A Probabilistic Programming Language for Biology. In: Gupta, A., Henzinger, T.A. (eds.) CMSB 2013. LNCS, vol. 8130, pp. 150–163. Springer, Heidelberg (2013)
13. Goodman, N.D., Mansinghka, V.K., Roy, D.M., Bonawitz, K., Tenenbaum, J.B.: Church: a language for generative models. In: McAllester, D.A., Myllymäki, P. (eds.) UAI, pp. 220–229. AUAI Press (2008)
14. Hermanns, H.: Interactive Markov Chains: and the quest for quantified quality. Springer (2002)
15. Hillston, J.: A Compositional Approach to Performance Modelling. CUP (1996)
16. Jha, S.K., Clarke, E.M., Langmead, C.J., Legay, A., Platzer, A., Zuliani, P.: A Bayesian Approach to Model Checking Biological Systems. In: Degano, P., Gorrieri, R. (eds.) CMSB 2009. LNCS, vol. 5688, pp. 218–234. Springer, Heidelberg (2009)
17. Marco, D., Cairns, D., Shankland, C.: Optimisation of process algebra models using evolutionary computation. In: 2011 IEEE Congress on Evolutionary Computation (CEC), pp. 1296–1301 (2011)
18. Marco, D., Shankland, C., Cairns, D.: Evolving Bio-PEPA process algebra models using genetic programming. In: Proceedings of the Fourteenth International Conference on Genetic and Evolutionary Computation Conference, GECCO 2012, New York, NY, USA, pp. 177–184 (2012)
19. Minka, T., Winn, J., Guiver, J., Knowles, D.: Infer.NET 2.5, Microsoft Research Cambridge (2012), http://research.microsoft.com/infernet
20. de Nicola, R., Latella, D., Loreti, M., Massink, M.: A Uniform Definition of Stochastic Process Calculi. ACM Comput. Surv. 46(1), 5:1–5:35 (2013)
21. Pfeffer, A.: The Design and Implementation of IBAL: A General-Purpose Probabilistic Language. In: Getoor, L., Taskar, B. (eds.) Introduction to Statistical Relational Learning. The MIT Press (2007)
22. Pfeffer, A.: CTPPL: A Continuous Time Probabilistic Programming Language. In: IJCAI, pp. 1943–1950 (2009)
23. Sciacca, E., Spinella, S., Calcagno, C., Damiani, F., Coppo, M.: Parameter Identification and Assessment of Nutrient Transporters in AM Symbiosis through Stochastic Simulations. ENTCS 293, 83–96 (2013), Proceedings of CS2Bio 2012
24. Sen, K., Viswanathan, M., Agha, G.: Model-Checking Markov Chains in the Presence of Uncertainties. In: Hermanns, H., Palsberg, J. (eds.) TACAS 2006. LNCS, vol. 3920, pp. 394–410. Springer, Heidelberg (2006)
25. Toni, T., Welch, D., Strelkowa, N., Ipsen, A., Stumpf, M.P.: Approximate Bayesian computation scheme for parameter inference and model selection in dynamical systems. Journal of The Royal Society Interface 6(31), 187–202 (2009)

PALOMA: A Process Algebra
for Located Markovian Agents

Cheng Feng and Jane Hillston

LFCS, School of Informatics, University of Edinburgh
s1109873@sms.ed.ac.uk, jane.hillston@ed.ac.uk
http://www.quanticol.eu

Abstract. We present a novel stochastic process algebra that allows the expression of models representing systems comprised of populations of agents distributed over space, where the relative positions of agents influence their interaction. This language, PALOMA, is given both discrete and continuous semantics and it captures multi-class, multi-message Markovian agent models (M^2MAM). Here we present the definition of the language and both forms of semantics, and demonstrate the use of the language to model a flu epidemic under various quarantine regimes.

1   Introduction

Collective systems, comprised of many communicating entities and without centralised control, are becoming pervasive. Without any global knowledge, entities interact locally to create a system with discernible characteristics at the global level; a phenomenon sometimes termed *emergence*.

The notion of *locality* has spatial relationship implicit within it, and thus to faithfully capture these systems we have to be able to represent the spatial arrangement of entities and the constraints that this places on their communication. For example, interactions may only be allowed for entities which are co-located or within a certain physical distance of each other, or space may be segmented in such a way that even physically close entities are unable to communicate. Furthermore movement can be a crucial aspect of the behaviour of entities within the system. Therefore it becomes essential to develop modelling formalisms in which space is captured explicitly, and in which the same entity in different locations can be distinguished. Meanwhile, given the scale of collective systems, which often rely on large populations of entities in order to meet their objectives, we must also find efficient mechanisms both to express and to analyse the developed models.

Multi-class Multi-message Markovian Agents Models (M^2MAM) have recently been proposed by Cerrotti *et al.* as a suitable framework for modelling collective systems comprised of populations of agents which are spatially distributed [1]. Several case studies [2–4] demonstrated that this is a powerful and useful framework. However the model specification is in terms of matrices which capture the possible interactions and influences between agents. This form of specification

G. Norman and W. Sanders (Eds.): QEST 2014, LNCS 8657, pp. 265–280, 2014.
© Springer International Publishing Switzerland 2014

is highly demanding on the modeller and prone to error. In this paper we propose a process algebra to capture models within the M^2MAM framework and circumvent the rather cumbersome matrix specification. The process algebra, PALOMA, is equipped with both discrete event and differential semantics. This high-level specification is human-readable, less error-prone, amenable to automated checking and supports automated derivation of the executable models defined by the semantics. More specifically, the discrete semantics provides the theoretical foundation for discrete event simulation whilst the differential semantics allows us to automatically derive the matrices, and thus the underlying mean-field model of the M^2MAM in the form of initial value problems.

The paper is structured as follows. We briefly introduce the concepts of M^2MAMs in the next section. Section 3 presents the syntax and discrete semantics of PALOMA. This is followed by the differential semantics in Section 4. In Section 5, a case study in which we apply PALOMA to the modelling of the spread of flu in a multi-community society is presented. Finally, Sections 6 and 7 discuss related work, future research and draw final conclusions.

2 M^2MAM

In this section, we briefly introduce the key concepts of M^2MAMs, originally presented by Cerrotti *et al.* in [1]. M^2MAMs consist of a collection of Markov agents (MAs) distributed over space, which is represented by a finite set of locations. Each MA has a location attribute and can be denoted by a finite state machine in which two types of transitions can happen: *local* transitions and *induced* transitions. Local transitions occur whenever the MA changes its state spontaneously with a delay governed by an exponential distribution. Local transitions can also possibly emit messages that can cause the occurrence of induced transitions in MAs at the same or other locations. This enables location-based asynchronous interaction between MAs. The reception of a message is governed by the *perception function*, which depends on both the location and state of the sender and receiver MAs. When a MA receives a message, it can either ignore or accept it. In the latter case, the agent will change its state immediately by performing an induced transition.

Following [1], we use $MA^c(\ell)$ to denote a MA of class c in location ℓ. A $MA^c(\ell)$ can be defined as a tuple $\{Q^c(\ell), \Lambda^c(\ell), G^c(\ell, m), A^c(\ell, m), \pi_0^c(\ell)\}$, in which:

- $Q^c(\ell) = [q_{ij}^c(\ell)]$ is a $n_c \times n_c$ matrix, in which each element $q_{ij}^c(\ell)$ represents the rate of the local transition from state i to state j, with $q_{ii}^c(\ell) = -\sum_{j \neq i}^{n_c} q_{ij}^c(\ell)$ where n_c is the number of states of a MA of class c.
- $\Lambda^c(\ell) = [\lambda_i^c(\ell)]$ is a vector, in which each element $\lambda_i^c(\ell)$ denotes the rate of a self-jump transition which reenters the same state i, for a MA of class c.
- $G^c(\ell, m) = [g_{ij}^c(\ell, m)]$ is a $n_c \times n_c$ matrix in which each element $g_{ij}^c(\ell, m)$ describes the probability of $MA^c(\ell)$ generating a message of type m during a local transition from state i to state j.
- $A^c(\ell, m) = [a_{ij}^c(\ell, m)]$ is a $n_c \times n_c$ matrix, in which each element $a_{ij}^c(\ell, m)$ $(i \neq j)$ describes the acceptance probability of message type m for the $MA^c(\ell)$,

with induced transition from state i to state j whereas $a_{ii}^c(\ell, m)$ denotes the probability of dropping this message, and $a_{ii}^c(\ell, m) = 1 - \sum_{j \neq i} a_{ij}^c(\ell, m)$.

- $\pi_0^c(\ell)$ is the initial state probability distribution of an agent of class c in location ℓ.

2.1   Model Analysis

The density of agents of class c in state i, in location ℓ at time t is denoted $p_i^c(\ell, t)$. In M^{2}MAMs, the density of agents of each class in a location is assumed to remain constant, i.e. the value of $\sum_{i=1}^{n_c} p_i^c(\ell, t) = P^c(\ell)$ is invariant. We use a vector $p^c(\ell, t) = [p_i^c(\ell, t)]$ to denote the state density distribution of agents of class c in location ℓ and at time t. The analysis of interest is the transient evolution of $p^c(\ell, t)$. It can be computed by solving a set of coupled ODEs.

First of all, the total rate at which messages of type m are generated by a MA of class c in state j and location ℓ can be computed by:

$$\beta_j^c(\ell, m) = \lambda_j^c(\ell) g_{jj}^c(\ell, m) + \sum_{k \neq j} q_{jk}^c(\ell) g_{jk}^c(\ell, m) \tag{1}$$

where the first term on the right hand side of the above equation gives the rate at which messages of type m are generated by the MA in state j by a self-jump transition, whereas the second term denotes the rate of message generation by the MA during a local transition from state j to another state.

With $\beta_j^c(\ell, m)$, we can compute $\gamma_{ii}^c(\ell, m, t)$, the total reception rate of messages of type m by a MA of class c in state i and location ℓ, at time t. The rate $\gamma_{ii}^c(\ell, m, t)$ is contributed to by all the messages of type m generated by MAs of all classes in all states and all locations, as long as they can be perceived by the receiver MA. Thus, $\gamma_{ii}^c(\ell, m, t)$ is obtained by the following equation:

$$\gamma_{ii}^c(\ell, m, t) = \sum_{\ell' \in \mathcal{V}} \sum_{c'=1}^{C} \sum_{j=1}^{n_{c'}} u_m(\ell, c, i, \ell', c', j) \beta_j^{c'}(\ell', m) p_j^{c'}(\ell', t) \tag{2}$$

where $\mathcal{C} = \{1, \ldots, C\}$ is the set of agent classes in the model, $\mathcal{V}$ is the location set $u_m(\ell, c, i, \ell', c', j)$ is the perception function of message m, whose value represents the probability that an agent of class c, in state i, and in location ℓ perceives a message m sent by an agent of class c' in state j and in position ℓ'.

Finally, we use a diagonal matrix, $\Gamma^c(\ell, m, t) = diag(\gamma_{ii}^c(\ell, m, t))$ to collect the rates in Equation (2), and the infinitesimal generator matrix $K^c(\ell, t)$ for the population CTMC of agents of class c in location ℓ at time t can be obtained by:

$$K^c(\ell, t) = Q^c(\ell) + \sum_m \Gamma^c(\ell, m, t)[A^c(\ell, m) - I] \tag{3}$$

where I is the identity matrix, the first term on the right hand side of (3) is the infinitesimal generator matrix of the CTMC for local transitions, and the second term gives the infinitesimal generator matrix for induced transitions which uses the averaged effect to approximate the effect of all other agents' interactions with an agent of class c in location ℓ at time t.

Shifting to a mean field view, the transient evolution of $p^c(\ell, t)$ is captured by standard Kolmogorov equations with initial conditions $p^c(\ell, t_0) = P^c(\ell)\pi_0^c(\ell)$ for all ℓ and c:

$$\frac{dp^c(\ell, t)}{dt} = p^c(\ell, t)K^c(\ell, t) \quad \forall(\ell, c) \tag{4}$$

3   PALOMA

PALOMA, the Process Algebra of Located Markovian Agents, is intended to provide a simple process algebra-based formalism which can be used to generate models in the M^2MAM framework. M^2MAM is used to generate a mean field model, but being based on Markovian agents it is also amenable to a discrete interpretation. As mentioned previously, PALOMA is equipped with both discrete and differential semantics. In this section, we first introduce the discrete interpretation, considering individual agents. We will then make the shift to population CTMC and ultimately a mean field model in the next section.

3.1   Syntax

In keeping with the M^2MAM framework, in PALOMA each agent is a finite state machine and the language is *conservative* in the sense that no agents are spawned or destroyed during the evolution of a model (although they can cease to change state). Thus the language has a two level grammar:

$$X(\ell) ::= !(\alpha, r).X(\ell) \mid ?(\alpha, p).X(\ell) \mid X(\ell) + X(\ell) \qquad P ::= X(\ell) \mid P \parallel P$$

Agents are parameterised by a *location*, here denoted by ℓ. Agents can undertake two types of actions, *spontaneous* actions, denoted $!(\alpha, r)$, and *induced* actions, denoted $?(\alpha, p)$. When an agent performs a spontaneous action, it does so with a given rate r, which is taken to be the parameter of an exponential distribution, where $1/r$ is the expected duration of the action. Spontaneous actions are broadcast to the entire system, and can induce change in any other agent which enables an induced action with the matching type α. An induced action has an associated probability p, which records the probability that the agent responds to a spontaneous action of the same type. In the style of the Calculus of Broadcasting Systems [5], this can be thought of as the probability that the agent *listens* as opposed to simply *hearing*. Alternative behaviours are represented by the standard choice operator, $+$. A choice between spontaneous actions is resolved via the race policy, based on their corresponding rates. We assume that there is never a choice between induced actions of the same type.

A model, P, consists of a number of agents composed in parallel. There is no direct communication between agents, for example in the style of shared actions in PEPA [6]. Instead, all interaction is via spontaneous/induced actions. When an action is induced in an agent the extent of its impact is specified by a perception function, $u(\alpha, \ell, X, \ell', X')$[1]. This is a further probability which, given the locations of the two agents, their current states and action type involved, determines the likelihood that the induced action occurs. For example, the perception function might have value 1 when the two agents are within a communication radius r of each other, but a value of 0 whenever the distance between them is greater than r. Obviously this gives a rich set of possible styles of interaction, but note that each agent with an induced action chooses independently whether to respond or not.

3.2   Semantics

The semantics proceeds in sequences of alternating steps. This can be regarded as a semi-Markov process: the first step, corresponding to the spontaneous action, determines a delay, whilst the second step is probabilistic and determines what the next state will be, as each possible induced action makes the choice of whether to respond, based both on its inherent probability of "listening" and the perception function. Since each agent makes such a decision independently, the probabilities can be multiplied to obtain the overall probability of a given next state. Correspondingly we define two transition relations $\longrightarrow$ and $\longrightarrow_{\mathcal{P}}$. These are shown in Figures 1 and 2 respectively.

In order to keep track of which agents have "heard" the messages which are broadcast by spontaneous actions we associate an *ether* element with the system, which provides the environment for *all* agents. This has a distinguished empty state E_0. As shown in rule SpA, a spontaneous action can only be initiated if the ether is currently empty, and no probabilistic transitions are enabled ($\not\longrightarrow_{\mathcal{P}}$). The resulting local state records that the ether contains the message α which originated with rate r at location ℓ from the state $!(\alpha, r).X(\ell)$, and that the continuation is subject to a probabilistic resolution. Any state awaiting probabilistic resolution is denoted $S^{\mathcal{P}}$. Note that $S^{\mathcal{P}}$ states will not be in the CTMC.

If the ether contains a message α then an agent enabling an induced α action may progress to a probabilistic state in which, with probability p, the continuation $X(\ell)$ is chosen, and with probability $1 - p$, the continuation $?(\alpha, p).X(\ell)$ is chosen (rule InA). For other agents, their spontaneous actions are blocked until the current one has been fully broadcast and probabilistically resolved (rule NoSp). If the ether contains a message of type α then an agent enabling a spontaneous action of any type (including α) witnesses the ongoing action, enters a probabilistic state and awaits resolution. This ensures that only one spontaneous action can be in progress at a time. Note that there is no possibility of an agent "sharing" the α action as would be possible in a language such as CSP or PEPA. Similarly, if the ether contains a message of type α then an agent that enables an induced message of any other type simply witnesses the ongoing action, enters a probabilistic state and awaits resolution (rule NoIn). In some cases a spontaneous

---

[1] Here we do not need to explicitly specify the class of sender and receiver agents as it can be deduced by the state and the location attributes.

SpA $E_0, !(\alpha, r).X(\ell) \xrightarrow{(\alpha,r)} [\alpha, r, \ell, X], X(\ell)^{\mathcal{P}} \quad (\not\rightarrow_{\mathcal{P}})$

InA $[\alpha, r, \ell', X'], ?(\alpha, p).X(\ell) \xrightarrow{(\alpha,r)} [\alpha, r, \ell', X'], (?(\alpha, p).X(\ell) +^{p} X(\ell))^{\mathcal{P}}$

NoSp $[\alpha, r, \ell', X'], !(\beta, s).X(\ell) \xrightarrow{(\alpha,r)} [\alpha, r, \ell', X'], !(\beta, s).X(\ell)^{\mathcal{P}}$

NoIn $[\alpha, r, \ell', X'], ?(\beta, p).X(\ell) \xrightarrow{(\alpha,r)} [\alpha, r, \ell', X'], ?(\beta, p).X(\ell)^{\mathcal{P}} \quad (\beta \neq \alpha)$

Ch1 $\dfrac{E, X_1(\ell) \xrightarrow{(\alpha,r)} E', X_1'(\ell')^{\mathcal{P}}}{E, X_1(\ell)+X_2(\ell) \xrightarrow{(\alpha,r)} E', X_1'(\ell')^{\mathcal{P}}}$ Ch2 $\dfrac{E, X_2(\ell) \xrightarrow{(\alpha,r)} E', X_2'(\ell')^{\mathcal{P}}}{E, X_1(\ell)+X_2(\ell) \xrightarrow{(\alpha,r)} E', X_2'(\ell')^{\mathcal{P}}}$

Par $\dfrac{E_1, X_1(\ell_1) \xrightarrow{(\alpha,r)} E', X_1'(\ell_1')^{\mathcal{P}} \quad E_2, X_2(\ell_2) \xrightarrow{(\alpha,r)} E', X_2'(\ell_2')^{\mathcal{P}}}{(E_1, X_1(\ell_1)) \parallel (E_2, X_2(\ell_2)) \xrightarrow{(a,r)} E', (X_1'(\ell_1') \parallel X_2'(\ell_2'))^{\mathcal{P}}}$

Fig. 1. The delay transition relation for PALOMA

action may not induce any actions in other agents. If this is the case the message will propagate, without impacting any other agents, except to put them into the trivial probabilistic state. Choice behaves as we would anticipate. We assume that within a choice both elements are in the same location as they correspond to a single agent. Parallel agents must agree on the single spontaneous action to take place, and consequently update the ether in the same way. A spontaneous action is deemed to be complete when all agents have moved to a probabilistic state. In this case a probabilistic resolution must be made to determine the next state. This is defined by the probabilistic transition relation, which will clear the ether and create the opportunity for the next spontaneous message.

Probabilistic resolutions are determined by a second transition relation $\longrightarrow_{\mathcal{P}}$, shown in Figure 2. The only probabilistic states which genuinely have different possible outcomes are those which resulted from an induced action. In this case there are two different resolutions according to whether the induced action is "listened to" or simply "heard". In either case the ether is emptied when the probabilistic resolution is made (rule PR1). First, a choice is made whether to hear the message or not, but secondly, if the message is heard, its impact is adjusted according to the perception function. This is consistent with the M^2MAM formalism. For other states the probabilistic resolution will simply clear the ether and return the agent to an active state again (rule PR2). Parallel agents undergo probabilistic resolution independently and their probabilities are multiplied (rule ParP).

4   Differential Semantics of PALOMA Models

Obtaining performance metrics via discrete event simulation can become very expensive or even infeasible for PALOMA models when there is a large number of

$$\text{PR1 } [\alpha, r, \ell', X'], (?(\alpha,p).X(\ell) +^p X(\ell))^{\mathcal{P}} \begin{cases} \xrightarrow{(\alpha, p \times u(\alpha, \ell, X, \ell', X'))}_{\mathcal{P}} E_0, X(\ell) \\ \xrightarrow{(\alpha, 1-p \times u(\alpha, \ell, X, \ell', X'))}_{\mathcal{P}} E_0, ?(\alpha,p).X(\ell) \end{cases}$$

$$\text{PR2 } [\alpha, r, \ell', X'], X(\ell)^{\mathcal{P}} \xrightarrow{(\alpha,1)}_{\mathcal{P}} E_0, X(\ell)$$

$$\text{ParP } \frac{E, X_1(\ell_1)^{\mathcal{P}} \xrightarrow{(\alpha,p)}_{\mathcal{P}} E_0, X_1'(\ell_1') \qquad E, X_2(\ell_2)^{\mathcal{P}} \xrightarrow{(\alpha,q)}_{\mathcal{P}} E_0, X_2'(\ell_2')}{E, (X_1(\ell_1) \parallel X_2(\ell_2))^{\mathcal{P}} \xrightarrow{(\alpha, p \times q)}_{\mathcal{P}} (E_0, X_1'(\ell_1')) \parallel (E_0, X_2'(\ell_2'))}$$

Fig. 2. The probabilistic transition relation for PALOMA

agents in the model. Thus, it is advantageous to define the differential semantics for PALOMA which can automatically derive the mean-field model in the form of initial value problems (a set of coupled ODEs with initial values) as done for M^2MAM. As solving the mean-field model is independent of the number of agents in the system, this enables scalable analysis of PALOMA models. In this section, we introduce the differential semantics of PALOMA models, first developing a population-based structured operational semantics which lifts the individual-based PALOMA model to a population-level view. This serves as an intermediate tool for the generation of the mean-field model.

4.1 Population-Based Structured Operational Semantics

In PALOMA, as agents in the same state and location are identical, it is advantageous to use a population-based state vector to represent the state of the model in which symmetric states are aggregated to mitigate the well-known state space explosion problem. For example, consider the following simple two-location SIS model in PALOMA, which we refer to as Example 1:

$$S(\ell) = ?(contact, p).I(\ell) + !(move, q).S(\ell')$$
$$I(\ell) = !(contact, \beta).I(\ell) + !(recover, \gamma).S(\ell) + !(move, q).I(\ell')$$
$$S(\ell') = ?(contact, p).I(\ell') + !(move', q').S(\ell)$$
$$I(\ell') = !(contact, \beta).I(\ell') + !(recover, \gamma).S(\ell') + !(move', q').I(\ell)$$

$$u(contact, \ell, X, \ell', X') = \begin{cases} \dfrac{1}{S_\ell + I_\ell} & \text{if } (\ell = \ell' \wedge X = S) \\ 0 & \text{otherwise} \end{cases}$$

$$S(\ell)[N_{S(\ell)}] \parallel I(\ell)[N_{I(\ell)}] \parallel S(\ell')[N_{S(\ell')}] \parallel I(\ell')[N_{I(\ell')}]$$

The model captures a disease spread scenario, in which an infective agent (I) makes a *contact* action spontaneously at the rate β. A susceptible agent (S) gets infected by accepting a *contact* message with the probability p. The perception function of the *contact* message can be explained as follows. If the message is received by a susceptible agent in the *same location* as the message sender, it can be perceived with probability $\frac{1}{S_\ell + I_\ell}$, where X_ℓ denotes the number of agents in state X in location ℓ. Otherwise, the message cannot be perceived. Intuitively, the perception function of message *contact* means that an infective agent contact one arbitrary agent in its current location. Thus, a susceptible agent in the same location as the infective agent can perceive the *contact* message with probability $1/N_\ell$, where $N_\ell = S_\ell + I_\ell$ is the total number of agents in the location.

Agents in location ℓ move to location ℓ' by performing a spontaneous action *move* at the rate q, and move back by a spontaneous action *move'* at the rate q'. The spontaneous actions *move* and *move'* do not have any corresponding induced actions and so can be thought of as not emitting a message. An infective agent can also do a *recover* action spontaneously without message emission at rate γ.

Lastly, the final equation gives the initial populations of agents, where for example, $S(\ell)[N_{S(\ell)}]$ denotes $N_{S(\ell)}$ agents in the state $S(\ell)$ in parallel. This is syntactic sugar to ease the definition of large population models.

Now suppose we use a counting abstraction, constructing a state vector $\mathbf{X} = (x_1, x_2, x_3, x_4)$ to represent the current state of the model, in which x_1 denotes the number of agents in state $S(\ell)$, x_2, the number of agents in state $I(\ell)$, x_3, the number of agents in state $S(\ell')$ and x_4, the number of agents in state $I(\ell')$. Then the size of the state space is reduced from $O(4^N)$ to $O(N^4)$, where $N = \sum x_i$.

Using this notation, the x_2 enabled transitions caused by the spontaneous actions *contact* made by the x_2 agents in state $I(\ell)$ can be aggregated to a population-level transition as follows:

$$E_0, \mathbf{X} \xrightarrow{(contact, \beta \times x_2)} [contact, \beta \times x_2, \ell, I], \mathbf{X}^\mathcal{P} \;\not\mapsto_\mathcal{P} \qquad (5)$$

The probabilistic resolutions following from this transition can be aggregated:

$$[contact, \beta x_2, \ell, I], \mathbf{X}^\mathcal{P} \left\{ \begin{array}{c} \cdots \\ \xrightarrow{(contact, \binom{x_1}{i} \times (pu)^i \times (1-pu)^{x_1-i})}_\mathcal{P} E_0, \mathbf{X} + (-i, i, 0, 0) \\ \cdots \end{array} \right. \qquad (6)$$

where $i = (0, 1, ..., x_1)$, u is the value of the perception function. Note that $\binom{x_1}{i} \times (pu)^i \times (1-pu)^{x_1-i}$ is the probability that there are i out of x_1 induced transitions of the form $S(\ell) =?(contact, p).I(\ell)$ actually fired, where $(-i, i, 0, 0)$ is the associated net change on the state vector caused by these transitions.

Furthermore, as the probabilistic resolutions occur immediately and finish instantaneously after a spontaneous transition fired, we can use a new transition relation $\longrightarrow_*$, which we call the *population-based* transition relation to form the following transitions to represent the transitions in equations (5) and (6):

$$\mathbf{X} \xrightarrow{(\tau, \beta \times x_2)}_* \mathbf{X} + (0, 0, 0, 0) \qquad (7)$$

$$\mathbf{X} \left\{ \begin{array}{c} \cdots \\ \xrightarrow{(\tau_i, \beta \times x_2 \times \binom{x_1}{i} \times (pu)^i \times (1-pu)^{x_1-i})}_* \mathbf{X} + (-i, i, 0, 0) \\ \cdots \end{array} \right. \qquad (8)$$

where τ, τ_i are the transition names, $\beta \times x_2$, $\beta \times x_2 \times \binom{x_1}{i} \times (pu)^i \times (1-pu)^{x_1-i}$ are the rates of the transitions, $(0, 0, 0, 0)$ and $(-i, i, 0, 0)$ are the net change of the elements in the state vector caused by the transitions. Note that in *population-based* transitions, induced transitions have rates derived from the spontaneous

$$\textsf{PbSp} \quad \frac{E_0, !(\alpha, r).X(\ell) \xrightarrow{(\alpha,r)} [\alpha, r, \ell, X], X(\ell)^{\mathcal{P}} \not\mapsto_{\mathcal{P}}}{\mathbf{X} \xrightarrow{(\tau, \mathbf{X}[i_1] \times r)}_* \mathbf{X}'\{\mathbf{X}'[i_1] = \mathbf{X}[i_1] - 1, \mathbf{X}'[i_2] = \mathbf{X}[i_2] + 1\}} \quad \mathbf{X}[i_1] > 0$$

$$\textsf{PbIn} \quad \frac{[\alpha, r, \ell, X], (?(\alpha, p).X'(\ell') +^{\mathcal{P}} X'(\ell'))^{\mathcal{P}} \begin{cases} \xrightarrow{(\alpha, p \times u)}_{\mathcal{P}} E_0, X'(\ell') \\ \xrightarrow{(\alpha, 1-p \times u)}_{\mathcal{P}} E_0, ?(\alpha, p).X'(\ell') \end{cases}}{\mathbf{X} \begin{cases} \cdots \\ \xrightarrow{(\tau_k, r_{\tau_k})}_* \mathbf{X}'\{\mathbf{X}'[i_2] = \mathbf{X}[i_2] - k, \mathbf{X}'[i_3] = \mathbf{X}[i_3] + k\} \\ \cdots \end{cases}}$$

with $\mathbf{X}[i_1] > 0$, $k = (0, ..., \mathbf{X}[i_2])$, $r_{\tau_k} = \mathbf{X}[i_1] \times r \times \binom{\mathbf{X}[i_2]}{k} \times (pu)^k \times (1 - pu)^{\mathbf{X}[i_2]-k}$

Fig. 3. The Population-based Structured Operational Semantics

transitions. By doing this, we can analyse the influence of the spontaneous transitions and induced transitions on the population level dynamics of the model separately. This simplifies the analysis of PALOMA models at the population level because there are now no probabilistic transitions at this level.

We formally define the population-based structured operational semantics with rules for population-based transitions for PALOMA in Figure 3. The premises of these two rules describe the behaviour of single agents whereas the conclusions gives the collective dynamics of the populations of agents. More specifically, the rule PbSp infers a population-based transition from a spontaneous transition of a single agent, in which $\mathbf{X}$ and $\mathbf{X}'$ are the state vectors representing the states of the model before and after the transition. i_1, i_2 are the indexes of count variables in the state vector for agents in states $!(\alpha, r).X(\ell)$ and $X(\ell)$ respectively. We do not need to explicitly specify the count of agents in other states because they remain invariant after the transition in the premise of rule PbSp. The rule PbIn infers a set of population-based transitions from a transition of a single agent induced by a sptontaneous action α fired at the rate r which is performed by an agent in state X and in location ℓ. i_1, i_2, i_3 are the indexes of count variables in the state vector for agents in states $X(\ell)$, $?(\alpha, p).X'(\ell')$ and $X'(\ell')$ respectively.

4.2   Population-Based CTMC Model for PALOMA

With the above state aggregation and population-based structured operational semantics, we can define the population-based CTMC model for PALOMA. Formally, the population-based CTMC model for PALOMA is defined as a tuple $\mathcal{P} = (\mathbf{X}, \mathcal{D}, \mathcal{T}, \mathbf{x_0})$, where:

- $\mathbf{X} = (x_1, ..., x_n)$ is a state vector format, where each vector element is the count variable of agents in a specific state and location.
- Each x_i takes a value in a finite domain $\mathcal{D}_i \subset \mathbb{Z}^+$. Thus, $\mathcal{D} = \mathcal{D}_1 \times ... \times \mathcal{D}_n$ is the state space of the model.

- $\mathcal{T}(\mathbf{X}) = \{\tau_1, ..., \tau_m\}$ is the set of population-based transitions enabled in state $\mathbf{X}$, of the form $\tau_i = (r(\mathbf{X}), \mathbf{d})$, where:
  1. $r : \mathcal{D} \to \mathbb{R}_{\geq 0}$ is the rate function which depends on the current state of the system.
  2. $\mathbf{d} \in \mathbb{Z}^n$, is the update vector which gives the net change on each element of $\mathbf{X}$ caused by the transition.
- $\mathbf{x_0}$ is the initial state of the model.

4.3   The Mean-Field Model

The population-based CTMC model for PALOMA is used to extract ODEs for the mean-field model similarly to [7]. We will explain how this can be done in this subsection. Firstly, if we are interested in the mean behaviour of the system dynamics, the set of population-based transitions in the conclusion of the rule Pbln can be aggregated by a single transition as:

$$\mathbf{X} \xrightarrow{(\tau, \mathbf{X}[i_1] \times r)}_* \mathbf{X} + \mathbf{d}_\tau \tag{9}$$

where $\mathbf{d}_\tau[i_2] = -pu \times \mathbf{X}[i_2]$, $\mathbf{d}_\tau[i_3] = pu \times \mathbf{X}[i_2]$ and for $j \notin \{i_2, i_3\}$, $\mathbf{d}_\tau[j] = 0$. Specifically, $\mathbf{X}[i_1] \times r$ is the rate at which the spontaneous transition α in the premise of Pbln occurs, and it is also treated as the rate of the above aggregated transition. $\binom{\mathbf{X}[i_2]}{k} \times (pu)^k \times (1-pu)^{\mathbf{X}[i_2]-k}$ is the probability that there are k out of $\mathbf{X}[i_2]$ enabled induced transitions actually fired in the probabilistic resolutions of the spontaneous transition, and in this case the net change in the count variables for agents in state $?(\alpha, p).X'(\ell')$ and $X'(\ell')$ is $(-k, k)$. Thus, by summing up the product of all the possible net changes and their associated probabilities, we can get the expected net change in the population level of agents in these two states caused by the transitions induced by the α action performed by the agents in state $X(\ell)$ as follows:

$$\mathbb{E}_{\mathbf{d}[i_2, i_3]} = \sum_{k=0}^{\mathbf{X}[i_2]} \binom{\mathbf{X}[i_2]}{k} (pu)^k \times (1 - pu)^{\mathbf{X}[i_2]-k} \times (-k, k) = (-pu \times \mathbf{X}[i_2], pu \times \mathbf{X}[i_2])$$

Now, consider a population-based CTMC model for PALOMA (in which all the population-based transitions in the conclusion of the rule Pbln are aggregated in the style of Equation (9)) which is currently in state $\mathbf{X}$ with an enabled transition τ. This means that in every $1/r_\tau$ time units on average, a change in the population level of some agents denoted by $\mathbf{X}' = \mathbf{X} + \mathbf{d}_\tau$ occurs. If we approximate such a discrete change in a continuous fashion, then the change in the population level of the agents over a finite time interval Δt is:

$$\mathbf{X}(t + \Delta t) = \mathbf{X}(t) + r_\tau \times \mathbf{d}_\tau \times \Delta t$$

Rearranging the above equation and taking the limit $\Delta t \to 0$, we obtain the ODE which describes the (approximated) transient evolution of the population level of the agents in the system caused by transition τ as: $\frac{d\mathbf{X}(t)}{dt} = r_\tau \times \mathbf{d}_\tau$

Taking all enabled transitions $\mathcal{T}(\mathbf{X}) = \{\tau_1, ..., \tau_m\}$ into account, the ODE which describes the (approximated) transient evolution of the complete population-level system dynamics, has initial condition $\mathbf{X}(0) = \mathbf{x_0}$ and is defined as:

$$\frac{d\mathbf{X}(t)}{dt} = \sum_{i=1}^{m} r_{\tau_i} \times \mathbf{d}_{\tau_i}.$$

The Motivating Example. We use Example 1 to illustrate our approach. From the induced transitions in each location $\hat{\ell} \in \{\ell, \ell'\}$:

$$S(\hat{\ell}) = ?(contact, p).I(\hat{\ell}) \quad \text{induced by} \quad I(\hat{\ell}) = !(contact, \beta).I(\hat{\ell})$$

we obtain $\tau_1 = (\beta x_2, (-pux_1, pux_1, 0, 0))$ and $\tau_2 = (\beta x_4, (0, 0, -pu'x_3, pu'x_3))$ respectively, where $u = \dfrac{1}{x_1 + x_2}$ and $u' = \dfrac{1}{x_3 + x_4}$.

From the following spontaneous transitions:

$$S(\ell) = !(move, q).S(\ell') \quad S(\ell') = !(move', q').S(\ell) \quad I(\ell) = !(move, q).I(\ell')$$
$$I(\ell') = !(move', q').I(\ell) \quad I(\ell) = !(recover, \gamma).S(\ell) \quad I(\ell') = !(recover, \gamma).S(\ell')$$
$$I(\ell) = !(contact, \beta).I(\ell) \quad I(\ell') = !(contact, \beta).I(\ell')$$

We get the following corresponding population-based transitions:

$$\tau_3 = (qx_1, (-1, 0, 1, 0)) \quad \tau_4 = (q'x_3, (1, 0, -1, 0)) \quad \tau_5 = (qx_2, (0, -1, 0, 1))$$
$$\tau_6 = (q'x_4, (0, 1, 0, -1)) \quad \tau_7 = (\gamma x_2, (1, -1, 0, 0)) \quad \tau_8 = (\gamma x_4, (0, 0, 1, -1))$$
$$\tau_9 = (\beta x_2, (0, 0, 0, 0)) \quad \tau_{10} = (\beta x_4, (0, 0, 0, 0))$$

Therefore, the mean-field model for Example 1 is $\frac{d\mathbf{X}(t)}{dt} = \sum_{i=1}^{i=10} r_{\tau_i} \times \mathbf{d}_{\tau_i}$. The associated ODE model for each state count variable is:

$$\frac{dx_1(t)}{dt} = -\beta \times x_2 \times p \times u \times x_1 - q \times x_1 + q' \times x_3 + \gamma \times x_2$$

$$\frac{dx_2(t)}{dt} = \beta \times x_2 \times p \times u \times x_1 - q \times x_2 + q' \times x_4 - \gamma \times x_2$$

$$\frac{dx_3(t)}{dt} = -\beta \times x_4 \times p \times u' \times x_3 + q \times x_1 - q' \times x_3 + \gamma \times x_4$$

$$\frac{dx_4(t)}{dt} = \beta \times x_4 \times p \times u' \times x_3 + q \times x_2 - q' \times x_4 - \gamma \times x_4$$

The mean-field model matches our intuition from the M²MAM definition.

5   Case Study: Modelling the Spread of Flu

In this section, we extend the PALOMA model in Example 1 to model the spread of flu in a multi-community society. The model captures a simplified scenario of the 1918-1919 flu epidemic in central Canada, which was originally described in [8]. Consider an isolated society which consists of m communities. Q is the rate matrix in which each element q_{ij} is the rate at which a resident travels from community i to community j. We use β_i to denote the number of contacts per person per day in community i. For various reasons, the number of contacts per person per day is not the same in all communities. For example, suppose community i is the business centre of the society, then the number of contacts per person per day in community i should be higher than other communities.

In the epidemic model, a resident has three states: State S for susceptible, I for infected and R for recovered. When a susceptible resident makes contact with an infected resident, he will be infected by the flu with the probability p. On average, it takes about $1/\gamma$ days for an infected resident to recover from the flu. Once a resident recovers, he will not be infected again. We are interested in how many residents are infected by the flu from the beginning to the end of the outbreak. This can be captured by the following PALOMA model:

$$S(\ell_i) = ?(contact, p).I(\ell_i) + \sum_{j \neq i}^{m} !(move_{ij}, q_{ij}).S(\ell_j)$$

$$I(\ell_i) = !(contact, \beta_i).I(\ell_i) + !(recover, \gamma).R(\ell_i) + \sum_{j \neq i}^{m} !(move_{ij}, q_{ij}).I(\ell_j)$$

$$R(\ell_i) = \sum_{j \neq i}^{m} !(move_{ij}, q_{ij}).R(\ell_j)$$

$$u(contact, \ell, X, \ell', X') = \begin{cases} \dfrac{1}{S_\ell + I_\ell + R_\ell} & \text{if } (\ell = \ell' \wedge X = S) \\ 0 & \text{otherwise} \end{cases}$$

$$S(\ell_1)[N_{S(\ell_1)}] \parallel I(\ell_1)[I_{I(\ell_1)}] \parallel \ldots \parallel S(\ell_m)[N_{S(\ell_m)}] \parallel I(\ell_m)[I_{I(\ell_m)}]$$

where the perception function of message *contact* also means that an infected resident can make contact with an arbitrary resident in their current community. Thus, a susceptible resident in the same location as the infective agent can perceive the *contact* message with probability $1/N_\ell$, where $N_\ell = S_\ell + I_\ell + R_\ell$ is the total number of residents in the community ℓ.

5.1   Investigate the Effect of Quarantine on the Spread of the Flu

In this subsection we present the results of some experiments run on the model to investigate the effect of different quarantine policies on the spread of the flu. For example, quarantine may be applied to a whole community which is believed to be the source of the outbreak, reducing the likelihood of travel to other communities. Alternatively, individuals who develop flu in any community may be individually isolated and prevented from travelling.

Community-Level Quarantine: Here we assume that the flu originates from community i. We model the effect of *community-level quarantine* by adding a quarantine factor $0 < \sigma < 1$ to the rate at which residents travel into and out of community i. More specifically, the value of $Q(i, j)$ and $Q(j, i)$ becomes $q_{ij} \times \sigma$ and $q_{ji} \times \sigma$ repectively.

Individual-Level Quarantine: Alternatively the focus may be on the individuals with the disease. Suppose that on average, an infected person exhibits symptoms $1/\eta$ days after infection. Once an infected person is discovered, they will be isolated immediately until recovery. To model this, we introduce a new state $ISO(\ell_i)$, which represents an isolated infected resident currently in community i. Note that not all infected individuals will exhibit symptoms. The modifications to the PALOMA model with individual-level quarantine are given as follows:

$$I(\ell_i) = !(contact, \beta_i).I(\ell_i) + !(recover, \gamma).R(\ell_i) + !(discovered, \eta).ISO(\ell_i)$$

$$+ \sum_{j \neq i}^{m} !(move_{ij}, q_{ij}).I(\ell_j)$$

$$ISO(\ell_i) = !(recover, \gamma).R(\ell_i)$$

$$u(contact, \ell, X, \ell', X') = \begin{cases} \dfrac{1}{S_\ell + I_\ell + R_\ell} & \text{if } (\ell = \ell' \wedge X = S) \\ 0 & \text{otherwise} \end{cases}$$

Note that an isolated resident cannot travel out of their current community or contact other residents. As a result, an agent in state $ISO(\ell_i)$ can only do a spontaneous action *recover* and go to the state $R(\ell_i)$. Moreover, ISO_ℓ is not included in the perception function of *contact* which also reflects that isolated residents do not have chances to contact other residents.

Model Analysis. We consider five communities located in a star topology as illustrated in Figure 4(d): we assume community 1 is the business centre of the society and the flu also originates in Community 1. Thus, the community-level quarantine is imposed on Community 1. We assume that there are 300 residents of Community 1, 10 of whom are infected at the start of the study; 150 residents of Community 2, 140 residents of Community 3, 100 residents of Community 4 and 100 residents of Community 5, all of whom are susceptible at the start of the study. The values of parameters used in our simulation are given in Table 1.

Our simulation tool can parse PALOMA models and run discrete event simulations. The corresponding mean-field model is automatically generated in the form of Matlab scripts when the model is parsed, and can be run directly in Matlab. Figure 4(a), 4(b), 4(c) show our simulation results with 95% confidence interval in three different senarios. The results generated by the discrete event simulation (taking the average of 100 simulation runs) match well with the results of the mean-field model. The run time of a discrete event simulation for this model is about 70 seconds on a dual CORE i5 machine whereas the mean-field model can generate results instantly.

The results also give us some interesting information. It can be seen that community-level quarantine only reduces the number of infected residents to a

Table 1. Parameters used in the simulation

$p = 0.5$	$\beta_1 = 1$	$\beta_i = 0.5 \ (i \neq 1)$	$\gamma = 0.2$	$\eta = 0.25$	$\sigma = 0.1$	$q_{12} = 0.1$
$q_{13} = 0.12$	$q_{14} = 0.13$	$q_{15} = 0.11$	$q_{21} = 0.4$	$q_{31} = 0.4$	$q_{41} = 0.3$	$q_{51} = 0.35$

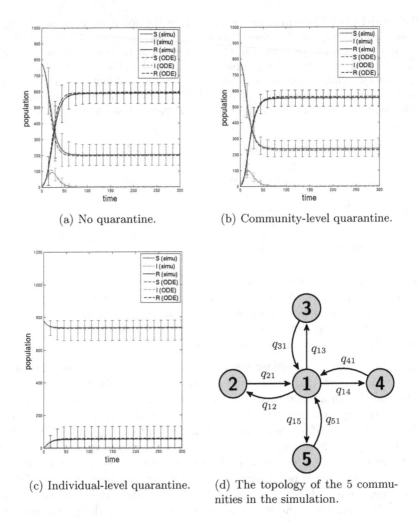

(a) No quarantine.

(b) Community-level quarantine.

(c) Individual-level quarantine.

(d) The topology of the 5 communities in the simulation.

Fig. 4. The simulation result and the community topology of the flu spread model

limited extent whereas individual-level quarantine has a more profound effect on controlling the spread of the flu.

6   Related Work

There have been many previous process algebras which encompass some spatial modelling, ranging from very abstract space and mobility in the π-calculus [9] to Cardelli and Gardner's elegant process algebra based around affine geometry [10]. Some also incorporate stochastic behaviour, such as stochastic π-calculus [11], Bio-PEPA [12], stochastic Bio-Ambients [13] and stochastic bigraphs [14]. But in each of these space is abstractly represented and most focus on a containment relationship between locations.

Closer to our work are the process algebras PALPS [15] and MASSPA [16]. PALPS, the Process Algebra with Location for Population Systems, is designed for building ecological models, and offers language primitives targeted at this application domain. Moreover, only an individual-based semantics is developed, severely restricting the scalability of the models which can be developed. Like PALOMA, MASSPA, the Markovian Agent Spatial Stochastic Process Algebra, takes the M^2MAM framework as its starting point. MASSPA emulates message broadcast by allowing each spontaneous action to emit a number of messages, loosely based on the likelihood that the single spontaneous action will trigger that number of induced actions. This multiplication of actions affects the dynamics of the system and no individual-based semantics is established. Instead MASSPA models are translated into systems of chemical reactions and population-level discrete event simulation based on mass action dynamics is developed. In contrast, PALOMA is equipped with both individual- and population-based semantics, supports dynamic perception functions (i.e. based on the current system state), which enables PALOMA to model adaptive behaviour. Furthermore, we slightly extend M^2MAM to allow agents to move between locations. The original M^2MAM framework requires the number of agents in each location to remain constant because the derived ODEs describe the evolution of the state density distribution of agents in each location over time, and then use that to derive the number of agents in different states. However, in the differential semantics of PALOMA, we use the ODEs to directly describe the evolution of number of agents in different states in different locations. Thus we are able to allow agents to move in PALOMA models, as demonstrated in the case study.

In [17] the authors apply mean-field models with locations (or classes more generally), which are close to the mean-field models developed in this paper, to the performance evaluation of network systems. PALOMA high-level language to define models of this kind.

7   Conclusion

PALOMA is a novel stochastic process algebra which treats location as a primary feature of each agent, and allows the interaction of agents to be adapted according to their locations. Location is just one possible interpretation of this parameter, and more generally, PALOMA can be seen as supporting *attribute-based communication*. This style of communication has previously been investigated in languages such as SCEL [18], but in that context has not been amenable to scalable analysis. Here we demonstrate how PALOMA may be equipped with scalable analysis through both discrete event and mean field interpretations. Future work will consider suitable notions of equivalence and logic, e.g. it could be useful to consider agents which exert influence of the same type over an analogous region of space, to be equivalent even though their exact locations differ. We will also investigate abstractions of space within PALOMA models.

Acknowledgement. This work is partially supported by the EU project QUAN-TICOL, 600708.

References

1. Cerotti, D., Gribaudo, M., Bobbio, A., Calafate, C.T., Manzoni, P.: A Markovian agent model for fire propagation in outdoor environments. In: Aldini, A., Bernardo, M., Bononi, L., Cortellessa, V. (eds.) EPEW 2010. LNCS, vol. 6342, pp. 131–146. Springer, Heidelberg (2010)
2. Gribaudo, M., Cerotti, D., Bobbie, A.: Analysis of on-off policies in sensor networks using interacting Markovian agents. In: 6th Annual IEEE International Conference on Pervasive Computing and Communications, pp. 300–305. IEEE (2008)
3. Bruneo, D., Scarpa, M., Bobbio, A., Cerotti, D., Gribaudo, M.: Markovian agent modeling swarm intelligence algorithms in wireless sensor networks. Performance Evaluation 69(3), 135–149 (2012)
4. Cerotti, D., Gribaudo, M., Bobbio, A.: Disaster propagation in heterogeneous media via markovian agents. In: Setola, R., Geretshuber, S. (eds.) CRITIS 2008. LNCS, vol. 5508, pp. 328–335. Springer, Heidelberg (2009)
5. Prasad, K.: A calculus of broadcasting systems. Science of Computer Programming 25(2), 285–327 (1995)
6. Hillston, J.: A Compositional Approach to Performance Modelling. CUP (2005)
7. Tribastone, M., Gilmore, S., Hillston, J.: Scalable differential analysis of process algebra models. IEEE Transactions on Software Engineering 38(1), 205–219 (2012)
8. Sattenspiel, L., Herring, D.A.: Simulating the effect of quarantine on the spread of the 1918–19 flu in central Canada. Bull. of Mathematical Biology 65(1), 1–26 (2003)
9. Sangiorgi, D., Walker, D.: The pi-calculus: a Theory of Mobile Processes. CUP (2003)
10. Cardelli, L., Gardner, P.: Processes in space. In: Ferreira, F., Löwe, B., Mayordomo, E., Mendes Gomes, L. (eds.) CiE 2010. LNCS, vol. 6158, pp. 78–87. Springer, Heidelberg (2010)
11. Priami, C.: Stochastic π-calculus. The Computer Journal 38(7), 578–589 (1995)
12. Ciocchetta, F., Hillston, J.: Bio-PEPA: A framework for the modelling and analysis of biological systems. Theoretical Computer Science 410(33), 3065–3084 (2009)
13. Brodo, L., Degano, P., Priami, C.: A stochastic semantics for bioAmbients. In: Malyshkin, V.E. (ed.) PaCT 2007. LNCS, vol. 4671, pp. 22–34. Springer, Heidelberg (2007)
14. Krivine, J., Milner, R., Troina, A.: Stochastic bigraphs. Electronic Notes in Theoretical Computer Science 218, 73–96 (2008)
15. Efthymiou, X., Philippou, A.: A process calculus for spatially-explicit ecological models. In: Application of Membrane Computing, Concurrency and Agent-based Modelling in Population Biology (AMCA-POP 2010), pp. 84–78 (2010)
16. Guenther, M.C., Bradley, J.T.: Higher moment analysis of a spatial stochastic process algebra. In: Thomas, N. (ed.) EPEW 2011. LNCS, vol. 6977, pp. 87–101. Springer, Heidelberg (2011)
17. Bakhshi, R., Endrullis, J., Endrullis, S., Fokkink, W., Haverkort, B.: Automating the mean-field method for large dynamic gossip networks. In: 7th International Conference on the Quantitative Evaluation of Systems, pp. 241–250. IEEE (2010)
18. De Nicola, R., Ferrari, G., Loreti, M., Pugliese, R.: A language-based approach to autonomic computing. In: Beckert, B., Damiani, F., de Boer, F.S., Bonsangue, M.M. (eds.) FMCO 2011. LNCS, vol. 7542, pp. 25–48. Springer, Heidelberg (2012)

On the Discriminating Power of Testing Equivalences for Reactive Probabilistic Systems: Results and Open Problems

Marco Bernardo[1], Davide Sangiorgi[2], and Valeria Vignudelli[2]

[1] Università di Urbino, Italy
[2] Università di Bologna, Italy and INRIA, France

Abstract. Testing equivalences have been deeply investigated on fully nondeterministic processes, as well as on processes featuring probabilities and internal nondeterminism. This is not the case with reactive probabilistic processes, for which it is only known that the discriminating power of probabilistic bisimilarity is achieved when admitting a copying capability within tests. In this paper, we introduce for reactive probabilistic processes three testing equivalences without copying, which are respectively based on reactive probabilistic tests, fully nondeterministic tests, and nondeterministic and probabilistic tests. We show that the three testing equivalences are strictly finer than probabilistic failure-trace equivalence, and that the one based on nondeterministic and probabilistic tests is strictly finer than the other two, which are incomparable with each other. Moreover, we provide a number of facts that lead us to conjecture that (i) may testing and must testing coincide on reactive probabilistic processes and (ii) nondeterministic and probabilistic tests reach the same discriminating power as probabilistic bisimilarity.

1 Introduction

Many relations have been defined in concurrency theory to capture the notion of "same behavior". They range from branching-time relations like (bi)simulations, which are very sensitive to branching points, to linear-time relations based on (decorated) traces, which in contrast abstract to different extents from those points. Most of these relations can be characterized in terms of testing scenarios. Two processes are testing equivalent if, when interacting with them by means of tests encompassing a success predicate, they result in the observation of the same outcomes. By varying the power of tests, it is possible to recover different behavioral relations in the linear-time/branching-time spectrum [15].

The formalization of testing equivalence that we consider in this paper was first introduced in [32] and then revisited in [20]. It is very general, in the sense that it is defined on processes featuring both internal nondeterminism and probabilities. We will describe such processes through a nondeterministic and probabilistic extension of labeled transition systems (LTS) [22], which we call NPLTS, where the target of each transition is a probability distribution over the set of states – in the style of [24,29] – rather than a single state.

G. Norman and W. Sanders (Eds.): QEST 2014, LNCS 8657, pp. 281–296, 2014.

The idea at the basis of this probabilistic testing equivalence, which we denote by $\sim_{\mathrm{PTe-}\sqcup\sqcap}$, is as follows. The interaction system, resulting from an NPLTS process under test and an NPLTS observer, does not have a unique probability of succeeding, but several success probabilities, one for each maximal resolution of nondeterminism. Only the supremum ($\sqcup$) and the infimum ($\sqcap$) of these success probabilities are taken into account in [32,20], so that two processes are deemed equivalent if they result, for each possible test, in the same suprema and infima. Following the terminology of classical testing equivalence [10], the constraint on suprema (resp. infima) – yielding $\sim_{\mathrm{PTe-}\sqcup}$ (resp. $\sim_{\mathrm{PTe-}\sqcap}$) – represents the may (resp. must) part; we know from [12] that $\sim_{\mathrm{PTe-}\sqcap}$ is strictly finer than $\sim_{\mathrm{PTe-}\sqcup}$ in the absence of divergence, i.e., infinite computations whose steps are all invisible.

The relation $\sim_{\mathrm{PTe-}\sqcup\sqcap}$ of [32,20] coincides, over processes and tests resulting in interaction systems with finitely many maximal resolutions, with a slightly finer variant comparing success probabilities of individual maximal resolutions, for which several characterizations were given. In [31], it was shown that $\sim_{\mathrm{PTe-}\sqcup}$ coincides with the coarsest congruence contained in the probabilistic trace-distribution equivalence of [30] and $\sim_{\mathrm{PTe-}\sqcap}$ coincides with the coarsest congruence contained in a probabilistic failure-distribution equivalence.[1] Besides providing logical and equational characterizations, in [11] it was later shown that $\sim_{\mathrm{PTe-}\sqcup}$ coincides with a probabilistic simulation equivalence akin to that of [25] and $\sim_{\mathrm{PTe-}\sqcap}$ coincides with a novel probabilistic failure-simulation equivalence. Such characterizations of $\sim_{\mathrm{PTe-}\sqcup\sqcap}$, together with its position in the spectrum of NPLTS behavioral equivalences studied in [4], reveals that this equivalence has a higher discriminating power with respect to the fully nondeterministic case.

When both the processes and the tests are fully nondeterministic, i.e., LTS models, $\sim_{\mathrm{PTe-}\sqcup\sqcap}$ boils down to the classical testing equivalence of [10]. In this case, as shown in [9] $\sim_{\mathrm{PTe-}\sqcup}$ coincides with trace equivalence and, in the absence of divergence, $\sim_{\mathrm{PTe-}\sqcap}$ coincides with failure equivalence [8]. Several subsequent works addressed how to make classical testing equivalence more powerful. In [27], a higher discriminating power – the one of failure-trace equivalence [15] – was reached by equipping tests with the possibility of expressing the refusal of performing certain actions (refusal testing). Then, it was illustrated in [1] that the discriminating power of bisimulation equivalence [26] can be achieved if, in addition to refusals, two further ingredients are introduced: making copies of intermediate states of the processes under test (copying capability) and enumerating all computations at some point inside a test and combining the related information (global testing). As later observed in [18,12,3], an alternative way of enhancing the discriminating power of classical testing equivalence consists of including probabilities within tests.

Unlike the NPLTS case and the LTS case, very little is known about the discriminating power of the relation $\sim_{\mathrm{PTe-}\sqcup\sqcap}$ of [32,20] over NPLTS models not admitting internal nondeterminism, i.e., Markov decision processes (MDP) [28] or, equivalently, reactive probabilistic labeled transition systems (RPLTS) [16].

---

[1] In [31], countably many different success actions are admitted but, as shown in [13], the single standard one suffices in the case of finitary processes and tests.

An analogous relation was investigated only in [23] for possibly replicated deterministic tests applied to RPLTS models extended with a form of internal choice; this relation is strictly coarser than the probabilistic bisimilarity of [24].

A testing approach for RPLTS models not concerned with extremal success probabilities was studied in [24,7]. It is based on tests formalized through a nonprobabilistic testing language, which allows a tuple of tests to be performed independently on as many copies of the current state of the process under test. The copying capability turns out to be sufficient for the resulting testing equivalence to coincide with the probabilistic bisimilarity of [24], as two RPLTS models that are not probabilistic bisimilar can be distinguished by some such test with probability arbitrarily close to one. As noticed in [6], this statistical approach cannot be exploited for classical bisimilarity, because there are bisimilar LTS models for which no pair of computable probabilizations in the form of RPLTS models renders them indistinguishable with respect to the considered tests.

The purpose of this paper is to examine the discriminating power of the relation $\sim_{\text{PTe-}\sqcup\sqcap}$ of [32,20] when the processes under test are RPLTS models. On the observer side, we consider three different classes of tests: RPLTS, LTS, and NPLTS. In all the three cases, $\sim_{\text{PTe-}\sqcup\sqcap}$ will turn out to be strictly finer than a probabilistic extension of failure-trace equivalence, thereby confirming the power of the interplay between probabilities and nondeterminism discussed in [18,12,3] even when testing RPLTS processes. We then show that the discriminating power of LTS tests and the discriminating power of RPLTS tests are not only below the discriminating power of NPLTS tests, but also incomparable with each other.

Finally, in the setting of testing RPLTS processes, we bring up two problems whose solution seems far from being trivial. The first one refers to may testing and must testing; while the latter is known to be strictly finer than the former for divergence-free LTS or NPLTS processes, we conjecture that they coincide in the case of RPLTS processes. The second one refers to the discriminating power of $\sim_{\text{PTe-}\sqcup\sqcap}$ under NPLTS tests; although no copying capability is admitted, we conjecture that the same identification power as the probabilistic bisimilarity of [24] is achieved. Our conjectures will be substantiated by a number of facts.

Some preliminary results for RPLTS testing are contained in [5]. However, that paper focusses on higher-order languages and addresses, for RPLTS processes, only the case of LTS-based tests generated by CCS-like calculi [26] with and without refusal. In contrast, here we systematically investigate the discriminating power of testing equivalence $\sim_{\text{PTe-}\sqcup\sqcap}$ when applied to RPLTS processes under each of the three classes of tests: RPLTS, LTS, and NPLTS.

This paper is organized as follows. In Sect. 2, we introduce the various LTS-like models that will be used throughout the paper. In Sect. 3, we present the spectrum of behavioral equivalences for RPLTS models by extending results over fully probabilistic processes proved in [21,17]. In Sect. 4, we define the three variants of $\sim_{\text{PTe-}\sqcup\sqcap}$. In Sect. 5, we place the three variants in the RPLTS spectrum and relate their respective discriminating powers. Finally, in Sect. 6 we discuss the two open problems mentioned above and motivate our conjectures about their solution.

2   Background

In this section, we provide definitions and notations for the various LTS-like models used in the paper to formalize processes, tests, and interaction systems.

2.1   Nondeterministic and Probabilistic Processes

The most expressive model that we need is the one that will be used to represent interaction systems, as well as the most powerful observers that we consider. Since it may contain both internal nondeterminism and probabilities, we start by defining it as a slight variation of simple probabilistic automata [29]. In the next two subsections, we derive the submodels employed to represent processes under test, as well as less powerful observers.

Definition 1. *A* nondeterministic and probabilistic labeled transition system, *NPLTS for short, is a triple* $(S, A, \longrightarrow)$ *where S is an at most countable set of states, A is a countable set of transition-labeling actions, and $\longrightarrow \subseteq S \times A \times Distr(S)$ is a transition relation, with $Distr(S)$ being the set of discrete probability distributions over S.* ∎

A transition (s, a, Δ) is written $s \xrightarrow{a} \Delta$. State $s' \in S$ is not reachable from s via that a-transition if $\Delta(s') = 0$, otherwise it is reachable with probability $p = \Delta(s')$. The reachable states form the support of Δ, i.e., $supp(\Delta) = \{s' \in S \mid \Delta(s') > 0\}$. The choice among all the outgoing transitions of s is nondeterministic and can be influenced by the external environment, while the choice of the target state for a specific transition is probabilistic and made internally.

In this setting, a computation is a sequence of state-to-state steps, each denoted by $s \xrightarrow{a} s'$ and derived from a state-to-distribution transition $s \xrightarrow{a} \Delta$.

Definition 2. *Let $\mathcal{L} = (S, A, \longrightarrow)$ be an NPLTS and $s, s' \in S$. A sequence c:*
$$s_0 \xrightarrow{a_1} s_1 \xrightarrow{a_2} s_2 \ldots s_{n-1} \xrightarrow{a_n} s_n$$
is a computation of $\mathcal{L}$ of length n from $s = s_0$ to $s' = s_n$ iff for all $i = 1, \ldots, n$ there exists a transition $s_{i-1} \xrightarrow{a_i} \Delta_i$ such that $s_i \in supp(\Delta_i)$, with $\Delta_i(s_i)$ being the execution probability of step $s_{i-1} \xrightarrow{a_i} s_i$ conditioned on the selection of transition $s_{i-1} \xrightarrow{a_i} \Delta_i$ of $\mathcal{L}$ at state s_{i-1}. Computation c is maximal iff it is not a proper prefix of any other computation. We denote by $\mathcal{C}_{fin}(s)$ the set of finite-length computations from s. ∎

A resolution of a state s of an NPLTS $\mathcal{L}$ is the result of a possible way of resolving nondeterminism starting from s. A resolution is a tree-like structure, whose branching points are probabilistic choices corresponding to target distributions of transitions. This is obtained by unfolding from s the graph structure of $\mathcal{L}$ and by selecting at each reached state at most one of its outgoing transitions.

Definition 3. *Let $\mathcal{L} = (S, A, \longrightarrow_{\mathcal{L}})$ be an NPLTS and $s \in S$. An NPLTS $\mathcal{Z} = (Z, A, \longrightarrow_{\mathcal{Z}})$ is a resolution of s iff there exists a state correspondence function $corr_{\mathcal{Z}} : Z \to S$ such that $s = corr_{\mathcal{Z}}(z_s)$ for some $z_s \in Z$, and for all $z \in Z$ it holds that:*

- If $z \xrightarrow{a}_{\mathcal{Z}} \Delta$, then $corr_{\mathcal{Z}}(z) \xrightarrow{a}_{\mathcal{L}} \Delta'$ with $corr_{\mathcal{Z}}$ being injective over $supp(\Delta)$ and $\Delta(z') = \Delta'(corr_{\mathcal{Z}}(z'))$ for all $z' \in supp(\Delta)$.
- If $z \xrightarrow{a_1}_{\mathcal{Z}} \Delta_1$ and $z \xrightarrow{a_2}_{\mathcal{Z}} \Delta_2$, then $a_1 = a_2$ and $\Delta_1 = \Delta_2$.

Resolution $\mathcal{Z}$ is maximal iff, for all $z \in Z$, whenever z has no outgoing transitions, then $corr_{\mathcal{Z}}(z)$ has no outgoing transitions either. We respectively denote by $Res(s)$ and $Res_{\max}(s)$ the sets of resolutions and maximal resolutions of s. ∎

Since $\mathcal{Z} \in Res(s)$ is fully probabilistic in that each of its states has at most one outgoing transition, the probability $prob(c)$ of executing $c \in \mathcal{C}_{\mathrm{fin}}(z_s)$ can be computed as the product of the (no longer conditional) execution probabilities of the individual steps of c. This notion is lifted to $\mathcal{C} \subseteq \mathcal{C}_{\mathrm{fin}}(z_s)$ by letting $prob(\mathcal{C}) = \sum_{c \in \mathcal{C}} prob(c)$ whenever none of the computations in $\mathcal{C}$ is a proper prefix of one of the others.

2.2   Reactive Probabilistic Processes

A reactive probabilistic process can be described as an RPLTS. This is an NPLTS $(S, A, \longrightarrow)$ in which, for all $s \in S$ and $a \in A$, whenever $s \xrightarrow{a} \Delta_1$ and $s \xrightarrow{a} \Delta_2$, then $\Delta_1 = \Delta_2$. This means that internal nondeterminism is not admitted.

Given a state $s \in S$ and a trace $\alpha \in A^*$, if no resolution of s contains computations labeled with α, then the probability of executing α from s is 0. Otherwise, due to the absence of internal nondeterminism, there exists a resolution of s containing the set $\mathcal{C}(s, \alpha)$ of all the computations from s labeled with α, in which case the probability of executing α from s is assumed to be the value $prob(\mathcal{C}(s, \alpha))$ computed in any of these resolutions containing $\mathcal{C}(s, \alpha)$.

2.3   Fully Nondeterministic Processes

The behavior of a fully nondeterministic process is usually represented through an LTS, which can be viewed as an NPLTS $(S, A, \longrightarrow)$ in which every transition leads to a Dirac distribution, i.e., a distribution that concentrates all the probability mass into a single target state. Formally, a Dirac transition $s \xrightarrow{a} \delta_{s'}$ fulfills $\delta_{s'}(s') = 1$ and $\delta_{s'}(s'') = 0$ for all $s'' \in S \setminus \{s'\}$. In these processes without probabilities, resolutions reduce to computations.

3   The Spectrum of Equivalences for RPLTS Processes

We know from [21,17,19] that the linear-time/branching-time spectrum of behavioral equivalences for fully probabilistic processes is narrower than the one for fully nondeterministic processes [15] as in the former many equivalences coincide. This is the case also with reactive probabilistic processes, as we now show.

Let $\mathcal{L} = (S, A, \longrightarrow)$ be an RPLTS and $s, s_1, s_2 \in S$. We introduce probabilistic trace-based equivalences on $\mathcal{L}$ as follows by analogy with [21,17]:

- $C(s, \alpha)$ is the set of computations from s labeled with trace $\alpha \in A^*$.
  $s_1 \sim_{\mathrm{PTr}} s_2$ iff $prob(C(s_1, \alpha)) = prob(C(s_2, \alpha))$ for all $\alpha \in A^*$.
- $CC(s, \alpha)$ is the set of completed computations from s labeled with $\alpha \in A^*$.
  $s_1 \sim_{\mathrm{PCTr}} s_2$ iff $s_1 \sim_{\mathrm{PTr}} s_2$ and $prob(CC(s_1, \alpha)) = prob(CC(s_2, \alpha))$ for all $\alpha \in A^*$.
- $FC(s, \varphi)$, where $\varphi = (\alpha, F)$ is a failure pair, is the set of computations from s labeled with α such that the last state of each computation cannot perform any action in F.
  $s_1 \sim_{\mathrm{PF}} s_2$ iff $prob(FC(s_1, \varphi)) = prob(FC(s_2, \varphi))$ for all $\varphi \in A^* \times 2^A$.
- $RC(s, \varrho)$, where $\varrho = (\alpha, R)$ is a ready pair, is the set of computations from s labeled with α such that the set of actions that can be performed by the last state of each computation is precisely R.
  $s_1 \sim_{\mathrm{PR}} s_2$ iff $prob(RC(s_1, \varrho)) = prob(RC(s_2, \varrho))$ for all $\varrho \in A^* \times 2^A$.
- $FTC(s, \phi)$, where $\phi = (a_1, F_1) \ldots (a_n, F_n)$ is a failure trace, is the set of computations from s labeled with $a_1 \ldots a_n$ such that the state reached by each computation after the i-th step, $1 \le i \le n$, cannot perform any action in F_i.
  $s_1 \sim_{\mathrm{PFTr}} s_2$ iff $prob(FTC(s_1, \phi)) = prob(FTC(s_2, \phi))$ for all $\phi \in (A \times 2^A)^*$.
- $RTC(s, \rho)$, where $\rho = (a_1, R_1) \ldots (a_n, R_n)$ is a ready trace, is the set of computations from s labeled with $a_1 \ldots a_n$ such that the set of actions that can be performed by the state reached by each computation after the i-th step, $1 \le i \le n$, is precisely R_i.
  $s_1 \sim_{\mathrm{PRTr}} s_2$ iff $prob(RTC(s_1, \rho)) = prob(RTC(s_2, \rho))$ for all $\rho \in (A \times 2^A)^*$.

Probabilistic bisimilarity $\sim_{\mathrm{PB}}$ for RPLTS processes was defined in [24], while probabilistic similarity $\sim_{\mathrm{PS}}$ can be introduced as follows by analogy with [19]. Given a binary relation $\mathcal{R}$ over S, its lifting $\mathcal{R}_{\mathrm{d}}$ to $Distr(S)$ is defined by letting $(\Delta_1, \Delta_2) \in \mathcal{R}_{\mathrm{d}}$ iff there exists a function $w : S \times S \to \mathbb{R}_{[0,1]}$ such that:

- $w(s_1, s_2) > 0 \implies (s_1, s_2) \in \mathcal{R}$ for all $s_1, s_2 \in S$;
- $\Delta_1(s_1) = \sum_{s' \in S} w(s_1, s')$ for all $s_1 \in S$;
- $\Delta_2(s_2) = \sum_{s' \in S} w(s', s_2)$ for all $s_2 \in S$.

A binary relation $\mathcal{R}$ on S is a probabilistic simulation iff, whenever $(s_1, s_2) \in \mathcal{R}$, then for all $a \in A$ it holds that $s_1 \xrightarrow{a} \Delta_1$ implies $s_2 \xrightarrow{a} \Delta_2$ with $(\Delta_1, \Delta_2) \in \mathcal{R}_{\mathrm{d}}$; the equivalence $\sim_{\mathrm{PS}}$ is the kernel of the largest probabilistic simulation. Relation $\mathcal{R}$ is a probabilistic bisimulation iff it is a symmetric probabilistic simulation; the equivalence $\sim_{\mathrm{PB}}$ is the largest probabilistic bisimulation.

It was shown in [2] that $\sim_{\mathrm{PB}}$ and $\sim_{\mathrm{PS}}$ coincide, hence the variants in between (ready similarity, failure similarity, completed similarity) collapse too. Moreover, the proofs of the results in [21,17] for fully probabilistic processes can be smoothly adapted to the RPLTS case, and also extended to deal with $\sim_{\mathrm{PRTr}}$ and $\sim_{\mathrm{PFTr}}$. As a consequence, we have the following spectrum under the assumption that every state has finitely many outgoing transitions.

Proposition 1. *On finitely-branching RPLTS processes, it holds that:*
$$\sim_{\mathrm{PB}} = \sim_{\mathrm{PS}} \subsetneq \sim_{\mathrm{PRTr}} = \sim_{\mathrm{PFTr}} \subsetneq \sim_{\mathrm{PR}} = \sim_{\mathrm{PF}} \subsetneq \sim_{\mathrm{PCTr}} = \sim_{\mathrm{PTr}} \qquad \blacksquare$$

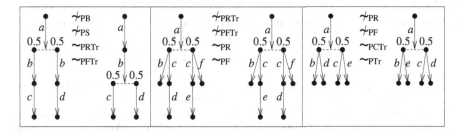

Fig. 1. Processes illustrating the strictness of the inclusions in Prop. 1

The strictness of all the inclusions above is witnessed by the counterexamples in Fig. 1. The graphical conventions for process descriptions are as follows. Vertices represent states and action-labeled edges represent action-labeled transitions. Given a transition $s \xrightarrow{a} \Delta$, the corresponding a-labeled edge goes from the vertex for state s to a set of vertices linked by a dashed line, each of which represents a state $s' \in supp(\Delta)$ and is labeled with $\Delta(s')$. The label $\Delta(s')$ is omitted when it is equal to 1, i.e., when Δ is the Dirac distribution $\delta_{s'}$.

4   Testing Equivalences for RPLTS Processes

In this section, we define a probabilistic testing equivalence for RPLTS processes under three different classes of observers respectively formalized as RPLTS, LTS, and NPLTS tests.

Given an RPLTS, we assume that the elements of its action set A are all visible. The action set of each considered test will be $\bar{A} \cup \{\omega\}$, where $\bar{A} = \{\bar{a} \mid a \in A\}$ is the set of coactions for A and $\omega \notin A$ is a distinguished action denoting success. Every coaction must synchronize with the corresponding action; when this happens, the invisible action $\tau \notin A$ is produced. Therefore, the resulting interaction system is an NPLTS with action set $\{\tau, \omega\}$, whose transition relation $\longrightarrow$ is derived from the transition relation $\longrightarrow_1$ of the RPLTS process under test and the transition relation $\longrightarrow_2$ of the observer, through the following two rules:

$$\frac{s \xrightarrow{a}_1 \Delta_1 \quad o \xrightarrow{\bar{a}}_2 \Delta_2}{(s,o) \xrightarrow{\tau} \Delta_1 \star \Delta_2} \qquad \frac{o \xrightarrow{\omega}_2 \Delta_2}{(s,o) \xrightarrow{\omega} \delta_s \star \Delta_2}$$

where $(\Delta \star \Gamma)(s', o') = \Delta(s') \cdot \Gamma(o')$.

A finite-length computation from the initial state (s, o) of the interaction system is successful iff its last state can perform ω, and no preceding state can perform ω. Given a resolution $\mathcal{Z}$ of (s, o), we denote by $\mathcal{SC}(z_{s,o})$ the set of successful computations from the state $z_{s,o}$ of $\mathcal{Z}$ corresponding to (s, o). We respectively denote by $\sqcup$ and $\sqcap$ the supremum and the infimum of the set of probability values $prob(\mathcal{SC}(z_{s,o}))$ computed in the various resolutions of the interaction system. To avoid infima to be trivially zero, in the next definition, which is inspired by [32,20,23], we restrict ourselves to maximal resolutions.

Definition 4. *Let* $\mathcal{L} = (S, A, \longrightarrow_{\mathcal{L}})$ *be an RPLTS. We say that* $s_1, s_2 \in S$ *are probabilistic* $\sqcup\sqcap$*-testing equivalent, written* $s_1 \sim_{\text{PTe-}\sqcup\sqcap} s_2$*, iff for every test* $\mathcal{T} = (O, A, \longrightarrow_{\mathcal{T}})$ *with initial state* $o \in O$ *it holds that:*

$$\bigsqcup_{\mathcal{Z}_1 \in Res_{\max}(s_1, o)} prob(\mathcal{SC}(z_{s_1, o})) = \bigsqcup_{\mathcal{Z}_2 \in Res_{\max}(s_2, o)} prob(\mathcal{SC}(z_{s_2, o}))$$

$$\bigsqcap_{\mathcal{Z}_1 \in Res_{\max}(s_1, o)} prob(\mathcal{SC}(z_{s_1, o})) = \bigsqcap_{\mathcal{Z}_2 \in Res_{\max}(s_2, o)} prob(\mathcal{SC}(z_{s_2, o}))$$

The equivalence is respectively denoted by $\sim_{\text{PTe-}\sqcup\sqcap,\text{rp}}$*,* $\sim_{\text{PTe-}\sqcup\sqcap,\text{nd}}$*, or* $\sim_{\text{PTe-}\sqcup\sqcap,\text{np}}$ *depending on whether the considered tests are all reactive probabilistic, fully non-deterministic, or nondeterministic and probabilistic.* ∎

We assume tests to be finite, i.e., finite state, finitely branching, and loop free. On the one hand, this entails that interaction systems will have finitely many maximal resolutions, thus ensuring the validity of our results also for a slightly finer variant of $\sim_{\text{PTe-}\sqcup\sqcap}$ that we could define following [31,11]. On the other hand, this restriction will be exploited in the proofs of some results.

5   Properties of the RPLTS Testing Equivalences

5.1   Placing the Testing Equivalences in the RPLTS Spectrum

Our first result is that the three relations $\sim_{\text{PTe-}\sqcup\sqcap,\text{rp}}$, $\sim_{\text{PTe-}\sqcup\sqcap,\text{nd}}$, and $\sim_{\text{PTe-}\sqcup\sqcap,\text{np}}$ are comprised between $\sim_{\text{PFTr}}$ and $\sim_{\text{PB}}$. This confirms the power of the interplay between probabilities and nondeterminism for discriminating purposes, which was already noticed in the testing theory for NPLTS processes [18,12,3].

The proof that each of the three equivalences is strictly finer than $\sim_{\text{PFTr}}$ benefits from an analogous result with respect to $\sim_{\text{PF}}$. Both proofs focus on tests that are deterministic LTS models (DLTS for short) as they admit neither internal nondeterminism nor probabilities. Since these tests constitute a submodel common to RPLTS, LTS, and NPLTS tests, the inclusion proofs relying on them scale to the three more expressive families of tests.

Lemma 1. *On RPLTS processes, for all* $* \in \{\text{rp}, \text{nd}, \text{np}\}$ *it holds that:*
$$\sim_{\text{PTe-}\sqcup\sqcap,*} \subsetneqq \sim_{\text{PF}}$$
∎

Theorem 1. *On RPLTS processes, for all* $* \in \{\text{rp}, \text{nd}, \text{np}\}$ *it holds that:*
$$\sim_{\text{PTe-}\sqcup\sqcap,*} \subsetneqq \sim_{\text{PFTr}}$$
∎

The inclusions in $\sim_{\text{PFTr}}$ are strict as shown by the two RPLTS processes, the DLTS test, and the two NPLTS interaction systems in Fig. 2, because we have $\sqcup = 1$ and $\sqcap = 0$ in the first system and $\sqcup = \sqcap = 0.5$ in the second one.

The proof that $\sim_{\text{PB}}$ is included in each of the three testing equivalences exploits the fact that $\sim_{\text{PB}}$ is a congruence with respect to parallel composition. Inclusion stems from showing that, under $\sim_{\text{PB}}$, for each maximal resolution of any of the two interaction systems, there exists a maximal resolution of the other interaction system, such that the two resolutions have the same success probability. The maximal resolutions to consider are those arising from randomized

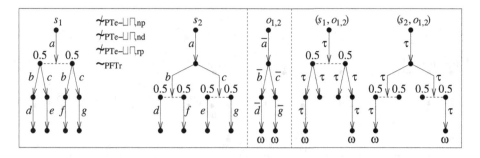

Fig. 2. Processes and test illustrating the strictness of the inclusions of Thm. 1

schedulers, as opposed to the deterministic ones used so far, which means that a convex combination of equally labeled transitions can be selected at each state. Formally, the first clause of Def. 3 changes by requiring that, if $z \xrightarrow{a}_{\mathcal{Z}} \Delta$, then there exist $n \in \mathbb{N}_{\geq 1}$, $(p_i \in \mathbb{R}_{]0,1]} \mid 1 \leq i \leq n)$, and $(corr_{\mathcal{Z}}(z) \xrightarrow{a}_{\mathcal{L}} \Delta_i \mid 1 \leq i \leq n)$ such that $\sum_{i=1}^{n} p_i = 1$ and $\Delta(z') = \sum_{i=1}^{n} p_i \cdot \Delta_i(corr_{\mathcal{Z}}(z'))$ for all $z' \in supp(\Delta)$. Given $s \in S$, we denote by $Res_{\max}^{rnd}(s)$ the set of maximal resolutions of s originated from randomized schedulers.

Lemma 2. *Let* $\mathcal{L} = (S, A, \longrightarrow_{\mathcal{L}})$ *be an RPLTS and* $s_1, s_2 \in S$. *If* $s_1 \sim_{PB} s_2$, *then for every test* $\mathcal{T} = (O, A, \longrightarrow_{\mathcal{T}})$ *with initial state* $o \in O$ *it holds that:*

- *For each* $\mathcal{Z}_1 \in Res_{\max}^{rnd}(s_1, o)$ *there exists* $\mathcal{Z}_2 \in Res_{\max}^{rnd}(s_2, o)$ *such that:*
$$prob(\mathcal{SC}(z_{s_1,o})) = prob(\mathcal{SC}(z_{s_2,o}))$$
- *For each* $\mathcal{Z}_2 \in Res_{\max}^{rnd}(s_2, o)$ *there exists* $\mathcal{Z}_1 \in Res_{\max}^{rnd}(s_1, o)$ *such that:*
$$prob(\mathcal{SC}(z_{s_2,o})) = prob(\mathcal{SC}(z_{s_1,o}))$$ ∎

Theorem 2. *On RPLTS processes, for all* $* \in \{rp, nd, np\}$ *it holds that:*
$$\sim_{PB} \subsetneq \sim_{PTe\text{-}\sqcup\sqcap,*}$$ ∎

5.2   Relationships among the RPLTS Testing Equivalences

Our second result is concerned with the relationships among the discriminating powers of $\sim_{PTe\text{-}\sqcup\sqcap,rp}$, $\sim_{PTe\text{-}\sqcup\sqcap,nd}$, and $\sim_{PTe\text{-}\sqcup\sqcap,np}$, which will help us investigating the strictness of the inclusions of Thm. 2.

First of all, we observe that $\sim_{PTe\text{-}\sqcup\sqcap,np}$ is included both in $\sim_{PTe\text{-}\sqcup\sqcap,rp}$ and in $\sim_{PTe\text{-}\sqcup\sqcap,nd}$, because RPLTS tests and LTS tests are special cases of NPLTS tests. Both inclusions are strict, as shown in the upper part of Fig. 3, where the NPLTS test yields $\sqcup = 0.75$ and $\sqcap = 0.25$ in the first interaction system and $\sqcup = \sqcap = 0.5$ in the second one. We remark the need of both internal nondeterminism and probabilities in the distinguishing test. A linear test succeeding after performing $\bar{a}$, $\bar{b}$, and $\bar{c}$ would not be able to tell apart s_3 and s_4. Likewise, those two states would not be distinguishable by a test obtained from the previous one by replacing the $\bar{c}$-transition with a probabilistic choice between that transition

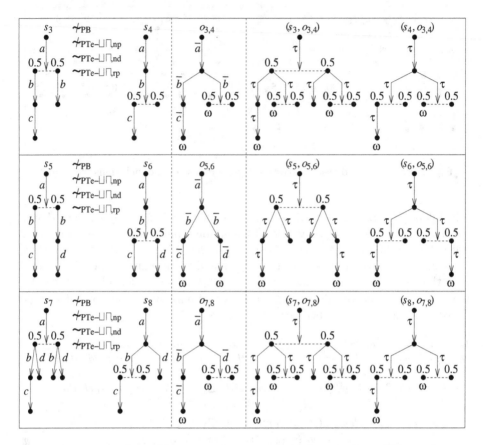

Fig. 3. Processes and tests illustrating the strictness of the inclusions of Thm. 3/Cor. 1

and a terminal/success state, or introducing a nondeterministic choice through a further $\bar{b}$-transition to a terminal/success state after the $\bar{a}$-transition.

Secondly, it turns out that, in general, $\sim_{\text{PTe-}\sqcup\sqcap,\text{rp}}$ and $\sim_{\text{PTe-}\sqcup\sqcap,\text{nd}}$ are incomparable with each other. For instance, in the middle part of Fig. 3 we have that $s_5 \sim_{\text{PTe-}\sqcup\sqcap,\text{rp}} s_6$, while $s_5 \not\sim_{\text{PTe-}\sqcup\sqcap,\text{nd}} s_6$ because the LTS test yields $\sqcup = 1$ and $\sqcap = 0$ in the first interaction system and $\sqcup = \sqcap = 0.5$ in the second one. Notice the necessity of internal nondeterminism in the distinguishing test. In contrast, in the lower part of Fig. 3 we have that $s_7 \sim_{\text{PTe-}\sqcup\sqcap,\text{nd}} s_8$, while $s_7 \not\sim_{\text{PTe-}\sqcup\sqcap,\text{rp}} s_8$ because the RPLTS test yields $\sqcup = 0.75$ and $\sqcap = 0.25$ in the first interaction system and $\sqcup = \sqcap = 0.5$ in the second one. Unlike the upper part of Fig. 3, here internal nondeterminism is not necessary in the distinguishing test.

Thirdly, if $\sim_{\text{PTe-}\sqcup\sqcap,\text{rp}}$ admitted only restricted RPLTS tests, then it would include $\sim_{\text{PTe-}\sqcup\sqcap,\text{nd}}$, with the inclusion being strict as shown in the middle part of Fig. 3. A restricted RPLTS (RRPLTS for short) test is such that its probabilistic choices, i.e., its non-Dirac transitions, are not preceded by nondeterministic choices. The proof of this fact is based on the deprobabilization of an

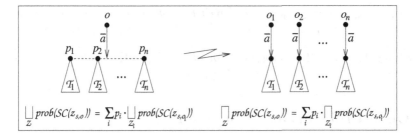

Fig. 4. Deprobabilization of an RRPLTS test (applies recursively to $\mathcal{T}_1, \mathcal{T}_2, \ldots, \mathcal{T}_n$)

RRPLTS test. This is an algorithm that performs a top-down traversal of the test until a set of DLTS subtests is generated, which preserves the extremal success probabilities induced by the original test.

When encountering a non-Dirac transition in the top-down traversal of the RRPLTS test, as shown in Fig. 4 the algorithm replaces the test with as many RRPLTS subtests – which are DLTS subtests in the final steps – as there are ways of resolving the probabilistic choice. For simplicity, only the non-Dirac transition, labeled with $\bar{a}$, originating the probabilistic choice is depicted in the figure, but in general it could be the last transition in a computation – traversing states where no nondeterministic choices occur – going from the initial state o of the test to the probabilistic choice. Given a state s of the process under test, the two formulas in Fig. 4 witness that the two convex combinations of the extremal success probabilities induced by the n subtests respectively coincide with the two extremal success probabilities induced by the original test.

Should a nondeterministic choice precede the considered probabilistic choice, it would not be appropriate to generate subtests by resolving both choices. The reason is that it would then be natural to focus on the maximum and the minimum of the extremal success probabilities induced by the various subtests arising from the resolution of the nondeterministic choice. This certainly works when the nondeterministic choice is originated from the initial state of the test, or from the state reached by a Dirac transition of the test that synchronizes with a Dirac transition of the process under test. However, the synchronization of a Dirac transition of the test with a non-Dirac transition of the process results in a non-Dirac transition in the interaction system, for which a convex combination (as opposed to maximum and minimum) of the extremal success probabilities of the various subtests needs to be computed.

Fourthly, if $\sim_{\text{PTe-}\sqcup\sqcap,\text{nd}}$ admitted only DLTS tests, then it would include $\sim_{\text{PTe-}\sqcup\sqcap,\text{rp}}$, with the inclusion being strict as shown in the lower part of Fig. 3. The reason is that a DLTS test is a special case of RPLTS test in which there are no probabilistic choices. In conclusion, we have:

Theorem 3. *On RPLTS processes, it holds that:*

1. $\sim_{\text{PTe-}\sqcup\sqcap,\text{np}} \subsetneq \sim_{\text{PTe-}\sqcup\sqcap,\text{nd}}$ *and* $\sim_{\text{PTe-}\sqcup\sqcap,\text{np}} \subsetneq \sim_{\text{PTe-}\sqcup\sqcap,\text{rp}}$.
2. $\sim_{\text{PTe-}\sqcup\sqcap,\text{nd}}$ *and* $\sim_{\text{PTe-}\sqcup\sqcap,\text{rp}}$ *are incomparable with each other.*

3. $\sim_{\text{PTe-}\sqcup\sqcap,\text{nd}} \subsetneq \sim_{\text{PTe-}\sqcup\sqcap,\text{rp}}$ *if only RRPLTS tests were admitted by* $\sim_{\text{PTe-}\sqcup\sqcap,\text{rp}}$.

4. $\sim_{\text{PTe-}\sqcup\sqcap,\text{rp}} \subsetneq \sim_{\text{PTe-}\sqcup\sqcap,\text{nd}}$ *if only DLTS tests were admitted by* $\sim_{\text{PTe-}\sqcup\sqcap,\text{nd}}$. ∎

Corollary 1. *On RPLTS processes, for all* $* \in \{\text{rp}, \text{nd}\}$ *it holds that:*
$$\sim_{\text{PB}} \subsetneq \sim_{\text{PTe-}\sqcup\sqcap,*}$$ ∎

6   Open Problems and Conjectures

In this section, we address further issues related to testing equivalences for RPLTS processes. Rather than proving new results, the value of this section consists of highlighting two problems that have not received attention in the literature so far, and then proposing two conjectures for them sustained by various arguments. We hope that these discussions will help other people finding solutions to the conjectures. We expect that their proof (or their refutation) will shed light on the subtle interplay between probabilities and nondeterminism.

6.1   May Testing vs. Must Testing

In the case of testing LTS or NPLTS processes, it is known that must testing equivalence is strictly finer than may testing equivalence in the absence of divergence, otherwise the two equivalences are incomparable [9,12]. When testing RPLTS processes, the relationships between $\sim_{\text{PTe-}\sqcup}$ (may testing) and $\sim_{\text{PTe-}\sqcap}$ (must testing) are not clear, even if we restrict ourselves to NPLTS tests and we admit τ-actions within them.

In that case, we could derive that $\sim_{\text{PTe-}\sqcap,\text{np}} \subseteq \sim_{\text{PTe-}\sqcup,\text{np}}$ by exploiting the construction used in [12] for proving an analogous result on NPLTS processes. The purpose of that construction is to build from a given NPLTS test a dual one, which generates all complementary success probabilities in the interaction system. The idea is to transform every state of the test having an outgoing ω-transition into a terminal state, and to add to any other state a τ-transition followed by an ω-transition.

The absence of internal nondeterminism within RPLTS processes would however prevent us from concluding that the above inclusion is strict. Indeed, the typical counterexample made out of a test succeeding after performing $\bar{a}$ followed by $\bar{b}$, which distinguishes a process that can perform either a followed by b, or a followed by c, from a process that can perform a and then has a choice between b and c, is not applicable because the first process is not an RPLTS.

Such considerations lead us to conjecture that, for each of the three variants of $\sim_{\text{PTe-}\sqcup\sqcap}$, its may part $\sim_{\text{PTe-}\sqcup}$ coincides with its must part $\sim_{\text{PTe-}\sqcap}$, and hence both coincide with $\sim_{\text{PTe-}\sqcup\sqcap}$ by virtue of the definition of the latter. This is certainly true when restricting attention to fully probabilistic tests – as they yield, when interacting with an RPLTS process, a single maximal resolution, in which $\sqcup$ and $\sqcap$ necessarily coincide – or tests having exactly one nondeterministic choice that occurs in the initial state – as can be easily proved by induction on the number of maximal resolutions of each such test.

Conjecture 1. On RPLTS processes, for all $* \in \{\text{rp}, \text{nd}, \text{np}\}$ it holds that:

$$\sim_{\text{PTe-}\sqcup,*} = \sim_{\text{PTe-}\sqcap,*} = \sim_{\text{PTe-}\sqcup\sqcap,*} \qquad \blacksquare$$

6.2   Characterizing RPLTS Testing Equivalences

Our findings in Sect. 5 leave open the question whether $\sim_{\text{PB}}$ is strictly finer than $\sim_{\text{PTe-}\sqcup\sqcap,\text{np}}$ or coincides with it. In the latter case, we would have that, in the RPLTS setting, testing equivalence reaches the same discriminating power as bisimilarity not only in the presence of an explicit copying capability within tests [24], but also in the absence of it, provided that tests are equipped with both internal nondeterminism and probabilities. We point out that this would be a peculiarity of RPLTS processes, because it is known that NPLTS tests are less powerful than bisimilarity in the case of NPLTS processes [4]. The numerous examples of RPLTS processes that we have examined lead us to the following:

Conjecture 2. On RPLTS processes, it holds that $\sim_{\text{PTe-}\sqcup\sqcap,\text{np}} = \sim_{\text{PB}}$. $\qquad \blacksquare$

As a consequence of Thm. 2, it suffices to prove that $\sim_{\text{PTe-}\sqcup\sqcap,\text{np}}$ is included in $\sim_{\text{PB}}$. This is equivalent to show that, given two states s_1 and s_2 of an RPLTS, if $s_1 \nsim_{\text{PB}} s_2$, then $s_1 \nsim_{\text{PTe-}\sqcup\sqcap,\text{np}} s_2$. The idea is to build a distinguishing NPLTS test from a distinguishing formula of PML, the modal logic characterizing $\sim_{\text{PB}}$ on RPLTS processes [24]. In its minimal form [14], PML comprises the constant true, logical conjunction $\cdot \wedge \cdot$, and the diamond operator $\langle a \rangle_p \cdot$ where a is an action and p is a probability lower bound. Formula $\langle a \rangle_p \phi$ is satisfied by an RPLTS state if an a-labeled transition is possible from that state, after which a set of states satisfying ϕ is reached with probability at least p.

The proof of the conjecture appears far from being trivial. The connection between PML and the testing approach of [24] is intuitively clear, as multiplying the success probabilities resulting from the application of independent choice-free tests to as many copies of the current state under test is analogous to taking the logical conjunction of a number of formulas each starting with a suitably decorated diamond. In contrast, our tests follow the classical theory of [10], hence do not admit any copying capability and, most importantly, may contain choices, which fit well together with logical disjunction rather than conjunction.

Nevertheless, on the basis of the examined examples, we have developed a procedure that, given an appropriate PML formula ϕ that is satisfied by s_1 but not s_2, builds an NPLTS test $\mathcal{T}(\phi)$ that should tell apart s_1 and s_2 (see Fig. 5). By appropriate PML formula, we mean that ϕ possesses the following three properties. First, among all the PML formulas distinguishing s_1 from s_2, ϕ is one of those with the minimum depth, where the depth of a formula is the maximum number of nested diamond operators occurring in the formula itself. Second, among all the distinguishing PML formulas of minimum depth, ϕ is one of those with the minimum number of conjunctions. Third, all the probability lower bounds in ϕ are maximal, in the sense that, as soon as one of them is increased, s_1 no longer satisfies the resulting formula.

If $depth(\phi) = 1$, then $\phi = \langle a \rangle_1 \text{true}$ in our RPLTS setting, and hence $\mathcal{T}(\phi)$ simply has an $\bar{a}$-transition followed by an ω-transition. If $depth(\phi) \geq 2$, then

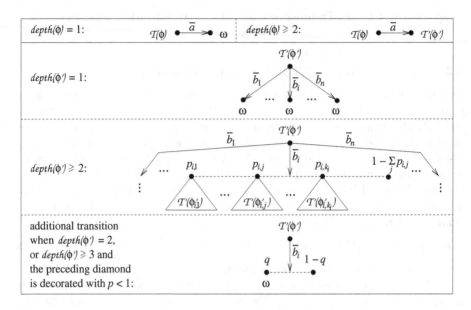

Fig. 5. Construction of the presumably distinguishing test $\mathcal{T}(\phi)$ based on $\mathcal{T}'(\phi')$

$\phi = \langle a \rangle_p \phi'$ because the initial state of an RPLTS has a nondeterministic choice among differently labeled transitions. As a consequence, $\mathcal{T}(\phi)$ has an $\bar{a}$-transition to the initial state of $\mathcal{T}'(\phi')$, which is recursively built as follows.

If $depth(\phi') = 1$, then $\phi' = \bigwedge_{1 \leq i \leq n} \langle b_i \rangle_1$ true, where $n \in \mathbb{N}_{\geq 1}$ and $b_{i_1} \neq b_{i_2}$ for $i_1 \neq i_2$. In this case, $\mathcal{T}'(\phi')$ has a nondeterministic choice among n transitions respectively labeled with $\bar{b}_1, \bar{b}_2, \ldots, \bar{b}_n$, each followed by an ω-transition. If $depth(\phi') \geq 2$, then $\phi' = \bigwedge_{1 \leq i \leq n} (\bigwedge_{1 \leq j \leq k_i} \langle b_i \rangle_{p_{i,j}} \phi'_{i,j})$, where $n \in \mathbb{N}_{\geq 1}$, $b_{i_1} \neq b_{i_2}$ for $i_1 \neq i_2$, and $k_i \in \mathbb{N}_{\geq 1}$ for all $i = 1, \ldots, n$ with $k_i > 1$ implying that $\phi'_{i,j} \neq$ true for all $j = 1, \ldots, k_i$. In this case, $\mathcal{T}'(\phi')$ has a nondeterministic choice among n transitions respectively labeled with $\bar{b}_1, \bar{b}_2, \ldots, \bar{b}_n$, with the i-th transition reaching a distribution Δ_i that, for each $j = 1, \ldots, k_i$, assigns probability $p_{i,j}$ to the initial state of $\mathcal{T}'(\phi'_{i,j})$; whenever the various probabilities $p_{i,j}$ do not sum up to 1, the residual probability is assigned by Δ_i to a terminal state. Test $\mathcal{T}'(\phi'_{i,j})$ simply has an ω-transition when $\phi'_{i,j} =$ true.

As far as the capability of discriminating s_1 and s_2 is concerned, there are two critical points in the construction of $\mathcal{T}(\phi)$. One of them is the last but one diamond operator occurring within each subformula of ϕ. Due to the minimality of ϕ with respect to diamond nesting depth, this is precisely a point in which a source of non-bisimilarity arises. Thus, when $depth(\phi') = 2$, we add to $\mathcal{T}'(\phi')$ a transition labeled with $\bar{b}_i$ for some subformula $(\bigwedge_{1 \leq j \leq k_i} \langle b_i \rangle_{p_{i,j}} \phi'_{i,j})$ having depth 2; the transition reaches with a suitable probability q a success state (i.e., a state having an ω-transition) and with probability $1 - q$ a terminal state.

To explain the role of this additional transition, consider the two $\sim_{\mathrm{PB}}$-inequivalent states s_3 and s_4 in the upper part of Fig. 3. The conjunction-free

PML formula $\phi = \langle a \rangle_{0.5} \langle b \rangle_1 \langle c \rangle_1$ true is satisfied by s_3 but not s_4. However, as argued at the beginning of Sect. 5.2, an additional transition that introduces both internal nondeterminism and a probabilistic choice between a success state and a terminal one is needed in the test to be able to distinguish s_3 and s_4.

The other critical point is any diamond operator, preceding the last but one, which is decorated with a probability lower bound less than 1. Due to the maximality of ϕ with respect to probability lower bounds, this is again a point in which a source of non-bisimilarity arises. Thus, when $depth(\phi') \geq 3$ and the diamond operator immediately preceding ϕ' is decorated with a probability lower bound less than 1, we add to $\mathcal{T}'(\phi')$ a transition labeled with $\bar{b}_i$ for some subformula $(\bigwedge_{1 \leq j \leq k_i} \langle b_i \rangle_{p_{i,j}} \phi'_{i,j})$ having depth at least 3; as before, the transition reaches with a suitable probability q a state equipped with an ω-transition and with probability $1 - q$ a terminal state.

We conclude by mentioning that an alternative proof strategy for Conj. 2 may exploit Prop. 1 ($\sim_{PB} = \sim_{PS}$), Conj. 1, and the characterization of may testing via simulation provided by [11]. However, we recall that in [11] τ-actions are admitted, the considered probabilistic simulation is not the standard one, and the focus is on preorders rather than equivalences.

Acknowledgment. We are grateful to the anonymous referees for their useful comments. This work is supported by MIUR-PRIN project CINA and ANR project 12IS02001 PACE.

References

1. Abramsky, S.: Observational equivalence as a testing equivalence. Theoretical Computer Science 53, 225–241 (1987)
2. Baier, C., Kwiatkowska, M.: Domain equations for probabilistic processes. Mathematical Structures in Computer Science 10, 665–717 (2000)
3. Bernardo, M., De Nicola, R., Loreti, M.: Revisiting trace and testing equivalences for nondeterministic and probabilistic processes. Logical Methods in Computer Science 10(1:16), 1–42 (2014)
4. Bernardo, M., De Nicola, R., Loreti, M.: Relating strong behavioral equivalences for processes with nondeterminism and probabilities. Theoretical Computer Science (2014)
5. Bernardo, M., Sangiorgi, D., Vignudelli, V.: On the discriminating power of passivation and higher-order interaction. In: Proc. of CSL/LICS. ACM Press (2014)
6. Bloom, B., Meyer, A.R.: A remark on bisimulation between probabilistic processes. In: Meyer, A.R., Taitslin, M.A. (eds.) Logic at Botik 1989. LNCS, vol. 363, pp. 26–40. Springer, Heidelberg (1989)
7. van Breugel, F., Mislove, M., Ouaknine, J., Worrell, J.: Domain theory, testing and simulation for labelled Markov processes. Theoretical Computer Science 333, 171–197 (2005)
8. Brookes, S.D., Hoare, C.A.R., Roscoe, A.W.: A theory of communicating sequential processes. Journal of the ACM 31, 560–599 (1984)
9. De Nicola, R.: Extensional equivalences for transition systems. Acta Informatica 24, 211–237 (1987)
10. De Nicola, R., Hennessy, M.: Testing equivalences for processes. Theoretical Computer Science 34, 83–133 (1984)

11. Deng, Y., van Glabbeek, R.J., Hennessy, M., Morgan, C.: Characterising testing preorders for finite probabilistic processes. Logical Methods in Computer Science 4(4:4), 1–33 (2008)
12. Deng, Y., van Glabbeek, R.J., Hennessy, M., Morgan, C., Zhang, C.: Remarks on testing probabilistic processes. In: Computation, Meaning, and Logic: Articles Dedicated to Gordon Plotkin. ENTCS, vol. 172, pp. 359–397. Elsevier (2007)
13. Deng, Y., van Glabbeek, R.J., Morgan, C., Zhang, C.: Scalar outcomes suffice for finitary probabilistic testing. In: De Nicola, R. (ed.) ESOP 2007. LNCS, vol. 4421, pp. 363–378. Springer, Heidelberg (2007)
14. Desharnais, J., Edalat, A., Panangaden, P.: Bisimulation for labelled Markov processes. Information and Computation 179, 163–193 (2002)
15. van Glabbeek, R.J.: The linear time – branching time spectrum I. In: Handbook of Process Algebra, pp. 3–99. Elsevier (2001)
16. van Glabbeek, R.J., Smolka, S.A., Steffen, B.: Reactive, generative and stratified models of probabilistic processes. Information and Computation 121, 59–80 (1995)
17. Huynh, D.T., Tian, L.: On some equivalence relations for probabilistic processes. Fundamenta Informaticae 17, 211–234 (1992)
18. Jonsson, B., Ho-Stuart, C., Yi, W.: Testing and refinement for nondeterministic and probabilistic processes. In: Langmaack, H., de Roever, W.-P., Vytopil, J. (eds.) FTRTFT 1994 and ProCoS 1994. LNCS, vol. 863, pp. 418–430. Springer, Heidelberg (1994)
19. Jonsson, B., Larsen, K.G.: Specification and refinement of probabilistic processes. In: Proc. of LICS, pp. 266–277. IEEE-CS Press (1991)
20. Jonsson, B., Yi, W.: Compositional testing preorders for probabilistic processes. In: Proc. of LICS, pp. 431–441. IEEE-CS Press (1995)
21. Jou, C.-C., Smolka, S.A.: Equivalences, congruences, and complete axiomatizations for probabilistic processes. In: Baeten, J.C.M., Klop, J.W. (eds.) CONCUR 1990. LNCS, vol. 458, pp. 367–383. Springer, Heidelberg (1990)
22. Keller, R.M.: Formal verification of parallel programs. Communications of the ACM 19, 371–384 (1976)
23. Kwiatkowska, M., Norman, G.: A testing equivalence for reactive probabilistic processes. In: Proc. of EXPRESS. ENTCS, vol. 16(2), pp. 114–132. Elsevier (1998)
24. Larsen, K.G., Skou, A.: Bisimulation through probabilistic testing. Information and Computation 94, 1–28 (1991)
25. Lynch, N.A., Segala, R., Vaandrager, F.: Compositionality for probabilistic automata. In: Amadio, R.M., Lugiez, D. (eds.) CONCUR 2003. LNCS, vol. 2761, pp. 208–221. Springer, Heidelberg (2003)
26. Milner, R.: Communication and Concurrency. Prentice Hall (1989)
27. Phillips, I.: Refusal testing. Theoretical Computer Science 50, 241–284 (1987)
28. Puterman, M.L.: Markov Decision Processes: Discrete Stochastic Dynamic Programming. John Wiley & Sons (1994)
29. Segala, R.: Modeling and Verification of Randomized Distributed Real-Time Systems. PhD Thesis (1995)
30. Segala, R.: A compositional trace-based semantics for probabilistic automata. In: Lee, I., Smolka, S.A. (eds.) CONCUR 1995. LNCS, vol. 962, pp. 234–248. Springer, Heidelberg (1995)
31. Segala, R.: Testing probabilistic automata. In: Sassone, V., Montanari, U. (eds.) CONCUR 1996. LNCS, vol. 1119, pp. 299–314. Springer, Heidelberg (1996)
32. Yi, W., Larsen, K.G.: Testing probabilistic and nondeterministic processes. In: Proc. of PSTV, pp. 47–61. North-Holland (1992)

Continuity Properties of Distances
for Markov Processes

Manfred Jaeger[1], Hua Mao[2], Kim Guldstrand Larsen[1], and Radu Mardare[1]

[1] Department of Computer Science, Aalborg University, Denmark
{jaeger,kgl,mardare}@cs.aau.dk
[2] College of Computer Science, Sichuan University, P.R. China
maohuamh@gmail.com

Abstract. In this paper we investigate distance functions on finite state
Markov processes that measure the behavioural similarity of non-bisimilar
processes. We consider both probabilistic bisimilarity metrics, and trace-
based distances derived from standard Lp and Kullback-Leibler distances.
Two desirable continuity properties for such distances are identified. We
then establish a number of results that show that these two properties are
in conflict, and not simultaneously fulfilled by any of our candidate natural
distance functions. An impossibility result is derived that explains to some
extent the fundamental difficulty we encounter.

1 Introduction

Markov processes are widely used as formal system models in the presence of uncertainty. In the formal analysis of such models, notions of equivalence traditionally play a key role [9]. However, there is an increasing interest in approximate
models, such as simplified models obtained by model abstraction, or models that
are automatically learned by statistical inference from empirical data [12,10].
When analysing the relationship between a true model and its approximation,
then equivalence clearly is too strong a criterion. Therefore, concepts of approximate equivalence that generalize probabilistic bisimulation equivalence via the
introduction of bisimulation distances have received some attention [5,18,17,3,1].

It turns out however, that some of these distances violate some natural properties one would expect from a distance function that in a meaningful sense
measures the quality of approximation. As an example, consider the automaton
$\mathcal{M}_\epsilon$ shown in Figure 1 representing a process where a biased ($\epsilon \neq 0$) or unbiased
($\epsilon = 0$) coin is tossed repeatedly. For a small $\epsilon > 0$ the model $\mathcal{M}_\epsilon$ would be
considered a good approximation of the model $\mathcal{M}_0$, and if a distance measure d
represents quality of approximation, then $d(\mathcal{M}_0, \mathcal{M}_\epsilon)$ should go to zero as $\epsilon \to 0$.
This property, which we will formalize as *parameter continuity*, is not satisfied
by the original bisimulation distances (though it turns out to be satisfied by the
discounted versions of these distances).

Parameter continuity is not the only requirement we have on a distance function. It should also be the case that when one model is a good approximation of
another model according to a given distance function, then some upper bounds

G. Norman and W. Sanders (Eds.): QEST 2014, LNCS 8657, pp. 297–312, 2014.
© Springer International Publishing Switzerland 2014

are implied on the error incurred by using the approximation instead of the real model. We can, thus, formulate two high-level objectives for the design of a distance function:

O1 If $(\mathcal{M}_n)_n$ is a sequence of approximations for a target model $\mathcal{M}$, and for increasing n, $\mathcal{M}_n$ is obtained by applying an increasing amount of resources to obtain a good approximation, then $d(\mathcal{M}, \mathcal{M}_n) \rightarrow 0$.

O2 In a particular use scenario for an approximate model $\mathcal{M}'$, an upper distance bound $d(\mathcal{M}, \mathcal{M}') < \delta$ between $\mathcal{M}'$ and the correct model $\mathcal{M}$ should imply an upper bound on the error, or loss, incurred when using $\mathcal{M}'$ instead of $\mathcal{M}$ in the given scenario.

We here have formulated these two objectives in a deliberately vague manner in order to emphasize that they can give rise to a variety of more concrete, formal conditions. One aspect of objective O1 will be captured by the parameter continuity condition illustrated by Figure 1, and formally defined in Section 4.1 below. Parameter continuity matches the informal description of O1 in the sense that if the correct model is $\mathcal{M}_0$, then models obtained by an increasing amount of approximation effort (e.g., learned or constructed from an increasing amount of empirical data) will be of the form $\mathcal{M}_\epsilon$ with decreasing ϵ.

Objective O2 was the main design criterion in the development of the probabilistic bisimulation metrics: a bound on the probabilistic bisimulation distance implies the same bound on the difference in probability for all properties definable in certain formal languages. We follow the same approach, and partly capture the broad objective O2 by a formal condition we will call *property continuity*.

O1 and O2 are conflicting objectives. Each one, individually, has a trivial solution: O1 will be satisfied by a "minimal" distance that is constant zero (we will allow distances that are not metrics, and where non-identical models can have distance zero). O2, on the other hand, is satisfied by any "maximal" distance, where any two non-identical models have the maximal possible distance, typically 1 or ∞. The challenge, then, is to find distances that in a meaningful manner balance O1 and O2.

In this paper we investigate how a number of different distance functions perform with regard to the criteria of parameter and property continuity. Besides the existing bisimulation distances, our main interest is with *trace-based* distances that measure the distance between automata only as a function of the

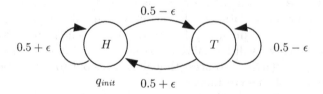

Fig. 1. Biased coin model $\mathcal{M}_\epsilon$

probability distributions over infinite sequences defined by the automata. We here study several constructions of distance measures derived from the standard L_p and Kullback-Leibler distances. It will turn out that the conflict between O1 and O2 is not fully resolved by any of our candidate distance measures, and we will derive an impossibility result that explains to some extent the fundamental difficulty we encounter.

2   Preliminaries

Throughout, Σ denotes a finite alphabet; $\Sigma^n, \Sigma^*, \Sigma^\omega$ denote the sets of all strings of length n, all finite strings, and all infinite strings, respectively. A finite string $w \in \Sigma^*$ defines the *cylinder set* $w\Sigma^\omega \subseteq \Sigma^\omega$. This is just the set of all infinite strings with prefix w. Let Cyl denote the set of all cylinder sets. Cyl is the basis of the standard topology $\mathcal{O}(\Sigma^\omega)$ on Σ^ω, i.e., open sets in this topology are just unions of cylinder sets. Furthermore, the cylinder sets generate the σ-algebra $\mathcal{A}(\Sigma^\omega)$ on Σ^ω.

The basic automaton model we use in this work is that of a *Labeled Markov Chain*, or, more specifically, state-labelled, discrete time Markov chain.

Definition 1. *LMC A* labeled Markov chain (LMC) *over* Σ *is a tuple* $\mathcal{M} = \langle Q, \Sigma, \Pi, \pi, L \rangle$, *where*

- *Q is a finite set of states,*
- *$\Pi : Q \rightarrow [0,1]$ is an* initial probability distribution *with* $\sum_{q \in Q} \Pi(q) = 1$,
- *$\pi : Q \times Q \rightarrow [0,1]$ is the* transition probability function *such that for all* $q \in Q$, $\sum_{q' \in Q} \pi(q, q') = 1$.
- *$L : Q \rightarrow \Sigma$ is a* labeling function

If $\Pi(q_{\text{init}}) = 1$ for some unique initial state q_{init} *of Q, then we denote $\mathcal{M}$ also as* $\langle Q, \Sigma, q_{\text{init}}, \pi, L \rangle$. *In contexts where the initial distribution Π does not matter, we also simply consider the structure* $\langle Q, \Sigma, \pi, L \rangle$ *as a LMC.*

An LMC is deterministic *if a state $q_{\text{init}} \in Q$ as described above exists, and for all $q \in Q$, $\sigma \in \Sigma$ there exists at most one state $q' \in Q$ with $\pi(q, q') > 0$ and $L(q') = \sigma$.*

According to the preceding definition, we assume that each state is labelled with a unique symbol from Σ, not by a subset of a set of atomic propositions AP, as, e.g., in [2]. Clearly, by taking $\Sigma = 2^{AP}$, Definition 1 also accommodates this alternative view of labeled Markov chains.

An LMC defines for each n a probability distribution over Q^n, which induces via the mapping $q_{i_1} \ldots q_{i_n} \mapsto L(q_{i_1}) \ldots L(q_{i_n})$ a probability distribution on Σ^n. Via standard measure-theoretic constructions, these distributions define a unique distribution on $\mathcal{A}(\Sigma^\omega)$, which we denote by $P_\mathcal{M}$. When the initial distribution Π of $\mathcal{M}$ is re-defined to assign probability one to $q \in Q$, then we denote the distribution defined by the resulting LMC by $P_{\mathcal{M},q}$. This can be simplified to P_q, when the underlying structure $\langle Q, \Sigma, \cdot, \pi, L \rangle$ is clear from the context.

Linear-time properties related to traces of the model can be expressed in linear-time temporal logic (LTL) enriched also with the derived temporal operators □ (always) and ◊ (eventually). The fragment of LTL obtained by omitting the until operator $\varphi_1 U \varphi_2$ is called *bounded LTL (BLTL)*.

3 From Equivalence to Distance

The most fundamental approach to comparing system models is by means of concepts of system equivalence. For non-probabilistic system models, the basic tools here are bisimulation and trace equivalence. Adapted to probabilistic system models, this gives rise to the following two notions of equivalence.

Definition 2 (Probabilistic Bisimulation [9]). *Let* $\mathcal{M} = \langle Q, \Sigma, \Pi, \pi, L \rangle$ *be an LMC. A probabilistic bisimulation on* $\mathcal{M}$ *is an equivalence relation* R *on* Q *such that for all states* $(q_1, q_2) \in R$:

- $L(q_1) = L(q_2)$.
- $\pi(q_1, C) = \pi(q_2, C)$ *for each equivalence class* $C \in Q/R$.

States q_1 *and* q_2 *are bisimulation-equivalent (or bisimilar), denoted* $q_1 \sim q_2$, *if there exists a bisimulation* R *on* $\mathcal{M}$ *such that* $(q_1, q_2) \in R$.

Definition 3 (Probabilistic Trace Equivalence). *Two states* $q_1 \in \mathcal{M}_1, q_2 \in \mathcal{M}_2$ *are probabilistic trace equivalent, denoted* $q_1 \overset{T}{\sim} q_2$, *if* $P_{\mathcal{M}_1, q_1} = P_{\mathcal{M}_2, q_2}$.

Equivalence often is too strong a condition when comparing system models. We therefore also need quantitative measures that allow us to determine whether one system very closely resembles another system, without being completely indistinguishable in the sense of an equivalence relation. We study such measures given in the form of distance functions, where small distance indicates similarity, and zero distance means equivalence.

Thus, we consider *distance functions* d that map pairs of states to non-negative numbers: $d : (q_1, q_2) \to \mathbb{R}^{\geq 0} \cup \{\infty\}$. The only condition we always require is that $d(q, q) = 0$. If $d < \infty$, d is symmetric and satisfies the triangle inequality, then d is called a *pseudo-metric*. If also $q_1 \neq q_2 \Rightarrow d(q_1, q_2) > 0$, then d is a *metric*. We note that as a measure of approximation quality, non-symmetric distances can be quite natural, because here the two arguments of the distance function can have distinct roles: one being the approximation, and one being the "real" model that is approximated. For example, if ϕ expresses a crucial safety property, and $\mathcal{M}_1, \mathcal{M}_2$ are LMCs with $P_{\mathcal{M}_1}(\neg\phi) = 0$ and $P_{\mathcal{M}_2}(\neg\phi) = 10^{-5}$, then $\mathcal{M}_2$ may be considered a good (i.e., safe) approximation of $\mathcal{M}_1$, but not vice-versa.

A distance function d is *consistent with bisimilarity* if $d(q_1, q_2) = 0 \Leftrightarrow q_1 \sim q_2$; it is *consistent with trace equivalence* if $d(q_1, q_2) = 0 \Leftrightarrow q_1 \overset{T}{\sim} q_2$. If a distance is consistent with trace equivalence, then the implication $q_1 \sim q_2 \Rightarrow d(q_1, q_2) = 0$ still holds, but not the converse.

We next introduce two types of distance functions. First we consider distance functions that are quantitative extensions of bisimulation equivalence, and then distance functions that extend trace equivalence.

3.1   Probabilistic Bisimilarity Metric

The bisimilarity pseudometric was originally introduced by means of logical expressions that are evaluated to real numbers at system states according to a functional semantics [5,18]. The distance of two states then is defined as the supremum over all logical expressions of the differences of function values. Alternative characterizations as the fixedpoint of monotone operators on pseudometrics have been developed in [17,3,1].

However, there are some differences in the assumed underlying system models in these papers, and the literature does not fully establish an equivalence of all available versions of bisimilarity distance for the LMC models we here use. In the following we therefore review one particular formalization of the bisimilarity pseudometrics in terms of *couplings*.

The definitions below follow the style traditionally used in the bisimulation context in that a distance is defined between states q_1, q_2 in a single underlying model $\mathcal{M}$. There is just a minor conceptual difference with no technical implications between this perspective, and the view that each q_i is embedded in its own model $\mathcal{M}_i$.

Given two probability measures μ, ν on Q, we use the notation $J(\mu, \nu)$ to denote the set of all probability measures on $Q \times Q$ that have μ and ν as the marginals on the first, respectively second, component.

Definition 4 (Coupling). *Let* $\mathcal{M} = \langle Q, \Sigma, \pi, L \rangle$ *be a finite LMC. The Markov chain* $\mathcal{C} = \langle Q \times Q, \Sigma \times \Sigma, \omega, L \rangle$ *is called a* coupling *for* $\mathcal{M}$ *if, for all* $q_1, q_2 \in Q$,

1. $\omega((q_1, q_2), \cdot) \in J(\pi(q_1, \cdot), \pi(q_2, \cdot))$, *and*
2. $L(q_1, q_2) = (L(q_1), L(q_2))$.

A coupling for $\mathcal{M}$ can be seen as a probabilistic pairing of two copies of $\mathcal{M}$ running synchronously, although not necessarily independently.

Given a coupling $\mathcal{C}$ for $\mathcal{M}$, and a *discount factor* $\lambda \leq 1$, we define $\Gamma_\lambda^{\mathcal{C}} : [0, 1]^{Q \times Q} \to [0, 1]^{Q \times Q}$ for $d : Q \times Q \to [0, 1]$ and $q_1, q_2 \in Q$, as follows:

$$
\Gamma_\lambda^{\mathcal{C}}(d)(q_1, q_2) = \begin{cases} 1 & \text{if } L(q_1) \neq L(q_2) \\ \lambda \cdot \displaystyle\sum_{u,v \in Q} d(u, v) \cdot \omega((q_1, q_2), (u, v)) & \text{if } L(q_1) = L(q_2) \end{cases}
$$

The operator $\Gamma_\lambda^{\mathcal{C}}$ has a unique least fixedpoint [1], which we denote by $\gamma_\lambda^{\mathcal{C}}$. Each $\gamma_\lambda^{\mathcal{C}}$ is a distance function on Q. The bisimulation distance is obtained by taking the minimum over all possible couplings:

$$
d_{b,\lambda} := \min\{\gamma_\lambda^{\mathcal{C}} \mid \mathcal{C} \text{ coupling for } \mathcal{M}\}. \tag{1}
$$

The minimum here is taken pointwise at each argument (q_1, q_2). It is shown in [1] that $d_{b,\lambda}$ is well-defined, as the minimum on the right of (1) is attained. Furthermore, there is a coupling that minimizes $\gamma_\lambda^{\mathcal{C}}(q_1, q_2)$ simultaneously for all (q_1, q_2). We here use the extra subscript b to distinguish this bisimilarity distance more clearly from other distance functions we will also consider in the sequel. $d_{b,\lambda}$ is consistent with probabilistic bisimilarity.

3.2   Trace-Based Distances

A distance d is trace-based, if $d(q_1, q_2)$ is a function only of P_{q_1} and P_{q_2}. The measure-theoretic construction of distributions P_q on Σ^ω is essentially a limit of finite-dimensional distributions on Σ^n ($n \in \mathbb{N}$). In a similar manner, it is natural to construct distances between distributions on Σ^ω as a limit of distances on distributions on Σ^n. There are, however, several possible ways of doing this. We consider the following three canonical constructions. If $d^{(n)}$ is a distance function for distributions on Σ^n ($n \in \mathbb{N}$), we define induced distance functions for distributions on Σ^ω as

- (limit) $d^\infty := \lim_{n \to \infty} d^{(n)}$
- (per-symbol distance; limit average) $d^{ps} := \lim_n \frac{1}{n} d^{(n)}$
- (discounted sum) $d^\lambda := \sum_{n \geq 1} \lambda^n d^{(n)}$ ($\lambda < 1$)

For all three constructions it holds that symmetry and triangle inequality are preserved, i.e., if all the $d^{(n)}$ possess these properties, then so do d^λ, d^∞, and d^{ps} (provided the limits exist).

The limit and the per-symbol distances are opposite in nature to the discounted sum distances: the latter emphasizes the differences in the distribution of initial segments $w \in \Sigma^n$ of $s = ws' \in \Sigma^\omega$, whereas the first two are most sensitive to the distribution of the infinite tail s'.

In the following P_1, P_2 always denote probability distributions on Σ^ω. We are mostly concerned with distributions P_i that are defined by states q_i in LMCs $\mathcal{M}_i$, i.e., $P_i = P_{q_i}$. However, many of our general considerations also apply to arbitrary distributions P_i. If P_i is of the form P_{q_i} for some $q_i \in \mathcal{M}_i$, then we say that P_i is generated by an LMC.

P_i induces for each $n \in \mathbb{N}$ a distribution $P_i^{(n)}$ on Σ^n. In order to avoid notational clutter, we suppress the superscript (n) to distinguish $P_i^{(n)}$ from P_i. Which probability space we assume for P_i in a given context will be implicit from the arguments of $P_i()$.

In this paper, we consider the following standard distance functions between distributions P_1, P_2 on Σ^n:

- (Kullback-Leibler distance) $d_{KL}^{(n)}(P_1, P_2) := \sum_{w \in \Sigma^n} P_1(w) \log \frac{P_1(w)}{P_2(w)}$
- (L_p-distance) $d_{L_p}^{(n)}(q_1, q_2) := (\sum_{w \in \Sigma^n} | P_1(w) - P_2(w) |^p)^{1/p}$

For L_p-distances we focus our attention on $p = 1$ (total variation distance), $p = 2$ (Euclidean distance), and $p = \infty$ (Maximum distance). The distance d_{KL}^{ps} is well-known in information theory, and there usually called the *divergence-rate*.

An important tool in the analysis of the Kullback-Leibler distance is an additivity property [8, Chapter 2], which adapted to our context can be stated as:

$$d_{KL}^{(n+1)}(P_1, P_2) = d_{KL}^{(n)}(P_1, P_2) + \sum_{w \in \Sigma^n} P_1(w) \sum_{\sigma \in \Sigma} P_1(\sigma|w) \log \frac{P_1(\sigma|w)}{P_2(\sigma|w)}. \quad (2)$$

Also important is the following relationship between d_{KL} and d_{L_1}:

$$d_{KL}^{(n)} \geq \frac{(d_{L_1}^{(n)})^2}{2}. \tag{3}$$

(see [16] for this and further, sharper, bounds). A direct implication is that for all $A \subseteq \Sigma^n$

$$|P_1(A) - P_2(A)| \leq \sqrt{d_{KL}^{(n)}(P_1, P_2)/2}. \tag{4}$$

For all definitions of distances as limits one needs to verify that the limits actually exist in order to ensure that the distances are well-defined. The following table summarizes some relevant facts:

	KL	L_1	L_2	L_∞
d^λ	Lemma 1	$< \infty$	$< \infty$	$< \infty$
d^∞	Lemma 1	$< \infty$	?	?
d^{ps}	Prop. 1	$\equiv 0$	$\equiv 0$	$\equiv 0$

Here '$< \infty$' means that the distance is well defined and finite. For the $d_{L_p}^\lambda$ distances this is the case because the $d_{L_p}^{(n)}$ are bounded by a common constant for all n. For L_1 one furthermore has that $d_{L_1}^{(n)}$ is monotonically increasing in n, which entails $d_{L_1}^\infty < \infty$. $d_{L_2}^{(n)}$ and $d_{L_\infty}^{(n)}$ are not monotone in n. From Proposition 2 below it follows that $d_{L_2}^\infty, d_{L_\infty}^\infty$ will be not very useful even if guaranteed to be well-defined. We therefore do not analyse their exact status further.

The $d_{L_p}^{(n)}$ being bounded, it is also immediate that the $d_{L_p}^{ps}$ are identically zero, denoted $\equiv 0$ in the table. We will not consider these distances further.

We now turn to the limiting behavior of $d_{KL}^{(n)}$, where the situation is a little more intricate.

Lemma 1. (i) $d_{KL}^{(n)}(P_1, P_2)$ *is monotonically increasing for all* P_1, P_2.
If P_1, P_2 *are generated by LMCs, then one of the following cases holds:*

(iia) *There exists an* $n > 0$ *and* $w \in \Sigma^n$ *with* $0 = P_2(w) < P_1(w)$, *so that*
$d_{KL}^{(m)}(P_1, P_2) = \infty$ *for all* $m \geq n$.
(iib) $d_{KL}^{(n)}(P_1, P_2) \in O(n)$

From this Lemma it follows that d_{KL}^λ and d_{KL}^∞ are well-defined, but possibly infinite.[1] Furthermore, if case (iib) holds, then d_{KL}^λ is finite.

Turning to the per-symbol distance, we first obtain from Lemma 1 that if the P_i are generated by LMCs, and case (iia) of the lemma does not hold, then

$$0 \leq \liminf_n \frac{1}{n} d_{KL}^{(n)}(P_1, P_2) \leq \limsup_n \frac{1}{n} d_{KL}^{(n)}(P_1, P_2) < \infty.$$

To ensure that d_{KL}^{ps} is well-defined, one has to establish that the lim inf and lim sup are equal in this equation. The question of this equality, i.e., the problem

---

[1] The proof of this lemma and subsequent results can be found in the online appendices for this paper available at `people.cs.aau.dk/~jaeger/publications.html`

of the existence of the divergence rate, is non-trivial, and has received considerable attention in the literature. [13] gives examples of stochastic processes for which the divergence rate does not exist, but also states that it exists when P_1, P_2 are generated by Hidden Markov Models (HMMs). Since LMCs are a special type of Hidden Markov Models, this would provide the solution to our problem. However, no proof of this statement is given in [13]. Positive results on the existence of the divergence rate for several classes of Markov processes can be found in [11] and [7, Chapter 10]. These results do not cover the case of HMMs or LMCs, however. In contrast, [6,14] specifically consider the class of HMMs, but the results of [6] applied to our problem will only lead to the trivial bound $\limsup_n \frac{1}{n} d_{KL}^{(n)}(P_1, P_2) \leq \infty$, and [14] is concerned with models with continuous observation spaces.

We will not solve the question of the existence of d_{KL}^{ps} in full generality here. In the following, we only consider the case of deterministic LMCs. This case not only greatly facilitates the theoretical analysis, but the proof of the following Proposition also leads to an efficient way of computing d_{KL}^{ps}. [2]

Proposition 1. *Let P_1, P_2 be defined by deterministic LMCs $\mathcal{M}_1, \mathcal{M}_2$. Then $\lim_n 1/n \, d_{KL}^{(n)}(P_1, P_2)$ exists.*

In light of [13] it is strongly conjectured that the existence of $d_{KL}^{ps}(P_1, P_2)$ also holds for nondeterministic LMCs. In the following, all statements relating to d_{KL}^{ps} are implicitly restricted to those cases where d_{KL}^{ps} is well-defined.

Having defined several candidate trace-based distances, we first check which ones are consistent with trace equivalence.

Proposition 2. *Distances are or are not consistent with trace equivalence, as indicated by y (yes), respectively n (no), in the following table:*

	KL	L_1	L_2	L_∞
d^λ	y	y	y	y
d^∞	y	y	n	n
d^{ps}	n			

For $d_{L_2}^\infty$ and $d_{L_\infty}^\infty$ the proposition is shown by considering the automata of Figure 1: denote by q_ϵ the initial state of automaton $\mathcal{M}_\epsilon$. Then one obtains that for all ϵ $d_{L_2}^\infty(\mathcal{M}_0, \mathcal{M}_\epsilon) = d_{L_\infty}^\infty(\mathcal{M}_0, \mathcal{M}_\epsilon) = 0$. Not being able to measure any distance between different $\mathcal{M}_\epsilon$ models makes these distance measure clearly unsuitable for our purpose, and we will not consider them any further.

According to Proposition 2, also d_{KL}^{ps} is not consistent with trace equivalence. An example illustrating this point is given by Figure 2. It shows a (deterministic) LMC $\mathcal{M}_{k,p}$ parameterized by k (length of an initial sequence of a-labeled states),

---

[2] We note that the efficient computability of the finite-dimensional $d_{KL}^{(n)}$, as well as the limits d_{KL}^∞ and d_{KL}^{ps} is a different problem than the computation of relative entropies for probabilistic automata, as investigated by [4]. The automata investigated in this latter work define probability distributions over Σ^*, and the Kullback-Leibler distance therefore becomes the infinite sum $\sum_{w \in \Sigma^*} P_1(w) \log(P_1(w)/P_2(w))$.

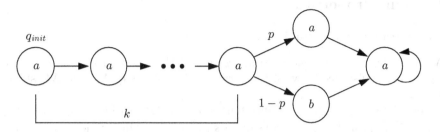

Fig. 2. The automata $\mathcal{M}_{k,p}$

and p (the indicated transition probability). Consider the case $k = 1$. Let $p \neq p'$, and q, q' the initial states of $\mathcal{M}_{1,p}$ and $\mathcal{M}_{1,p'}$, respectively. Then one obtains that $d_{KL}^{(n)}(q, q') = p \log(p/p') + (1-p) \log((1-p)/(1-p'))$ is constant for all n, so that $d_{KL}^{ps}(q, q') = 0$.

Even though d_{KL}^{ps} here also fails to distinguish different models $\mathcal{M}_{k,p}$ and $\mathcal{M}_{k,p'}$, this failure is much less significant than the failure of $d_{L_2}^{\infty}$ and $d_{L_\infty}^{\infty}$ for the models $\mathcal{M}_\epsilon$. The $\mathcal{M}_{k,p}$ are indeed only distinguishable by their initial behavior, but indistinguishable in their infinitary, ergodic behavior. If one is primarily concerned with the limiting behavior of systems, then $d(q, q') = 0$ is appropriate. For $\mathcal{M}_0$ and $\mathcal{M}_\epsilon$ of Figure 1 the ergodic behaviors are characterized by a different frequency of H and T, and $d_{KL}^{ps}(q_0, q_\epsilon) = 0.5 \log(0.5/(0.5+\epsilon)) + 0.5 \log(0.5/(0.5-\epsilon))$ appropriately reflects this.

We therefore still consider d_{KL}^{ps} as a meaningful distance. Even if we insist on consistency with trace equivalence as a necessary property for a distance, d_{KL}^{ps} remains relevant for the following reason: if d is a distance that is consistent with trace equivalence, then any mixture $\alpha d + (1-\alpha)d'$ $(0 < \alpha < 1)$ of d with another distance d' still is consistent with trace equivalence. Thus, even if d_{KL}^{ps} may not satisfy our demands for a stand-alone distance, it can still be a very useful component in a distance defined as a mixture. We will return to the construction of distances as mixtures in Section 6.

We here have considered constructions of distance functions for distributions on Σ^ω from distance functions on Σ^n. Of course, one may also directly define distances on Σ^ω using integrals rather than sums. For example, one may define

$$d_{KL}(P_1, P_2) = \int_{\Sigma^\omega} f_1(s) \log \frac{f_1(s)}{f_2(s)} d\mu(s),$$

where the f_i are density functions for P_i relative to the reference measure μ. For this, however, according to the Radon-Nikodym theorem, we first need a reference measure μ, so that the P_{q_i} are both absolutely continuous with respect to μ. In general, it will be impossible to find a natural μ that serves this purpose for all relevant P_i. However, one can work around this problem by letting $\mu = 1/2(P_1 + P_2)$. Distances defined in this manner, however, will fail our first desirable property, introduced in the following section.

4   Main Properties

4.1   Parameter Continuity

We begin by giving a general formalization of the intuition that as $\epsilon \to 0$ in Figure 1, the distance between the corresponding states of $\mathcal{M}_0$ and $\mathcal{M}_\epsilon$ should go to zero.

Let π be a transition probability function on a state set Q. A sequence $(\pi_n)_n$ of transition probability functions on Q *s-converges* against π, denoted $\pi_n \overset{s}{\to} \pi$, if

(i) $\forall n \forall q, q' \in Q : \ \pi_n(q, q') = 0 \Leftrightarrow \pi(q, q') = 0$
(ii) $\forall q, q' \in Q : \ \pi_n(q, q') \to \pi(q, q') \ (n \to \infty)$

We call this s-convergence, because condition (i) requires that the functions in the sequence (π_n) all have the same set of support as π. In other words, we do not allow a sequence of non-zero transition probabilities to converge to zero.

Definition 5. *A distance function d is* parameter continuous, *if for any labeled Markov chain $\mathcal{M} = \langle Q, \Sigma, q_{\text{init}}, \pi, L \rangle$, and any sequence $\pi_n \overset{s}{\to} \pi$ the following holds: for $\mathcal{M}_n := \langle Q, \Sigma, q_{\text{init}}, \pi_n, L \rangle$, $P := P_{\mathcal{M}, q_{\text{init}}}$, and $P_n := P_{\mathcal{M}_n, q_{\text{init}}}$ it holds that $\lim_{n \to \infty} d(P_n, P) = \lim_{n \to \infty} d(P, P_n) = 0$.*

Note that we are considering potentially non-symmetric distance functions, which is why we have the requirements both for the limit of $d(P_n, P)$ and $d(P, P_n)$.

Parameter continuity in the sense of this definition captures an important aspect of the informal objective O1 from Section 1. We only consider s-convergent sequences of transition probabilities in this definition, because a stronger requirement that also applies to sequences of transition probabilities $\epsilon_n \to 0$ would be immediately inconsistent with objective O2, as formalized by LTL-continuity below: consider the coin model of Figure 1, but now let the transition probabilities into the H state be $1 - \epsilon$, and the transition probabilities into T be ϵ. For the LTL property $\Diamond T$ we then have $P(\Diamond T) = 1$ in all $\mathcal{M}_\epsilon$ with $\epsilon > 0$, and $P(\Diamond T) = 0$ in $\mathcal{M}_0$. Thus, if we required that $d(\mathcal{M}_\epsilon, \mathcal{M}_0) \to 0$ as $\epsilon \to 0$, then an upper bound on the distance between models could not imply an upper bound on the probability difference for LTL formulas.

The following Proposition summarizes parameter continuity properties of selected distances. We do not consider any more those trace-based distances that from Proposition 2 turned out to be uninteresting.

Proposition 3. *Distances are or are not parameter continuous, as indicated by y (yes), respectively n (no), in the following tables:*

	KL	L_1	L_2	L_∞
d^λ	y	y	y	y
d^∞	n	n		
d^{ps}	y			

	$d_{b,\lambda}$
$\lambda = 1$	n
$\lambda < 1$	y

The negative result for $d_{b,1}$ is obtained from a characterization of $d_{b,1}$ in terms of the reachability probability in couplings $\mathcal{C}$ of a state (q_1, q_2) with $L(q_1) \neq L(q_2)$ [3,1]. Applied to the models $\mathcal{M}_\epsilon$ of Figure 1, this characterization shows that $d_{b,1}(q_0, q_\epsilon) = 1$ for all $\epsilon > 0$.

The negative results for d_{KL}^∞ and $d_{L_1}^\infty$ are also obtained by considering the models $\mathcal{M}_\epsilon$, where one again obtains that distances between $\mathcal{M}_0$ and $\mathcal{M}_\epsilon$ are given by the maximal possible values: for all $\epsilon > 0$ $d_{KL}^\infty(q_0, q_\epsilon) = \infty$, $d_{L_1}^\infty(q_0, q_\epsilon) = 2$.

4.2   Property Continuity

In the following, we call any measurable subset $\varphi \subseteq \Sigma^\omega$ a *property*. Thus, "property" is the same as "event" in standard probability theoretic language. We prefer the term property here, because in the present context we view φ rather as a property of a system behavior than as an observed event, and it will later be more natural to speak about LTL-definable properties, than LTL-definable events.

Definition 6 (Φ-continuity). *Let $\varphi \subseteq \Sigma^\omega$ be a property. A distance d is φ-continuous, if*

$$\forall \epsilon > 0 \; \exists \delta > 0 \; \forall P_1, P_2 : d(P_1, P_2) \leq \delta \Rightarrow |P_1(\varphi) - P_2(\varphi)| \leq \epsilon. \qquad (5)$$

If $\Phi \subset 2^{\Sigma^\omega}$ is a class of properties, then d is Φ-continuous, if d is φ-continuous for all $\varphi \in \Phi$.

If d is *Φ-continuous*, then the δ-bound on $d(P_1, P_2)$ needed to ensure that $|P_1(\varphi) - P_2(\varphi)| \leq \epsilon$ will depend on φ. In the following definition these bounds are required to be uniform for all φ.

Definition 7 (Uniform Φ-continuity). *Let $\Phi \subset 2^{\Sigma^\omega}$ be a class of properties. A distance d is* uniformly Φ-continuous, *if*

$$\forall \epsilon > 0 \; \exists \delta > 0 \; \forall \varphi \in \Phi, \forall P_1, P_2 : d(P_1, P_2) \leq \delta \Rightarrow |P_1(\varphi) - P_2(\varphi)| \leq \epsilon. \qquad (6)$$

The following lemma is a straightforward, but useful observation.

Lemma 2. *Let d_1, d_2 be two distance function, such that there exists a continuous function f with $f(0) = 0$, and $d_1 \leq f(d_2)$. Then, for any Φ: (uniform) Φ-continuity of d_1 implies (uniform) Φ-continuity of d_2.*

According to (3) Lemma 2 applies to $d_1 = d_{L_1}^{(n)}$ and $d_2 = d_{KL}^{(n)}$ with $f(x) = \sqrt{2x}$. Since f does not depend on n, the same also is true for $d_1 = d_{L_1}^\infty$ and $d_2 = d_{KL}^\infty$. Thus, proving (uniform) Φ-continuity for $d_{L_1}^\infty$ is sufficient to also prove it for d_{KL}^∞.

Lemma 3. *A BLTL-definable property $\phi \subseteq \Sigma^\omega$ is a finite union of cylinder sets. A distance d is (uniformly) BLTL continuous iff it is (uniformly) Cyl-continuous.*

The first statement in this lemma follows from a straightforward induction on BLTL formulas. The second statement then is a direct consequence of the definitions. Combining Lemma 3 with the fact that measures on Σ^ω are uniquely defined by the measures of cylinder sets, one obtains:

Lemma 4. *BLTL-continuity implies consistency with trace equivalence.*

We now formulate our main results on property continuity.

Proposition 4. *Distances are or are not uniformly BLTL continuous, BLTL continuous, or not BLTL continuous as indicated by uy, y, respectively n, in the following tables:*

	KL	L_1	L_2	L_∞
d^λ	y	y	y	y
d^∞	uy	uy		
d^{ps}	n			

	$d_{b,\lambda}$
$\lambda = 1$	uy
$\lambda < 1$	y

Proposition 5. *Distances are or are not uniformly LTL continuous, LTL continuous, or not LTL continuous, as indicated by uy, y, respectively n, in the following tables:*

	KL	L_1	L_2	L_∞
d^λ	n	n	n	n
d^∞	uy	uy		
d^{ps}	n			

	$d_{b,\lambda}$
$\lambda = 1$	uy
$\lambda < 1$	n

The negative results for the d^λ distances are established by again considering the automata $\mathcal{M}_{k,p}$ of Figure 2, and the LTL sentence $\Diamond b$. Let $p_1 \neq p_2$, and $P_{k,i}$ the distribution defined by $\mathcal{M}_{k,p_i}$ $(i = 1, 2)$ Then, for all k: $|P_{k,1}(\Diamond b) - P_{k,2}(\Diamond b)| = |p_1 - p_2|$. On the other hand, for all discounted distances, and all $\delta > 0$, there exists a k such that $d(P_{k,1}, P_{k,2}) < \delta$.

According to Proposition 5, $d^\infty_{KL}, d^\infty_{L_1}$ and $d_{b,1}$ have very strong property continuity characteristics. However, according to Proposition 3, this comes at the price of not fulfilling objective O1.

Comparison of Propositions 3 and 5 shows that so far we have failed to construct a distance function implementing both our objectives. In the following section we will see that to some extent this is due to a fundamental limitation.

5   Impossibility Results

The proofs of the positive results expressed by Propositions 4 and 5 are not based on the concrete logical characterizations of property classes LTL and BLTL, but on the underlying topological and measure-theoretic structure of these properties. This is not surprising, since the definitions of the distance measures we have been considering also are based on general measure-theoretic concepts, without reference to linear temporal logic.

It is therefore tempting to try to construct on a slightly broader topological basis also a distance measure that is both parameter- and LTL-continuous. This approach also suggests itself because of the fact that LTL-definable properties still have a quite simple topological structure: any LTL-definable set $A \subseteq \Sigma^\omega$ is a Boolean combination of G_δ-sets, where a G_δ-set is a countable intersection of open sets [15, Theorem 5.2]. From this it follows that for LTL-continuity of d it would be enough to show that d is continuous for G_δ-sets.

However, as we now show, it is even impossible to obtain continuity for all open sets in conjunction with parameter continuity.

Proposition 6. *There exists an open set O, such that there does not exist a distance function d that is parameter continuous and O-continuous.*

The open set O constructed in the proof of the preceding theorem is not LTL-definable. The theorem, therefore, only delimits the possibilities of obtaining parameter continuous and LTL-continuous distance functions by purely topological and measure-theoretic constructions. However, the proof of Proposition 6 also directly leads to the following:

Proposition 7. *There does not exist a distance function d that is parameter continuous and uniformly BLTL-continuous.*

Thus, we find that uniform (B)LTL-continuity is inconsistent with parameter continuity. However, uniform continuity is a very strong demand to begin with, so the main objective of combining parameter continuity and LTL-continuity still could be feasible. In the next section we show that this is indeed the case. Before we give a concrete example, we establish some general results about mixtures of distance functions.

6 Mixture Constructions

In Section 3 we justified our continued interest in the d_{KL}^{ps} distance in spite of the fact that it is not consistent with trace equivalence by its possible use as a component in a mixture of distances.

Definition 8. *Let $n \in \mathbb{N} \cup \{\infty\}$, and for $i = 1, \ldots, n$: $\alpha_i \in [0, 1]$ with $\sum_i \alpha_i = 1$, and d_i a distance function. Then $d := \sum_i \alpha_i d_i$ is called a mixture of the d_i. If $n < \infty$, then d is a finite mixture.*

It is well-known that mixtures of distances preserve essential metric properties such as symmetry and the triangle-inequality. In the following we summarize to what extent the distance properties we are studying are preserved. We say that a property of a distance is *preserved* under mixtures, if a mixture d has the property whenever all its constituent d_i have the property. A property is *strongly preserved* if d has the property whenever at least one d_i has the property. The following Lemma summarizes the relevant preservation properties.

Lemma 5. *The following properties are preserved under mixtures:*

- *The left to right direction $d(P_1, P_2) = 0 \Leftarrow q_1 \equiv q_2$ ($\equiv \in \{\sim, \overset{T}{\sim}\}$) of consistency with bisimulation- or trace-equivalence.*
- *Parameter continuity*

The following properties are strongly preserved under mixtures:

- *The right to left direction $d(P_1, P_2) = 0 \Rightarrow q_1 \equiv q_2$ ($\equiv \in \{\sim, \overset{T}{\sim}\}$) of consistency with bisimulation- or trace-equivalence.*
- *Φ-continuity and uniform Φ-continuity.*

We next investigate two different distances that are constructed as mixtures.

6.1 Expected LTL Distance

Definition 9. *For $\phi \in$ LTL define*

$$d_\phi(P_1, P_2) := |P_1(\phi) - P_2(\phi)|.$$

Let $\phi_1, \phi_2, \dots$ be an enumeration of LTL, $\alpha_i \in (0, 1)$ with $\sum_i \alpha_i = 1$, and define

$$d_Q := \sum_i \alpha_i d_{\phi_i}.$$

$d_Q(P_1, P_2)$ can be interpreted as the expected difference $|P_1(\phi) - P_2(\phi)|$ for LTL formulas that are randomly generated according to probabilities α_i. As an empirical evaluation measure for how well a learned system model approximates the LTL properties of a true system model $\mathcal{M}_1$, this distance was used in [10]. The following proposition now states that with d_Q we have the first distance that satisfies both of our main objectives.

Proposition 8. *d_Q is parameter and LTL continuous.*

Even though d_Q satisfies our main objectives, it clearly still has some significant shortcomings. First, the concrete values of $d_Q(P_1, P_2)$ depend very much on the coefficients α_i. When the α_i are just more or less arbitrarily set in a synthetic construction of d_Q, then the actual values of d_Q will lack a meaningful interpretation. If, however, α_i represents a meaningful probability of ϕ_i (for example, the expected frequency with which ϕ_i will be checked in a given application context), then $d_Q(P_1, P_2)$ is interpretable as the expected deviation between LTL-probabilities computed in $\mathcal{M}_1$ and $\mathcal{M}_2$.

Second, d_Q poses computational problems. The only currently available approach to (approximately) computing d_Q is to compute d_{ϕ_i} for a sample $i = i_1, \dots, i_k$. If α_i can be computed for a given ϕ_i, and the ϕ_{i_j} in the sample are all distinct, then d_Q is bounded by $[\sum_{j=1}^k \alpha_{i_j} d_{\phi_{i_j}}, \sum_{j=1}^k \alpha_{i_j} d_{\phi_{i_j}} + (1 - \sum_{j=1}^k \alpha_{i_j})]$. If the α_i are only implicitly given by a random generator for LTL formulas, then d_Q can be estimated by the empirical distance $1/k \sum_{j=1}^k d_{\phi_{i_j}}$.

6.2   KL Mixture

A second mixture construction we consider is

$$d_{KL}^{mix} := \alpha d_{KL}^{\lambda} + (1 - \alpha)d_{KL}^{ps}.$$

The motivation for d_{KL}^{mix} is that it combines a distance function that is mostly sensitive to differences in the initial behavior of a system (d_{KL}^{λ}), and a distance that measures differences in the long-run, ergodic behavior (d_{KL}^{ps}).

d_{KL}^{mix} is consistent with trace-equivalence, parameter continuous, and inherits the BLTL-continuity of d_{KL}^{λ}. However, d_{KL}^{mix} is not LTL continuous (as expected from Proposition 6, since d_{KL}^{mix} is a purely measure theoretic construction). Concretely, d_{KL}^{mix} still is subject to the counterexample described for the d^{λ} in connection with Proposition 5.

7   Conclusion

In this paper we have investigated a number of distances on finite state Markov Processes, which measure the behavioural similarity of non-bisimilar processes. We have considered both bisimulation distances and trace-based distances. In particular, we focused on several constructions derived from the standard L_p and Kullback-Leibler distances. The continuity aspects for which we have tested the distances are natural properties one would expect from a distance that in a meaningful sense measures the relationship between a true model and its approximations. On one hand we study the parameter continuity, which guarantees that the distances are continuous in the transition probabilities. On the other hand we analyzed the concept of a good approximation of a system in the light of a given distance function. We expect from a good distance to provide us some bounds on the error incurred by using the approximation of a model instead of the real model in given contexts.

We demonstrated that none of the considered distances fully respects the continuity properties that we considered. This failure is partially explained by an impossibility result that reveals to some extent the fundamental difficulties that one encounters when trying to achieve such complex goals.

References

1. Bacci, G., Bacci, G., Larsen, K.G., Mardare, R.: On-the-fly exact computation of bisimilarity distances. In: Piterman, N., Smolka, S.A. (eds.) TACAS 2013. LNCS, vol. 7795, pp. 1–15. Springer, Heidelberg (2013)
2. Baier, C., Clarke, E.M., Hartonas-Garmhausen, V., Kwiatkowska, M., Ryan, M.: Symbolic model checking for probabilistic processes. In: Degano, P., Gorrieri, R., Marchetti-Spaccamela, A. (eds.) ICALP 1997. LNCS, vol. 1256, pp. 430–440. Springer, Heidelberg (1997)
3. Chen, D., van Breugel, F., Worrell, J.: On the complexity of computing probabilistic bisimilarity. In: Birkedal, L. (ed.) FOSSACS 2012. LNCS, vol. 7213, pp. 437–451. Springer, Heidelberg (2012)

4. Cortes, C., Mohri, M., Rastogi, A., Riley, M.: On the computation of the relative entropy of probabilistic automata. Int. J. Found. Comput. Sci. 19(1), 219–242 (2008)
5. Desharnais, J., Gupta, V., Jagadeesan, R., Panangaden, P.: Metrics for labeled Markov systems. In: Baeten, J.C.M., Mauw, S. (eds.) CONCUR 1999. LNCS, vol. 1664, pp. 258–273. Springer, Heidelberg (1999)
6. Do, M.N.: Fast approximation of Kullback-Leibler distance for dependence trees and hidden Markov models. IEEE Signal Processing Letters 10(4), 115–118 (2003)
7. Gray, R.M.: Entropy and Information Theory, 2nd edn. Springer (2011)
8. Kullback, S.: Information Theory and Statistics. Wiley (1959)
9. Larsen, K.G., Skou, A.: Bisimulation through probabilistic testing. Inf. Comput. 94(1), 1–28 (1991)
10. Mao, H., Chen, Y., Jaeger, M., Nielsen, T.D., Larsen, K.G., Nielsen, B.: Learning probabilistic automata for model checking. In: Proceedings of the 8th International Conference on Quantitative Evaluation of SysTems (QEST), pp. 111–120 (2011)
11. Rached, Z., Alajaji, F., Campbell, L.L.: The Kullback-Leibler divergence rate between Markov sources. IEEE Transactions on Information Theory 50(5), 917–921 (2004)
12. Sen, K., Viswanathan, M., Agha, G.: Learning continuous time Markov chains from sample executions. In: Proceedings of International Conference on Quantitative Evaluation of Systems (QEST), pp. 146–155 (2004)
13. Shields, P.C.: Two divergence-rate counterexamples. Journal of Theoretical Probability 6(3), 521–545 (1993)
14. Silva, J., Narayanan, S.: Upper bound Kullback-Leibler divergence for transient hidden Markov models. IEEE Transactions on Signal Processing 56(9), 4176–4188 (2008)
15. Thomas, W.: Automata on infinite objects. In: van Leeuwen, J. (ed.) Handbook of Theoretical Computer Science, vol. 2. Elsevier/MIT Press (1990)
16. Toussaint, G.T.: Sharper lower bounds for discrimination information in terms of variation (corresp.). IEEE Transactions on Information Theory 21(1), 99–100 (1975)
17. van Breugel, F., Sharma, B., Worrell, J.: Approximating a behavioural pseudometric without discount for probabilistic systems. Logical Methods in Computer Science 4(2), 1–23 (2008)
18. van Breugel, F., Worrell, J.: A behavioural pseudometric for probabilistic transition systems. Theoretical Computer Science 331, 115–142 (2005)

Deciding the Value 1 Problem for Reachability in 1-Clock Decision Stochastic Timed Automata[*]

Nathalie Bertrand[1], Thomas Brihaye[2], and Blaise Genest[3]

[1] Inria, Team SUMO, UMR IRISA, Rennes, France
[2] Université de Mons, Mons, Belgium
[3] CNRS, Team SUMO, UMR IRISA, Rennes, France

Abstract. We consider reachability objectives on an extension of stochastic timed automata (STA) with nondeterminism. Decision stochastic timed automata (DSTA) are Markov decision processes based on timed automata where delays are chosen randomly and choices between enabled edges are nondeterministic. Given a reachability objective, the value 1 problem asks whether a target can be reached with probability arbitrary close to 1. Simple examples show that the value can be 1 and yet no strategy ensures reaching the target with probability 1. In this paper, we prove that, the value 1 problem is decidable for single clock DSTA by non-trivial reduction to a simple almost-sure reachability problem on a finite Markov decision process. The ε-optimal strategies are involved: the precise probability distributions, even if they do not change the winning nature of a state, impact the timings at which ε-optimal strategies must change their decisions, and more surprisingly these timings cannot be chosen uniformly over the set of regions.

1 Introduction

Stochastic timed automata (STA) were originally defined in [2,3] as a probabilistic semantics for timed automata, with the motivation to rule out 'unrealistic' paths in timed automata, and therefore alleviate some drawbacks of the mathematical model such as infinite precision of the clocks and instantaneous events. Of course, STA also form a new stochastic timed model, interesting on its own. Informally, the semantics of a stochastic timed automaton consists of an infinite-state infinitely-branching Markov chain whose underlying graph is the timed transition system associated with a timed automaton. The transitions between states are governed by the following: first, a delay is sampled randomly among possible delays, and second, an enabled transition is chosen randomly among enabled ones.

[*] The research leading to these results has received funding from the ARC project (number AUWB-2010-10/15-UMONS-3), a grant "Mission Scientifique" from the F.R.S.-FNRS, the FRFC project (number 2.4545.11), the European Union Seventh Framework Programme (FP7/2007-2013) under Grant Agreement number 601148 (CASSTING), the project ANR-13-BS02-0011-01 STOCH-MC, and the *professeur invité ISTIC-Rennes 1* programme.

G. Norman and W. Sanders (Eds.): QEST 2014, LNCS 8657, pp. 313–328, 2014.

Several models combining dense-time, continuous probabilities, and nondeterminism have been studied [7,8,11] and most result focus on qualitative questions, such as deciding the existence of a strategy ensuring a reachability objective with probability 1 (see the related work section).

A model that extends stochastic timed automata with nondeterminism was defined in [5]: the delays are random, but the choice between enabled transitions is nondeterministic. For this model, optimal strategies always exist for the *time-bounded* reachability problem. Yet, a simple example also shows that optimal strategies do not always exist for the reachability problem: there might be strategies to ensure a probability arbitrary close to 1 to reach a target location, and no strategy achieving probability 1.

More generally, the value 1 problem asks whether for every $\varepsilon > 0$ there exists a strategy ensuring a given objective with probability at least $1 - \varepsilon$. It can be defined in various game-like contexts, ranging from probabilistic finite automata (PFA) to concurrent games. In most models where the agent has full information, the value 1 problem coincides with the almost-sure problem, that is, whether there exists a strategy to ensure a given objective with probability 1. For partial observation models however, the value 1 problem and the almost-sure problem often differ: for concurrent games, both are decidable [12,9], whereas the value 1 problem is undecidable for PFA [14], and decidable only for subclasses [13,10].

In this paper, we consider a probabilistic and nondeterministic version of stochastic timed automata, called decision stochastic timed automata (DSTA), in which delays are random but edges are selected by the player. Contrary to most existing frameworks on stochastic and timed models, we do not assume the distributions over delays to be exponential. We consider (time-unbounded) reachability objectives on DSTA with a single clock. The restriction to 1-clock DSTA derives from the fact that even for purely stochastic models without decisions (*i.e.* STA), the decidability of the almost-sure reachability problem is open, for models with at least two clocks. Using the classical region abstraction we show that the existence of an almost-surely winning strategy is decidable for reachability objectives on 1-clock DSTA. Interestingly, in our context, the value 1 problem does not coincide with the almost-sure problem, although the agent has full information. The main contribution of the paper is then to prove that the value 1 problem is decidable too. To do so, we build an *ad hoc* abstraction based on a refinement of regions, and reduce to an almost-sure reachability question in the derived finite-state Markov decision process (MDP). The correctness proof is complex, and ε-optimal strategies are involved: first they are not uniform within a region as actions they dictate depend on the comparison of the precise clock value with some cutpoint. Second, and more surprisingly, these cutpoints cannot be chosen uniformly over the set of regions.

Related Work

In stochastic timed games [7], locations are partitioned into locations owned by three players, a reachability player (who has a time-bounded reachability objective), a safety player (who has the opposite objective), and an environment player (who makes random moves). In a location of the reachability or safety player, the respective player decides both the sojourn time and the edge to fire, whereas in the environment's locations, the delay as well as the edge are chosen randomly. For this model, it was shown that, assuming there is a single player and the underlying timed automaton has only one clock, the existence of a strategy for a reachability goal almost-surely (resp. with positive probability) is PTIME-complete (resp. NLOGSPACE-complete). For two-player games, quantitative questions are undecidable. Simple examples show that even for one player and 1-clock timed automata, the value 1 and probability 1 problems differ. This is due to strict inequalities in guards, that prevent the player to choose an optimal delay. We believe that our proof techniques can be adapted to solve the value 1 problem in 1-player stochastic timed games over 1-clock timed automata.

In stochastic real-time games [8], environment nodes (in which the behaviour is similar to continuous time Markov decision processes (CTMDPs)) and control nodes (where players choose a distribution over actions) induce a probability distribution on runs. The objective for player 0 is to maximise the probability that a run satisfies a specification given by a deterministic timed automaton (DTA). The main result states that if player 0 has an almost-sure winning strategy, then she also has a simple one which can be described by a DTA.

Markovian timed automata (MTA) consist in an extension of timed automata with exponentially distributed sojourn time. Optimal probabilities can be approximated for time-(un)bounded reachability properties in MTA [11].

2 Definitions and Problem Statement

2.1 Timed Automata

Timed automata [1] were introduced in the early nineties. We recall the definition and semantics of one-clock timed automata. Given a clock x, a *guard* is a finite conjunction of expressions of the form $x \sim c$ where $c \in \mathbb{N}$ is an integer, and $\sim$ is one of the symbols $\{<, \leq, =, \geq, >\}$. We denote by $\mathcal{G}(x)$ the set of guards over clock x. Often, for $g \in \mathcal{G}(x)$ a guard and t a clock value, we will write $t \in g$ to express that t satisfies the constraints expressed in g.

Definition 1. *A one-clock timed automaton is a tuple $(L, \ell_0, E, \mathcal{I})$ such that: L is a finite set of locations, $\ell_0 \in L$ is the initial location, $E \subseteq L \times \mathcal{G}(x) \times 2^{\{x\}} \times L$ is a finite set of edges, and $\mathcal{I} : L \to \mathcal{G}(x)$ assigns an invariant to each location.*

In the following, we assume all timed automata to be *well-formed*: for every location $\ell \in L$, $\mathcal{I}(\ell) = \bigcup_{(\ell,g,a,r,\ell') \in E} g$, that is, the invariant in a location coincides with the union of the guards on its outgoing edges. This implies in particular that the union of guards outgoing a location is an interval.

The semantics of a one-clock timed automaton $(L, \ell_0, E, \mathcal{I})$ is a timed transition system $\mathcal{T} = (S, s_0, \delta)$ where $S = L \times \mathbb{R}_{\geq 0}$, $s_0 = (\ell_0, 0)$ and the transition function $\delta \subseteq S \times (\mathbb{R}_{\geq 0} \cup E) \times S$ is composed of

- Delay transitions: $\big((\ell, t), \tau, (\ell, t + \tau)\big) \in \delta$ whenever $[t, t + \tau] \subseteq \mathcal{I}(\ell)$

- Discrete transitions: $((\ell, t), e, (\ell', t')) \in \delta$ as soon as the edge $e = (\ell, g, r, \ell') \in E$ satisfies $t \in g$ and if $r = \{x\}$, $t' = 0$ else $t' = t$.

When convenient, we will use the alternative notations $(\ell, t) \xrightarrow{\tau} (\ell, t + \tau)$ and $(\ell, t) \xrightarrow{e} (\ell', t')$. Edge e is said *enabled* in state $s = (\ell, t)$, whenever there exists $s' \in S$ such that $s \xrightarrow{e} s'$.

2.2   Decision Stochastic Timed Automata

We now introduce the concept of *decision stochastic timed automaton* (DSTA). Roughly speaking, a decision stochastic timed automaton is a one-clock timed automaton equipped with probability distributions over delays. The semantics of DSTA is provided by an infinite-state MDP, in the spirit of [5].

In the following, given $X \subseteq \mathbb{R}_{\geq 0}$, we denote by $\mathsf{Dist}(X)$ the set of probability distributions on X.

Definition 2. *A decision stochastic timed automaton is a tuple $\mathcal{A} = (L, \ell_0, E, \mathcal{I}, \mu)$ where $(L, \ell_0, E, \mathcal{I})$ is a one-clock timed automaton and $\mu = (\mu_{\ell,t})$ is a family of distributions, one for each state $(\ell, t) \in L \times \mathbb{R}_{\geq 0}$, and such that $\mu_{\ell,t} \in \mathsf{Dist}(\mathcal{I}(\ell) \cap [t, +\infty[)$.*

Intuitively, for every state $(\ell, t) \in S$, for every interval $I \subseteq \mathbb{R}_{\geq 0}$, $\mu_{\ell,t}(I)$ is the probability that from (ℓ, t) a delay d_0, such that $t + d_0 \in I$, is chosen by μ.

We make some reasonable assumptions on the distributions. For every location ℓ, the function must satisfy the following sanity conditions:

(c_1) for every $t \in \mathcal{I}(\ell)$, and any non-punctual interval $I \subseteq \mathcal{I}(\ell) \cap [t, +\infty[$, $\mu_{\ell,t}(I) > 0$; also if $[t, +\infty[\cap \mathcal{I}(\ell) \neq \{t\}$, then for any $a \in \mathbb{R}_{\geq 0}$, $\mu_{\ell,t}(\{a\}) = 0$;
(c_2) for every $t < t' \in \mathcal{I}(\ell)$, and $I \subseteq [t', +\infty[$, $\mu_{\ell,t}(I) \leq \mu_{\ell,t'}(I)$;
(c_3) if $|\mathcal{I}(\ell)| = \infty$, then for every $t, t' \in \mathcal{I}(\ell)$, for every $a, b \in \mathbb{R}_{\geq 0}$, $\mu_{\ell,t}([t + a, t + b]) = \mu_{\ell,t'}(t' + a, t' + b)$;
if $|\mathcal{I}(\ell)| < \infty$, and $m = \sup\{t \mid t \in \mathcal{I}(\ell)\}$, then for every $t, t' \in \mathcal{I}(\ell)$, for every $a, b \in \mathbb{R}_{\geq 0}$, $\mu_{\ell,t}(t + \frac{a}{m-t}, t + \frac{b}{m-t}) = \mu_{\ell,t'}(t' + \frac{a}{m-t'}, t' + \frac{b}{m-t'})$.

Let us comment on these conditions. First, (c_1) states that the distributions are equivalent to the Lebesgue measure: they do not assign 0 measure to interval with non-empty interior, and do not assign positive probability to points. Then, (c_2) is a monotonicity condition: the higher the clock value, the more likely a fixed interval is to be sampled. Last, with (c_3) one assumes that distributions depend only on the location, not on the precise clock value. More precisely, in case the invariant is not bounded, the distributions should be equal in all states; and if the invariant is bounded, they should coincide up to normalisation. It is important to notice the classical exponential and uniform distributions satisfy these three hypotheses.

Notice that stochastic timed automata (STA) [2,3] and DSTA share the same syntax, and only differ in their semantics: STA are interpreted as purely stochastic system whereas DSTA are interpreted as stochastic and nondeterministic systems. Let $\mathcal{A}$ be a decision stochastic timed automaton. Its semantics is given in terms of an infinite state MDP (or equivalently a 1-1/2 player game), based on the timed transition system $\mathcal{T}$ of the underlying timed automaton. The set of states is composed of two copies $S_\square$ and $S_\Diamond$ of S: stochastic states $S_\Diamond = \{\langle s \rangle \mid s \in S\}$ and player states $S_\square = \{[s] \mid s \in S\}$. The transitions are of the form:

- *stochastic transition:* $\langle s \rangle \xrightarrow{\tau} [s']$ if $(s, \tau, s') \in \delta$;
- *player transition:* $[s] \xrightarrow{e} \langle s' \rangle$ if $(s, e, s') \in \delta$.

The result of each transition is thus deterministic. However stochastic transitions are not played in an arbitrary way, but follow the family of probability distributions $(\mu_{\ell,t})$. Precisely, for $I \subseteq \mathbb{R}_{\geq 0}$ an interval, the probability from $\langle \ell, t \rangle$ to reach a clock value in I is given by $\mathbb{P}(\langle \ell, t \rangle \xrightarrow{\tau} [\ell, t'] \wedge t' \in I) = \mu_{\ell,t}(I)$.

Decisions of the nondeterministic player are specified through the notion of strategy. A *history* is a finite path in the MDP, ending in a player state: $\langle s_0 \rangle \xrightarrow{\tau_0} [s_0'] \xrightarrow{e_0} \langle s_1 \rangle \xrightarrow{\tau_1} [s_1'] \cdots \langle s_n \rangle \xrightarrow{\tau_n} [s_n']$. The set of all histories is denoted Hist. A strategy dictates the decision in states of $S_\square$, given the history so far. Formally, a *strategy* is a function $\sigma : $ Hist $\to E$ such that $\sigma(\langle s_0 \rangle \xrightarrow{\tau_0} [s_0'] \xrightarrow{e_0} \langle s_1 \rangle \xrightarrow{\tau_1} [s_1'] \cdots \langle s_n \rangle \xrightarrow{\tau_n} [s_n'])$ is enabled in s_n'.

As pointed out in [16] in the context of continuous-time Markov decision processes, not all strategies are meaningful. The same phenomenon appears for DSTA, and in the following we thus restrict to so-called *measurable strategies* that induce measurable sets of runs for reachability objectives.

For a fixed measurable strategy σ, and an initial state $s_0 \in S_\Diamond \cup S_\square$, the decision stochastic timed automaton $\mathcal{A}$ gives rise to a stochastic process. For a measurable event $\mathcal{E}$, we write $\mathbb{P}_\sigma^{\mathcal{A}}(s_0 \models \mathcal{E})$ for the probability of $\mathcal{E}$ starting from s_0 and under strategy σ. Given a target set $F \subseteq S_\Diamond \cup S_\square$ in the DSTA, the event $\Diamond F$, denotes the set of paths that eventually visit F.

2.3 Problem Definition

Let $\mathcal{A}$ be a decision stochastic timed automaton, Goal $\subseteq L$ and $s \in S_\square \cup S_\Diamond$. We define $F = \{\langle \ell, t \rangle \mid \ell \in$ Goal$\}$. The value of s, with respect to the objective Goal, is the supremum, over all strategies, of the probability from s to reach F.

Definition 3. *The* value *of state s is* $\mathrm{val}_{\mathcal{A}}(s) = \sup_\sigma \mathbb{P}_\sigma^{\mathcal{A}}(s \models \Diamond F)$.
The value 1 problem *asks, given a decision stochastic timed automaton $\mathcal{A}$, a target set* Goal $\subseteq L$ *and an initial state $s \in S_\square \cup S_\Diamond$, whether* $\mathrm{val}_{\mathcal{A}}(s) = 1$.
F is said limit-surely *reachable from s if* $\mathrm{val}_{\mathcal{A}}(s) = 1$.

Notice that this definition is different from the *almost-sure* reachability problem, which asks whether there exists a strategy σ such that $\mathbb{P}_\sigma^{\mathcal{A}}(s \models \Diamond F) = 1$. From a state with value 1, for every ε, there exists a strategy achieving probability $1 - \varepsilon$ to reach F, yet it does not imply that some strategy realises the objective with probability 1. For finite-state Markov decision processes, and in many simple frameworks, value 1 and probability 1 coincide. However, this is untrue for DSTA, as shown in the example below.

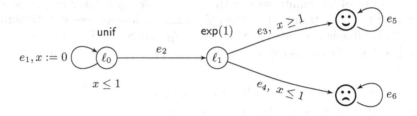

Fig. 1. A simple example of decision stochastic timed automaton

Example 1. Figure 1 shows a basic example of a DSTA, where the distributions μ are uniform in ℓ_0 and exponential with rate 1 in location ℓ_1. The smiley location is limit-surely reachable from the initial location ℓ_0 with clock value 0. Indeed, the idea, in order to ensure a high probability $1 - \varepsilon$ to reach the target is to loop on ℓ_0 until a player state $[\ell_0, 1 - \tau]$ is reached (this happens almost surely) for a small τ, and then to move to location ℓ_1. Now, the probability from $\langle \ell_1, 1 - \tau \rangle$ to reach the target converges to 1 as τ converges to 0. Yet, no strategy can ensure to reach the target with probability 1. This is thus a simple example where limit-sure reachability and almost-sure reachability differ. Such phenomena are not due to invariants, and already occur in DSTA where only exponential distributions are allowed. Indeed, one can adapt the above example and consider an exponential distribution in ℓ_0, while transferring the invariant $x \leq 1$ to the guard of e_2.

2.4   Limit Corner-Point MDP

As an extension of timed automata, DSTA have infinitely many states, because of continuous time. The usual technique to deal with this issue for timed automata, is to resort to the region abstraction [1], which we recall here. For one-clock timed automata, the number of regions is linear [15]: they all are intervals, either punctual $\{c\}$, open and bounded (c, d), or open and unbounded $(c, +\infty)$, for $c, d \in \mathbb{N}$.

We write R for the set of such regions, and r denotes a typical element of R.

Beyond these classical regions, we will use *pointed regions*, similar to the notion introduced for the corner-point abstraction by [6]: every bounded open region (c, d) is duplicated into $(\underline{c}, d)$ and $(c, \underline{d})$, with the intuitive meaning of being close to the left limit, or to the right limit of the interval. Other regions (unbounded or punctual) are kept as is. When a timed automaton is fixed, $\mathbf{R}$ denotes the set of pointed regions, with $\mathbf{r}$ a typical element, and it is partitioned into: $\mathbf{R}_{\text{right}}$ (resp. $\mathbf{R}_{\text{left}}$) for the set of pointed regions of the form $(c, \underline{d})$ (resp. $(\underline{c}, d)$), and $\mathbf{R}_{\text{plain}}$ for punctual regions or the unbounded region. Pointed regions are equipped with a natural total order $<$; for example $\{0\} < (\underline{0}, 1) < (0, \underline{1})$. We say that pointed region $\mathbf{r}'$ is a *successor* of $\mathbf{r}$ if $\mathbf{r} < \mathbf{r}'$. The *immediate open successor* of $\mathbf{r}$ is the least region $\mathbf{r}'$, for the order $<$, that is different from $\mathbf{r}$ and open. The set of all successors of a region $\mathbf{r}$ is denoted $\overrightarrow{\mathbf{r}}$. A pointed region $\mathbf{r}'$ is said *negligible* with respect to location ℓ and region $\mathbf{r}$, if $\mathbf{r}' \in \overrightarrow{\mathbf{r}}$, $\mathbf{r}'$ is punctual and $\mathcal{I}(\ell) \cap \overrightarrow{\mathbf{r}}$ is not.

We now define the limit corner-point MDP associated with a DSTA.

Definition 4. *Given* $\mathcal{A} = (L, \ell_0, E, \mathcal{I}, \mu)$ *a DSTA, its* limit corner-point MDP *is* $\mathcal{A}_{cp} = (\mathcal{S}, \mathbf{s}_0, \mathsf{Act}, \Delta)$, *where*

- $\mathcal{S} = \mathcal{S}_\square \cup \mathcal{S}_\lozenge$ *is partitioned into player states and stochastic states:*
  $\mathcal{S}_\square = \{[\ell, \mathbf{r}] \mid \ell \in L, \ \mathbf{r} \in \mathbf{R}\}$ *and* $\mathcal{S}_\lozenge = \{\langle \ell, \mathbf{r}\rangle \mid \ell \in L, \ \mathbf{r} \in \mathbf{R}\}$;
- $\mathbf{s}_0 = \langle \ell_0, \{0\}\rangle$;
- $\mathsf{Act} = E \cup E^{limit}$, *where* E^{limit} *is a copy of* E;
- Δ *consists of the following transitions:*
  - $\langle \ell, \mathbf{r}\rangle \xrightarrow{\tau} [\ell, \mathbf{r}']$ *as soon as* $\mathbf{r}' \geq \mathbf{r}$ *and* $\mathbf{r}'$ *is not negligible w.r.t.* ℓ, *and the probabilities are uniform over all* τ-*successors;*
  - $[\ell, \mathbf{r}] \xrightarrow{e} \langle \ell', \{0\}\rangle$ *as soon as* $e = (\ell, g, \{x\}, \ell') \in E$, *and* $\mathbf{r} \models g$;
  - $[\ell, \mathbf{r}] \xrightarrow{e} \langle \ell', \mathbf{r}\rangle$ *as soon as* $e = (\ell, g, \emptyset, \ell') \in E$, *and* $\mathbf{r} \models g$;
  - $[\ell, \mathbf{r}] \xrightarrow{e^{limit}} \langle \ell', \mathbf{r}'\rangle$ *as soon as* $\mathbf{r} \in \mathbf{R}_{right}$, $e = (\ell, g, \emptyset, \ell') \in E$, $\mathbf{r} \models g$, $\mathbf{r}'$ *is the immediate open successor of* $\mathbf{r}$, *and* $\mathbf{r}' \models \mathcal{I}(\ell')$.

With the exception of limit-edges, the definition of the limit corner-point MDP is natural since it mimics the behaviour of the DSTA, at the region level and abstracting precise probabilities. Limit-edges are particular to the value 1 problem. Roughly speaking, they offer the player, from region $[\ell, r]$, the possibility to play as if the clock value was *arbitrarily close* to the right border of r, therefore as if it was in r' the immediate open successor of r. In particular, there cannot be two consecutive transitions starting with a limit edge and staying in the same pointed region $[\ell, \mathbf{r}] \xrightarrow{e^{limit}} \xrightarrow{\tau} [\ell', \mathbf{r}]$.

Example 2. Let us illustrate Definition 4 on the example of Fig. 2 below. Its limit corner-point MDP is represented in Fig. 3. For readability reasons we only represented states with left-pointed region for $(1, 2)$, since the behaviour is exactly the same from right-pointed regions.

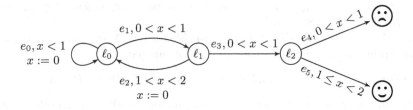

Fig. 2. The first illustrating example

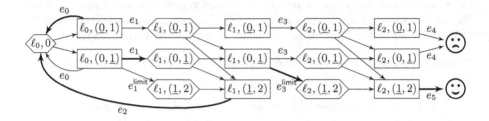

Fig. 3. The limit corner-point MDP for the example from Fig. 2

We let $\mathcal{F} = \{[\ell, \mathbf{r}] \mid \ell \in \mathsf{Goal}\}$ in the rest of the paper. $\mathcal{A}_{\mathsf{cp}}$ is a finite MDP, and one can define strategies in the usual way. In the following, for $\mathbf{s}$ a state of $\mathcal{A}_{\mathsf{cp}}$, and $\mathcal{F} \subseteq \mathcal{S}$ we write $\mathbb{P}^{\mathcal{A}_{\mathsf{cp}}}_{\max}(\mathbf{s} \models \Diamond \mathcal{F})$ for the maximum probability, over all strategies, to reach $\mathcal{F}$ from $\mathbf{s}$.

Last, we introduce some notations: First, for any region $r \in R$, we define $\bullet r \in \mathbf{R}_{\mathsf{left}} \cup \mathbf{R}_{\mathsf{plain}}$ (resp. $r \bullet \in \mathbf{R}_{\mathsf{right}} \cup \mathbf{R}_{\mathsf{plain}}$) with: $\bullet(c, d) = (\underline{c}, d)$ (resp. $(c, d) \bullet = (c, \underline{d})$), $\bullet\{c\} = \{c\} = \{c\}\bullet$ and $\bullet(c, +\infty) = (c, +\infty) = (c, +\infty)\bullet$. Now, given $t \in \mathbb{R}_{\geq 0}$, $\mathbf{r}_{\mathsf{left}}(t)$ (resp. $\mathbf{r}_{\mathsf{right}}(t)$) represents the left (resp. right) pointed region t belongs to: if $t \in r$, then $\mathbf{r}_{\mathsf{left}}(t) = \bullet r$ (resp. $\mathbf{r}_{\mathsf{right}}(t) = r\bullet$).

3   Main Results

We now state the main results of our paper. We start with an expected result:

Proposition 1. *The almost-sure reachability problem is decidable in* PTIME *for DSTA.*

Proposition 1 is not a consequence of the decidability result in [7]. Although our model of DSTA can be encoded into the stochastic timed games of [7], the naive encoding requires an additional clock, in order to prevent players from letting time elapse. Since their decidability result applies only to stochastic timed games with a single clock, this simple reduction is of no help here. We believe their techniques can be adapted though. An alternative, which we take here, is to use the region abstraction, in order to solve the almost sure reachability problem. Details are provided Section 5.

As value 1 and probability 1 do not coincide for DSTA, the following theorem is non trivial, and is the main contribution of this paper.

Theorem 1. *The value 1 problem is decidable in* PTIME *for DSTA.*

To obtain Theorem 1, we reduce the value 1 problem for DSTA to the almost-sure reachability problem in the limit corner-point abstraction.

Proposition 2. *Let $\mathcal{A}$ be a decision stochastic timed automaton, $\mathcal{A}_{cp}$ its limit corner point abstraction, and $\ell \in L$ a location, and $t \in \mathbb{R}_{\geq 0}$ a clock value. Then*

$$\mathsf{val}_{\mathcal{A}}([\ell,t]) = 1 \iff \mathbb{P}^{\mathcal{A}_{cp}}_{\max}([\ell, \mathbf{r}_{left}(t)]) \models \Diamond \mathcal{F}) = 1 \text{ and}$$
$$\mathsf{val}_{\mathcal{A}}(\langle\ell,t\rangle) = 1 \iff \mathbb{P}^{\mathcal{A}_{cp}}_{\max}(\langle\ell, \mathbf{r}_{left}(t)\rangle) \models \Diamond \mathcal{F}) = 1 .$$

Example 3. Let us illustrate the result of Proposition 2 on the example of Fig. 2, whose limit corner-point MDP is represented on Fig. 3. Bold edges give a winning strategy in the MDP for the almost-sure reachability of the smiley state. According to Proposition 2, the set of states with value 1 for the target ☺ in the stochastic timed automaton, is thus $(\ell_0, [0, 1)) \cup (\ell_1, [1, 2)) \cup (\ell_2, [1, 2))$. (Here, we use brackets as a short-cut, rather than square brackets or angle brackets, not to distinguish stochastic and player states.) Intuitively, an ε-optimal strategy from $\langle\ell_0, 0\rangle$ to reach ☺ is the following: stay in ℓ_0 until a large clock value is sampled, then move to ℓ_1; if then the sampled clock value is above 1, move back to ℓ_0 and iterate the same process, otherwise, proceed to ℓ_2; finally, reach ☺ or ☹ from ℓ_2 depending on the last sampled clock value.

Theorem 1 is a consequence of Proposition 2. To obtain a polynomial-time algorithm, one exploits that for 1-clock decision stochastic timed automaton, the number of regions, and thus the number of states in the limit corner-point MDP is linear [15], and almost-sure reachability properties can be checked in polynomial-time for finite MDP.

In order to show Proposition 2 we proceed in two steps: first, we prove it for player states $[\ell, t] \in S_\square$. This suffices to prove it as well for stochastic states $\langle\ell, t\rangle \in S_\Diamond$ thanks to the structure of the limit corner-point MDP. Given $\mathcal{A}_{cp}$, we write $W \subseteq S$ for the set of states from which there exists an almost-sure winning strategy for the reachability objective. We now detail what having left and/or right pointed region winning in the limit corner-point MDP abstraction implies:

Proposition 3. *Let $r \in R$ be a region and $\ell \in L$ a location.*

- *If $[\ell, \bullet r] \in W$, then $[\ell, r\bullet] \in W$.*
- *If $[\ell, \bullet r] \in W$, then for every $t \in r$, $\mathsf{val}_{\mathcal{A}}([\ell, t]) = 1$;*
- *Else, if $[\ell, r\bullet] \in W$, then for every $t \in r$, $\mathsf{val}_{\mathcal{A}}([\ell, t]) < 1$;*
- *Else, there exists $\varepsilon > 0$ such that for every $t \in r$, $\mathsf{val}_{\mathcal{A}}([\ell, t]) \leq 1 - \varepsilon$.*

The first item is a simple observation: for every location ℓ and region $r \in R$, any winning strategy from $[\ell, \bullet r]$ can be mimicked from $[\ell, r\bullet]$, and is also winning from there. Section 4 is devoted to the rest of the proof of Proposition 3.

Proposition 3 suffices to prove Proposition 2, in the case of player states. Indeed, the second item (whose proof is in Section 4.2) shows the right-to-left implication of Proposition 2, and the third and fourth items show the other implication by contraposition (the proof is in Section 4.1). Remark that we can be more specific in the third case and state: when $[\ell, \bullet r] \notin W$ and $[\ell, r\bullet] \in W$, then $\sup_{t \in r} \mathsf{val}_{\mathcal{A}}([\ell, t]) = 1$. This explains the difference between the third and fourth items.

Before moving to the proofs, we show an example where ε-optimal strategies need to be conceptually complex: the cutpoint inside regions where the strategy changes decision cannot be chosen independently of the location.

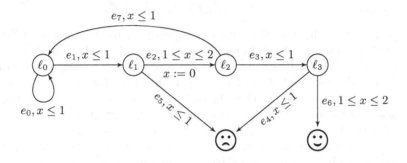

Fig. 4. An example where non-uniform strategies are needed

Consider the example from Fig. 4, where the implicit probability distributions over delays are all uniformly distributed. Decisions can only be taken in locations ℓ_0 and ℓ_2, where transitions with overlapping guards are possible. Intuitively, from ℓ_0 to reach ☺, transition e_1 needs to be taken, with a risk that once ℓ_1 is reached, transition e_5 is triggered. Let t_0 be the cutpoint in ℓ_0 such that if $t_0 < t < 1$, the player decides to take e_1 from (ℓ_0, t). In the same way, let $t_2 \in (0, 1)$ be the cutpoint in ℓ_2 such that if $t_2 < t < 1$, the player decides to take e_3 from (ℓ_2, t). To reach a contradiction, we assume $t_0 = t_2$, and write τ for this value. From $[\ell_2, t]$ with $1 - \tau < t < 1$, a simple calculation shows that the probability to lose is $p_{\text{lose}}^T(\ell_2, t) = (1 - t)/(2 - t)$. Also, from $[\ell_0, t]$, the losing probability is lower bounded by the probability to lose in two steps, directly from ℓ_1, hence $p_{\text{lose}}^T(\ell_0, t) \geq (1 - t)/(2 - t)$. Moreover, $p_{\text{win}}^T(\ell_0, t) \leq p_{\text{win}}^T(\ell_1, t) \leq (1 - t)p_{\text{win}}^T(\ell_3, t) \leq (1 - t)$. Hence, $p_{\text{lose}}^T(\ell_0, t) > p_{\text{win}}^T(\ell_0, t)/2$ for all $t > 1 - \tau$, that is $\mathbb{P}_T(\langle \ell_0, t \rangle \models \Diamond F) < 2/3$. This shows that F is not limit-surely reachable under simple strategies, defined by constant mappings. Yet, the ☺-state is limit-surely reachable from $[\ell_0, 0]$. However, to achieve this, t_0 needs to be set to a much lower value than t_2, *e.g.* $t_2 = \tau$ and $t_0 = \tau^2$.

4   Deciding the Value 1 Problem

The goal of this section is to provide a proof of Proposition 3 (and thus of Proposition 2 and of Theorem 1). For the sake of completeness, we recall the algorithm to compute the set of states of an MDP from which there exists a strategy to reach a target set $\mathcal{F}$ almost-surely (see *e.g.* [4]). We denote by $\mathcal{W}$ the winning states for this objective, and $\mathcal{W}_\square$ the subset of winning player states. The algorithm that computes $\mathcal{W}$ computes at the same time for every player state $\mathbf{w} \in \mathcal{W}_\square$, the largest set of safe actions $\mathsf{Safe}(\mathbf{w})$, *i.e.* the set of all actions that ensure staying in $\mathcal{W}$.

- Initially: $\mathcal{L} = \emptyset$, and for every $\mathbf{s} \in \mathcal{S}_\square$, $\mathsf{Safe}(\mathbf{s}) = \{e \in E \cup E^{\mathsf{limit}} \mid \exists \mathbf{s} \xrightarrow{e} \mathbf{s}'\}$.
- Perform steps 1 and 2 until convergence.
  - Step 1: Move to $\mathcal{L}$ every $[\ell, \mathbf{r}]$ from which there is no path to $\mathcal{F}$ via states in $\mathcal{S} \setminus \mathcal{L}$ only.
  - Step 2: Remove from $\mathsf{Safe}([\ell, \mathbf{r}])$ any e such that if $[\ell, \mathbf{r}] \xrightarrow{e} \langle \ell', \mathbf{r}' \rangle$, there exists $[\ell', \mathbf{r}''] \in \mathcal{L}$ with $\langle \ell', \mathbf{r}' \rangle \to [\ell', \mathbf{r}'']$.
    Move to $\mathcal{L}$ every $[\ell, \mathbf{r}]$ such that $\mathsf{Safe}([\ell, \mathbf{r}]) = \emptyset$.
- Return $(\mathcal{W} = \mathcal{S} \setminus \mathcal{L}, \mathsf{Safe})$.

The rest of the section is organised as follows. Subsection 4.1 establishes the third and fourth items of Proposition 3: states whose left region is not winning in the limit corner-point MDP, do not have value 1 in the decision stochastic timed automaton. Then, Subsection 4.2 shows its second item: states whose left region is winning in the limit corner-point MDP do have value 1 in the decision stochastic timed automaton.

4.1   Non Limit-Surely Winning States

We first aim at showing the right-to-left implication in Proposition 2, by contraposition: $\mathbb{P}_{\max}^{\mathcal{A}_{\mathsf{cp}}}((\ell, \mathbf{r}_{\mathsf{left}}(t)) \models \Diamond \mathcal{F}) < 1$ implies $\mathsf{val}_{\mathcal{A}}([\ell, t]) < 1$. This corresponds to proving the third and fourth items of Proposition 3.

Lemma 1. *Let $\ell \in L$ be a location and $r \in R$ a region.*

- *If $[\ell, r\bullet] \in \mathcal{L}$, then there exists $\varepsilon_{\ell,r} > 0$ such that for every $t \in r$, $\mathsf{val}_{\mathcal{A}}([\ell, t]) \leq 1 - \varepsilon_{\ell,r}$;*
- *If $[\ell, \bullet r] \in \mathcal{L}$, then for every $t \in r$, there exists $\varepsilon_{\ell,t} > 0$ such that $\mathsf{val}_{\mathcal{A}}([\ell, t]) \leq 1 - \varepsilon_{\ell,t}$; Moreover, one can pick non-increasing values for the $\varepsilon_{\ell,t}$'s, that is, $\varepsilon_{\ell,t} \leq \varepsilon_{\ell,t'}$ as soon as $t \geq t'$.*

Proof (Sketch). The proof is by induction on the moment in the MDP algorithm at which $[\ell, r]$ has been moved to $\mathcal{L}$. We thus define $\emptyset = \mathcal{L}_0 \subseteq \mathcal{L}_1 \subseteq \cdots \mathcal{L}_n = \mathcal{L}$ to describe the evolution of $\mathcal{L}$ during time with $|\mathcal{L}_i| = i$. Notice that this decomposition is finer than steps (this is important for step 2 of the MDP algorithm).

We show one important subcase here. Assume $[\ell, (\underline{a}, b)] \in \mathcal{L}$ because of step 2. Every transition e in the DSTA are associated with a transition e in the MDP which leads from $[\ell, (\underline{a}, b)]$ to $\langle \ell', \mathbf{r}' \rangle$, and there exists $\langle \ell', \mathbf{r}' \rangle \xrightarrow{\tau} [\ell', \mathbf{r}'']$ with $[\ell', \mathbf{r}''] \in \mathcal{L}_i$. The hardest case is when $\mathbf{r}'' = \mathbf{r}' = \mathbf{r} = (\underline{a}, b)$. Other cases are actually easier to treat and lead to a uniform bound ε over t. Let $\nu_{\ell,t} = \mu_{\ell,t}(t, (t+b)/2)$ for every $t \in (a, b)$. Observe that $\nu_{\ell,t} > 0$ and is non increasing with $t \in (a, b)$, by assumption (c_3) on the measure functions μ. We then set $\varepsilon_e^{\ell,t} = \nu_{\ell,t} \cdot \varepsilon_{\ell',(t+b)/2} > 0$ for all $t \in (a, b)$. Note that $\varepsilon_e^{\ell,t}$ depends upon t. Last, we define $\varepsilon_{\ell,t} = \min_e \varepsilon_e^{\ell,t}$, the minimum over all transitions e outgoing from $[\ell, t]$. So defined, $\varepsilon_{\ell,t}$ is positive and non increasing because $\nu_{\ell,t}$ and $\varepsilon'_{\ell',(t+b)/2}$ are positive and non increasing. $\qquad\square$

4.2 Limit-Surely Winning States

We now prove that $\mathbb{P}_{\max}^{\mathcal{A}_{cp}}((\ell, \mathbf{r}_{\text{left}}(t)) \models \Diamond \mathcal{F}) = 1$ implies $\text{val}_{\mathcal{A}}([\ell, t]) = 1$. This amounts to show the second item of Proposition 3.

Covering Forest and Golden Paths. Let $\mathcal{A}$ be a decision stochastic timed automaton, and $\mathcal{A}_{cp}$ the associated limit corner-point MDP. From the almost sure winning set of states and actions $(\mathcal{W}, \text{Safe})$ of $\mathcal{A}_{cp}$, we extract a *covering forest* whose roots are elements of $\mathcal{F}$. Each edge of the forest from a player state $\mathbf{w} \in \mathcal{W}_{\square}$ to its unique parent is a transition of $\mathcal{A}_{cp}$, labelled with action $\text{sel}(\mathbf{w}) \in \text{Safe}(\mathbf{w})$. Globally, for every $\mathbf{w} \in \mathcal{W}_{\square}$, the unique path to a root of the forest $\mathbf{w}'_n \in \mathcal{F}$ is a path in $\mathcal{A}_{cp}$: there are $\mathbf{w}_1 \cdots \mathbf{w}_n \in \mathcal{W}_{\square}$ and $\mathbf{w}'_1 \cdots \mathbf{w}'_n \in \mathcal{W}_{\Diamond}$ with

$$\mathbf{w} \xrightarrow{\text{sel}(\mathbf{w})} \mathbf{w}' \xrightarrow{\tau} \mathbf{w}_1 \xrightarrow{\text{sel}(\mathbf{w}_1)} \mathbf{w}'_1 \cdots \mathbf{w}_{n+1} \xrightarrow{\text{sel}(\mathbf{w}_n)} \mathbf{w}'_n \ ,$$

and such a path is called a *golden path* in the following. Notice that many edges emanating from stochastic states do not appear in this forest. They may lead to states that are further from $\mathcal{F}$. The intuition is that this forest represents the ideal situation. Even if it is not guaranteed to take these ideal edges from stochastic states, there is a chance to follow the forest towards $\mathcal{F}$, and we will show it is sufficient.

Example 4. Fig. 5 represents the covering forest, here a tree, on our running example from Fig. 2. Remark here that e_0 is selected in $[\ell_0, (\underline{0}, 1)]$ and e_1 in $[\ell_0, (0, \underline{1})]$. The fact that they differ reflects that the decision for an optimal strategy should not be uniform within region $(0, 1)$ in location ℓ_0.

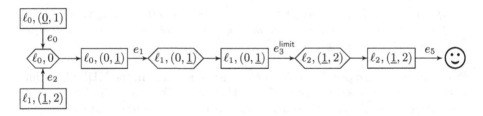

Fig. 5. The covering forest on our running example

Assume now that $\mathcal{F}$ is almost-surely reachable from $(\ell, \mathbf{r}_{\text{left}}(t))$ in the limit corner-point $\mathcal{A}_{cp}$, and let us exhibit a family of ε-optimal strategies, showing that F is limit-surely reachable from $\langle \ell, t \rangle$ in $\mathcal{A}$. More precisely, these ε-optimal strategies are positional but not region uniform: for every location ℓ and bounded open region $r = (c, d)$, there can be a cut-point $c + \tau \in (c, d)$ such that the decisions in the left part of r, $(c, c+\tau)$ and in its right part $(c+\tau, d)$ differ, but are uniform over each of these sub-intervals. These cut-points can be defined through a mapping $T : L \times R \to \mathbb{R}_{\geq 0}$. When T is fixed, we write $\text{left}_T(\ell, r)$ for the sub-region of (ℓ, r), to the left of its cutpoint $T(\ell, r)$. Similarly, $\text{right}_T(\ell, r)$ denotes the states to the right of the cutpoint. For $t \in r$, we write $\mathbf{r}_T([\ell, t]) = [\ell, \mathbf{r}_{\text{left}}(t)]$ if $t \in \text{left}_T(\ell, r)$, and $\mathbf{r}_T([\ell, t]) = [\ell, \mathbf{r}_{\text{right}}(t)]$ if $t \in \text{right}_T(\ell, r)$. We will abuse the notation and call a path $s_1 \cdots s_n$ in the DSTA $\mathcal{A}$ a golden path if for $(\ell_i, \mathbf{r}_i)$ the golden path associated with $\mathbf{r}_T(s_1)$, we have $\mathbf{r}_T(s_{2i+1}) = [\ell_{2i+1}, \mathbf{r}_{2i+1}]$ for all i (if a limit edge is played, it is not possible to agree on the stochastic state).

Based on the covering forest of selected actions, we build a family of strategies $(\sigma_T)_{T:L \times R \to \mathbb{R}_{\geq 0}}$, parametrised by a cutpoint function $T : L \times R \to \mathbb{R}_{\geq 0}$: if $s \in \text{left}_T(\ell, (c, d))$, then $\sigma_T(s) = \text{sel}((\ell, (\underline{c}, d)))$, else $s \in \text{right}_T(\ell, (c, d))$ and $\sigma_T(s) = \text{sel}((\ell, (c, \underline{d})))$. In short, when T is fixed, $\sigma_T(s) = \text{sel}(\mathbf{r}_T(s))$. Since limit edges do not exist in the DSTA, by $\sigma_T(s) = e^{\text{limit}}$, we implicitly mean $\sigma_T(s) = e$, where $e \in E$ is the unique edge associated with $e^{\text{limit}} \in E^{\text{limit}}$.

We define the following subset of states in the DSTA $\mathcal{A}$: $S_{\text{win}} = \{[\ell, t] \in S_\square \mid [\ell, \mathbf{r}_{\text{left}}(t)] \in \mathcal{W}\}$. We show that all states in S_{win} have value 1 for the reachability objective in $\mathcal{A}$. Further, $S_T^{\text{right}} = \{[\ell, t] \in S_\square \mid \mathbf{r}_T([\ell, t]) = [\ell, \mathbf{r}_{\text{right}}(t)] \in \mathcal{W}\}$ is the set of states belonging to the right-part (as specified by the cut-point function T) of limit winning bounded open regions; Note that S_T^{right} and S_{win} are not necessarily disjoint. Our objective is to show that for every $\varepsilon > 0$, there exists a mapping T such that, from any state $s \in S_{\text{win}}$, the probability to reach F under σ_T is at least $1 - \varepsilon$.

Lemma 2. *For every $s \in S_{\text{win}}$, $\mathbb{P}_{\sigma_T}^{\mathcal{A}}(s \models \Diamond(F \cup S_T^{\text{right}})) = 1$.*

When the mapping T is fixed, we let $[\![\mathcal{W}]\!]_T = \{s \mid \mathbf{r}_T(s) \in \mathcal{W}\}$, that is, states whose pointed-region (relatively to T) is winning in the limit corner point. Notice that $[\![\mathcal{W}]\!]_T = S_T^{\text{right}} \cup S_{\text{win}}$. We show that the probability to leave $[\![\mathcal{W}]\!]_T$ can be made arbitrarily small.

Lemma 3. *For every $\varepsilon > 0$, there exists a function $T : L \times R \to \mathbb{R}_{\geq 0}$ such that, writing $\bigcirc$ for the next-step operator, for every $s \in [\![\mathcal{W}]\!]_T$*

- *if $\sigma_T(s) \in E^{\text{limit}}$ then $\mathbb{P}_{\sigma_T}^s(\bigcirc [\![\mathcal{W}]\!]_T) \geq 1 - \varepsilon$,*
- *else $\mathbb{P}_{\sigma_T}^s(\bigcirc [\![\mathcal{W}]\!]_T) = 1$.*

Taking a limit edge in the limit corner-point MDP is the only case where a losing state can be reached in the concrete DSTA. However, staying in the current region when the decision in the abstraction was a limit-edge may not necessarily lead to a losing state. Actually, limit-edges are not always the best choice: the only way to reach F might be to stay in the current region (and therefore avoid limit-edges) even if the probability to stay there is very small. This is illustrated on the example from Fig. 3 in which from $[\ell_0, (0, \underline{1})]$, in order to eventually reach $\mathcal{F}$, one should pick e_1 rather than e_1^{limit}.

We now explain that the probability to follow the covering forest towards $\mathcal{F}$, although small, will be arbitrarily bigger than the probability to follow the forest until reaching a losing state.

Given T a mapping assigning cut-points to each bounded open region, and $[\ell, t]$ a player state in $[\![\mathcal{W}]\!]_T$, we write $p_{\text{win}}^T(\ell, t)$ for the probability, from $[\ell, t]$ and under σ_T, to execute a golden path (which therefore reach F). Also, $p_{\text{lose}}^T(\ell, t)$

is the probability to execute a golden path until a losing state in $S \setminus [\![\mathcal{W}]\!]_T$ is reached. If from a stochastic state, a golden path is not executed, and yet $\mathcal{L}$ is not immediately reached either (this corresponds to behaviours not "counted" in p_{lose}^T and p_{win}^T), then a winning region will be reached, possibly further away from $\mathcal{F}$.

Lemma 4. *For every $\varepsilon > 0$, there exists a function $T : L \times R \to \mathbb{R}_{\geq 0}$ such that for every $[\ell, t] \in [\![\mathcal{W}]\!]_T$*

$$p_{\mathsf{win}}^T(\ell, t) \cdot \varepsilon \geq p_{\mathsf{lose}}^T(\ell, t) .$$

Given a tolerance ε, the proof of Lemma 4 details how to define a mapping, denoted $T(\varepsilon)$ to make the dependency explicit, under which the probability to reach $\mathcal{F}$ by progressing in the covering forest is arbitrarily bigger than the probability to reach a losing state. The definition of $T(\varepsilon)$ is non trivial and is done by induction on the distance to $\mathcal{F}$ in the covering forest. Recall that once a cut-point mapping T is fixed, the strategy σ_T is perfectly defined. It now remains to justify that, the strategies $(\sigma_{T(\varepsilon)})$ form a family of limit-sure strategies.

Lemma 5. *For every $s \in S_{\mathsf{win}}$, $\mathbb{P}_{\sigma_{T(\varepsilon)}}^{\mathcal{A}}(s \models \Diamond F) \geq 1 - \varepsilon$.*

Proof. Let $\varepsilon > 0$, $T(\varepsilon) : L \times R \to (0, 1)$ the mapping as defined in Lemma 4, and $\sigma_{T(\varepsilon)}$ the corresponding strategy. To establish Lemma 5 we provide a lower bound on the probability, under $\sigma_{T(\varepsilon)}$ to reach F from winning states.

To do so, we consider the set X of runs under $\sigma_{T(\varepsilon)}$ that stay forever in $[\![\mathcal{W}]\!]_T \setminus F$. Such runs never reach the target, and also stay away from the losing states. We will show that $\mathbb{P}_{\sigma_{T(\varepsilon)}}^{\mathcal{A}}(s \models X) = 0$. To do so, we again partition X into three categories: X_1 gathers runs with infinitely many resets; X_2 consists of runs with finitely many resets and ending in the unbounded clock region (M, ∞); and X_3 is the set of runs with finitely many resets eventually staying in a bounded region (c, d).

Let us first consider X_1. Runs in X_1 necessarily visit some some state $(\ell_0, 0)$ infinitely often. Since, at each visit of $(\ell_0, 0)$, there is a strictly positive probability to execute a golden path and thus reach F, we reach a contradiction. Thus $\mathbb{P}_{\sigma_{T(\varepsilon)}}^{\mathcal{A}}(s \models X_1) = 0$. We now consider runs in X_2, that ultimately stay in the unbounded region. As explained earlier, almost surely F will be reached, a contradiction: $\mathbb{P}_{\sigma_{T(\varepsilon)}}^{\mathcal{A}}(s \models X_2) = 0$.

Last, for runs in X_3, that ultimately stay in a bounded region (c, d), the reasoning is exactly the same as for runs of X_2. Thus, $\mathbb{P}_{\sigma_{T(\varepsilon)}}^{\mathcal{A}}(s \models X_3) = 0$.

We now exploit Lemma 4 to conclude. Since almost all runs leave $[\![\mathcal{W}]\!]_T \setminus F$, it must be that either a losing states or F is reached, and it suffices to compare the probabilities in each case. Thanks to Lemma 4, for all states in $[\![\mathcal{W}]\!]_T$, it is much more likely to reach F than to reach a losing state. More precisely,

$$\frac{p_{\mathsf{win}}^T(\ell, t)}{p_{\mathsf{win}}^T(\ell, t) + p_{\mathsf{lose}}^T(\ell, t)} \geq \frac{p_{\mathsf{win}}^T(\ell, t)}{p_{\mathsf{win}}^T(\ell, t) + p_{\mathsf{win}}^T(\ell, t) \cdot \varepsilon} \geq \frac{1}{1 + \varepsilon} \geq 1 - \varepsilon .$$

As a consequence $\mathbb{P}_{\sigma_{T(\varepsilon)}}^{\mathcal{A}}(s \models \Diamond F) \geq (1 - \varepsilon)$. □

This ends the proof of the left-to-right implication in Proposition 2.

5   Deciding the Probability 1 Problem

If one wants to solve the probability 1 problem, rather than the more difficult value 1 problem, it suffices to consider the region MDP $\mathcal{A}_R$. This MDP $\mathcal{A}_R$ is equivalent to the fragment of $\mathcal{A}_{\mathsf{cp}}$ restricted to left and plain regions, and hence without limit edges. As for the value 1 problem, the decidability of the probability 1 problem is given thanks to the following reduction:

Lemma 6. *Let $\mathcal{A}$ be a decision stochastic timed automaton, $\mathcal{A}_R$ its region MDP, $r \in R$ a region, and $\ell \in L$ a location and $t \in \mathbb{R}_{\geq 0}$ a clock value with $t \in r$. Then*

$$\exists \sigma, \ \mathbb{P}^{\mathcal{A}}_{\sigma}([\ell, t] \models \Diamond F) = 1 \quad \Longleftrightarrow \quad \mathbb{P}^{\mathcal{A}_R}_{\max}([\ell, r] \models \Diamond \mathcal{F}) = 1 \ and$$
$$\exists \sigma, \ \mathbb{P}^{\mathcal{A}}_{\sigma}(\langle \ell, t \rangle \models \Diamond F) = 1 \quad \Longleftrightarrow \quad \mathbb{P}^{\mathcal{A}_R}_{\max}(\langle \ell, r \rangle \models \Diamond \mathcal{F}) = 1 \ .$$

Proof. The proof is not different for player and stochastic states, so we treat them indistinctly, and use brackets in place of square or angle brackets.

Let (ℓ, t) with $t \in R$, be a state of $\mathcal{A}$ such that $\mathbb{P}^{\mathcal{A}_R}_{\max}((\ell, r) \models \Diamond \mathcal{F}) < 1$. One can easily adapt the inductive proof of Lemma 1, showing that if (ℓ, r) is losing in the MDP $\mathcal{A}_R$, then for every $t \in r$, there exists $\varepsilon_{\ell,t}$ with $\mathbb{P}^{\mathcal{A}}_{\sigma}((\ell, t) \models \Diamond F) < 1 - \varepsilon_{\ell,t}$ whatever the strategy σ.

For the other implication, it suffices to mimic faithfully in $\mathcal{A}$ the positional winning strategy $\sigma_{\mathcal{A}_R}$ from $\mathcal{A}_R$. For every (ℓ, r) a winning state in the region MDP $\mathcal{A}_R$, for every $t \in r$, we let $\sigma(\ell, t) = \sigma_{\mathcal{A}_R}(\ell, r)$. Now that the strategy is fixed, we recover the purely probabilistic framework of Stochastic Timed Automata, and can apply the results of [3] to conclude that $\mathbb{P}^{\mathcal{A}}_{\sigma}((\ell, t) \models \Diamond F) = 1$. Alternatively, partitioning runs into three categories, as we did for the proofs of Lemmas 2 and 5, allows one to conclude that σ is almost-surely winning in $\mathcal{A}$.   □

Lemma 6, and precisely its right-to-left implication, does not hold in DSTA with at least two clocks. We emphasise here again that the decomposition of runs we use is only valid for 1-clock timed automata.

As an immediate consequence, we obtain the decidability in PTIME of the probability 1 problem for reachability objectives in DSTA (see Proposition 1).

6   Conclusion

This paper shows the decidability in PTIME of the probability 1 and value 1 problems for reachability objectives on an extension of 1-clock timed automata with random delays, and in which edges are chosen according to a strategy. It would be natural to allow for more general objectives (*e.g.* Büchi or parity). We could also investigate the extension of our framework to 2 players, taking decisions in turn or concurrently. Moving to more quantitative questions, such as computing the value would probably require a finer abstraction than the limit corner-point MDP. Last, the class of 1-clock DSTA can seem restrictive, and it is definitely a challenge to tackle already stochastic timed automata without decisions, for which the almost-sure model checking of reachability properties is still open.

References

1. Alur, R., Dill, D.L.: A Theory of Timed Automata. Theoretical Computer Science 126(2), 183–235 (1994)
2. Baier, C., Bertrand, N., Bouyer, P., Brihaye, T., Größer, M.: Probabilistic and Topological Semantics for Timed Automata. In: Arvind, V., Prasad, S. (eds.) FSTTCS 2007. LNCS, vol. 4855, pp. 179–191. Springer, Heidelberg (2007)
3. Baier, C., Bertrand, N., Bouyer, P., Brihaye, T., Größer, M.: Almost-Sure Model Checking of Infinite Paths in One-Clock Timed Automata. In: Proceedings of LICS 2008, pp. 217–226. IEEE Comp. Soc. Press (2008)
4. Baier, C., Katoen, J.-P.: Principles of model checking. MIT Press (2008)
5. Bertrand, N., Schewe, S.: Playing optimally on timed automata with random delays. In: Jurdziński, M., Ničković, D. (eds.) FORMATS 2012. LNCS, vol. 7595, pp. 43–58. Springer, Heidelberg (2012)
6. Bouyer, P., Brinksma, E., Larsen, K.G.: Optimal infinite scheduling for multi-priced timed automata. Formal Methods in System Design 32(1), 3–23 (2008)
7. Bouyer, P., Forejt, V.: Reachability in Stochastic Timed Games. In: Albers, S., Marchetti-Spaccamela, A., Matias, Y., Nikoletseas, S., Thomas, W. (eds.) ICALP 2009, Part II. LNCS, vol. 5556, pp. 103–114. Springer, Heidelberg (2009)
8. Brázdil, T., Krčál, J., Křetínský, J., Kučera, A., Řehák, V.: Stochastic Real-Time Games with Qualitative Timed Automata Objectives. In: Gastin, P., Laroussinie, F. (eds.) CONCUR 2010. LNCS, vol. 6269, pp. 207–221. Springer, Heidelberg (2010)
9. Chatterjee, K., de Alfaro, L., Henzinger, T.A.: Qualitative concurrent parity games. ACM Transactions on Computation Logic 12(4), 28 (2011)
10. Chatterjee, K., Tracol, M.: Decidable problems for probabilistic automata on infinite words. In: Proceedings of LICS 2012, pp. 185–194. IEEE (2012)
11. Chen, T., Han, T., Katoen, J.-P., Mereacre, A.: Reachability Probabilities in Markovian Timed Automata. In: Proceedings of CDC-ECC 2011, pp. 7075–7080. IEEE (2011)
12. de Alfaro, L., Henzinger, T.A., Kupferman, O.: Concurrent reachability games. Theoretical Computer Science 386(3), 188–217 (2007)
13. Fijalkow, N., Gimbert, H., Oualhadj, Y.: Deciding the value 1 problem for probabilistic leaktight automata. In: Proceedings of LICS 2012, pp. 295–304. IEEE (2012)
14. Gimbert, H., Oualhadj, Y.: Probabilistic automata on finite words: Decidable and undecidable problems. In: Abramsky, S., Gavoille, C., Kirchner, C., Meyer auf der Heide, F., Spirakis, P.G. (eds.) ICALP 2010. LNCS, vol. 6199, pp. 527–538. Springer, Heidelberg (2010)
15. Laroussinie, F., Markey, N., Schnoebelen, P.: Model checking timed automata with one or two clocks. In: Gardner, P., Yoshida, N. (eds.) CONCUR 2004. LNCS, vol. 3170, pp. 387–401. Springer, Heidelberg (2004)
16. Wolovick, N., Johr, S.: A Characterization of Meaningful Schedulers for Continuous-Time Markov Decision Processes. In: Asarin, E., Bouyer, P. (eds.) FORMATS 2006. LNCS, vol. 4202, pp. 352–367. Springer, Heidelberg (2006)

Decidable Problems for Unary PFAs

Rohit Chadha[1], Dileep Kini[2], and Mahesh Viswanathan[2]

[1] University of Missouri, USA
[2] University of Illinois at Urbana-Champaign, USA

Abstract. Given a PFA A and a cut-point λ, the isolation problem asks if there is a bound $\epsilon > 0$ such that the acceptance probability of every word is bounded away from λ by ϵ. In this paper we show that the isolation problem for PFAs with a unary input alphabet is (a) coNP-complete, if the cut-point is 0 or 1, and (b) is in coNP^{RP} and coNP-hard, if the cut-point is in $(0, 1)$. We also show that the language containment problem, language equivalence problem, the emptiness problem and the universality problem for unary PFAs with limit isolated cut-points is in the fourth level of counting hierarchy C_4P (and hence in PSPACE).

1 Introduction

Probabilistic finite automata (PFA), introduced by Rabin [18], are a generalization of deterministic finite automata that model finite state, randomized algorithms that process an input string one-way. Given a cut-point $\lambda \in [0, 1]$, an input string w is accepted by a PFA A iff the probability of reaching a final/accept state of A on input w from the initial state is $> \lambda$, and $L_{>\lambda}(A)$ denotes the collection of all strings accepted by A with cut-point λ. The emptiness problem for PFAs is not only an important mathematical problem, but it has applications in verifying useful properties of open, probabilistic, reactive systems like sensor networks, biochemical reactions, and software [14,15,13,1]. For example, checking emptiness is equivalent to checking if drug concentrations in certain organs is always below toxicity levels [13]. In such contexts, one would, in fact, like to ensure a stronger property, namely, that the drug concentrations are well below acceptable levels, and not just barely acceptable.

Checking such "robust" properties of a system is closely related to another important problem of PFAs, namely, the *isolation* problem. A cut-point λ is said to be isolated for an automaton A if there is an $\epsilon > 0$ (called a *degree of isolation*) such that A's probability of reaching an accepting state on any word is either $> \lambda + \epsilon$ or is $< \lambda - \epsilon$. PFAs with isolated cut-points enjoy many nice properties. First, they represent algorithms with bounded error, for which, the error probability can be driven down below a fixed level using repeated experiments [18]. Second, λ being isolated ensures that $L_{>\lambda}(A)$ is regular; note, if λ is not isolated then $L_{>\lambda}(A)$ may not be regular [25,18] (even for unary alphabet [17,20]). Finally, certain PFAs with isolated cut-points are *stable*, in that small changes to the transition probabilities don't change the language with respect to cut-point λ [18].

G. Norman and W. Sanders (Eds.): QEST 2014, LNCS 8657, pp. 329–344, 2014.
© Springer International Publishing Switzerland 2014

Even though the emptiness and isolation problems have important practical applications, these problems have been shown to be computationally very hard. The emptiness problem is co-r.e.-complete [17,6] and the isolation problem is Σ_2^0-complete [4,9,5]. However, these lower bounds only apply when the input alphabet of the PFA has at least two symbols. The decidability of the isolation problem for PFAs over the unary input alphabet was claimed to be true in Bertoni's original paper [4] but was not proved. Furthermore, there was no complexity analysis for the isolation problem. The decidability of the emptiness problem for PFAs over the unary alphabet is a long-standing open problem.

In this paper we consider these decision problems for PFAs over a unary input alphabet. Unary PFAs are nothing but (finite state) Markov chains, which are a standard model to define the semantics of probabilistic systems, and have been used in a number of contexts. Our main result is about the complexity of the isolation problem for unary PFAs. When $\lambda \in (0, 1)$, we show that the isolation problem is decidable and is in $\mathsf{coNP^{RP}}$.[1] Note that since RP is contained in NP (see [3]), this implies that the problem of checking isolation when $\lambda \in (0, 1)$ is in the second level of polynomial hierarchy. Furthermore, given that RP is believed to be P, this would imply the problem to be in coNP which matches the lower bound of coNP-hardness mentioned ahead. Our procedure also gives a way to compute a degree of isolation if the PFA is isolated. Our result is proved as follows. Let us call a PFA A isolated in the limit if there is a $n_0 > 0$ such that the probability of accepting any string a^n, with $n > n_0$ is bounded away from λ. Thus λ is an isolated cut-point for a PFA A iff A is isolated in the limit and the probability of accepting any "short" string (i.e., one whose length is less than n_0) is bounded away from λ. We first prove that the problem of checking if A is isolated in the limit is in coNP. Next, we show that if A is isolated in the limit, then the bound n_0 is "small". More precisely, we show that this number n_0 can be represented in binary using polynomially (in the size of A) many bits. Using this observation, we can conclude that if a PFA A is isolated in the limit, then λ is not isolated if there is some string (of exponential length) that is accepted with probability λ. The check of whether a string of length ℓ is accepted with probability λ can be reduced to checking if a straight-line program of length ℓ using addition, multiplication, and subtraction computes a real number that is equal to 0. Based on this observation, and results on the complexity of the EquSLP problem [21], we conclude that checking the isolation of a PFA over the unary alphabet is in $\mathsf{coNP^{RP}}$. Next, we show that if the cut-point λ is either 0 or 1 the isolation problem is easier. We show that for these extremal cut-points, the isolation problem is in coNP. The proof uses observations about the complexity of the universality problem for NFAs [24]. We also show that the isolation problem is coNP-hard.

Our techniques for checking isolation for unary PFAs have a few consequences. One can show that if A and B are limit isolated PFAs then the problem of checking $L_{>\lambda}(A) \subseteq L_{>\lambda}(B)$ is in $\mathsf{coNP^{C_3P}}$. The complexity class $\mathsf{coNP^{C_3P}}$ lies in

---

[1] RP is the set of decision problems that can be decided by randomized polynomial-time Turing Machines with one-sided error (see Section 2.3).

the fourth level of counting hierarchy and hence in PSPACE [27]. That means that the language equivalence problem, the emptiness problem and the universality problem for limit isolated PFAs are decidable in $coNP^{C_3P}$ and hence in PSPACE. These results need to be contrasted with the fact that the decidability of the emptiness problem for unary PFAs (when λ is not necessarily isolated) is still open. Similarly, the decidability of the emptiness problem for PFAs with isolated cut-points, with input alphabet having at least 2 symbols, is also open.

The rest of the paper is organized as follows. We discuss preliminary definitions and results in Section 2. Our results on complexity of checking limit isolation and isolation for unary PFAs are discussed in Section 3 and Section 4. We discuss the problems of language emptiness, containment and universality for limit isolated unary PFAs in Section 5. We present our conclusions in Section 6.

Related Work. As pointed out in the introduction, the undecidability of the emptiness and isolation problems for PFAs was established in [17,6,4,9,5]. The efficient decidability of bisimulation for probabilistic systems can be exploited to efficiently check a strong version of equivalence of PFAs [26,7,11,12] — here PFAs A and B are said to be equivalent if the acceptance probability of each input string is the same in both A and B. The decidability of language equivalence is a consequence of a more general result on minimizing weighted automata [22]. Model checking of Markov chains with respect to PCTL properties is a mature technology [19]. However, such tools cannot answer emptiness, containment, and language equivalence of unary PFAs because such properties are not expressible in PCTL. Convergence properties of Markov chains have been widely studied (see [8,16]) but questions of complexity of isolation pertain to both transient and asymptotic behavior of Markov chains which, to the best of our knowledge, has not been studied. The decidability of checking emptiness for isolated unary PFAs can also be derived as a consequence of the results in [5] which establishes that the problem of checking emptiness of isolated eventually weakly ergodic PFAs is decidable. However, [5] does not contain any complexity analysis. Furthermore, our results on complexity of emptiness checking applies to limit isolated unary PFAs which are a strict super-set of isolated unary PFAs.

2   Preliminaries

We introduce some notation and recall some standard notation. We will fix some notational conventions used in this paper. We will assume that for any finite S of k elements, we have a *fixed* enumeration $\{1, 2, \ldots, k\}$ of the elements of S. We will identify elements of S with the corresponding numeral.

2.1   Distributions, Stochastic Matrices and Markov Chains

Distributions. Given a finite set S, a distribution over S is any function $\mu : S \mapsto [0,1]$ such that $\sum_{s \in S} \mu(s) = 1$. Since S is finite, μ can be thought of as a vector with $|S|$ coordinates. The set of all distributions over S is represented by

$dist(S)$. For a set $S' \subseteq S$, we write $\mu(S') = \sum_{i \in S'} \mu(i)$. For any two distributions $\mu, \nu \in dist(S)$ the distance between them is defined as

$$d(\mu, \nu) = \sum_{i \in S} \frac{|\mu(i) - \nu(i)|}{2} = \max_{S' \subseteq S} |\mu(S') - \nu(S')|.$$

The distance d is a metric.

Stochastic matrices. A stochastic matrix δ is a square matrix with non-negative entries such that each row of the matrix sums up to one. This distance between $n \times n$ matrices δ_1, δ_2 is defined as:

$$d(\delta_1, \delta_2) = \max_i \sum_j |\delta_1(i, j) - \delta_2(i, j)|.$$

We use $\delta(; , j)$ to represent the jth column of the matrix δ. We use $\sigma_t(\delta)$ to denote the sequence of matrices $\delta^t, \delta^{2t}, \delta^{3t}, \ldots$ and use $\widehat{\delta^t}$ to denote the limit $\lim_{r \to \infty} \delta^{rt}$ if it exists. When clear from the context, we shall drop δ from $\sigma_t(\delta)$ and just write σ_t. A stochastic matrix δ is called *positive* if all of its entries are strictly positive.

A stochastic matrix of dimension $n \times n$ can be represented as a directed graph with n vertices, and an edge from i to j if $\delta(i, j) > 0$. A maximally strongly connected component is called a *Bottom Strongly Connected Component* (BSCC) if it has no outgoing edges. A *transient state* is a state which is not in a BSCC, and a *terminal state* is one which is within a BSCC. δ is said to be *irreducible* if it has only one BSCC and no transient states. The collection of all BSCCs of a δ will be represented by $\mathcal{C}_\delta$. The set of all transient states of δ will be denoted by T_δ. Lower case c_δ will be used to denote individual BSCCs. When clear from the context, we shall drop the subscript δ.

The *period* of a vertex is defined as the g.c.d (greatest common divisor) of all the cycle lengths going through the vertex. For a SCC, the periods of all the vertices in that component will be the same and will be defined as the period of that component. A component is called *aperiodic* if its period is 1. δ is said to be *aperiodic* if all vertices have period 1. The *ultimate period* of δ is the l.c.m (least common multiple) of the periods of its BSCCs. Since BSCCs and their related periods can be computed in polynomial time we have the following:

Proposition 1. *The ultimate period of a $n \times n$ matrix δ can be computed in polynomial time and is a number with $O(n \log n)$ bits.*

The following Lemma is proved in [23]

Lemma 1. *For any $n \times n$ stochastic matrix δ, if δ is an aperiodic and irreducible stochastic matrix then δ^{n^2} is positive.*

A stochastic matrix γ with dimensions $n \times n$ is called a *contraction map with contracting factor* $\alpha < 1$ if for all distributions μ and ν of dimension n it is the case that $d(\mu\gamma, \nu\gamma) < \alpha d(\mu, \nu)$.

Proposition 2. *For any $n \times n$ stochastic matrix δ if δ is positive then δ is contracting with contracting factor $1 - n \min_{i,j} \delta(i,j)$.*

Proof. Let s be $\min_{i,j} \delta(i,j)$. Observe that $s \leq \frac{1}{n}$.

$$
\begin{aligned}
2d(\mu\delta, \nu\delta) &= \sum_j |\mu\delta(j) - \nu\delta(j)| = \sum_j \left| \sum_i (\mu(i)\delta(i,j) - \nu(i)\delta(i,j)) \right| \\
&= \sum_j \left| \sum_i (\mu(i)(\delta(i,j) - s) - \nu(i)(\delta(i,j) - s) + s(\mu(i) - \nu(i))) \right| \\
&= \sum_j \left| \sum_i (\mu(i)(\delta(i,j) - s) - \nu(i)(\delta(i,j) - s)) + s\left(\sum_i \mu(i) - \sum_i \nu(i) \right) \right| \\
&= \sum_j \left| \sum_i ((\mu(i) - \nu(i))(\delta(i,j) - s)) + s(1 - 1) \right| \\
&\leq \sum_j \sum_i |(\mu(i) - \nu(i))|(\delta(i,j) - s) = \sum_i \sum_j |(\mu(i) - \nu(i))|(\delta(i,j) - s) \\
&\leq \sum_i |(\mu(i) - \nu(i))| \sum_j (\delta(i,j) - s) = \sum_i |(\mu(i) - \nu(i))|(1 - ns) \\
&\leq 2(1 - ns)d(\mu, \nu).
\end{aligned}
$$

$\square$

Markov Chains. A Markov chain M is a tuple (Q, δ, μ_0) where Q is a finite set of states, δ is stochastic matrix of dimension $|Q| \times |Q|$ and $\mu_0 \in dist(Q)$. δ is referred to as the transition matrix and μ_0 denotes the initial distribution. The Markov chain represents an infinite sequence of distributions $\mu_0, \mu_1, ..$ where $\mu_i = \mu_0 \delta^i$.

2.2 Probabilistic Finite Automata

A PFA [18] is like a deterministic automaton except that the transition on an input symbol is probabilistic.

Definition 1. *A Probabilistic Finite Automaton (PFA) A is a tuple $(Q, \Sigma, (\delta_\sigma)_{\sigma \in \Sigma}, \mu_0, Q_F)$, where Q is a finite set of states, Σ is a finite alphabet, $\mu_0 \in dist(Q)$ is the initial distribution, $Q_F \subseteq Q$ is the set of final states, and $(\delta_\sigma)_{\sigma \in \Sigma}$ is an indexed set of stochastic matrices with dimension $|Q| \times |Q|$.*

For a symbol a, $\delta_a(s,t)$ represents the probability of going from state s to t on input symbol a. For any input word $w \in \Sigma^*$ of length n the probability of going from s to t along $w = a_1 a_2 \cdots a_n$ is then given by $\delta_w(s,t)$ where δ_w is the matrix $(\delta_{a_1} \cdot \delta_{a_2} \cdots \delta_{a_n})$. The distribution reached on input $w \in \Sigma^*$ in A is then given by $\mu_0 \delta_w$.

Definition 2. *The acceptance probability of a word $w \in \Sigma^*$ on PFA A is given by $\sum_{q \in Q_F} \mu_0 \delta_w(q)$ or $\mu_0 \delta_w \eta_F$ where η_F is the column vector such that $\eta_F(j) = 1$ if $j \in Q_F$ and $\eta_F(j) = 0$ otherwise.*

We will say that η_F is the vector corresponding to Q_F.

Languages defined by PFAs need not be regular (even over unary alphabet [17][20]). Emptiness checking turns out to be undecidable in general [6] but is still open for the unary case.

Since this paper only considers PFAs over a unary alphabet, the remaining definitions only apply to unary PFAs.

We will assume that the unique letter in the unary alphabet is a, and we will drop the index a in the transition δ_a. Thus, the probability of accepting the string of length a^ℓ is given by $\mu_0 \delta^\ell \eta_F$. We will often use μ_k to denote $\mu_0 \delta^k$. A unary PFA is essentially a Markov chain with a subset of the states designated as final states. The language of PFA is defined with respect to a cut-point λ:

Definition 3. *Given a cut-point $\lambda \in [0, 1]$ the language accepted by a unary PFA A with respect to λ denoted by $L_{>\lambda}(A)$ is*

$$\{a^n \mid \mu_0 \delta^n \eta_F > \lambda\}.$$

We are interested in special kinds of cut-points, called *isolated* cut-points [18]. A cut-point λ is isolated for a PFA A if the acceptance probabilities of all the words are bounded away from λ.

A cut-point is said to be extremal if it is either 0 or 1, and non-extremal if it is the open interval $(0, 1)$.

Definition 4. *The cut-point λ is said to be isolated for A if there exists an $\epsilon > 0$ such that for all $n > 0$,*

$$|\mu_0 \delta^n \eta_F - \lambda| > \epsilon.$$

ϵ *is known as a degree of isolation.*

When λ is isolated, the language recognized $L_{>\lambda}(A)$ is known to be regular [18]. We introduce the notion of *limit isolation*, which generalizes the notion of isolated cut-points. We say that a cut-point is limit isolated if it is isolated for asymptotically large inputs. Formally,

Definition 5. *The cut-point λ is said to be limit isolated for A if there exists $\epsilon > 0$ and $n_0 > 0$ and such that for all $n > n_0$,*

$$|\mu_0 \delta^n \eta_F - \lambda| > \epsilon.$$

ϵ *is known as a degree of limit isolation.*

Note 1. A PFA A is isolated at λ iff it is limit isolated at λ and there is no word that is accepted with probability exactly λ. It is also easy to see that the language recognized $L_{>\lambda}(A)$ is regular if λ is limit isolated.

The two definitions lead us to the problems of checking if a given rational cut-point is limit isolated or isolated, which we will tackle in sections 3 and 4 respectively.

The following proposition implies that checking whether 0 is isolated (limit isolated, respectively) is as hard as checking if 1 is isolated (limit isolated, respectively) and vice-versa.

Proposition 3. *For any PFA A, there is another PFA B such that 0 is an isolated (limit isolated respectively) cut-point of A iff 1 is an isolated (limit isolated respectively) cut-point of B.*

Proof. We just need to interchange final and non-final states. That is if $A = (Q, \Sigma, (\delta_\sigma)_{\sigma \in \Sigma}, \mu_0, Q_F)$ then we can take $B = (Q, \Sigma, (\delta_\sigma)_{\sigma \in \Sigma}, \mu_0, Q \setminus Q_F)$.

2.3 Complexity

The complexity class RP consists of problems which can be solved using a randomized polynomial time algorithm that always returns "yes" on yes-instances, and returns "no" with probablity at least $\frac{1}{2}$ on no-instances. We know that RP is contained in NP.

The counting hierarchy CH is a class of decision problems contained within PSPACE, which was introduced by Wagner [27]. The 0-th level, C_0P, is defined as P. The k-th level of the hierarchy is denoted by C_kP and is defined recursively as $C_{k+1}P = PP^{C_kP}$. Here PP denotes the class of decision problems for which there are polynomial time randomized algorithms which answer "yes" with probability $> \frac{1}{2}$ on yes-instances, and answer "no" with probability $\geq \frac{1}{2}$ on no-instances. The whole counting hierarchy is contained in PSPACE.

In this paper we will assume every rational number is represented as $\frac{p}{q}$ where p and q are integers in binary. So, when we say a rational r can be computed in polynomial time given rationals $r_1, \ldots, r_k$, it implies that r can also be represented using polynomially many bits in the inputs $r_1, \ldots, r_k$.

2.4 Straight Line Programs

We will use *straight line programs* (SLP) to represent the computation of quantities such as acceptance probability of a word. A SLP over a set of variables V is a sequence of statements of the form $x := E$ where $x \in V$; E is either a constant in $\{0, 1\}$, a variable in V, or an expression of the form $e_1 \circ e_2$ where the operator $\circ \in \{+, -, *\}$ and $e_i \in \{0, 1\} \cup V$. Furthermore, each variable occurring on the right hand side of an assignment must occur in the left hand side of (some) earlier assignment. The value of a SLP is defined as the value assigned in its last statement. EquSLP is the problem of deciding if the value returned by the SLP is 0. PosSLP is defined as the problem of determining whether the value of the given SLP is positive. EquSLP was shown to be in coRP in [21]. A recent result [2] shows that PosSLP is in P^{C_3P} and hence in the 4-th level of counting hierarchy.

3 Limit Isolation

We prove that the problem of checking if a cut-point (extremal or non-extremal) is limit-isolated for a unary PFA is coNP-complete. In order to prove these results, we recall some standard facts about Markov chains. The proofs of these facts can be found in [8] and hence are omitted.

Theorem 1. *Let $c \in C$ be a BSCC of a Markov Chain $M = (Q, \delta, \mu_0)$, p be the period of c, then for any state j in c:*

1. *If i is a transient state of M then $\lim\limits_{r \to \infty} \delta^{pr}(i, j)$ exists and can be calculated in time polynomial in the size of δ.*
2. *If i is in c, then $\lim\limits_{r \to \infty} \delta^{pr}(i, j)$ exists and can be calculated in time polynomial in the size of δ.*
3. *If i is neither a transient state of M nor in c then $\lim\limits_{r \to \infty} \delta^{pr}(i, j) = 0$ (in fact $\delta^\ell(i, j) = 0$ for all ℓ).*

This leads to the following corollary (recall that the ultimate period of δ is the l.c.m of the period of its BSCCs).

Corollary 1. *For any stochastic matrix δ, with ultimate period p, $\widehat{\delta^p} = \lim\limits_{r \to \infty} \delta^{pr}$ exists.*

We are ready to show that limit isolation is coNP-complete.

Theorem 2. *The problem of checking given a unary PFA A and a rational cut-point λ whether λ is limit-isolated for A is coNP-complete.*

Proof. (**Upper Bound**). Let $A = (Q, \Sigma, \delta, \mu_0, Q_F)$ and let p be the ultimate period of δ. According to Corollary 1, there are possibly p different limits towards which the Markov chain approaches in a cyclic manner. That is for each $0 \leq k < p$, we have that $\lim_{r \to \infty} \delta^{k+pr}$ exists.

If λ is not a limit isolated cut-point then it is easy to see that there is a $0 \leq k < p$ such that $\lim_{r \to \infty} \mu_0 \delta^{k+pr} \eta_F$ is λ. The witness for a *no* answer to our problem is therefore going to be this number k which requires only $n \log n$ bits to be represented.

The result will follow if we can compute the distribution $\mu_k \widehat{\delta^p} = \lim_{r \to \infty} \mu_0 \delta^{k+pr}$ in polynomial time. This can be achieved as follows. Note that $\mu_k \widehat{\delta^p}(i) = 0$ for any transient state i. We only have to compute $\mu_k \widehat{\delta^p}(i)$ for terminal states i.

Consider a BSCC c_i of δ. Let its period be p_i, let k_i be k mod p_i. Note that p can be exponentially large but each of the p_is at most n. Although $\sigma_{p_i} = \delta^{p_i}, \delta^{2p_i}, \ldots$ need not converge, it follows from Theorem 1 that the columns corresponding to c_i do converge to a limit. Now $\widehat{\delta^{p_i}}(;, j) = \widehat{\delta^p}(;, j)$ for any state $j \in c_i$ because σ_p is a subsequence of σ_{p_i}. So the entire matrix $\widehat{\delta^p}$ can be calculated in polynomial time. Essentially the jth column of $\widehat{\delta^p}$ is identical to the jth column of $\widehat{\delta^{p_i}}$. In order to calculate $\delta^k \widehat{\delta^p}$ observe that its jth column $\delta^k \widehat{\delta^p}(;, j) = \delta^k \widehat{\delta^{p_i}}(;, j) = \delta^{k_i} \widehat{\delta^{p_i}}(;, j)$ where again p_i is the period of the BSCC c_i that contains j. Note that Now $k_i < p_i \leq n$ and so δ^{k_i} can be calculated in polynomial time. The upper bound follows.

(**Lower Bound**). In order to prove hardness we use the reduction in [10,24] which is used to show coNP-hardness of the universality problem for unary non-deterministic finite automata (NFA). We briefly describe the salient features of the reduction; for further details the reader should refer to [10]. The original reduction is from 3SAT to non-universality of unary NFA. Given a 3SAT formula ϕ with n variables and m clauses, [10] constructs a NFA N_ϕ as a union of m cyclic automata. Intuitively, each cycle corresponds to a clause, has an initial state and a cycle accepts if and only if the input encodes an assignment that does not satisfy that clause. So N_ϕ accepts every input iff ϕ is unsatisfiable. The only non-determinism in N_ϕ is from having to choose a cycle at the beginning, so we can transform it into a PFA P_ϕ by choosing amongst the cycles uniformly at random. Since there are only m cycles any word that is accepted by N_ϕ is accepted by P_ϕ with probability at least $\frac{1}{m}$; otherwise it is accepted with probability 0. So 0 is an isolated cut-point for P_ϕ iff ϕ is unsatisfiable. Finally we observe that P_ϕ is isolated at 0 iff it is limit isolated at 0, which proves that 0 is a limit-isolated cut-point for P_ϕ iff ϕ is unsatsifiable. We have already observed that if a cut-point is isolated then it is also limit-isolated. For the converse observe that the constructed unary NFA N_ϕ is a disjoint union of cycles. Let d be the lcm of all the cycles of N_ϕ. Now it is easy to see that for each j, the probability distribution on the states of the unary PFA P_ϕ on input a^j is the same as the probability distribution on input $a^{j \bmod d}$. So if there is some word a^j that is accepted with 0 probability then 0 cannot be a limit-isolated cut-point.   □

Remark 1. Please note that the lower bound proof of Theorem 2 can be modified if the cut-point λ is not extremal; simply add an additional state with a self loop, which you choose initially with probability λ. Also, we could have taken the cut-point to be 1 thanks to Proposition 3. Thus, complexity of limit-isolation does not depend on whether the cut-point is extremal or not. Also, note that the lower bound proof also establishes the coNP-hardness of the isolation problem.

4   Complexity of Isolation Checking

We will prove that the problem of checking whether λ is isolated is in coNP^{RP} (see Theorem 3). For extremal cut-points, i.e., when λ is 0 or 1, we will show the problem to be coNP-complete (see Theorem 4). We start by discussing non-extremal cut-points.

Non-Extremal Cut-Points. Broadly speaking, the proof for showing that isolation is in coNP^{RP} is as follows:

- We can use Theorem 2 to check if the cut-point λ is limit isolated for A. If it is not limit-isolated then we know that the cut-point is not isolated.
- If it is limit isolated, then λ will be isolated iff there is no word a^ℓ accepted with probability λ. We will show that that this word cannot be too long (see Lemma 3).
- We can then guess this word, construct a straight-line program such that its value is 0 iff the probability of accepting this word is λ, and check if it evaluates to 0 or not (see Lemma 2).

We start by showing that the problem of deciding given a PFA A and a number n in binary, whether a word a^n is accepted with probability $=\lambda$ is in coRP and $> \lambda$ is in the counting hierarchy.

Lemma 2. *Given unary PFA A, a non-negative integer n in binary and a rational number λ, the problem of checking:*

1. *if a^n is accepted with probability equal to λ is in coRP.*
2. *if a^n is accepted with probability greater than λ lies in $\mathsf{P}^{\mathsf{C_3 P}}$.*

Proof. The word a^n is accepted with probability $\mu_0 \delta^n \eta_F$ where μ_0 is the initial distribution, δ the transition matrix and η_F the vector corresponding to the final states. In order to find out if this quantity is equal to λ, one can write a straight line program p that calculates $\mu_0 \delta^n \eta_F - \lambda$. The program is the usual square-and-multiply algorithm for exponentiation and it is going to be $O(\log_2 n)$ long because the number of iterations in the algorithm is equal to the number of bits required to represent n. The value of the program p is equal to (greater than) 0 iff a^n is accepted with probability exactly (greater than) λ. Now, we can check if $val(p) = 0$ in coRP [21] and $val(p) > 0$ in $\mathsf{P}^{\mathsf{C_3 P}}$ [2]. The result follows.   □

We will now show that if a limit isolated PFA accepts a word with probability exactly λ then this word cannot be too long. This fact is proved in Lemma 3 with the help of auxiliary Propositions 4, 5 and 6. We start by proving a result about irreducible stochastic matrices. Recall that $\widehat{\delta^t}$ is used to denote the limit of the sequence $\lim_{r \to \infty} \delta^{rt}$.

Proposition 4. *Given an irreducible stochastic matrix δ with period p and rational $\epsilon \in (0, 1)$ there exists a number k, computable in polynomial time, such that for all $\ell \geq k : d(\delta^{p\ell}, \widehat{\delta^p}) \leq \epsilon$.*

Proof. A stochastic matrix γ with all positive entries acts as a contraction map on the set of distributions. The associated contraction factor α is $(1 - ns)$ where s is the smallest entry in γ (see Proposition 2). So we have

$$d(\mu\gamma^i, \mu\widehat{\gamma}) = \lim_{j \to \infty} d(\mu\gamma^i, \mu\gamma^j) \leq \lim_{j \to \infty} \sum_{i'=i}^{j-1} d(\mu\gamma^{i'}, \mu\gamma^{i'+1})$$

$$\leq \lim_{j \to \infty} \sum_{i'=i}^{j-1} \alpha^{i'} d(\mu, \mu\gamma) \leq \frac{\alpha^i}{1 - \alpha} = \frac{(1 - ns)^i}{ns} \leq \frac{e^{-nsi}}{ns}.$$

Choosing $i > \frac{1}{ns} \log \frac{2}{ns\epsilon}$ will give us $d(\mu\gamma^i, \mu\widehat{\gamma}) \leq \frac{\epsilon}{2}$ and because the μ is arbitrary we also have $d(\gamma^i, \widehat{\gamma}) \leq \epsilon$.

Coming back to δ, the graph of δ^p consists of p disjoint irreducible and aperiodic components. It is enough to show the above bound on each of the individual components (because the distance between the matrices takes maximum across rows), so consider δ^p to be irreducible and aperiodic. From Lemma 1, we know that δ^{pn^2} has all positive entries. The smallest entry of δ^{pn^2}, say s, requires only polynomially many bits to be represented. According to the above observation, for $i \geq \frac{1}{ns} \log \frac{2}{ns\epsilon}$ we have $d(\delta^{pn^2 i}, \widehat{\delta^p}) \leq \epsilon$. If $\frac{1}{ns\epsilon} = \frac{x}{y}$, and j represents the number of bits of y then we can choose $k = \lceil \frac{n}{s}(j+1) \rceil$, which is computable in polynomial time.   □

We now bound the number of steps required so that the probability of being in a transient state is small.

Proposition 5. *Given a stochastic matrix δ and rational $\epsilon \in (0,1)$ there exists a number k, computable in polynomial time such that for all $\ell \geq k$ it is the case that for all distributions μ_0, $\sum_{j \in T_\delta} \mu_0 \delta^\ell(j) \leq \epsilon$ where T_δ is the set of transient states of δ.*

Proof. Here we are required to show that after k steps the probability of being in a transient state is small. Every transient state has a path of length at most n to at least one terminal state, so choose one for each transient state. Let u be the minimum probability associated with any of those paths. So after every n steps each transient state loses at least u fraction of its probability to a terminal state, or in other words the probability of being in any transient state reduces by a factor of u. Hence after $k'n$ steps the probability of being in a transient state is at most $(1-u)^{k'}$, and choosing $k' \geq \frac{1}{u} \log \frac{1}{\epsilon}$ makes $(1-u)^{k'} \leq \epsilon$. So choosing k to be a number bigger then $\frac{n}{u} \log \frac{1}{\epsilon}$ we have our required number.   □

We now bound the length of input needed to be close to the limit distribution $\mu \widehat{\delta^p}$ where p is the ultimate period of δ.

Proposition 6. *Given a stochastic matrix δ, a distribution μ and rational $\epsilon \in (0,1)$ there exists a k, computable in polynomial time such that for all $\ell \geq k$: $d(\mu \delta^{p\ell}, \mu \widehat{\delta^p}) \leq \epsilon$ where p is the ultimate period of δ.*

Proof. (Sketch.) First we use Proposition 5 to get a k_1 such that it suffices to take k_1 steps to get to a distribution where the probability of being in any transient state is less than $\frac{\epsilon}{4}$. This ensures that for $l \geq k_1$, the probability of being in any BSCC c after pl steps is at least $1 - \frac{\epsilon}{4}$. This means that taking any more steps beyond k_1 can only perturb the probability in terminal states by a small amount which adds up to $\frac{\epsilon}{4}$ across all BSCCs. Let us focus on one BSCC c. Taking, k_2 steps beyond the k_1 will do two things to c:
 i) bring in more probability from the transient states
 ii) distribute the probability already present in c (i.e., the probability of being in c) at step k_1 according to μ_c, the stationary distribution of c.

The first effect can only result in pumping at most a small probability into c, which adds at most $\frac{\epsilon}{4}$ to the distance. The probability already present in c after k_1 steps is close to the limiting probability, and hence the contribution of the second effect into the distance can be made small by choosing k_2 according to Proposition 4 for the BSCC c with the bound $\frac{\epsilon}{4}$. Instead of choosing k_2 for a particular c, we can choose it to be the maximum across all c which will give us the desired result. We formalize these ideas in the calculations included in the Appendix.   □

We can prove a similar result about the matrix products as well.

Lemma 3. *Given a stochastic matrix δ and a rational $\epsilon \in (0,1)$, there exists a number k, computable in polynomial time, such that for all $\ell \geq k : d(\delta^{p\ell}, \widehat{\delta^p}) \leq \epsilon$ where p is the ultimate period of δ.*

Proof. The distance between the matrices can be broken into

$$d(\delta^{p\ell}, \widehat{\delta^p}) = \max_i \sum_j |\delta^{p\ell}(i,j) - \widehat{\delta^p}(i,j)| = \max_i \sum_j |\nu_i \delta^{p\ell}(j) - \nu_i \widehat{\delta^p}(j)|$$

$$= \max_i (2\, d(\nu_i \delta^{p\ell}, \nu_i \widehat{\delta^p})). \qquad \text{Here } \nu_i \text{ represents the distribution with probability 1 at state } i$$

Proposition 6 tells us we can choose a k of appropriate size such that for any μ, the distance $d(\mu\delta^{p\ell}, \mu\widehat{\delta^p})$ for $\ell \geq k$ is below $\frac{\epsilon}{2}$. $\qquad\square$

We are ready to establish the complexity of the problem of checking if λ is an isolated cut-point for a unary PFA A.

Theorem 3. *Given a unary PFA A and a rational λ, the problem of checking if A is isolated at λ is in $\mathsf{coNP^{RP}}$ and is coNP-hard.*

Proof. The lower bound follows from the proof of Theorem 2. For the $\mathsf{coNP^{RP}}$ upper bound, let us consider the complement of the problem where A is not limit isolated at λ. In this case either λ is not a limited isolated cut-point or there is some string which is accepted with probability λ. If the given PFA is not limit isolated then we guess this fact and check if it is true in NP (Thanks to Theorem 2). So assume that the PFA is limit isolated. We now need to check if there is any "short" string accepted with probability λ.

Let $A = (Q, \Sigma, \delta, \mu_0, Q_F)$ and let p be the ultimate period of A. Let $\widehat{\delta^p} = \lim_{t\to\infty} \delta^{pt}$. For each $r > 0$, let $\mu_r = \mu_0 \delta^r$.

Consider $\epsilon_r = |\mu_r \widehat{\delta^p}\eta - \lambda|$. Since any $\mu_r \widehat{\delta^p}$ can be computed in polynomial time (see proof of Theorem 2), it is the case that ϵ_r can be computed in polynomial time (given μ_0, δ, r). Suppose the length of the string accepted with probability λ is ℓ. Let $\ell = pq + r$ where $r = \ell \bmod p$. According to Lemma 3, there exists a k_r (computable in polynomial time) such that if $q > k_r$ then $d(\mu_r \delta^{pq}, \mu_r \widehat{\delta^p}) \leq \frac{\epsilon_r}{2}$. Since $d(\mu_r \delta^{pq}, \mu_r \widehat{\delta^p}) \geq |\mu_r \delta^{pq}\eta - \mu_r \widehat{\delta^p}\eta|$, we get that a^ℓ will not be accepted with probability λ if $q > k_r$.

Now, the decision procedure for checking non-isolation proceeds as follows. It first guesses $0 \leq r < p$, then it computes ϵ_r and subsequently computes k_r. Now, it guesses $q \leq k_r$ and then it computes $\ell = pq + r$. ℓ requires only polynomially many bits (because k_r is computable in polynomial time from r). Hence we can use the procedure of Lemma 2 as an oracle to check if a^ℓ is accepted with probability exactly λ. Note that this final check is done by a coRP algorithm and hence the non-isolation is in $\mathsf{NP^{coRP}}$. Note that $\mathsf{NP^{coRP}}$ is exactly the class $\mathsf{NP^{RP}}$ since we can always switch the yes/no answer from the oracle-calls. This results in a $\mathsf{coNP^{RP}}$ upper bound for the limit isolation problem in the non-extremal case. $\qquad\square$

Extremal cut-points. For extremal cut-points, the upper bound matches the lower bound.

Theorem 4. *Given a unary PFA A, the problem of checking if 0 is isolated is* coNP-*complete. Similarly checking if 1 is isolated is also* coNP-*complete.*

Proof. The lower bound follows from the proof of Theorem 2. For upper bound, first thing to note is that the coNP upper bound for limit isolation proved in Theorem 2 also holds for the cut-point 0. So in case it is limit isolated, we need to check if there is a string accepted with probability 0. If μ_0 is the initial distribution, let $Q_I = \{q|\mu_0(q) > 0\}$. A word is accepted by A with probability 0 iff it has no path from a state in Q_I to a final state. So checking if 0 is isolated reduces to the universality checking of unary NFA which is known to be in coNP [24]. □

5 Other Decidable Problems

In this section, we observe that the problems of language containment, equality, emptiness, and universality are all in counting hierarchy for unary PFAs with limit isolated cut-points. We need one proposition.

Proposition 7. *Given a PFA A with ultimate period p, a number $0 \leq r < p$ and a rational cut-point λ such that λ is limit isolated for A, there is a number k computable in polynomial time s.t $\forall q \geq k$, $a^{pq+r} \in L_{>\lambda}(A)$ iff $a^{pk+r} \in L_{>\lambda}(A)$.*

Proof. Let $A = (Q, \Sigma, (\delta_\sigma)_{\sigma \in \Sigma}, \mu_0, Q_F)$. Let $\mu_r = \mu_0 \delta^r$, let $\widehat{\delta^p} = \lim_{t \to \infty} \delta^{pt}$ and let η_F be the vector corresponding to F. Consider $\epsilon_r = |\mu_r \widehat{\delta^p} \eta_F - \lambda|$. ϵ_r can be computed in polynomial time (see proof of Theorem 2). According to Lemma 3, there is a k computable in polynomial time such that if $q > k$ then $\boldsymbol{d}(\mu_r \delta^{pq}, \mu_r \widehat{\delta^p}) \leq \frac{\epsilon_r}{2}$. Since $\boldsymbol{d}(\mu_r \delta^{pq}, \mu_r \widehat{\delta^p}) \geq |\mu_r \delta^{pq}(Q_f) - \mu_r \widehat{\delta^p}(Q_f)|$, we get that a^{pq+r} has acceptance probability $> \lambda$ iff a^{pk+r} has acceptance probability $> \lambda$.

Theorem 5. *Given two unary PFAs A and B and rational cut-points λ_1 and λ_2, such that λ_1 and λ_2 are limit isolated for A and B respectively, the following problems are in* coNP^{C_3P}.
  1. $L_{>\lambda_1}(A) \subseteq L_{>\lambda_2}(B)$.
  2. $L_{>\lambda_1}(A) = L_{>\lambda_2}(B)$.
  3. $L_{>\lambda_1}(A) = \emptyset$.
  4. $L_{>\lambda_1}(A) = \Sigma^*$.

Proof. Without loss of generality, we can assume that the ultimate periods of A and B are the same (since we can always add unreachable cycles). Let the ultimate period be p. The algorithm for checking containment proceeds as follows. The algorithm is going to guess a number ℓ such that a^ℓ is accepted by A and rejected by B. Note, ℓ can be written as $\ell = pq + r$ where $q = \ell \text{ div } p$ and $r = \ell \bmod p$. Hence we have to guess q and r. First, the algorithm guesses the offset $0 \leq r < p$ which is a polynomial-sized number.

Thanks to Proposition 7, there is a k_A such for all $q_A \geq k_A$, a^{pq_A+r} is accepted by A iff the string a^{pk_A+r} is accepted. Furthermore, k_A can be computed in polynomial time from A and r. Similarly, there is a k_B such that for all $q_B \geq k_B$, a^{pq_B+r} is accepted by B iff the string a^{pk_B+r} is accepted. Let $k = \max(k_A, k_B)$.

By construction, we can conclude that if a^ℓ with $\ell = pq + r$ is in the language of A but not in the language of B then we can take $q \leq k$. So, now the algorithm guesses $q \leq k$ and then checks that i) $a^\ell \in L_{>\lambda_1}(A)$ and ii) $a^\ell \notin L_{>\lambda_2}(B)$. These checks can be carried out by $\mathsf{P}^{\mathsf{C}_3\mathsf{P}}$ algorithms as in Lemma 2 and the result follows. The other problems of language equality, emptiness, and universality follow immediately from the result for language containment.   □

6   Conclusions

In this paper we established the complexity of a variety of decision problems for unary PFAs. In particular, we showed that the isolation problem is in coNP-complete, when the cut-point is extremal, and is in $\mathsf{coNP}^{\mathsf{RP}}$ when the cut-point is not extremal. We also show that limit isolation of unary PFAs allows us to conclude that language, containment, equality, emptiness, and universality are decidable within PSPACE.

Acknowledgements. We thank anonymous referees for pointing out that EquSLP is in coRP, and observing that Bertoni had claimed the decidability of the isolation problem in the conclusions of [4]. Rohit Chadha was supported by NSF grant CNS 1314338. Dileep Kini was supported by NSF grant SHF 1016989. Mahesh Viswanathan was supported by NSF grant CNS 1314485.

References

1. Agrawal, M., Akshay, S., Genest, B., Thiagarajan, P.S.: Approximate verification of the symbolic dynamics of markov chains. In: LICS, pp. 55–64 (2012)
2. Allender, E., Bürgisser, P., Kjeldgaard-Pedersen, J., Bro Miltersen, P.: On the complexity of numerical analysis. SIAM J. Comput. 38(5), 1987–2006 (2009)
3. Arora, S., Barak, B.: Computational Complexity: A Modern Approach. Cambridge University Press (2009)
4. Bertoni, A.: The solution of problems relative to probabilistic automata in the frame of formal language theory. In: Siefkes, D. (ed.) GI 1974. LNCS, vol. 26, pp. 107–112. Springer, Heidelberg (1975)
5. Chadha, R., Sistla, A.P., Viswanathan, M.: Probabilistic automata with isolated cut-points. In: Chatterjee, K., Sgall, J. (eds.) MFCS 2013. LNCS, vol. 8087, pp. 254–265. Springer, Heidelberg (2013)
6. Condon, A., Lipton, R.J.: On the complexity of space bounded interactive proofs (extended abstract). In: 30th Annual Symposium on Foundations of Computer Science, pp. 462–467 (1989)
7. Doyen, L., Henzinger, T.A., Raskin, J.-F.: Equivalence of labeled markov chains. Inernational Journal of Foundations of Computer Science 19(3), 549–563 (2008)
8. Gantmacher, F.R.: Applications of the Theory of Matrices. Dover Books on Mathematics Series, Dover (2005)

9. Gimbert, H., Oualhadj, Y.: Probabilistic automata on finite words: Decidable and undecidable problems. In: Abramsky, S., Gavoille, C., Kirchner, C., Meyer auf der Heide, F., Spirakis, P.G. (eds.) ICALP 2010. LNCS, vol. 6199, pp. 527–538. Springer, Heidelberg (2010)
10. Gramlich, G.: Probabilistic and nondeterministic unary automata. In: Rovan, B., Vojtáš, P. (eds.) MFCS 2003. LNCS, vol. 2747, pp. 460–469. Springer, Heidelberg (2003)
11. Kiefer, S., Murawski, A.S., Ouaknine, J., Wachter, B., Worrell, J.: Language equivalence for probabilistic automata. In: Gopalakrishnan, G., Qadeer, S. (eds.) CAV 2011. LNCS, vol. 6806, pp. 526–540. Springer, Heidelberg (2011)
12. Kiefer, S., Murawski, A.S., Ouaknine, J., Wachter, B., Worrell, J.: On the complexity of the equivalence problem for probabilistic automata. In: Birkedal, L. (ed.) FOSSACS 2012. LNCS, vol. 7213, pp. 467–481. Springer, Heidelberg (2012)
13. Korthikanti, V.A., Viswanathan, M., Agha, G., Kwon, Y.: Reasoning about mdps as transformers of probability distributions. In: QEST, pp. 199–208 (2010)
14. Kwon, Y., Agha, G.: Linear Inequality LTL (iLTL): A Model Checker for Discrete Time Markov Chains. In: Davies, J., Schulte, W., Barnett, M. (eds.) ICFEM 2004. LNCS, vol. 3308, pp. 194–208. Springer, Heidelberg (2004)
15. Kwon, Y.M., Agha, G.: A Markov Reward Model for Software Reliability. In: IEEE International Parallel and Distributed Processing Symposium, IPDPS 2007, pp. 1–6 (2007)
16. Norris, J.R.: Markov Chains. Cambridge University Press (1997)
17. Paz, A.: Introduction to probabilistic automata. Academic Press (1971)
18. Rabin, M.O.: Probabilistic automata. Information and Computation 6(3), 230–245 (1963)
19. Rutten, J.M., Kwiatkowska, M., Norman, G., Parker, D.: Mathematical Techniques for Analyzing Concurrent and Probabilistic Systems. AMS (2004)
20. Salomaa, A., Soittola, M., Bauer, F.L., Gries, D.: Automata-theoretic aspects of formal power series. Texts and monographs in computer science. Springer (1978)
21. Schönhage, A.: On the power of random access machines. In: Maurer, H.A. (ed.) ICALP 1979. LNCS, vol. 71, pp. 520–529. Springer, Heidelberg (1979)
22. Schützenberger, M.P.: On the definition of a family of automata. Information and Control 4(2-3), 245–270 (1961)
23. Senata, E.: Non-negative Matrices and Markov Chains. George Allen & Unwin Ltd. (1973)
24. Stockmeyer, L.J., Meyer, A.R.: Word problems requiring exponential time: Preliminary report. In: Proc. of the 5th Ann. ACM Symposium on Theory of Computing, pp. 1–9 (1973)
25. Turakainen, P.: On stochastic languages. Information and Control 12(4), 304–313 (1968)
26. Tzeng, W.-G.: A polynomial-time algorithm for the equivalence of probabilistic automata. SIAM J. Comput. 21(2), 216–227 (1992)
27. Wagner, K.W.: Some observations on the connection between counting and recursion. Theoretical Computer Science 47(3), 131–147 (1986)

Appendix

Calculations from Proposition 6: A *sub-distribution* over a finite set S is a function $\mu : S \mapsto [0, 1]$ such that $\sum_{s \in S} \mu(s) \leq 1$. The distance between sub-distributions can be defined in the same way we do for distributions.

We first describe the notation we will use in the following calcualtions: μ_T and μ_C denote the sub-distributions on the transient states and the BSCCs after pk_1 steps. For any (sub-)distribution μ , $\mu \upharpoonright c$ denotes the vector with the entries in the states in c alone. The matrix δ restricted to the states of c is written as δ_c. Starting from μ, and having taken pk_1 steps, the probability of being in component c is denoted by $\pi_c^{pk_1}$, and the relative distribution on a component c is given by $\mu_c^{pk_1}$, i.e for any $i \in c$, $\mu_c^{pk_1}(i) = \mu\delta^{pk_1}(i)/\pi_c^{pk_1}$. Starting from μ the probability of being in c in the limit is given by $\hat{\pi}_c$ and the relative distribution on c in the limit is give by $\hat{\mu}_c$. Now we are ready to proceed.

$$d\big(\mu\delta^{p\ell}, \widehat{\mu\delta^p}\big) = d\big(\mu\delta^{pk_1}\delta^{pk_2}, \widehat{\mu\delta^p}\big) = d\big((\mu_T + \mu_C)\delta^{pk_2}, \widehat{\mu\delta^p}\big)$$

$$=d\big(\mu_T\delta^{pk_2} + \mu_C\delta^{pk_2}, \widehat{\mu\delta^p}\big) \leq \underbrace{\frac{\sum_j \mu_T\delta^{pk_2}(j)}{2}}_{\text{Apply Prop 5}} + d\big(\mu_C\delta^{pk_2}, \widehat{\mu\delta^p}\big)$$

Let us focus on $d\big(\mu_C\delta^{pk_2}, \widehat{\mu\delta^p}\big)$

$$= \sum_{c\in C} d\big((\mu_C\delta^{pk_2})\upharpoonright c, (\widehat{\mu\delta^p})\upharpoonright c\big) = \sum_{c\in C} d\big(\pi_c^{pk_1}\mu_c^{pk_1}\delta_c^{pk_2}, \hat{\pi}_c\,\widehat{\hat{\mu}_c\delta_c^P}\big)$$

We have $(\mu_C\delta^{pk_2})\upharpoonright c = \pi_c^{pk_1}\mu_c^{pk_1}\delta_c^{pk_2}$ because when we start from μ_C there is no probability of being in any transient state, so we can ignore the transient states, and then the BSCCs cannot communicate so they evolve independently.

$$= \sum_{c\in C}\sum_{j\in c} \big|\pi_c^{pk_1}\mu_c^{pk_1}\delta_c^{pk_2}(j) - \hat{\pi}_c\,\widehat{\hat{\mu}_c\delta_c^P}(j)\big|$$

$$= \sum_{c\in C}\sum_{j\in c} \big|\pi_c^{pk_1}\big(\mu_c^{pk_1}\delta_c^{pk_2}(j) - \widehat{\hat{\mu}_c\delta_c^P}(j)\big) + (\pi_c^{pk_1} - \hat{\pi}_c)\widehat{\hat{\mu}_c\delta_c^P}(j)\big|$$

$$\leq \Big(\sum_{c\in C}\pi_c^{pk_1}\sum_{j\in c}\big|\mu_c^{pk_1}\delta_c^{pk_2}(j) - \widehat{\hat{\mu}_c\delta_c^P}(j)\big|\Big) + \Big(\sum_{c\in C}(\hat{\pi}_c - \pi_c^{pk_1})\sum_{j\in c}\widehat{\hat{\mu}_c\delta_c^P}(j)\Big)$$

$$\leq \Big(\sum_{c\in C}\pi_c^{pk_1}\overbrace{d\big(\mu_c^{pk_1}\delta_c^{pk_2}(j), \widehat{\hat{\mu}_c\delta_c^P}(j)\big)}^{\text{Apply Prop 4}}\Big) + \overbrace{\sum_{c\in C}(\hat{\pi}_c - \pi_c^{pk_1})}^{\text{Apply prop 5}}$$

$$\leq \Big(\sum_{c\in C}\pi_c^{pk_1}\frac{\epsilon}{2}\Big) + \frac{\epsilon}{4} \leq \frac{3\epsilon}{4}$$

Therefore $d\big(\mu\delta^\ell, \widehat{\mu\delta^p}\big) \leq \epsilon$.

A Scalable Approach
to the Assessment of Storm Impact
in Distributed Automation Power Grids

Alberto Avritzer[1], Laura Carnevali[4], Lucia Happe[3], Anne Koziolek[3],
Daniel Sadoc Menasche[2], Marco Paolieri[4], and Sindhu Suresh[1]

[1] Siemens Corporation, Corporate Technology, Princeton, USA
[2] Federal University of Rio de Janeiro, Rio de Janeiro (UFRJ), Brazil
[3] Karlsruhe Institute of Technology (KIT), Germany
[4] University of Florence, Florence, Italy

Abstract. We present models and metrics for the survivability assessment of distribution power grid networks accounting for the impact of multiple failures due to large storms. The analytical models used to compute the proposed metrics are built on top of three design principles: state space factorization, state aggregation, and initial state conditioning. Using these principles, we build scalable models that are amenable to analytical treatment and efficient numerical solution. Our models capture the impact of using reclosers and tie switches to enable faster service restoration after large storms. We have evaluated the presented models using data from a real power distribution grid impacted by a large storm: Hurricane Sandy. Our empirical results demonstrate that our models are able to efficiently evaluate the impact of storm hardening investment alternatives on customer affecting metrics such as the expected energy not supplied until complete system recovery.

Keywords: Survivability, cyber-physical systems, smart-grid.

1 Introduction

In this paper, we introduce a new approach to the modeling and analysis of large power distribution networks, with the goal of supporting the evaluation of investment alternatives for storm hardening of overhead transmission facilities. Specifically, the focus of this work is on the modeling and analysis of US Northeast power distribution network outages that result from mid-Atlantic hurricanes and tropical storms.

Hurricane Sandy hit the US northeast overhead power distribution network with strong winds on October 29 and 30, 2012. The impact on the overhead power network in New York City and Westchester county was so severe that about 70% of the 868,347 non-network customers (i.e., customers served by overhead lines) in these areas were interrupted. In Westchester county alone 320,926 customer outages were reported [11]. The total number of interruptions of non-network customers in the Con Edison territory was 1,115,294. These customer

G. Norman and W. Sanders (Eds.): QEST 2014, LNCS 8657, pp. 345–367, 2014.
© Springer International Publishing Switzerland 2014

outages were a consequence of the loss of nearly 1,000 utility poles and over 900 transformers. In the Bronx/Westchester area, 699 poles and 718 transformers were replaced [11]. One of the most important reported causes of these customer outages in Westchester county was the fall-down of trees and branches due to the co-habitation of power distribution with Westchester county's forests [26]. Therefore, over 1,000 roads in Westchester county were blocked by trees and branches after Hurricane Sandy.

As a result of the damage to overhead power distribution network caused by Hurricane Sandy, a study was commissioned by the City of New York to assess the feasibility of undergrounding parts of the overhead network [27]. The total cost of replacement of the overhead network in the Bronx/Westchester county is estimated at $27.2 billion. A total replacement covering all of New York City and Westchester county is estimated at $42.9 billion. In [12] a list of the storm hardening initiatives for the electric overhead distribution system was presented. These initiatives include, among others, the use of additional reclosers and sectionalizer switches, tree trimming, and selective undergrounding [12].

The utility uses a coarse grained risk model to identify the relationship between the required capital investment for storm hardening of a specific asset and the risk reduction achieved in terms of asset outage durations using wind damage probabilities [12]. Unfortunately, the coarse grain risk assessment approach is not detailed enough to assess the customer impact of large storms. The risk assessment approach used by the utility can be improved by using a metric that captures the evolution of the repair process (both automated and manual), and the energy not supplied from the start of the outage event to the completion of all required repairs. Therefore, there is a need to improve the utility risk prioritization of storm hardening investment approaches by modeling the impact of sectionalizing, undergrounding, and tree trimming on a metric of interest to the utility. We call this metric the *customer affecting metric*. In this paper, the customer affecting metric used is the *average energy not supplied from the time of the emergency to full restoration of service to all customers*. The level of accuracy required in the power grid model needs to be sufficient to compare design alternatives; the model has to be accurate enough to properly distinguish between the investment options.

We model the impact of using reclosers and tie switches to enable faster service restoration after large storms. The use of reclosers and tie switches provides the following benefits to the power utility: (1) enables sectionalization of customers reducing the impact of outages, (2) reduces the number of energized down wires, and (3) enables the automated and remote reconfiguration of the overhead distribution network during the several phases of the storm emergency (preparation and restoration) [11].

Survivability is the ability of the system to continue functioning during and after a failure or disturbance event [18]. In our previous work, we developed survivability models accounting for single failures in distribution automation power grids [21,24,1]. The analytical models used to compute the proposed metrics are built on top of three design principles: state space factorization, state aggregation, and initial state conditioning. In [2], we extended the survivability model to account for multiple-failures.

The main contributions of this paper are the following.

- A *scalable model* to assess survivability related metrics of smart power-grids. The model allows us to efficiently compute survivability related metrics in networks consisting of hundreds of loops. The model captures the smart-grid/cyber-physical interconnection as well as automatic restoration/manual restoration, and allows for general distributions for the automatic and manual restoration, adopting recently proposed techniques of non-Markovian analysis.
- A *characterization of hurricanes*, which accounts for historical data and can make use of geographical information. Each hurricane is characterized by a hurricane model, which indicates the wind strength in knots at each section of the grid. The hurricane model is then used to obtain a global survivability related metric for the whole network as a function of the network topology and the hurricane pattern.
- *What-if analysis*, which allows to quickly identify the impact of different strategies for power grid storm hardening, such as distributed generation, tree trimming, and moving lines underground. This analysis can be used for planning and optimization purposes.

We illustrate the practicality of our approach by evaluating the impact of Hurricane Sandy on a model of a overhead distribution network of the scale of the Con Edison overhead distribution network in New York, which serves the areas of Staten Island, the Bronx, Brooklyn, Queens, and Westchester county.

The remainder of this paper is organized as follows. In Section 2, we describe the cyber-physical system under study. Section 3 and 4 present the failure model used for hurricane characterization and an overview of the survivability model, respectively. In Section 5, we present the evaluation of the survivability model for the cyber-physical system. In Section 6, we give a brief summary of related work. Section 7 contains our conclusions and suggestions for further research.

2   Modeled Power Grid Topology

In this section, we first introduce terminology (Section 2.1) and describe the key features of the smart-grid topology that are used to derive the proposed survivability model (Section 4). In addition, we also introduce storm hardening strategies whose effect is evaluated quantitatively in Section 5.

2.1   Terminology

For the sake of clarity, we define here the key terms used throughout this paper.
- **Wind gust**. The *wind gust* is measured in knots and classified in small, medium, large and catastrophic.

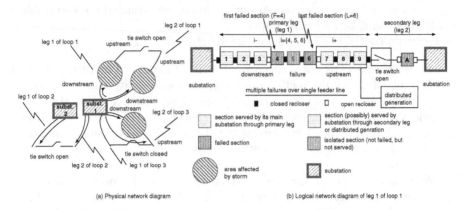

(a) Physical network diagram   (b) Logical network diagram of leg 1 of loop 1

Fig. 1. Physical and logical diagrams of the network topology after the hurricane. Note that in the logical diagram we represent two substations at the two ends of the loop, irrespective of whether they are the same physical substation or different physical substations. In autoloops, as autoloops 1 and 3 in the figure, the two ends of the loop are connected to the same substation. In loop 2, in contrast, the two ends of the loop are connected to different substations.

- **Incidence matrix.** Given a physical smart-grid topology, different factors such as the geography of the terrain and the distance from the sea will affect the probability that a given storm will directly hit a section, resulting in the need of manual repair. The *incidence matrix* characterizes, for each section, the wind gust at that section after the occurrence of a storm of a given type.
- **Loop.** The power grid topology is divided in *loops*, which consist of sections. Each loop is connected to a substation at each of its ends.
- **Autoloop.** An autoloop is a special loop that is connected to the same substation at each of its ends. For all practical purposes, in this paper we do not distinguish between loops and autoloops, as the logical network diagrams that are built on top of the physical network diagram are the same for loops and autoloops (see Figure 1). Therefore, we use the terms *loop* and *autoloop* indistinguishably.
- **Legs.** Each loop comprises two *legs*, which are separated by a tie switch. Each leg is a set of contiguous sections: the first section in the leg is directly connected to the substation, and the last section is directly connected to the tie switch. The leg under study is referred to as the *primary leg* and the additional leg of the loop is referred to as the *secondary leg* (the distinction being clear from the context).
- **Tie switch.** The *tie switch* is a switch that controls the flow of energy in a loop. When open, the two separate legs in a loop are fed with energy that flows from the substation up to the tie switch. When the tie switch is closed, the substation feeds energy to the loop through one of its legs, which then relays energy to the other leg.

- **Isolated sections.** The sections in a loop are indexed based on their distance from the substation. After a storm, the set of contiguous sections in a leg between the first and last failed sections are referred to as *isolated sections* (which include the first and last failed sections). The isolated sections will be restored after manual repair.
- **Upstream sections.** The *upstream sections* are the set of contiguous sections farther from the substation, which are not damaged but might be indirectly affected by the storm due to loss of connectivity. The sections in this set are amenable to automatic restoration after the isolated sections are set aside if either (1) distributed generation is available to supply them, or (2) there exists a secondary path from the substation up to the upstream sections, making use of the secondary leg.
- **Downstream sections.** The *downstream sections* are a set of contiguous sections closer to the substation, which are not affected by the storm. The sections in this set are automatically fed by the substation after the isolated sections are set aside.
- **Phased recovery model.** The *phased recovery model* associated with a given leg of a loop is a state machine that characterizes the dynamic state of the leg, from a failure up to full recovery. The transition rates between states, as well as the reward rates associated with each state, depend on the distribution of isolated sections, which in turn depends on the incidence matrix. Although the reward rates and the transition rates of the phased recovery model may vary across legs, the number of states and the possible transitions between states in the phased recovery model are assumed to be fixed. Fixing the structure of the phased recovery model allows us to pre-compute solutions in a scalable fashion.
- **Reward table.** We associate a set of *reward rates* with each state of the phased recovery model characterizing a given leg. The *reward table* characterizes the expected reward rate (e.g., energy not supplied per time unit) associated with each state for each leg. The expected reward rate depends on the distribution of isolated sections, which in turn depends on the incidence matrix. Note that we use the term *reward rate* even if the corresponding metric of interest represents a cost that should be minimized.

2.2   Con Edison Overhead System in New York

The cyber-physical system under study is based on the Con Edison overhead distribution power grid in New York City and Westchester county, covering an area of 604 square miles. The overhead network consists of 37,000 miles of overhead cable lines that supply power to Westchester County, Staten Island, and parts of the Bronx, Brooklyn, and Queens [27]. The considered power grid includes 154 auto-loops with 219 substations. The 154 auto-loop line feeders are supported structurally by about 284,000 poles and use 47,119 overhead transformers to convert medium voltage (33kV–4kV) to low voltage (120V–240V) supplied to customers [11].

Table 1. Interval of measured maximum wind gust for locations close to the Con Edison overhead power grid, per county. Maximum wind gust values of sections are sampled from these intervals.

County	Maximum wind gust intervals (knots)
Brooklyn	[57, 68]
Queens	[60, 74]
Bronx	[57, 62]
Westchester County	[56, 64]
Staten Island	[59, 64]

Fig. 2. Wind gust values in knots for Hurricane Sandy. The parts of the Con Edison network that consist of overhead lines are taken from [27].

2.3 Hurricane or Tropical Storm Event

Our work is concerned with power distribution network outages that result from a typical hurricane or tropical storm event. We used Hurricane Sandy wind data reported by the U.S. Government National Hurricane Center [13]. Figure 2 shows the maximum wind gusts measured at different locations in New York City and Westchester County, as reported in [13, p.55 et seqq.]. The parts of the Con Edison network that consist of overhead lines are taken from [27] and are shown in light gray. For example, the east and south-east of Queens is predominantly fed by overhead lines, while Manhattan is completely served by an underground network (and thus not considered in this paper).

From this data, we approximate the maximum wind gust at each section required for our model as follows. We derive an interval of observed maximum wind gust for each of the different counties served by the Con Edison overhead network and report these intervals in Table 1. To sample the maximum wind gust at a section, we randomly draw a maximum wind gust speed from the interval associated with the corresponding county.

2.4 Storm Hardening Strategies

The main vulnerability of an overhead distribution system during a typical storm event is wind and tree damage to power distribution and support equipment (e.g., poles, wires, transformers). The storm hardening strategies considered in our model are: (1) undergrounding certain sections of the overhead system, and (2) tree trimming. The utility is deploying several other strategies (e.g., pole hardening) that are out of scope of this work [12]. The overhead distribution system resilience can be improved by replacing portions of the power line with underground equipment. However, due to high cost of undergrounding, the cost effectiveness of the approach needs to be evaluated. The phased recovery model introduced in Section 4 can be used to support such evaluations and to include other less expensive alternatives such as tree trimming.

2.5 Input Data for the Analysis of the Cyber-Physical System

We model the Con Edison overhead distribution power grid by extracting its most important properties as necessary to evaluate the impact of storm events. We model each autoloop as a sequence of sections and we associate each section s (where $s = 1, 2, \ldots, 1542$) with its (1) average load ℓ_s, (2) distributed generation capacity g_s, and (3) maximum wind gust, as shown in Table 1. We allocated to each county a number of loops proportional to the length of overhead distribution power lines in that county (from [27]). The loads in different sections were set based on the load profile benchmark proposed by Rudion et al. [29]. The aggregated input data is presented in Table 2.

Table 2 shows that the 154 auto-loops are supplied with 2,276 MW. We assume the distribution of the load among the counties as indicated in the last column of the table. In addition to being supplied by the substations, we additionally assume that the auto-loops are also fed by distributed generation from two different renewable energy sources: solar and biomass power systems. For solar, we assume that roughly 30% of the load can be provided by solar generation and distributed irregularly over the sections. For biomass, we assume that 4 biomass generators are available, each producing 20 MW.

Table 2. Model of the Con Edison distribution power grid, including all 154 auto-loops. Each auto-loop comprises a minimum of 8 and a maximum of 12 sections. Net load is the average load minus the load amenable to reduction due to Distributed Generation (DG): the average net load per section (sixth column) is obtained after subtracting the average DG per sections (fifth column) from the average load per section (fourth column).

County	Loops	Sections	Average load per section (kW)	Average DG per section (kW)	Average net load per section (kW)	Total net load (kW)
Brooklyn	16	158	1,479.33	525.32	954.01	150,734.58
Queens	32	317	1,452.16	375.39	1,076.77	341,335.60
Bronx	12	117	1,500.63	559.83	940.80	110,073.20
Westchester	62	634	1,472.39	435.75	1,036.64	658,267.96
Staten Island	32	317	1,488.01	374.45	1,113.56	352,998.18

3 Failure Model

The damage caused by a hurricane at a given section depends on the susceptibility of the section, characterized by the incidence matrix, and the hurricane strength. The susceptibility of the section depends on a number of adjustment factors such as whether the section is underground and trees were trimmed (Table 3). The hurricane strength depends on the geography (see Table 4).

Table 3. Adjustment factor θ

Storm hardening	Factor θ
Underground	0.0
Trees trimmed	0.8

Table 4. Probability of failure $\psi(w)$ as a function of wind strength w

w	Classification	Knots	$\psi(w)$
1	Small	< 34	0.1
2	Medium	$[34, 64)$	0.3
3	Large	$[64, 74)$	0.7
4	Catastrophic	$\geqslant 74$	1.0

Let W_s be a discrete random variable that characterizes the wind strength level at section s. Table 4 shows the different wind strength levels considered in this paper and the corresponding probabilities of failure $\psi(w)$ as a function of wind strength w. At each section, the probability of failure must be adjusted to account for the fact that sections are underground and/or trees were trimmed. Let θ_s be the adjustment factor corresponding to section s. The adjustment factors considered in this paper are presented in Table 3.

Let D_s be the random variable that characterizes the state of section $s \in S$ immediately after a failure in leg S. If section s has failed, $D_s = 1$. Otherwise, $D_s = 0$. Then, $P(D_s = 1)$ denotes the probability that s has been damaged by the hurricane and $P(D_s = 0) = 1 - P(D_s = 1)$ denotes the probability that s is still operational. The probability that a section s is directly affected by a storm is given by

$$P(D_s = 1 | W_s = w) = \theta_s \psi(w). \tag{1}$$

Given the distribution of wind strengths, we obtain the probability $P(D_s = 1)$ that section s is affected as

$$P(D_s = 1) = \sum_{w=1}^{4} P(D_s = 1 | W_s = w) P(W_s = w). \tag{2}$$

Equations (1) and (2) are used in Section 4 to derive key parameters of the survivability model.

4 Survivability Model Overview

In this section, we characterize the principles that have guided the formalization of the recovery procedure, we introduce the assumptions made at design stage, and we present the Markovian and non-Markovian version of the phased recovery model, discussing its properties.

4.1 Modeling Principles

The modeling principles are discussed with reference to the physical and logical diagrams of the network topology after the hurricane passage, which are shown in Figure 1.

State space factorization into legs of loops. We consider the system sectionalized into legs of loops, where each loop is divided into two legs. Each leg is composed of a set of sections, which are separated by reclosers; a leg starts at a substation, and ends at a tie switch.

State aggregation. We aggregate the sections around a failure into upstream sections and downstream sections. The downstream sections are still served through the primary leg. Conversely, the upstream sections are served by the secondary leg, if reachable, or by the distributed generation sources, if available.

Initial state conditioning. We condition the initial state to be a failure state. This allows to avoid dealing with different time scales and characterizing the failure rate.

4.2 Model Properties

The principles followed in the modeling phase as well as the assumptions made on failures and their effects permit to develop a separate survivability model for each loop, sharing the same structure while exposing different parameter values. In so doing, the model structure turns out to be independent of the topology of the power distribution grid, guaranteeing not only simplicity and flexibility of modeling, but also scalability of the overall approach. Moreover, the initial state conditioning permits to characterize the recovery actions given the occurrence of a failure, thus making the model independent of the failure rate.

4.3 Phased Recovery Model

Markovian model. The phased recovery model is characterized by the following states and events. After a section failure, the model is initially in state 0. The sojourn time in state 0 corresponds to the time required for the recloser to isolate the section, which takes an average of $1/\epsilon$ time units. A recloser isolates a section within 10-50 ms after a failure, so in the remainder of this paper we assume $\epsilon = \infty$. Let p be the probability that the communication network is still operational after a section failure, and q be the probability that there is a secondary path to supply energy for sections $i+$. After the isolation of section i is completed, the model transitions to one of following three states:

1. With probability pq the model transitions to state 1, where the distribution network is amenable to automatic restoration.
2. With probability $p(1 - q)$ the model transitions to state 3, where the effectiveness of distributed generation will determine if the system is amenable to automatic restoration.

3. With probability $1 - p$, the model transitions to state 4, where the communication system requires manual repair, which occurs with rate γ.

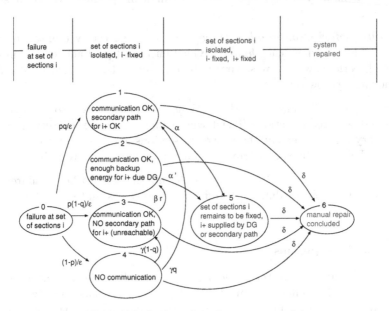

Fig. 3. Markovian phased recovery model

In state 3, distributed generation is activated after a period of time with average duration $1/\beta$. Let r be the probability that distributed generation can effectively be used. In this case, the model transitions from state 3 to state 2 with rate βr. When the model is in states 1 or 2, the distribution network is amenable to automatic restoration, which occurs after a period of time with average duration $1/\alpha$ and $1/\alpha'$, respectively. A manual repair of section i takes on average $1/\delta$ units of time (and can occur while the system is in states 1-5). After a manual repair, the model transitions to state 6, which corresponds to a fully repaired system.

Non-Markovian Model. The phased recovery procedure is modeled through a stochastic Time Petri Net (sTPN) [31,7], which extends Time Petri Nets (TPN) [25] by associating each transition with a Probability Density Function (PDF) supported over its static firing interval and with a weight used to resolve random switches.

A transition is enabled if each of its input places contains a token and none of its inhibitor places contains any tokens. Each enabled transition takes a time-to-fire sampled according to its PDF. We provide here a straightforward description of the sTPN phased recovery model shown in Figure 4, and refer the reader to [31,7] for a formal treatment of sTPN syntax and semantics. The specific

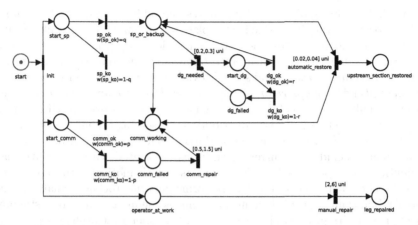

Fig. 4. Non-Markovian phased recovery model. IMM and GEN transitions are represented by thin bars and thick black bars, respectively.

distributions associated with model transitions refer to the case study analyzed in the experiments.

The immediate (IMM) transition *start* represents the beginning of the recovery procedure (IMM transitions fire in zero time). Its firing enables:

1. The general (GEN) transition *manual_repair*, modeling manual restoration of a set of sections.
2. The IMM transition *comm_ok*, with probability p, accounting for the cases where the communication network is working after a section failure. With probability $1 - p$ the communication network is not working after a section failure, which triggers the IMM transition *comm_ko*. When the transition *comm_ko* fires, it enables the GEN transition *comm_repair*, which characterizes the time for restoration of the communication system. After the restoration of the communication system, a token is added to *comm_working*.
3. The IMM transition *sp_ok*, with probability q, accounting for the cases where backup power is sufficient to supply energy to the upstream sections. With probability $1 - q$ the backup power does not suffice, which triggers the IMM transition *sp_ko*.

Automatic restoration of upstream sections occurs if the communication network is working and either (1) there is a secondary path to supply for the upstream sections (which fires the *sp_ok* transition, then placing a token in *sp_or_backup*), or (2) distributed generation suffices to supply the upstream sections (which fires the *dg_ok* transition, then placing a token in *sp_or_backup*). In any of these two cases, there will be a token in place *sp_or_backup* and the GEN transition *automatic_restore* will be enabled. After automatic restoration of the upstream sections occurs, the remaining isolated sections must be manually repaired.

Note that the firing of *dg_needed* removes and adds a token to place *comm_working* so as to maintain a token in that place if communication is available, and thus distinguish logical states where communication is up from those where it is down. For the same reason, the firing of *automatic_restore* removes and adds a token to places *comm_working* and *sp_or_backup*. Inhibitor arcs from *sp_or_backup* to *dg_needed* and from *upstream_section_restored* to *automatic_restore* are used to prevent multiple firings of *dg_needed* and *automatic_restore*. All GEN transitions are associated with uniform distributions preserving the mean value of the corresponding EXP distributions in the Markovian model of Figure 3.

The model includes concurrent transitions associated with non-Markovian distributions over possibly bounded supports, which motivates the use of the solution technique proposed in [19] to perform transient stochastic analysis. The approach builds an embedded chain that samples the underlying stochastic process after each firing, maintaining an additional timer that evaluates the time elapsed since the failure event. In Section 5 we use the techniques proposed in [19] and the model presented in Figure 4 to assess storm impact accounting for general residence time distributions in the phased recovery model.

4.4   Parameterization of the Phased Recovery Model

In this section, we characterize the effects of the hurricane on the infrastructure under consideration, which depends on its topology as well as on the characteristics and strength of the hurricane. Together, the infrastructure and the hurricane strength will determine the model parameters p, q, and r.

Disrupted Sections. Each leg of the smart grid infrastructure consists of n sections $s_1, \ldots, s_n$. As mentioned in Section 2.1, we consider three regions in an affected leg S: downstream sections, failed sections, and upstream sections. These regions are characterized by the first failed section s_f (section 4 in Figure 1) and the last failed section s_l (section 6 in Figure 1) with $1 \leq f \leq l \leq n$. If no section fails, we set $f = l = 0$.

Probability of Available Communication. At each loop, the probability of available communication between a substation and the tie switch after the hurricane depends on whether the communication is established through radio, wire, or through the power lines. The value of p depends on the technology adopted for communication, and to capture the different levels of investment we vary p between 0 and 1 in our numerical experiments.

Probability of Secondary Path Available. The probability that a secondary path to restore energy to the upstream sections is available depends on many factors, including the recovery of failed sections. Let Γ' be the indicator random variable that characterizes if the secondary leg is operational. Then,

Definition 1. *The probability that there exists a secondary path to provide energy for the upstream sections after a failure is*

$$q = P(\Gamma' = 1). \tag{3}$$

Leg S includes n sections, i.e., $n = |S|$. Recall that D_s is the random variable that characterizes the state of section $s \in S$ immediately after a failure in leg S for $s = 1, 2, \ldots, n$. If section s has failed, $D_s = 1$; otherwise, $D_s = 0$ (see Section 3). Furthermore, let S' be the other leg (secondary path) in the current loop, with n' sections, that might be used to provide energy for the upstream sections of S. In this paper, to simplify the presentation, we assume that all sections fail independently. This is clearly a simplifying assumption, which can be relaxed without compromising the general methodology presented in this work.

Then, the probability that the secondary path to the failed region in S is operational is

$$q = \prod_{s' \in S'} P(D_{s'} = 0). \tag{4}$$

Table 5 summarizes the notation used throughout this paper.

Table 5. Table of notation. All variables are a function of the leg under study, which must be clear from the context.

Variable	Description
p	probability that communication is working after failure
q	probability that there is a secondary path to upstream sections
r	probability that distributed generation suffices to provide for upstream
$\mathcal{S}_u$	average energy supplied per time-unit at state u of the phased recovery model
F	index of the first failed section
L	index of the last failed section
n	number of sections in primary leg, $n = |S|$
S	set of sections in primary leg
S'	set of sections in secondary leg
Γ'	indicator random variable, equals 1 if secondary path is available
$\mathcal{I}$	set of contiguous sections between first and last failed section (including them)
$\mathcal{I}+$	set of upstream sections
$\mathcal{I}-$	set of downstream sections
D_j	indicator random variable, equals 1 if section j failed
ℓ_j	load at section j
g_j	distributed generation at section j
$\mathcal{U}(l)$	upstream surplus when last failed section is section l

Characterizing Isolated Sections. Next, our goal is to characterize the set of sections that are isolated from the network due to failures. We order the sections in a leg increasingly as a function of their distance from the substation. In what follows, we refer to the *first* and *last* failed sections in a leg with respect to that order. Let F be the index of the first failed section in S (see Figure 1). Then, the probability that the first failed section is section f is given by

$$P(F = f) = \begin{cases} P(D_f = 1) \prod_{j=1}^{f-1} P(D_j = 0) & \text{if } f > 0, \\ \prod_{j=1}^{n} P(D_j = 0) & \text{if } f = 0. \end{cases} \quad (5)$$

Equation (5) indicates that section f is the first failed section in the leg if it is damaged ($D_f = 1$) and the sections before section f have not been affected ($D_j = 0$ for $j = 1, \ldots, f-1$). Note that if the hurricane did not affect any section in the leg, $F = 0$, which occurs when $D_j = 0$ for $j = 1, \ldots, n$.

Let L be the index of the last failed section in S. Then, the probability that the last failed section is section l is given by

$$P(L = l) = \sum_{f=0}^{n} P(L = l | F = f) P(F = f) \quad (6)$$

where

$$P(L = l | F = f) = \begin{cases} P(D_l = 1) \prod_{j=l+1}^{n} P(D_j = 0) & \text{if } f > 0 \text{ and } l > f, \\ \prod_{j=l+1}^{n} P(D_j = 0) & \text{if } f > 0 \text{ and } l = f, \\ 1 & \text{if } f = l = 0, \\ 0 & \text{otherwise}, \end{cases} \quad (7)$$

with the convention that $\prod_{j=m}^{n} P(D_j = 0) = 1$ if $m > n$. According to Eq. (7), section l is the last failed section in the leg if it is damaged ($D_l = 1$) and the sections after section l have not been affected ($D_j = 0$ for $j = l+1, \ldots, n$). If the hurricane affects only section f in the leg, then $F = L = f$, which occurs if section f is the first failed section and the sections afterwards have not been affected ($D_j = 0$ for $j = l+1, \ldots, n$). Note that, if the hurricane did not affect any section in the leg, $F = 0$, which implies that $L = 0$.

Rewards. Let ℓ_s denote the average load at section s. Next, our goal is to determine the average energy not supplied after a failure.

States 1, 2, 3, 4. Let $\mathcal{S}_u$ be the energy supplied per time unit when the phased recovery model is in state u, $1 \le u \le 4$. Then, the expected energy supplied per unit time, in state u, is given by

$$E[\mathcal{S}_u] = \sum_{f=0}^{n} E[\mathcal{S}_u | F = f] P(F = f) \quad (8)$$

where, for $1 \le u \le 4$,

$$E[\mathcal{S}_u | F = f] = \begin{cases} \sum_{1 \le j < f} \ell_j & \text{if } f > 0, \\ \sum_{1 \le j \le n} \ell_j & \text{if } f = 0. \end{cases} \quad (9)$$

Equation (9) indicates that the expected energy supplied at states 1-4 is the total load supplied to the downstream sections in case failures occur (see Figure 1), and is the total demanded load otherwise.

State 5. Next, our goal is to compute the expected energy supplied in state 5 of the phased recovery model. To this aim, we consider two cases, depending on whether the secondary leg is operational. Recall that $q = P(\Gamma' = 1)$ is the probability that the secondary leg is operational. Then, conditioning on whether the secondary leg is operational in state 5 of the phased recovery model, the expected energy supplied per time unit is given by

$$E[\mathcal{S}_5] = E[\mathcal{S}_5|\Gamma' = 1]\, q + E[\mathcal{S}_5|\Gamma' = 0]\,(1-q). \tag{10}$$

In what follows, we compute $E[\mathcal{S}_5|\Gamma' = 1]$ and $E[\mathcal{S}_5|\Gamma' = 0]$. The expected energy supplied in state 5 of the phased recovery model, given that a secondary path is available, is given by

$$E[\mathcal{S}_5|\Gamma' = 1] = \sum_{0 \le f \le n} \sum_{0 \le l \le n} E[\mathcal{S}_5|F = f, L = l, \Gamma' = 1] P(L = l|F = f, \Gamma' = 1) P(F = f|\Gamma' = 1)$$
$$\tag{11}$$

where $P(L = l|F = f, \Gamma' = 1)$ and $P(F = f|\Gamma' = 1)$ are given by Eqs. (7) and (5), respectively, and $E[\mathcal{S}_5|F = f, L = l, \Gamma' = 1]$ is given by

$$E[\mathcal{S}_5|F = f, L = l, \Gamma' = 1] = \sum_{1 \le j < f} \ell_j + \sum_{l < j \le n} \ell_j. \tag{12}$$

Next, we compute $E[\mathcal{S}_5|\Gamma' = 0]$. Let g_s denote the average distributed energy generated at section s.

Definition 2. *Let* $\mathcal{U}(l)$ *be the surplus generation at the upstream of the current leg when* $L = l$,

$$\mathcal{U}(l) = \sum_{l < j \le n} (g_j - \ell_j). \tag{13}$$

Once a storm hits a leg of a loop, we assume that the leg is broken into *isolated sections, upstream sections* and *downstream sections* (see Section 3). The isolated sections are restored through manual repair, and energy is supplied to them only after manual repair concludes. Upstream sections, in contrast, are amenable to automatic restoration through Distributed Generation (DG) or making use of a secondary path (secondary leg). We do not consider isolated restoration of sections within the failed region, i.e., we assume that either there is enough backup energy from distributed generation to supply to the upstream sections (in which case they will be automatically recovered), or distributed generation will not be used. In addition, we only consider distributed generation capacities up to the tie switch.

If the surplus generation is zero or positive, the upstream sections can be restored using distributed generation, even if the secondary leg is not operational.

Definition 3. *The probability* r *of whether DG can restore the isolated upstream sections is*

$$r = \sum_{1 < l \le n} 1_{\mathcal{U}(l) > 0} P(L = l) \tag{14}$$

where

$$1_{\mathcal{U}(l) > 0} = \begin{cases} 1 & \text{if } \mathcal{U}(l) > 0, \\ 0 & \text{otherwise.} \end{cases} \tag{15}$$

The expected energy supplied in state 5 of the phased recovery model, given that a secondary path is not available, is given by

$$E[\mathcal{S}_5|\Gamma' = 0] = \sum_{1 < l \le n} P(L = l|\Gamma' = 0)E[\mathcal{S}_5|L = l, \Gamma' = 0] \tag{16}$$

where

$$E[\mathcal{S}_5|L = l, \Gamma' = 0] = \begin{cases} \sum_{l < j \le n} \ell_j & \text{if } \mathcal{U}(l) \ge 0, \\ 0 & \text{otherwise.} \end{cases} \tag{17}$$

Equation (17) indicates that the expected energy supplied, given that a secondary path is not available, is equal to the energy load of the upstream sections in case DG suffices. Replacing Eqs. (11) and (16) into Eq. (10), we obtain the expected energy supplied in state 5 of the phased recovery model. To simplify the presentation, in the remainder of this paper, when evaluating Eq. (16) we will additionally assume that the probability $P(L = l)$ that l is the last failed section in the primary leg is independent of whether the secondary leg is available, i.e., $P(L = l|\Gamma' = 0) = P(L = l)$.

State 6. In state 6, the expected energy supplied equals the total system load, and is given by

$$E[\mathcal{S}_6] = \sum_{1 \le j \le |S|} \ell_j . \tag{18}$$

Summary. Table 6 summarizes the results presented in this section. It shows how the probabilities and reward rates of different states of the phased recovery model are computed, and their meaning.

Table 6. Summary of probabilities and reward rate semantics and expressions

State	Reward rate	Equations
1-4	Energy supplied per unit time to downstream sections	(8)–(9)
5	Energy supplied per unit time to downstream and upstream sections	(10)–(17)
6	Energy supplied per unit time to whole system	(18)

Variable	Event probability	Equations
q	There is a secondary path to upstream sections	(3)–(4)
r	Distributed generation suffices for upstream	(14)–(15)

5   Evaluation

In this section, we present the results from the analysis of the Con Edison network described in Section 2. Section 5.1 illustrates the investment options under evaluation, the experimental setup, and information on the execution. Section 5.2 presents the results of Markovian and non-Markovian analysis.

5.1 Setup and Execution

To analyze the survivability of the Con Edison network N, we both derive the parameters and solve the survivability models described in Section 4 separately for each autoloop $S \in N$. The survivability models yield the expected energy not supplied at each point in time for each autoloop. To aggregate the results, we sum up the expected energy not supplied at each point in time over all considered autoloops. As an additional metric, we consider the Accumulated Expected Energy Not Supplied (AEENS) until system recovery.

In order to evaluate the ability of the proposed approach to quantify storm hardening strategies, we consider three different investment strategies and quantify their effect on the expected energy not supplied:

1. *Investment strategy 1 (INV1): Trim the trees along all sections.* Under this investment strategy, we multiply the failure probability of each section by the "trees trimmed" adjustment factor $\theta = 0.8$ (see Table 3).
2. *Investment strategy 2 (INV2): For each autoloop, place the first section and the last section underground.* Under this investment strategy, we set the failure probability of each section neighboring the substation to zero (adjustment factor $\theta = 0$, see Table 3).
3. *Investment strategy 3 (INV3): Combine strategies 1 and 2.* Under this investment strategy, we set the probability of each section neighboring the substation to zero and multiply the failure probability of the remaining sections by the "trees trimmed" adjustment factor $\theta = 0.8$.

All investment options in place reduce the failure probability of some or all sections and thus affect the rewards of the model as well as the probabilities q and r. In the base model, we expect 575 sections to fail as a result of $\sum_{S \in N} \sum_{s \in S} P(D_s = 1)$. For the investment options, the expected number of failed sections is reduced to 460 (INV1), 458 (INV2), and 366 (INV3).

Table 7 shows the average rewards for each state over all autoloops. Note that this is averaged over all autoloops, while we actually solve the model separately for each autoloop. Thus, the different states of the models are not reached at the same time for different autoloops. Table 8 characterizes the parameters r and q over all legs of all autoloops. We use expert knowledge to set the different model

Table 7. Rewards in kW averaged over all legs of all autoloops, for each investment option

Investment	State 1	State 2	State 3	State 4	State 5	State 6
None	7.4936	7.4936	7.4936	7.4936	6.6600	0
INV1	6.5089	6.5089	6.5089	6.5089	5.2990	0
INV2	6.3918	6.3918	6.3918	6.3918	5.1499	0
INV3	5.5224	5.5224	5.5224	5.5224	3.9104	0

Table 8. For parameters r and q and each investment option, the number of sections with parameter equal to zero, greater than zero, and the average value over sections with parameter greater than zero

Investment	Sections with $r = 0$	Sections with $r > 0$	Average of r over sections with $r > 0$	Sections with $q = 0$	Sections with $q > 0$	Average of q over sections with $q > 0$
None	298	14	0.159	18	294	0.094
INV1	296	16	0.157	0	312	0.139
INV2	298	14	0.186	20	292	0.140
INV3	296	16	0.183	0	312	0.190

parameters. We let the mean manual repair time be 4 hours ($\delta = 1/4$) and the mean automatic repair time be 2 minutes ($\alpha = \alpha' = 30$). Distributed generation takes an average of 15 minutes to be activated ($\beta = 4$) and communication takes an average of 1 hour to be repaired ($\gamma = 1$). Throughout the evaluation, we let $p = 0.5$. We implemented the reward calculations and the calculation of parameters q and r in Matlab as described in Section 4.4.

Deriving the rewards for the case study setup takes about 5 seconds on a commercial off-the-shelf machine. The solution of the Markovian model was implemented in Matlab as well, and solving the Markovian phased recovery model takes about 10 seconds. Using the recent release of the ORIS Tool based on the Sirio framework [8,9,6], regenerative transient analysis of the non-Markovian phased recovery model up to time 16 h can be performed in nearly 3 seconds with a time step of 0.1 h. Evaluating the EENS of the non-Markovian model takes roughly 15 minutes for each investment option.

We assumed an infinite number of repair trucks and repair teams, i.e., we assumed that the mean manual repair time $1/\delta$ of each autoloop is independent of the number of overall failures. This is a simplifying assumption that we will relax in future work.

Note that loops operate autonomously from each other as far as distribution automation features are concerned. There might be dependencies between loop repair times due to geographical closeness, global availability of required resources, and so on. The geographical closeness was taken into account by analyzing the wind gusts per location. Nevertheless, the analysis of manual repair times as a function of geographical closeness and number of trucks available is out of scope of the paper.

5.2   Evaluation Results

Table 9 shows the accumulated EENS until the complete system recovery for the base network and the three investment options.

Figure 5 shows the results for the EENS over time for the base network with no investment (red) and the three networks resulting from the three investment strategies (blue, green, black). The first four curves in the figure key are the results of the Markovian analysis and have the typical exponential form. The fifth to eighth curves are the results of the non-Markovian analysis; the EENS value at time zero is the same as in Markovian analysis. With both solution approaches,

Table 9. Accumulated EENS in gigawatt (GW) until complete system recovery for different investment options

	Investment option		
None	INV1	INV2	INV3
4.5862	3.7915	3.8762	3.1268

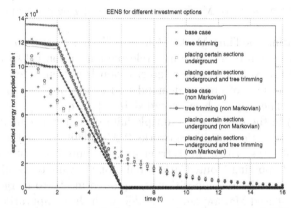

Fig. 5. Expected energy not supplied over time: base network and different investment options

the base model has the highest EENS at all points in time. Investment strategy 2 (some sections underground) performs slightly better than investment strategy 1 (trim trees) for the considered setup. Combining the two strategies, i.e., strategy 3, yields the best results.

Comparing the behavior of the EENS over time, we observe that the non-Markovian EENS results have a different behavior, exactly reaching a null value at time 6 h; this corresponds to the completion of the manual repair, which is uniformly distributed over the bounded support [2, 6] h. Before time 2 h, the decrease of the EENS rate is due to the repair of upstream sections; given the low probability that distributed generation is sufficient to provide energy to upstream sections (r parameter), the expected reduction is limited and the overall dynamics is dominated by the manual recovery operation. Different investment options are distinguished more significantly by guaranteeing different initial EENS immediately after a failure.

The results demonstrate how our models can be used to quantitatively assess investment options for storm hardening of distribution grids. Note that the numerical results of our analysis are by no means general recipes for the suitability of storm hardening strategies. Instead, for each power network under study, each considered storm scenario, and each set of storm hardening investment options, the input data (cf. Section 2) has to be determined and fed into our tool chain to quantify the effects of storm hardening strategies. Then, the calculation of the survivability parametrization and the solution of the survivability models can be done automatically by our tools.

6   Related Work

Survivability models of distribution automation power grids were first introduced in [2,21,24,1]. These models were solved analytically using multiple techniques, such as transient analysis of Markov chains, state aggregation, and hierarchical modeling.

In [1], a Continuous Time Markov Chain (CTMC) is used to model the actions taken in reaction to a failure in a telecommunication network, evaluating an extension of the System Average Interruption Duration Index (SAIDI) that accounts for variations of energy demand and supply during a multi-step recovery process. The approach is extended in [2] to quantify the Energy Not Supplied (ENS) in the presence of multiple failures under specific independence assumptions. Stochastic Activity Networks (SANs) [30] are used in [3] to model the operation of large critical infrastructures, encompassing interdependencies among them and applying Monte Carlo simulation to evaluate the distribution of cascade sizes. Hierarchical composition is exploited in [23] to merge the expressiveness of state-based Markov reward models with the computational efficiency of combinatorial methods, deriving transient availability and performability measures for telecommunication systems.

Unit commitment scheduling for coordination of energy demand and supply is studied in [5]. The authors model renewable energy resources through Hidden Markov Models (HMMs) [28] and power demand loads as a Markov-Modulated Poisson Process (MMPP) [14]. The problem is formulated as a partially observable Markov decision process and a distributed scheme is presented such that the most suitable generation unit is dynamically scheduled based on system parameters including demand loads, utility costs, reliability, and pollution emissions of generation units. In [10], a probabilistic model checker based on the PRISM tool [22] is developed and used to evaluate demand-side management in microgrids. The authors consider a decentralized infrastructure which allows users to oversee demand while dissuading them from abuse and incentivizing cooperation among them. The approach leverages the model of turn-based stochastic multi-player games, where players can either collaborate or compete to achieve a specific goal, and is used to detect potential weaknesses and unexpected behaviors in smart energy management algorithms.

The approach of [16,17] discusses elementary mechanisms for distributed runtime control of power grids with a substantial share of renewable energy sources (especially photovoltaic power generators), which make electric power production much more subject to unpredictable and significant fluctuations. To this end, non-Markovian models specified in MODEST [4] are used to evaluate production control algorithms and demand-side mechanisms, especially in terms of stability, availability, quality of service, and fairness.

Heegaard and Trivedi [18] study the survivability of the Internet and computer networks. Similarities between the operation of the Internet and the working principles of future power grids are leveraged in [15] to support design of distributed and decentralized power grid control appliances. Keshav and Rosenberg [20] also point out how concepts pioneered by the Internet are applicable to the design of smart grids.

The papers mentioned above are related to ours as they consider the survivability of critical infrastructures. Nonetheless, the work presented here significantly differs from previous work as we (1) combine the survivability model with a model to characterize a hurricane, (2) propose a scalable way to assess the survivability of large infrastructures, and (3) consider and compare investment options that have not been analyzed before using survivability models.

7   Conclusion

In this paper, we have introduced an innovative approach to model failures and recoveries resulting from large hurricanes. To this end, a scalable survivability model has been developed to assess the evolution of the failure recovery process on a real distribution automation network. More specifically, we have used as a case study Hurricane Sandy impacts on the overhead distribution of New York metropolitan region.

We have created a scalable survivability model based on a phase-recovery Markov chain with rewards. The reward rates characterize metrics such as the expected energy not supplied per time unit. They are parameterized using information about the geography and the network topology. Our model can be evaluated efficiently because each distribution loop is modeled independently by a separate phased recovery Markov chain. We have also developed a non-Markovian phased recovery model that allowed us to better approximate repair distributions. We have presented evaluations of the Markovian and non-Markovian phased recovery models and we are encouraged by the efficiency at which we obtained our initial results.

As topics for further research, we envision the development of heuristics to evaluate investment alternatives for distribution automation reliability improvement, by assessing customer affecting metrics such as energy not supplied up to full system recovery. The validation to specific environments requires engagement of the target utilities and possible model refinements. The proposed model is general enough to allow for topology generalizations and to incorporate historical data from different environments.

References

1. Avritzer, A., Suresh, S., Menasché, D.S., Leão, R.M.M., de Souza e Silva, E., Diniz, M.C., Trivedi, K., Happe, L., Koziolek, A.: Survivability models for the assessment of smart grid distribution automation network designs. In: Proceedings of the International Conference on Performance Engineering, pp. 241–252. ACM (2013)

2. Avritzer, A., Suresh, S., Menasché, D.S., Leão, R.M.M., de Souza e Silva, E., Diniz, M.C., Trivedi, K., Happe, L., Koziolek, A.: Survivability models for the assessment of smart grid distribution automation network designs. Concurrency and Computation Practice and Experience (2014)

3. Bloomfield, R., Buzna, L., Popov, P., Salako, K., Wright, D.: Stochastic modelling of the effects of interdependencies between critical infrastructure. In: Rome, E., Bloomfield, R. (eds.) CRITIS 2009. LNCS, vol. 6027, pp. 201–212. Springer, Heidelberg (2010)

4. Bohnenkamp, H.C., D'Argenio, P.R., Hermanns, H., Katoen, J.-P.: Modest: A compositional modeling formalism for hard and softly timed systems. IEEE Trans. Softw. Eng. 32(10), 812–830 (2006)

5. Bu, S., Yu, F.R., Liu, P.X.: Stochastic unit commitment in smart grid communications. In: 2011 IEEE Conference on Computer Communications Workshop, pp. 307–312. IEEE (2011)

6. Bucci, G., Carnevali, L., Ridi, L., Vicario, E.: Oris: a tool for modeling, verification and evaluation of real-time systems. Int. Journal of SW Tools for Technology Transfer 12(5), 391–403 (2010)

7. Carnevali, L., Grassi, L., Vicario, E.: State-Density Functions over DBM Domains in the Analysis of Non-Markovian Models. IEEE Trans. on Software Engineering 35(2), 178–194 (2009)

8. Carnevali, L., Ridi, L., Vicario, E.: A framework for simulation and symbolic state space analysis of non-Markovian models. In: Flammini, F., Bologna, S., Vittorini, V. (eds.) SAFECOMP 2011. LNCS, vol. 6894, pp. 409–422. Springer, Heidelberg (2011)

9. Carnevali, L., Ridi, L., Vicario, E.: Sirio: A framework for simulation and symbolic state space analysis of non-Markovian models. In: QEST 2011, pp. 153–154 (2011)

10. Chen, T., Forejt, V., Kwiatkowska, M.Z., Parker, D., Simaitis, A.: Automatic verification of competitive stochastic systems. Formal Methods in System Design 43(1), 61–92 (2013)

11. Consolidated Edison of New York, Inc. Report on the preparation and system restoration performance (January 2013)

12. Consolidated Edison of New York, Inc. Storm Hardening and Resiliency Collaborative Report (December 2013)

13. Blake, E.S., et al.: Tropical Cyclone Report - Hurricane Sandy (AL182012). Nat'l Hurricane Center (February 2013)

14. Fischer, W., Meier-Hellstern, K.: The Markov-modulated Poisson process (mmpp) cookbook. Performance Evaluation 18(2), 149–171 (1993)

15. Hahn, E.M., Hartmanns, A., Hermanns, H., Katoen, J.-P.: A compositional modelling and analysis framework for stochastic hybrid systems. Formal Methods in System Design 43(2), 191–232 (2013)

16. Hartmanns, A., Hermanns, H.: Modelling and decentralised runtime control of self-stabilising power micro grids. In: Margaria, T., Steffen, B. (eds.) ISoLA 2012, Part I. LNCS, vol. 7609, pp. 420–439. Springer, Heidelberg (2012)

17. Hartmanns, A., Hermanns, H., Berrang, P.: A comparative analysis of decentralized power grid stabilization strategies. In: Winter Simulation Conference, p. 158 (2012)

18. Heegaard, P.E., Trivedi, K.S.: Network survivability modeling. Computer Networks 53(8), 1215–1234 (2009)

19. Horváth, A., Paolieri, M., Ridi, L., Vicario, E.: Transient analysis of non-Markovian models using stochastic state classes. Performance Evaluation 69(7), 315–335 (2012)

20. Keshav, S., Rosenberg, C.: How internet concepts and technologies can help green and smarten the electrical grid. ACM SIGCOMM Computer Communication Review (2011)
21. Koziolek, A., Happe, L., Avritzer, A., Suresh, S.: A common analysis framework for smart distribution networks applied to survivability analysis of distribution automation. In: 2012 International Workshop on Software Engineering for the Smart Grid (SE4SG), pp. 23–29 (June 2012)
22. Kwiatkowska, M., Norman, G., Parker, D.: PRISM 4.0: Verification of probabilistic real-time systems. In: Gopalakrishnan, G., Qadeer, S. (eds.) CAV 2011. LNCS, vol. 6806, pp. 585–591. Springer, Heidelberg (2011)
23. Lanus, M., Yin, L., Trivedi, K.S.: Hierarchical composition and aggregation of state-based availability and performability models. IEEE Transactions on Reliability 52(1), 44–52 (2003)
24. Menasché, D., Leão, R.M.M., de Souza e Silva, E., Avritzer, A., Suresh, S., Trivedi, K., Marie, R.A., Happe, L., Koziolek, A.: Survivability analysis of power distribution in smart grids with active and reactive power modeling. In: GreenMetrics Workshop (2012)
25. Merlin, P., Farber, D.J.: Recoverability of communication protocols. IEEE Trans. on Comm. 24(9), 1036–1043 (1976)
26. NYC Department of Environmental Protection, Fred Gliesing, CF. Challenges to westchester's forests
27. Office of Long-Term Planning and Sustainability, City of New York. Utilization of Underground and Overhead Power Lines in the City of New York. City of New York (December 2013)
28. Rabiner, L., Juang, B.-H.: An introduction to hidden markov models. IEEE ASSP Magazine 3(1), 4–16 (1986)
29. Rudion, K., Orths, A., Styczynski, Z.A., Strunz, K.: Design of benchmark of medium voltage distribution network for investigation of dg integration. In: Power Engineering Society General Meeting. IEEE (2006)
30. Sanders, W.H., Meyer, J.F.: Stochastic activity networks: Formal definitions and concepts. In: Brinksma, E., Hermanns, H., Katoen, J.-P. (eds.) FMPA 2000. LNCS, vol. 2090, pp. 315–343. Springer, Heidelberg (2001)
31. Vicario, E., Sassoli, L., Carnevali, L.: Using stochastic state classes in quantitative evaluation of dense-time reactive systems. IEEE Trans. on Software Engineering 35(5), 703–719 (2009)

Compositionality Results
for Quantitative Information Flow[*]

Yusuke Kawamoto[1,2], Konstantinos Chatzikokolakis[2,3],
and Catuscia Palamidessi[1,2]

[1] INRIA, France
[2] École Polytechnique, France
[3] CNRS, France

Abstract. In the min-entropy approach to quantitative information flow, the leakage is defined in terms of a minimization problem, which, in case of large systems, can be computationally rather heavy. The same happens for the recently proposed generalization called g-vulnerability. In this paper we study the case in which the channel associated to the system can be decomposed into simpler channels, which typically happens when the observables consist of several components. Our main contribution is the derivation of bounds on the g-leakage of the whole system in terms of the g-leakages of its components.

1 Introduction

The problem of preventing confidential information from being leaked is a fundamental concern in the modern society, where the pervasive use of automatized devices makes it hard to predict and control the *information flow*. While early research focussed on trying to achieve *non-interference* (i.e., no leakage), it is nowadays recognized that, in practical situations, some amount of leakage is unavoidable. Therefore an active area of research on information flow is dedicated to the development of theories to *quantify* the amount of leakage, and of methods to minimize it. See, for instance, [15,5,20,18,10,11,25,6].

Among these theories, min-entropy leakage [25,7] has become quite popular, partly due to its clear operational interpretation in terms of one-try attacks. This quite basic setting has been recently extended to the *g-leakage* framework [2]. The main novelty consists in the introduction of gain functions, that permit to quantify the vulnerability of a secret in terms of the gain of the adversary, thus allowing to model a wide variety of operational scenarios.

While g-leakage is appealing for its generality and flexible operational interpretation, its computation is not trivial. Like most of the quantitative approaches, its definition is based on the probabilistic correlation between the secrets and

[*] This work has been partially supported by the project ANR-12-IS02-001 PACE, by the INRIA Equipe Associée PRINCESS, by the INRIA Large Scale Initiative CAPPRIS, and by EU grant agreement no. 295261 (MEALS). The work of Y. Kawamoto has been supported by a postdoc grant funded by the IDEX Digital Society project.

G. Norman and W. Sanders (Eds.): QEST 2014, LNCS 8657, pp. 368–383, 2014.

Table 1. Computation of information leakage measures in various scenarios

Kinds of systems	small systems	large systems	large unknown systems
Input distribution π	known	known	known
Component channels C_i	known	known	approx. statistically
Leakage of C_i with π_i	computable	computable	approx. statistically
Composed channel C	computable	unfeasible	unfeasible
Leakage of C with π	computable	unfeasible	unfeasible

the observables. Such correlation is usually expressed in terms of an *information-theoretic channel*, where the secrets constitute the input and the observables the output. The channel is characterized by the *channel matrix*, namely the conditional probabilities of each output for any given input. The computation of the channel matrix from the system can be performed via model checking (see, e.g., [3]), if a system is completely specified and it is not too complicated. Once the matrix is known, the computation of the g-leakage involves solving an optimization problem. This can be quite costly when the matrix is large.

Worse yet, in many cases it is not possible to compute the channel matrix exactly, for instance because the system may be too complicated, or because the conditional probabilities are partially determined by unknown factors. Fortunately, there are statistical methods that allow to approximate the channel matrix and the leakage [9,12]. There is also a tool, leakiEst [14], which allows to estimate min-entropy leakage from a set of trial runs [13]. However, if the cardinality of secrets and observables is large, such estimation becomes computationally heavy, due to the huge amount of trial runs that need to be performed.

In this paper we determined bounds on g-leakages in compositional terms. More precisely, we consider the parallel composition of channels, defined on the cross-products of the inputs and of the outputs. Then, we derive lower and upper bounds on the g-leakage of the whole channel in terms of the g-leakages of the components. Since the size of the whole channel is the product of the sizes of the components, there is an evident benefit in terms of computational cost. Table 1 illustrates the situation for the various kinds of channel matrices (small, large, unknown): the first three rows characterize the situation, and the last three express the feasibility of computing the leakage of the components, the matrix of the whole system, and the leakage of the whole system, respectively. This computation is meant to be exact in the first two columns, and statistical in the last one. The number of components is assumed to be huge. Note that the size of the whole channel increases exponentially with the number of the components.

We evaluate our compositionality results on randomly generated channels and on Crowds, a protocol for anonymous communication, run on top of a mobile ad-hoc network (MANET). In such a network users are mobile, can communicate only with nearby nodes, and the network topology changes frequently. As a result, Crowds routes can become invalid forcing the user to re-execute the protocol to establish a new route. These protocol repetitions, modeled by

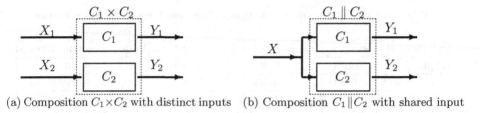

(a) Composition $C_1 \times C_2$ with distinct inputs   (b) Composition $C_1 \| C_2$ with shared input

Fig. 1. The two kinds of parallel compositions on channels, $\times$ and $\|$

the composition of the corresponding channels, lead to more information being leaked. Although the composed channel quickly becomes too big to compute the leakage directly, our compositionality results allow to obtain bounds on it.

The rest of the paper is organized as follows: Section 2 introduces basic notions of information theory, defines compositions of channels, and presents information leakage measures. Section 3 presents lower/upper bounds for g-leakages in compositional terms. Section 4 instantiates these results to min-entropy leakages. Section 5 introduces a transformation technique which improves the precision of our method. Section 6 evaluates our results by experiments.

All proofs can be found in the report version [17] of this paper.

2   Preliminaries

In this section we recall the notion of information-theoretic channels, define channel compositions, and recall some information leakage measures.

2.1   Channels

A *discrete channel* is a triple $(\mathcal{X}, \mathcal{Y}, C)$ consisting of a finite set $\mathcal{X}$ of secret input values, a finite set $\mathcal{Y}$ of observable output values, and an $|\mathcal{X}| \times |\mathcal{Y}|$ matrix C, called *channel matrix*, where each element $C[x, y]$ represents the conditional probability $p(y|x)$ of obtaining the output $y \in \mathcal{Y}$ given the input $x \in \mathcal{X}$. The input values have a probability distribution, called *input distribution* or *prior*. Given a prior π on $\mathcal{X}$, the joint distribution for X and Y is defined by $p(x, y) = \pi[x]C[x, y]$. The output distribution is given by $p(y) = \sum_{x \in \mathcal{X}} \pi[x]C[x, y]$.

2.2   Composition of Channels

We now introduce the two kinds of composition which will be considered in the paper. We assume that the channels are *independent*, in the sense that, given the respective inputs, the outcome of one channel does not influence the outcome of the other. We start with defining *parallel composition with separate inputs* $\times$ (*parallel composition* for short). Note that the term "parallel" here does not carry a temporal meaning: the actual execution of the corresponding systems could take place simultaneously or in any order.

Definition 1 (Parallel composition (with distinct inputs)). Given two discrete channels $(\mathcal{X}_1, \mathcal{Y}_1, C_1)$ and $(\mathcal{X}_2, \mathcal{Y}_2, C_2)$, their *parallel composition (with distinct inputs)* is the discrete channel $(\mathcal{X}_1 \times \mathcal{X}_2, \mathcal{Y}_1 \times \mathcal{Y}_2, C_1 \times C_2)$ where $C_1 \times C_2$ is the $(|\mathcal{X}_1| \cdot |\mathcal{X}_2|) \times (|\mathcal{Y}_1| \cdot |\mathcal{Y}_2|)$ matrix such that $(C_1 \times C_2)[(x_1, x_2), (y_1, y_2)] = C_1[x_1, y_1] \cdot C_2[x_2, y_2]$ for each $x_1 \in \mathcal{X}_1$, $x_2 \in \mathcal{X}_2$, $y_1 \in \mathcal{Y}_1$ and $y_2 \in \mathcal{Y}_2$.

The condition $(C_1 \times C_2)[(x_1, x_2), (y_1, y_2)] = C_1[x_1, y_1] \cdot C_2[x_2, y_2]$ is what we mean by "the channels are independent". Note that, although the output distributions Y_1 and Y_2 may be correlated, they are *conditionally independent*, in the sense that $p(y_1, y_2 | x_1, x_2) = p(y_1 | x_1) p(y_2 | x_2)$.

Next, we define the parallel composition with shared input $\|$.

Definition 2 (Parallel composition with shared input). Given two discrete channels $(\mathcal{X}, \mathcal{Y}_1, C_1)$ and $(\mathcal{X}, \mathcal{Y}_2, C_2)$, their *parallel composition with shared input* is the discrete channel $(\mathcal{X}, \mathcal{Y}_1 \times \mathcal{Y}_2, C_1 \| C_2)$ where $C_1 \| C_2$ is the $|\mathcal{X}| \times (|\mathcal{Y}_1| \cdot |\mathcal{Y}_2|)$ matrix such that $(C_1 \| C_2)[x, (y_1, y_2)] = C_1[x, y_1] \cdot C_2[x, y_2]$ for each $x \in \mathcal{X}$, $y_1 \in \mathcal{Y}_1$ and $y_2 \in \mathcal{Y}_2$.

Note that $\|$ is a special case of $\times$. In fact, $(C_1 \| C_2)[x, (y_1, y_2)] = C_1[x, y_1] \cdot C_2[x, y_2] = (C_1 \times C_2)[(x, x), (y_1, y_2)]$.

Fig. 1 illustrates these definitions. These two kinds of compositions are used to represent different situations. For example, in the Crowds protocol (explained in Section 6.1) repeated executions of the protocol with different senders are described by the parallel composition ($\times$), while repeated executions with the same sender are described by the parallel composition with shared input ($\|$).

2.3   Quantitative Information Leakage Measures

The *information leakage* of a channel is measured as the difference between the *prior uncertainty* about the secret value of the channel's input and the *posterior uncertainty* of the input after observing the channel's output. The uncertainty is defined in terms of an attacker's operational scenario. In this paper we will focus on *min-entropy leakage*, in which such measure, min-entropy, represents the difficulty for an attacker to guess the secret inputs in a single attempt.

Definition 3. Given a prior π on $\mathcal{X}$ and a channel $(\mathcal{X}, \mathcal{Y}, C)$, the *prior vulnerability* and the the *posterior vulnerability* are defined respectively as

$$V(\pi) = \max_{x \in \mathcal{X}} \pi[x] \qquad \text{and} \qquad V(\pi, C) = \sum_{y \in \mathcal{Y}} \max_{x \in \mathcal{X}} \pi[x] C[x, y].$$

Definition 4. Given a prior π on $\mathcal{X}$ and a channel $(\mathcal{X}, \mathcal{Y}, C)$, the *min-entropy* $H_\infty(\pi)$ and *conditional min-entropy* $H_\infty(\pi, C)$ are defined by:

$$H_\infty(\pi) = -\log V(\pi) \qquad \text{and} \qquad H_\infty(\pi, C) = -\log V(\pi, C)$$

and the *min-entropy leakage* $I_\infty(\pi, C)$ and *min-capacity* $C_\infty(C)$ are defined by:

$$I_\infty(\pi, C) = H_\infty(\pi) - H_\infty(\pi, C) \qquad \text{and} \qquad C_\infty(C) = \sup_{\pi'} I_\infty(\pi', C).$$

Min-entropy leakage has been generalized by *g-leakage* [2], which allows a wide variety of operational scenarios. These are modeled using a set $\mathcal{W}$ of possible *guesses*, and a *gain function* $g : \mathcal{W} \times \mathcal{X} \rightarrow [0,1]$ such that $g(w,x)$ represents the gain of the attacker when the secret value is x and he makes a guess w on x.

Then *g-vulnerability* is defined as the maximum expected gain of the attacker:

Definition 5. Given a prior π on $\mathcal{X}$ and a channel $(\mathcal{X}, \mathcal{Y}, C)$, the *prior g-vulnerability* and the *posterior g-vulnerability* are defined respectively by

$$V_g(\pi) = \max_{w \in \mathcal{W}} \sum_{x \in \mathcal{X}} \pi[x] g(w,x) \ \text{ and } \ V_g(\pi, C) = \sum_{y \in \mathcal{Y}} \max_{w \in \mathcal{W}} \sum_{x \in \mathcal{X}} \pi[x] C[x,y] g(w,x).$$

We now extend Definition 4 to the *g*-setting:

Definition 6. Given a prior π on $\mathcal{X}$ and a channel $(\mathcal{X}, \mathcal{Y}, C)$, the *g-entropy* $H_g(\pi)$, *conditional g-entropy* $H_g(\pi, C)$, *g-leakage* $I_g(\pi, C)$ and *g-capacity* $C_g(C)$ are defined by: $H_g(\pi) = - \log V_g(\pi)$, $H_g(\pi, C) = - \log V_g(\pi, C)$, $I_g(\pi, C) = H_g(\pi) - H_g(\pi, C)$, $C_g(C) = \sup_{\pi'} I_g(\pi', C)$.

The min-entropy notions are particular cases of the *g*-entropy ones, obtained by instantiating g to the identity function g_{id} defined as $g_{id}(w, x) = 1$ if $w = x$ and $g_{id}(w, x) = 0$ otherwise. Then we have $H_\infty = H_{g_{id}}$, $I_\infty = I_{g_{id}}$ and $C_\infty = C_{g_{id}}$.

3   Compositionality Results on *g*-Leakage

In this section we introduce joint gain functions for composed channels and present compositionality results for *g*-leakage.

3.1   Joint Gain Functions for Composed Channels

To formalize the *g*-leakages of composed channels, we need to know in advance a *joint gain function* g that is defined as a function from $(\mathcal{W}_1 \times \mathcal{W}_2) \times (\mathcal{X}_1 \times \mathcal{X}_2)$ to $[0,1]$. When a joint secret input is $(x_1, x_2) \in \mathcal{X}_1 \times \mathcal{X}_2$ and the attacker's joint guess is $(w_1, w_2) \in \mathcal{W}_1 \times \mathcal{W}_2$, the attacker's joint gain from the guesses is represented by $g((w_1, w_2), (x_1, x_2))$.

For the sake of generality, we do not assume any relation between g and the two gain functions g_1 and g_2, except for the following: a joint guess is worthless iff at least one of the single guesses is worthless. Formally: $g((w_1, w_2), (x_1, x_2)) = 0$ iff $g_1(w_1, x_1) g_2(w_2, x_2) = 0$. [1]

We say that g_1 and g_2 are *independent* if $g((w_1, w_2), (x_1, x_2)) = g_1(w_1, x_1) g_2(w_2, x_2)$ for all x_1, x_2, w_1 and w_2.

---

[1] This property holds, for example, when g, g_1, g_2 are the identity gain functions.

3.2   Jointly Supported Input Distributions

Given a joint prior π on $\mathcal{X}_1 \times \mathcal{X}_2$, the *marginal distribution* π_1 on $\mathcal{X}_1$ is defined as $\pi_1[x_1] = \sum_{x_2 \in \mathcal{X}_2} \pi[x_1, x_2]$ for all $x_1 \in \mathcal{X}_1$. The *marginal distribution* π_2 on $\mathcal{X}_2$ is defined analogously. Note that $\pi_1[x_1] \cdot \pi_2[x_2] = 0$ implies $\pi[x_1, x_2] = 0$. The converse does not hold in general, but occasionally we will assume it:

Definition 7. A prior π on $\mathcal{X}_1 \times \mathcal{X}_2$ is *jointly supported* if, for all $x_1 \in \mathcal{X}_1$ and $x_2 \in \mathcal{X}_2$, $\pi_1[x_1] \cdot \pi_2[x_2] \neq 0$ implies $\pi[x_1, x_2] \neq 0$.

Essentially, this condition rules out all the distributions in which there exist two events that happen with a non-zero probability, but that never happen together, i.e., events that are incompatible with each other.

If π_1 and π_2 are independent, i.e., $\pi[x_1, x_2] = \pi_1[x_1] \cdot \pi_2[x_2]$ for all $x_1 \in \mathcal{X}_1$ and $x_2 \in \mathcal{X}_2$, then we denote π by $\pi_1 \times \pi_2$. Note that $\pi_1 \times \pi_2$ is jointly supported.

3.3   The g-Leakage of Parallel Composition

In this section we present a lower and an upper bound for the g-leakage of $C_1 \times C_2$ in terms of the g-leakages of C_1 and C_2. We first introduce some notation.

Definition 8. Let π be a prior on $\mathcal{X}_1 \times \mathcal{X}_2$, and $g : (\mathcal{W}_1 \times \mathcal{W}_2) \times (\mathcal{X}_1 \times \mathcal{X}_2) \to [0,1]$ be a joint gain function. For $w_1 \in \mathcal{W}_1$ and $w_2 \in \mathcal{W}_2$, their *support with respect to g* is defined as: $\mathcal{S}_{w_1, w_2} = \{(x_1, x_2) \in \mathcal{X}_1 \times \mathcal{X}_2 \mid \pi[x_1, x_2] \cdot g((w_1, w_2), (x_1, x_2)) \neq 0\}$.

The lower and the upper bounds are based on the following two measures.

Definition 9. Let g be a joint gain function from $(\mathcal{W}_1 \times \mathcal{W}_2) \times (\mathcal{X}_1 \times \mathcal{X}_2)$ to $[0,1]$. Let g_1, g_2 be two gain functions from $\mathcal{W}_1 \times \mathcal{X}_1$ to $[0,1]$ and from $\mathcal{W}_2 \times \mathcal{X}_2$ to $[0,1]$ respectively. Given a prior π on $\mathcal{X}_1 \times \mathcal{X}_2$, we define $M_\pi^{\min}$ and $M_\pi^{\max}$:

$$M_\pi^{\min} = \min_{w_1 \in \mathcal{W}_1, w_2 \in \mathcal{W}_2} \min_{(x_1, x_2) \in \mathcal{S}_{w_1, w_2}} \frac{\pi_1[x_1] g_1(w_1, x_1) \cdot \pi_2[x_2] g_2(w_2, x_2)}{\pi[x_1, x_2] \cdot g((w_1, w_2), (x_1, x_2))}$$

$$M_\pi^{\max} = \max_{w_1 \in \mathcal{W}_1, w_2 \in \mathcal{W}_2} \sum_{(x_1, x_2) \in \mathcal{S}_{w_1, w_2}} \frac{\pi_1[x_1] g_1(w_1, x_1) \cdot \pi_2[x_2] g_2(w_2, x_2)}{\pi[x_1, x_2] \cdot g((w_1, w_2), (x_1, x_2))}.$$

When π_1 and π_2 are independent and g_1 and g_2 are independent, $M_\pi^{\min} = M_\pi^{\max} = 1$. In addition, for any prior π, $M_\pi^{\min}$ is strictly positive.

We now show compositionality results for generalized information measures.

Posterior g-Entropy of Parallel Composition

Lemma 1. *For any prior π on $\mathcal{X}_1 \times \mathcal{X}_2$ with marginals π_1 and π_2, and two channels $(\mathcal{X}_1, \mathcal{Y}_1, C_1)$, $(\mathcal{X}_2, \mathcal{Y}_2, C_2)$,*

- $H_g(\pi, C_1 \times C_2) \geq H_{g_1}(\pi_1, C_1) + H_{g_2}(\pi_2, C_2) + \log M_\pi^{\min}$
- *if π is jointly supported, then* $H_g(\pi, C_1 \times C_2) \leq H_{g_1}(\pi_1, C_1) + H_{g_2}(\pi_2, C_2) + \log M_\pi^{\max}$.

The equalities hold if the priors and the gain functions are independent:

Corollary 1. *If $g((w_1, w_2), (x_1, x_2)) = g_1(w_1, x_1) g_2(w_2, x_2)$ for all x_1, x_2, w_1 and w_2, then, for any π_1 and π_2, $H_g(\pi_1 \times \pi_2, C_1 \times C_2) = H_{g_1}(\pi_1, C_1) + H_{g_2}(\pi_2, C_2)$.*

The g-Leakage of Parallel Composition

Theorem 1. *Let π be a jointly supported prior on $\mathcal{X}_1 \times \mathcal{X}_2$ with marginals π_1 and π_2. Let $(\mathcal{X}_1, \mathcal{Y}_1, C_1)$, $(\mathcal{X}_2, \mathcal{Y}_2, C_2)$ be two channels. Then:*

$$I_{g_1}(\pi_1, C_1) + I_{g_2}(\pi_2, C_2) - \log \frac{M_\pi^{\max}}{M_\pi^{\min}} \le I_g(\pi, C_1 \times C_2) \le I_{g_1}(\pi_1, C_1) + I_{g_2}(\pi_2, C_2) + \log \frac{M_\pi^{\max}}{M_\pi^{\min}}$$

Again, the equality holds if the priors and the gain functions are independent:

Corollary 2. *If g_1 and g_2 are independent, then $I_g(\pi_1 \times \pi_2, C_1 \times C_2) = I_{g_1}(\pi_1, C_1) + I_{g_2}(\pi_2, C_2)$.*

These results can be naturally extended to the composition of n channels; this extension can be found in the report version of this paper [17].

3.4   The g-Leakage of Parallel Composition with Shared Input

In this section we present compositionality results for g-leakage when two channels share the same input value.

The parallel composition with shared input corresponds to the parallel composition with two identical inputs values: $(C_1 \| C_2)[x, (y_1, y_2)] = C_1[x, y_1]C_2[x, y_2] = (C_1 \times C_2)[(x, x), (y_1, y_2)]$. To give the same input value x to both C_1 and C_2, the prior $\pi^\dagger$ on $\mathcal{X} \times \mathcal{X}$ is defined from a prior π on $\mathcal{X}$ by:

$$\pi^\dagger[x, x'] = \begin{cases} \pi[x] & \text{if } x = x' \\ 0 & \text{otherwise} \end{cases}$$

Then $H_g(\pi, C_1 \| C_2) = H_g(\pi^\dagger, C_1 \times C_2)$. In addition, $\pi_1^\dagger[x] = \pi_2^\dagger[x] = \pi[x]$.

As we see in the definition, the attacker's gain is determined solely from a secret input x and his guess w on x (and independently of channels that receive x as input). Let g be a gain function from $\mathcal{W} \times \mathcal{X}$ to $[0, 1]$. Since C_1 and C_2 receive input from the same domain $\mathcal{X}$, we use the same gain function g to calculate both the g-leakages of C_1 and C_2. Since an identical input value x is given to C_1 and C_2 in the composed channel $C_1 \| C_2$ and the attacker makes a single guess w on the secret x, we define the joint gain function $g^\dagger \colon \mathcal{W} \times \mathcal{W} \times \mathcal{X} \times \mathcal{X} \to [0, 1]$ from g by: $g^\dagger((w, w'), (x, x')) = g(w, x)$ if $w = w'$ and $x = x'$ and $g^\dagger((w, w'), (x, x')) = 0$ otherwise. If $\pi^\dagger[x, x'] \cdot g^\dagger((w, w'), (x, x')) \ne 0$, then $w = w'$ and $x = x'$. Let $(\mathcal{W} \times \mathcal{X})^+ = \{ (w, x) \in \mathcal{W} \times \mathcal{X} \mid \pi[x]g(w, x) \ne 0 \}$. By $\pi_1^\dagger[x] = \pi_2^\dagger[x] = \pi[x]$, $M^{\min}(\pi^\dagger) = \min_{(w,x) \in (\mathcal{W} \times \mathcal{X})^+} \pi[x]g(w, x)$ and $M^{\max}(\pi^\dagger) = \max_{w \in \mathcal{W}} \sum_{x \in \mathcal{X}} \pi[x]g(w, x)$. Then $H_g(\pi) = -\log M^{\max}(\pi^\dagger)$.

To describe compositionality results, we introduce the following notation.

Definition 10. For any prior π on $\mathcal{X}$ and any gain function g, we define $H_g^{\min}(\pi)$ by: $H_g^{\min}(\pi) = -\log \min \{\pi[x]g(w, x) \colon x \in \mathcal{X}, w \in \mathcal{W}, \pi[x]g(w, x) \ne 0\}$.

Then, for any prior π, $H_g^{\min}(\pi) = -\log M^{\min}(\pi^\dagger)$ and $H_g^{\min}(\pi) \ge H_g(\pi)$.

Since $\pi^\dagger$ is *not* jointly supported, we can instantiate compositionality results in the previous sections only on a lower bound for the posterior g-entropy and upper bounds for g-leakage and g-capacity.

The posterior g-entropy $H_g(\pi, C_1 \parallel C_2)$ of a channel composed in parallel with shared inputs is lower-bounded by the summation of $\log H_g^{\min}(\pi)$ and the posterior g-entropies of its two components:

Theorem 2. *For any prior π on $\mathcal{X}$ and channels $(\mathcal{X}, \mathcal{Y}_1, C_1)$ and $(\mathcal{X}, \mathcal{Y}_2, C_2)$,*

$$H_g(\pi, C_1 \parallel C_2) \geq H_g(\pi, C_1) + H_g(\pi, C_2) - H_g^{\min}(\pi).$$

An upper bound of the g-leakage $I_g(\pi, C_1 \parallel C_2)$ of a channel composed in parallel with shared inputs is described using the g-leakages of its two components:

Theorem 3. *For any prior π on $\mathcal{X}$ and channels $(\mathcal{X}, \mathcal{Y}_1, C_1)$ and $(\mathcal{X}, \mathcal{Y}_2, C_2)$,*

$$I_g(\pi, C_1 \parallel C_2) \leq I_g(\pi, C_1) + I_g(\pi, C_2) + H_g^{\min}(\pi) - H_g(\pi).$$

We emphasize this result holds for any prior. Note that in the right-hand side of the above inequality, $H_g^{\min}(\pi) - H_g(\pi)$ is necessary as the following illustrates.

Example 1. Let us consider the channel $(\mathcal{X}, \mathcal{Y}, C)$ where $\mathcal{X} = \mathcal{Y} = \{0, 1\}$ and C is the 2×2 matrix defined by $C[0,0] = C[1,1] = 0.9$ and $C[0,1] = C[1,0] = 0.1$. Let g be the identity gain function g_{id} and π be the prior on $\mathcal{X}$ such that $\pi[0] = 0.1$ and $\pi[1] = 0.9$. Then $H_g(\pi) = H_g(\pi, C) = -\log 0.9$, $H_g(\pi, C \parallel C) = -\log 0.972$. Therefore $I_g(\pi, C \parallel C) = \log 1.08 > 0 = I_g(\pi, C) + I_g(\pi, C)$.

Note that the inequality of Theorem 3 does not give a useful upper bound when the prior π is far from the uniform distribution. In this example, by $H_g^{\min}(\pi) - H_g(\pi) = \log 9$, the left-hand side is $\log 1.08 \approx 0.111$ while the right-hand side is $\log 9 \approx 3.170$.

These compositionality results are naturally extended to n channels composed in parallel; the extension can be found in the report version of this paper [17]. On the other hand, the result may not hold when the composition of channels is done in a dependent way (i.e., it is not a parallel composition). The following is a counterexample:

Example 2. Let $\mathcal{X} = \mathcal{Y}_1 = \mathcal{Y}_2 = \{0, 1\}$, π be the uniform distribution on $\mathcal{X}$ and g be the identity gain function. We consider the channel that, given an input $x \in \mathcal{X}$, outputs a bit y_1 uniformly drawn from $\mathcal{Y}_1$ and the exclusive OR y_2 of x and y_1. Then the g-leakage of the channel is 1 while both of the g-leakages from $\mathcal{X}$ to $\mathcal{Y}_1$ and from $\mathcal{X}$ to $\mathcal{Y}_2$ are 0 and $H_g^{\min}(\pi) - H_g(\pi) = 0$. Hence the property expressed by Theorem 3 in general does not hold if we replace $\parallel$ with some other kind of composition.

4   Compositionality Results on Min-Entropy Leakage

In this section we present compositionality results for min-entropy leakage, which yield compositionality theorems for min-capacity.

4.1 Leakage of Parallel Composition

In this section we derive bounds for min-entropy, which, we recall, is a particular case of g-leakage obtained when g is the identity gain function.

We start by remarking that, when the gain functions are identity gain functions, $M_\pi^{\min}$ and $M_\pi^{\max}$ reduce to $M_{\infty,\pi}^{\min}$ and $M_{\infty,\pi}^{\max}$ defined as:

$$M_{\infty,\pi}^{\min} = \min_{(x_1,x_2)\in(\mathcal{X}_1\times\mathcal{X}_2)^+} \frac{\pi_1[x_1]\cdot\pi_2[x_2]}{\pi[x_1,x_2]}, \; M_{\infty,\pi}^{\max} = \max_{(x_1,x_2)\in(\mathcal{X}_1\times\mathcal{X}_2)^+} \frac{\pi_1[x_1]\cdot\pi_2[x_2]}{\pi[x_1,x_2]}$$

The next results are consequences of the results of Section 3:

Corollary 3. *For any prior π on $\mathcal{X}_1\times\mathcal{X}_2$ and channels $(\mathcal{X}_1,\mathcal{Y}_1,C_1)$, $(\mathcal{X}_2,\mathcal{Y}_2,C_2)$,*

- $H_\infty(\pi, C_1\times C_2) \geq H_\infty(\pi_1,C_1) + H_\infty(\pi_2,C_2) + \log M_{\infty,\pi}^{\min}$.
- *If π is jointly supported, $H_\infty(\pi,C_1\times C_2)\leq H_\infty(\pi_1,C_1)+H_\infty(\pi_2,C_2)+\log M_{\infty,\pi}^{\max}$.*
- *If $\pi = \pi_1\times\pi_2$, then $H_\infty(\pi_1\times\pi_2, C_1\times C_2) = H_\infty(\pi_1,C_1) + H_\infty(\pi_2,C_2)$.*

Corollary 4. *For a jointly supported prior π on $\mathcal{X}_1\times\mathcal{X}_2$, channels $(\mathcal{X}_1,\mathcal{Y}_1,C_1)$, $(\mathcal{X}_2,\mathcal{Y}_2,C_2)$ and $F = \log\frac{M_{\infty,\pi}^{\max}}{M_{\infty,\pi}^{\min}}$,*

- $I_\infty(\pi_1,C_1)+I_\infty(\pi_2,C_2) - F \leq I_\infty(\pi, C_1\times C_2) \leq I_\infty(\pi_1,C_1)+I_\infty(\pi_2,C_2) + F$
- *If $\pi = \pi_1\times\pi_2$, then $I_\infty(\pi_1\times\pi_2, C_1\times C_2) = I_\infty(\pi_1,C_1) + I_\infty(\pi_2,C_2)$.*

The min-entropy leakage coincides with the min-capacity when the prior π is uniform. Thus we re-obtain the following result from the literature [4]: $C_\infty(C_1\times C_2)=C_\infty(C_1) + C_\infty(C_2)$.

4.2 Leakage of Parallel Composition with Shared Input

As corollaries of Theorems 2 and 3 we obtain the compositionality results for the posterior min-entropy and the min-entropy leakage by taking g as the identity gain function g_{id}. For any prior π on $\mathcal{X}$, let $H^{\min}(\pi) = -\log\min\{\pi[x] \mid x \in \mathcal{X}, \pi[x] \neq 0\}$. Then $H^{\min}(\pi) \geq \log|\mathcal{X}| \geq H_\infty(\pi)$.

Corollary 5. *For any prior π on $\mathcal{X}$ and channels $(\mathcal{X},\mathcal{Y}_1,C_1)$ and $(\mathcal{X},\mathcal{Y}_2,C_2)$,*

- $H_\infty(\pi,C_1)+H_\infty(\pi,C_2)-H^{\min}(\pi)\leq H_\infty(\pi,C_1\|C_2)\leq\min\{H_\infty(\pi,C_1), H_\infty(\pi,C_2)\}$
- $\max\{I_\infty(\pi, C_1), I_\infty(\pi, C_2)\} \leq I_\infty(\pi, C_1 \| C_2) \leq I_\infty(\pi, C_1) + I_\infty(\pi, C_2) + H^{\min}(\pi) - H_\infty(\pi)$.

The min-entropy leakage coincides with the min-capacity when the prior π is uniform. If π is uniform we have $H^{\min}(\pi) = H_\infty(\pi)$. Thus we re-obtain the following result from the literature [16]: $C_\infty(C_1 \| C_2)\leq C_\infty(C_1) + C_\infty(C_2)$.

The following is an example of the above inequality.

Example 3. Consider the channel $(\mathcal{X},\mathcal{Y},C)$ shown in Example 1. Let π be the uniform prior on $\mathcal{X}$. Then $H_\infty(\pi) = 1$, $H_\infty(\pi, C) = H_\infty(\pi, C\|C) \approx 0.152$. Hence $C_\infty(C \| C) = H_\infty(\pi) - H_\infty(\pi, C\|C) \approx 0.848$ while $C_\infty(C) + C_\infty(C) \approx 1.696$.

5 Improving Leakage Bounds by Input Approximation

The compositionality results for g-leakage shown in a previous section may not give good bounds when the prior is far from the uniform distribution, as illustrated in Example 1. In particular, probabilities that are closer to 0 in priors make our leakage bounds much worse. Since such small probabilities do not affect true g-leakage values much, they can be removed from the priors while this may cause little error on g-leakage values. In the following we present a way of improving bad g-leakage bounds by removing small probabilities. We call it *input approximation* technique. We will only consider the case of min-entropy leakage, i.e., when g is the identity gain function.

The idea of removing small entropies is reminiscent of the notion of *smooth entropy* [8], although the motivation and technicalities are different.

5.1 Bounds for Known Channels

We first consider the case in which the channel components are known. Let π be a prior on $\mathcal{X}$. Let $\mathcal{X}'$ be a non-empty proper subset of $\mathcal{X}$ such that $\max_{x' \in \mathcal{X}'} \pi[x'] \leq \min_{x \in \mathcal{X} \setminus \mathcal{X}'} \pi[x]$. Then $\max_{x \in \mathcal{X} \setminus \mathcal{X}'} \pi[x] = \max_{x \in \mathcal{X}} \pi[x]$. Let $\epsilon = \sum_{x' \in \mathcal{X}'} \pi[x']$. We define a function $\pi|_{\mathcal{X} \setminus \mathcal{X}'}$ from $\mathcal{X}$ to $[0, 1]$ by:

$$\pi|_{\mathcal{X} \setminus \mathcal{X}'}[x] = \begin{cases} 0 & \text{if } x \in \mathcal{X}' \\ \pi[x] & \text{otherwise} \end{cases}$$

Then $\pi|_{\mathcal{X} \setminus \mathcal{X}'}$ is not a probability distribution, as it does not sum up to 1; i.e., $\sum_{x \in \mathcal{X}} \pi|_{\mathcal{X} \setminus \mathcal{X}'}[x] < 1$. However, the results in previous sections do not require π to be a probability distribution, and neither do the definitions of entropy and leakage. Errors caused by the above input approximation are bounded as follows:

Theorem 4. *For any prior π on $\mathcal{X}$ and channel $(\mathcal{X}, \mathcal{Y}, C)$,*

$$I_\infty(\pi|_{\mathcal{X} \setminus \mathcal{X}'}, C) \leq I_\infty(\pi, C) \leq I_\infty(\pi|_{\mathcal{X} \setminus \mathcal{X}'}, C) + \log(1 + \tfrac{\epsilon}{V(\pi|_{\mathcal{X} \setminus \mathcal{X}'}, C)}).$$

So, the idea is to remove very small probabilities in priors and then apply our compositional approach to derive bounds illustrated in a previous section. This will allow to obtain better bounds, as small probabilities affect dramatically the precision of our approach, while removing them produces only relatively small errors as shown in Theorem 4.

More precisely, the technique works as follows. Consider a channel C composed of C_1 and C_2 in parallel and a joint prior π on $\mathcal{X}_1 \times \mathcal{X}_2$. We take $\mathcal{X}_1 \times \mathcal{X}_2$ as $\mathcal{X}$ in the input approximation procedure and Theorem 4. Recall that the prior must be jointly supported in order to apply our compositional approach, therefore we take a $\mathcal{X}' \subseteq \mathcal{X}_1 \times \mathcal{X}_2$ so that $\pi|_{\mathcal{X} \setminus \mathcal{X}'}$ is jointly supported. Then we apply Corollary 4 to obtain a lower and an upper bound for $I_\infty(\pi|_{\mathcal{X} \setminus \mathcal{X}'}, C)$. Finally we apply Theorem 4 to obtain bounds for the original $I_\infty(\pi, C)$.

Example 4. Consider the channel $(\mathcal{X}, \mathcal{Y}, C)$ for $\mathcal{X} = \{x_0, x_1, x_2\}$, $\mathcal{Y} = \{y_0, y_1, y_2\}$ and C is given in Fig. 2. We assume the prior π such that $\pi(x_0) = 0.01$, $\pi(x_1) = 0.49$ and $\pi(x_1) = 0.50$, is shared among channels. Then the min-entropy leakage of the channel C^{10} composed of ten C's in parallel is 0.1319, while our upper bound is 0.7444 when $\epsilon = 0.01$. On the other hand, the upper bound obtained using min-capacity [4] is 4.114, which is much larger than ours.

	y_0	y_1	y_2
x_0	0.50	0.23	0.27
x_1	0.20	0.40	0.40
x_2	0.21	0.43	0.36

Fig. 2. Channel matrix

5.2 Bounds for Channels Composed of Unknown Channels

In some situations an analyst may not know the channel matrices C_1, C_2 and therefore cannot calculate $I_\infty(\pi|_{\mathcal{X}\setminus\mathcal{X}'}, C_i)$ or $V(\pi|_{\mathcal{X}\setminus\mathcal{X}'}, C_i)$ (necessary to apply Corollary 4), while he may know the information leakages $I_\infty(\pi_1, C_1)$ and $I_\infty(\pi_2, C_2)$. Our input approximation technique allows us to obtain bounds also in this case, although less precise than in the case of known channels. Hereafter we let $\pi' = \pi|_{\mathcal{X}\setminus\mathcal{X}'}$. From Theorem 4:

Theorem 5. $I_\infty(\pi_1, C_1) + I_\infty(\pi_2, C_2) - \log\dfrac{M^{\max}_{\infty,\pi'}}{M^{\min}_{\infty,\pi'}} - \log\dfrac{V(\pi_1,C_1)}{V(\pi_1,C_1)-\epsilon} - \log\dfrac{V(\pi_2,C_2)}{V(\pi_2,C_2)-\epsilon}$

$\leq I_\infty(\pi, C_1 \times C_2) \leq I_\infty(\pi_1, C_1) + I_\infty(\pi_2, C_2) + \log\dfrac{M^{\max}_{\infty,\pi'}}{M^{\min}_{\infty,\pi'}} + \log\dfrac{\max(V(\pi_1,C_1),V(\pi_2,C_2))}{\max(V(\pi_1,C_1),V(\pi_2,C_2))-\epsilon}.$

Theorem 6. $I_\infty(\pi, C_1 \| C_2) \leq I_\infty(\pi_1, C_1) + I_\infty(\pi_2, C_2) + \log\dfrac{\max(V(\pi_1,C_1),V(\pi_2,C_2))}{\max(V(\pi_1,C_1),V(\pi_2,C_2))-\epsilon}$
$+ H^{\min}(\pi') - H_\infty(\pi').$

When $\epsilon = 0$ these theorems coincide with Corollaries 4 and 5.

Note that $V(\pi_1, C_1)$ and $V(\pi_2, C_2)$ are calculated from $V(\pi_1)$, $V(\pi_2)$, $I_\infty(\pi_1, C_1)$ and $I_\infty(\pi_2, C_2)$. So it is sufficient for an analyst to know only π, $I_\infty(\pi_1, C_1)$ and $I_\infty(\pi_2, C_2)$ to calculate the above leakage bounds.

It is easy to see that these bounds are not as good as those in Section 5.1. Also they are more sensitive to the choice of ϵ. If we take a very small ϵ, the input approximation does not improve substantially, as neither $\dfrac{M^{\max}_{\infty,\pi'}}{M^{\min}_{\infty,\pi'}}$ nor $H^{\min}(\pi') - H_\infty(\pi')$ decreases much. If we take a very large ϵ, then the error caused by the input approximation is also very large, while $\dfrac{M^{\max}_{\infty,\pi'}}{M^{\min}_{\infty,\pi'}}$ and $H^{\min}(\pi') - H_\infty(\pi')$ are close to 0. We will later present experiments on the input approximation and illustrate that we should take ϵ as a value less than $\max\{V(\pi_1, C_1), V(\pi_2, C_2)\}$.

The input approximation techniques illustrated in Sections 5.1 and 5.2 can be extended to n-ary channel parallel composition. We refer to [17] for the details.

6 Experimental Evaluation

In this section we evaluate our bounds in two use-cases: first, on the Crowds protocol for anonymous communication, running on a mobile ad-hoc network (MANET), and second, on randomly generated channels.

6.1   Crowds Protocol on a MANET

Crowds [22] is a protocol for anonymous communication, in which participants achieve anonymity by forwarding messages through other users. A group of n users, called the Crowd, participate in the protocol, and one of them, called the *initiator* decides to send a message to some arbitrary recipient in the network, called the *server*. The protocol works as follows: first the initiator selects randomly (with uniform distribution) a member of the crowd, called the *forwarder*, and forwards the message to him. A forwarder, upon receiving a message, throws a (biased) probabilistic coin: with probability p_f (a parameter of the system) he randomly selects a new forwarder and advances the message to him, and with probability $1 - p_f$ he delivers the message directly to the server. Replies from the server follow the inverse path to arrive to the initiator and future requests use the already established route, to avoid repeating the protocol.

The goal of the protocol is to provide sender anonymity w.r.t. an attacker who does not control the whole network, but controls only some of the nodes and can only see traffic passing through them. Still, if the attacker controls some members of the crowd, strong anonymity is not satisfied. A forwarding request from user i is evidence that i is the initiator of the message. However, some anonymity is still provided since user i can always claim that he was in fact only forwarding a message from user j. If the number of corrupted users is relatively small, it is more likely that i is innocent (i.e. the initiator is user $j \neq i$) than guilty, offering a notion of anonymity called *probable innocence* [22].

In this section we consider an instance of Crowds running on a mobile ad-hoc network, in which users are mobile and can communicate only to neighbouring nodes hence the network topology changes frequently. Due to the network changes, routes become invalid and the initiator needs to rerun the protocol to establish a new route, which causes further information leakage. Our goal is to measure how quickly the leakage increases as a function of the number of re-executions. Concerning the attacker model, we assume that the attacker (i) knows the network topology (this could be achieved using known protocols for MANETs, e.g. [21]), (ii) controls some members of the crowd and (iii) controls the server. For a given network topology, the system is modeled by a channel with inputs init_i, meaning that user i is the initiator. The observable events are $\text{forw}_{j,k}$, meaning that user j forwarded the message to the corrupted node k (possibly the destination server). A matrix element $C[\text{init}_i, \text{forw}_{j,k}]$ gives the probability that $\text{forw}_{j,k}$ happens when i is the initiator. Finally, for channels C_1, C_2 modeling the protocol under different network topologies, the repetition of the protocol is modeled as $C_1 \parallel C_2$.

As anonymity metric, we use g-leakage with the 2-tries gain function g_{W_2}, modeling an attacker who can guess the initiator twice. Formally, W_2 is the set of all subsets of X with $\#X = 2$, and $g_{W_2}(w, x)$ is 1 if $x \in w$ and 0 otherwise.

We evaluate our compositionality results on a Crowds instance with 25 users, of which one is corrupted, and with $p_f = 0.7$. The network topology is generated by randomly adding a connection between any two users with probability 0.4. For a given topology, the matrix is computed by the PRISM model checker [19], using a model similar to one of [24]. Although executions in Crowds can be infinite, a finite state model can be employed, keeping track of only the current forwarder instead of the full route. Then each element of the channel matrix can be computed by PRISM as the probability of reaching the corresponding state.

The g-leakage of a single execution can be directly computed
from the channel; however, for
multiple executions, the channel
quickly becomes too big to be of
practical use (already at 5 repetitions). On the other hand, g-
leakage can be bounded using the
results in Section 4. The obtained
bounds for up to 9 protocol repetitions are shown in Fig. 3. Three
variations are given in which the
topology changes every 2 executions, every 3 executions or always

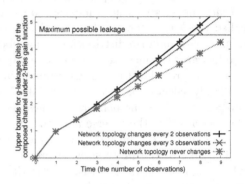

Fig. 3. Numbers of observations and bounds

stays the same. All bounds are computed using a uniform prior and some randomly generated channels. The experiments show that the compositionality technique allows us to obtain meaningful bounds when the system is too big to compute exact values.

Note that the assumption of uniformly chosen forwarders is standard for the Crowds protocol, however it would be interesting to study how our results would change if we considered non-uniform distributions. For instance, we could have a non-uniform distribution if the possible forwarders were equipped with a notion of trust, like in [23]. We leave this for future work.

6.2   Evaluation on Randomly Generated Channels

In this section we evaluate our bounds on min-entropy leakage using randomly generated channels. In particular, we evaluate the improvement on the bounds due to the input approximation technique, and the efficiency of our approach, which we have implemented as a library in leakiEst version 1.3 [1].

We first compare the exact leakage values with their upper bounds calculated using the input approximation technique in the case of shared input. Fig. 4 shows the average upper bounds obtained from Theorem 4, that can be applied when we know the channel matrix. Fig. 5 shows those obtained from Theorem 6 that we can apply when we *do not know* it. For both experiments we use randomly generated 10×10 channel matrices C and a prior π that contains some input with very small probabilities. We set $\epsilon = 0.1$ in the first case and $\epsilon = V(\pi, C)/3$ in the second one. We calculated the min-entropy leakage $I_\infty(\pi, C \parallel C \parallel C \parallel C \parallel C \parallel C)$ (composition of six C's), and its lower and upper bounds, using the n-ary generalizations of Theorems 4 and 6 (see [17] for the precise formulations.)

These cases give similar upper bounds as shown in Figs. 4 and 5. The x-
axis represents *noise levels* of randomly generated matrices, which we define as

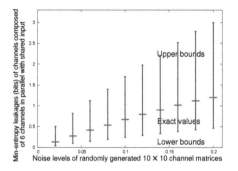

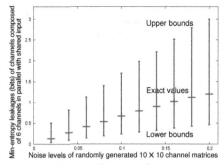

Fig. 4. Min-entropy leakages and their bounds for *known* channels

Fig. 5. Min-entropy leakages and their bounds for *unknown* channels

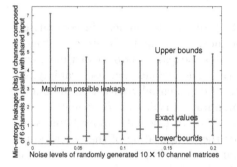

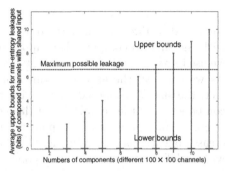

Fig. 6. Min-entropy leakages and their bounds when the analyst does *not* know the channels and chooses ϵ badly

Fig. 7. Upper bounds of min-entropy leakages as a function of the numbers of components

the maximum values (over rows of C) of the summations of the differences of probabilities from the uniform distributions. For instance, when the noise level is 0.10, the average upper bound is 1.699 in the first case (Fig. 4) while it is 1.701 in the second (Fig. 5).

These upper bounds depend on how we choose the parameter ϵ for the input approximation technique. In particular upper bounds strongly depend on ϵ in the case of unknown channels. In Fig. 5 we chose $\epsilon = V(\pi, C)/3$ which gives a relatively good upper bound. On the other hand, if we choose an ϵ too large we may obtain useless bounds. Indeed, if we set for instance $\epsilon = 0.2$, then we obtain upper bounds above the maximum possible leakage, which is the min-entropy, and is always $\log 10 \approx 3.322$ (as shown in Fig. 6) since the input is shared.

Fig. 7 shows average upper bounds of min-entropy leakages of randomly generated 100×100 channels, with randomly generated priors, noise level 0.1, and $\epsilon = 0.005$. As we can see from the figure, the gap between the lower and upper bounds increases with the number of components.

Finally we evaluate the efficiency of our method. We consider here the min-entropy leakage. Fig. 8 shows the execution time on a laptop (1.8 GHz Intel Core i5) for leakiEst to compute the exact min-entropy leakages of the channels

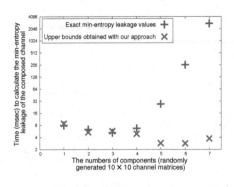

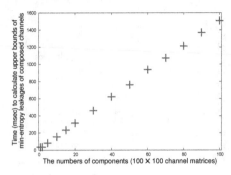

Fig. 8. Average time to calculate min-entropy leakages and their upper bounds

Fig. 9. Average time to calculate upper bounds as a function of the number of components

composed of randomly generated 10×10 component channels, in comparison with the time to compute their upper bounds. To compute the exact leakages, we used leakiEst with an option that calculates the leakages from exact matrices. As we can see, the execution time for the exact values increases rapidly. In fact, the size of composed channel increases exponentially with the number of components, so the complexity of this computation is at least exponential.

For a large number of components, the time to calculate upper bounds increases linearly as shown in Fig. 9. As for the computation of the exact values with leakiEst, we expected an exponential blow-up, but we could not check it since we run out of memory because of the size of the matrices.

7   Conclusion and Future Work

We have investigated compositional methods to derive bounds on g-leakage. To improve the precision of the bounds, we have proposed a technique based on the idea of approximating priors by removing small probabilities up to a parameter ϵ. From our experimental results we have found that the dependency of the precision on ϵ is not straightforward. We leave for future work the problem of determining optimal values for ϵ. We also want to explore a possible relation between our technique and the notion of smooth entropies from the information theory literature [8]. This could allow us to develop a more principled approach to the input approximation technique.

References

1. leakiEst, http://www.cs.bham.ac.uk/research/projects/infotools/leakiest/
2. Alvim, M.S., Chatzikokolakis, K., Palamidessi, C., Smith, G.: Measuring information leakage using generalized gain functions. In: Proc. of CSF, pp. 265–279. IEEE (2012)
3. Andrés, M.E., Palamidessi, C., van Rossum, P., Smith, G.: Computing the leakage of information-hiding systems. In: Esparza, J., Majumdar, R. (eds.) TACAS 2010. LNCS, vol. 6015, pp. 373–389. Springer, Heidelberg (2010)

4. Barthe, G., Köpf, B.: Information-theoretic bounds for differentially private mechanisms. In: Proc. of CSF, pp. 191–204. IEEE (2011)
5. Boreale, M.: Quantifying information leakage in process calculi. In: Bugliesi, M., Preneel, B., Sassone, V., Wegener, I. (eds.) ICALP 2006. LNCS, vol. 4052, pp. 119–131. Springer, Heidelberg (2006)
6. Boreale, M., Pampaloni, F., Paolini, M.: Asymptotic information leakage under one-try attacks. In: Hofmann, M. (ed.) FOSSACS 2011. LNCS, vol. 6604, pp. 396–410. Springer, Heidelberg (2011)
7. Braun, C., Chatzikokolakis, K., Palamidessi, C.: Quantitative notions of leakage for one-try attacks. In: Proc. of MFPS. ENTCS, vol. 249, pp. 75–91. Elsevier (2009)
8. Cachin, C.: Smooth entropy and rényi entropy. In: Fumy, W. (ed.) EUROCRYPT 1997. LNCS, vol. 1233, pp. 193–208. Springer, Heidelberg (1997)
9. Chatzikokolakis, K., Chothia, T., Guha, A.: Statistical Measurement of Information Leakage. In: Esparza, J., Majumdar, R. (eds.) TACAS 2010. LNCS, vol. 6015, pp. 390–404. Springer, Heidelberg (2010)
10. Chatzikokolakis, K., Palamidessi, C., Panangaden, P.: Anonymity protocols as noisy channels. Inf. and Comp. 206(2-4), 378–401 (2008)
11. Chatzikokolakis, K., Palamidessi, C., Panangaden, P.: On the Bayes risk in information-hiding protocols. J. of Comp. Security 16(5), 531–571 (2008)
12. Chothia, T., Kawamoto, Y., Novakovic, C., Parker, D.: Probabilistic point-to-point information leakage. In: Proc. of CSF, pp. 193–205. IEEE (June 2013)
13. Chothia, T., Kawamoto, Y.: Statistical estimation of min-entropy leakage (April 2014), http://www.cs.bham.ac.uk/research/projects/infotools/ (manuscript)
14. Chothia, T., Kawamoto, Y., Novakovic, C.: A tool for estimating information leakage. In: Sharygina, N., Veith, H. (eds.) CAV 2013. LNCS, vol. 8044, pp. 690–695. Springer, Heidelberg (2013)
15. Clark, D., Hunt, S., Malacaria, P.: Quantitative analysis of the leakage of confidential data. In: Proc. of QAPL. ENTCS, vol. 59(3), pp. 238–251. Elsevier (2001)
16. Espinoza, B., Smith, G.: Min-entropy as a resource. Information and Computation (2013)
17. Kawamoto, Y., Chatzikokolakis, K., Palamidessi, C.: Compositionality Results for Quantitative Information Flow. Tech. rep., INRIA (2014), http://hal.inria.fr/hal-00999723
18. Köpf, B., Basin, D.A.: An information-theoretic model for adaptive side-channel attacks. In: Proc. of CCS, pp. 286–296. ACM (2007)
19. Kwiatkowska, M.Z., Norman, G., Parker, D.: PRISM 2.0: A tool for probabilistic model checking. In: Proc. of QEST, pp. 322–323. IEEE (2004)
20. Malacaria, P.: Assessing security threats of looping constructs. In: Proc. of POPL, pp. 225–235. ACM (2007)
21. Nassu, B., Nanya, T., Duarte, E.: Topology discovery in dynamic and decentralized networks with mobile agents and swarm intelligence. In: Proc. of ISDA, pp. 685–690. IEEE (2007)
22. Reiter, M.K., Rubin, A.D.: Crowds: anonymity for Web transactions. ACM Trans. on Information and System Security 1(1), 66–92 (1998)
23. Sassone, V., Hamadou, S., Yang, M.: Trust in anonymity networks. In: Gastin, P., Laroussinie, F. (eds.) CONCUR 2010. LNCS, vol. 6269, pp. 48–70. Springer, Heidelberg (2010)
24. Shmatikov, V.: Probabilistic analysis of anonymity. In: Proc. of CSFW, pp. 119–128. IEEE (2002)
25. Smith, G.: On the foundations of quantitative information flow. In: de Alfaro, L. (ed.) FOSSACS 2009. LNCS, vol. 5504, pp. 288–302. Springer, Heidelberg (2009)

CyberSAGE: A Tool for Automatic Security Assessment of Cyber-Physical Systems

An Hoa Vu[1], Nils Ole Tippenhauer[1], Binbin Chen[1]
David M. Nicol[2], and Zbigniew Kalbarczyk[2]

[1] Advanced Digital Sciences Center, Singapore
[2] University of Illinois at Urbana-Champaign, IL, USA

Abstract. We present *CyberSAGE*, a Cyber Security Argument Graph Evaluation tool for cyber-physical systems. Specifically, CyberSAGE supports the automatic generation of *security argument graphs*, a graphical formalism that integrates diverse inputs—including workflow information for processes executed in the system, physical network topology, and attacker models—to argue about the level of security for the target system. Based on the generated graphs, CyberSAGE can combine numerical information to compute quantitative security assessment results. We illustrate the use of CyberSAGE through a power grid case study.

1 Introduction

Assessing the security of cyber-physical systems (CPS) in a holistic manner is challenging, since the results depend on a wide range of heterogeneous inputs: how the system is used, its network topology, which types of possible attacks one should consider, etc. In our previous work [1], we proposed a CPS security assessment framework that uses workflow—describing how a system provides its intended functionality—as a pillar for organizing different inputs. As shown in Figure 1a, our proposed framework suggests to first use the information about a security goal and the related workflow description to generate a high-level goal graph called *G-graph*, which can then be be used to generate a *GS-graph* by incorporating system information and finally a *GSA-graph* by further adding attacker information. We call the generated structures security argument graphs—they provide a graphical formalism that integrates diverse pieces of security-related inputs to argue about the security of the target system (more details in [2]). The graphs also support the combination of different pieces of numerical evidence (associated with different inputs) to produce quantitative assessment results.

While it is easy to explain the intuition behind the process, the manual construction of a holistic security argument graph for a complex CPS can be costly and error-prone. To better deal with the complexity, we have developed *CyberSAGE*, a Cyber Security Argument Graph Evaluation tool for CPS security assessment. Though still in its prototype stage, CyberSAGE can already automatically generate security argument graphs by putting together different types of inputs according to our methodology. It also supports a combinatorial

G. Norman and W. Sanders (Eds.): QEST 2014, LNCS 8657, pp. 384–387, 2014.

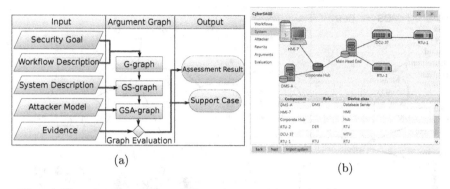

Fig. 1. The assessment framework implemented by CyberSAGE and its snapshot

approach to compute quantitative metrics over the graph. Figure 1b shows a snapshot of CyberSAGE. The rest of this paper will describe its main functionalities and illustrates its use in an example case study. More information about CyberSAGE can be found at our tool website [3].

2   Use of CyberSAGE

CyberSAGE can automatically evaluate a security goal that relates to the availability of specific processes. Those processes model the intended physical, cyber, and human interactions in the target CPS, and are provided to CyberSAGE as XML-based specifications. CyberSAGE converts the XML-based input into internal data structures and uses them to generate a security argument graph based on predefined *extension templates*. These templates are described in more details in [2], together with their definition and a set of CPS-specific templates. CyberSAGE performs the overall evaluation process in the following stages:

1) *Goal and workflow information input stage.* This stage loads the workflow for which the availability will be assessed. Since the workflow is typically modeled using UML activity diagrams, CyberSAGE supports XMI format inputs, as produced by UML modeling tools like Enterprise Architect[1].

2) *System information input stage.* This stage collects information about the deployed system. Currently, CyberSAGE can parse the topology information about a network, where each device plays one or more roles corresponding to the actors in the workflows. Devices are associated with properties such as availability, vulnerabilities, etc, according to their classes. CyberSAGE supports system inputs in an XML dialect used by the CSET tool [4].

3) *Attacker information input stage.* The next stage involves modeling potential threats to the system. Our attacker model contains a list of potential attack actions for different device classes and properties, and the required attacker properties to perform those actions. Currently, CyberSAGE has modeled

---

[1] http://www.sparxsystems.com/products/ea/

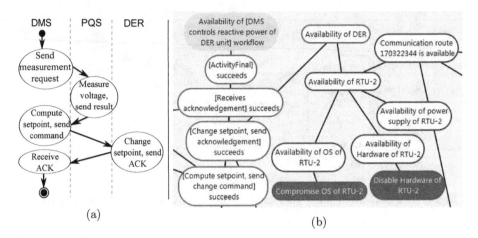

Fig. 2. The workflow input (a) and generated security argument graph in CyberSAGE (b, partially shown) for the example use case

attack actions that are targeted at the availability of software or hardware components in the system, via either remote or local access to the devices.

4) Argument graph generation stage. CyberSAGE then generates the argument graph using a built-in set of CPS-specific extension templates as defined in [2]. Users can inspect the generated argument graph and return to previous stages to change inputs and subsets of extension templates.

5) Evaluation stage. This stage performs a quantitative evaluation of the constructed security argument graph. Currently, CyberSAGE supports the labeling of the vertices by numerical evidence including component availability (when not under attack) and attack success probability, as well as the AND, OR, NEGATION operations for combining evidence. It then invokes the external libDAI [5] with a transformed form of the graph to compute the availability of the concerned process through Bayesian evaluation.

Example Use Case. We have used CyberSAGE in multiple use cases to assess the availability property of various CPS under attack. Due to space limitations, we focus on a concrete distributed energy resources control example (as adapted from [6]).

In the use case, the considered workflow (Figure 2a) captures the interactions among three main actors: a distribution management system (DMS) that manages the power quality and stability of a power grid; distributed energy resources (DER), such as solar power generators, that can adapt power generation based on the request from DMS; and a power quality sensor (PQS) that measures various power quality indicators, e.g., the voltage, and reports them to DMS. On a high level, the DMS controls the power generation output of DER based on the measurements from PQS. These three actors are implemented by distributed physical components, e.g., remote terminal units (RTU), that are not directly connected to each other. The system topology input captures the connectivity between the different physical components. Finally, we consider different types of attacks on different components and assign numerical evidence for the attack probability and component availability.

CyberSAGE applies a set of predefined extension templates [2] to incorporate the above inputs and generate a security argument graph, which consists of 42 vertices, as (partially) shown in Figure 2b. To interpret the graph, its root shows the security goal, and each vertex is expanded to one or several other vertices that it depends on. Based on the graph and numerical information provided at its vertices, CyberSAGE computes the availability of the modelled process. The runtime needed for generating the graph and evaluating the result is about 40ms.

We also tested other use cases with CyberSAGE, where the largest case had a security argument graph of 163 vertices and incurred a runtime of around 200ms. Since a security argument graph is meant to be human-readable (hence likely has no more than a few hundreds of vertices), we do not expect CyberSAGE to have performance issues for its graph generation and combinatorial computation.

3   Conclusion and Acknowledgements

In this paper, we introduced CyberSAGE, a tool that implements our workflow-oriented security assessment framework [1]. CyberSAGE supports automatic generation of security argument graphs and quantitative security assessment of CPS based on the generated graphs. We demonstrate how to use CyberSAGE to conduct an automatic security assessment for an electrical power grid use case.

This work is supported by Singapore's Agency for Science, Technology, and Research (A*STAR) under the Human Sixth Sense Programme (HSSP). We thank QEST reviewers and our shepherd David Parker for useful feedback, and William Temple, Sumeet Jauhar, and William Sanders for helpful discussions.

References

1. Chen, B., Kalbarczyk, Z., Nicol, D.M., Sanders, W.H., Tan, R., Temple, W.G., Tippenhauer, N.O., Vu, A.H., Yau, D.K.: Go with the flow: Toward workflow-oriented security assessment. In: New Security Paradigms Workshop (2013)
2. Tippenhauer, N.O., Temple, W.G., Vu, A.H., Chen, B., Nicol, D.M., Kalbarczyk, Z., Sanders, W.H.: Automatic generation of security argument graphs. Technical Report 1405.7475, CoRR (2014)
3. CyberSAGE: Tool Website, http://cybersagetool.com
4. CSET: The cyber security evaluation tool, http://ics-cert.us-cert.gov/satool.html
5. Mooij, J.M.: libDAI: A free and open source C++ library for discrete approximate inference in graphical models. Journal of Machine Learning Research 11, 2169–2173 (2010)
6. CEN-CENELEC-ETSI Smart Grid Coordination Group: Smart grid reference architecture (November 2012), http://ec.europa.eu/energy/gas_electricity/smartgrids/doc/xpert_group1_reference_architecture.pdf

Symbolic Approximation of the Bounded Reachability Probability in Large Markov Chains

Markus N. Rabe[1], Christoph M. Wintersteiger[2], Hillel Kugler[2],
Boyan Yordanov[2], and Youssef Hamadi[2]

[1] Saarland University, Germany
[2] Microsoft Research

Abstract. We present a novel technique to analyze the bounded reachability probability problem for *large* Markov chains. The essential idea is to incrementally search for sets of paths that lead to the goal region and to choose the sets in a way that allows us to easily determine the probability mass they represent. To effectively analyze the system dynamics using an SMT solver, we employ a finite-precision abstraction on the Markov chain and a custom quantifier elimination strategy. Through experimental evaluation on PRISM benchmark models we demonstrate the feasibility of the approach on models that are out of reach for previous methods.

1 Introduction

Probabilistic model checking is used in a wide array of applications, e.g., in reliability analysis, analysis of randomized algorithms, but also for analysis of system models that arise from the natural sciences like in computational biology [1]. Especially in the sciences, there has always been a large interest in the analysis of probabilistic models, as testified by countless applications of Markov chains, Markov decision process, and their associated analysis procedures. Versatile logics, such as PCTL [2], offer a flexible framework for specifying properties of probabilistic systems. We consider one of the main building blocks for the analysis of PCTL specifications: the bounded reachability probability problem. It asks for the probability that a given event, characterized by a set of states (the *goal region*), occurs within a given number of steps of the model.

The general area of probabilistic model checking has received increased interest recently, which is to a large degree due to advances made both in theory and in practical analysis tools, e.g., in model checkers like PRISM [3] and MRMC [4]. Like all model checkers, these tools face the state-space explosion problem, though the challenge of dealing with probabilities makes the analysis of even moderate size systems very difficult. Various strategies have been developed to manage the size of the state-space, including techniques like abstraction refinement (e.g., [5]), Stochastic SAT (SSAT) and Stochastic SMT (SSMT) [6, 7], generalized Craig interpolation [8], symmetry reduction [9], as well as bisimulation minimization [10, 11]. Other techniques use SAT-based path search to

G. Norman and W. Sanders (Eds.): QEST 2014, LNCS 8657, pp. 388–403, 2014.

```
1    module p1
2        x : [0..99] init 0;
3        [] x<50 -> 0.5:(x'=x+1) + 0.5:(x'=2*x);
4        [] x=50 -> (x'=x);
5    endmodule
```

Fig. 1. A symbolic DTMC in simplified PRISM syntax with one integer variable x with domain $[0, 99]$ and initial value 0. The first action is enabled if $x < 50$ and it offers 2 probabilistic choices: With probability 0.5, x is incremented or it is doubled. The second action is enabled when $x = 50$ and idles.

enumerate paths (possibly with cycles) that lead to the goal region [12, 13] and add up their probabilities; this approach was recently enhanced to enumerating path fragments in a BDD-representation [14]. Even though these approaches scale to models that cannot be solved by explicit state methods (i.e. numerical approaches), the number of states in these models is still fairly small compared to other symbolic techniques in automated (program) verification.

In this paper we present a new approach to approximate the bounded reachability probability in large Markov chains using solvers for satisfiability modulo theories (SMT-solvers). The approach starts with a new problem representation: Instead of focusing on probability distributions over states, we consider the whole probability space of *sequences of random decisions* up to the given step bound. We then iteratively approximate the bounded reachability probability through the SMT-based search for *sets of paths* that lead to the goal region.

Example. Consider the example system of Fig. 1. What is the probability that x is smaller than 20 after 8 steps? For example, the set of paths executing (x'=2*x) in the first 3 steps and anything in the 5 steps thereafter ensures that x is smaller than 20 until step 8 and it has a probability of 2^{-3}. A second set of paths starting with one execution of (x'=x+1), then four unrestricted steps, followed by three executions of (x'=x+1) is disjoint to the first set and also ensures to stay below 20. Hence its probability of 2^{-4} can be counted separately from the first set.

Organization. In Section 2 we characterize the probabilistic transition relation of a Markov chain *given in a symbolic representation* via an integral over a propositional formula, which enables a conceptually simple characterization of the bounded reachability probability. Next, we present the iterative approach to the approximation of the bounded reachability probability by searching for sets of paths that lead to the goal region. We choose the shape of sets in a way that allows us to easily determine their probability mass (Section 3). To effectively solve the resulting formulas, we discuss a finite-precision abstraction (Section 4) to obtain a purely discrete problem that we can effectively solve using SMT-based methods. To enhance the efficiency, we present a specialized quantifier elimination strategy that makes use of the convexity of the sets we search for (Section 5). In Section 6 we report on an experimental evaluation on a set of common benchmark models and discuss the findings. Section 7 discusses related work.

2   Preliminaries

We assume familiarity with the basic concepts of probability spaces and distributions and we start directly with the definition of the system model:

Definition 1 (Markov chain). *A (discrete-time) Markov chain* (S, s_{init}, P) *consists of a finite set of states* S, *an initial state* $s_{init} \in S$, *and a probabilistic transition function* $P : S \to Dist(S)$ *assigning each state* $s \in S$ *a probability distribution over successor states.*

An *execution* of a Markov chain is an infinite sequence of states. Although it is intuitively clear what the behavior of Markov chains is, we need to construct the probability space over executions carefully. We employ the theorem of Ionescu-Tulcea [15, Thm. 2.7.2] to build this probability space out of the infinite sequence of random experiments (random variables). Recursively, we define a random variable X_i over the sequences of states of length i as

$$X_i(s_0 \ldots s_i) = \sum_{s_{i-1} \in S} P(s_{i-1})(s_i) \cdot X_{i-1}(s_0 \ldots s_{i-1}) ,$$

and we define X_0 to assign probability 1 to s_{init}. The fact that each X_i, for $i \geq 0$, is a random variable, is easily verified.

This construction defines a σ-algebra over cylindrical sets of executions, i.e. sets of executions that are defined via a common prefix, and it yields a *unique* Borel-measurable probability space over the infinite executions of the Markov chain. For a given Markov chain M we denote this measure on executions (and on their prefixes) as $\Pr_M$, mapping Borel-measurable sets of finite and infinite executions to probabilities.

2.1   Bounded Reachability Probability

The analysis problem we consider in this paper is to determine the probability to reach a specified set of final states in a given number of steps. This problem is motivated by encodings of practical problems into Markov chains, where steps correspond to steps in time in the original system. To ask for the probability to reach the final states in a given number of steps is, therefore, often equivalent to asking what the probability is that a certain event happens at a certain time. The time bounded reachability probability problem is also a basic building block for model checking logics like PCTL [2].

Formally, for a given Markov chain $M = (S, s_{init}, P)$, a set of final states $F \subseteq S$, and a step number k we define the problem as computing

$$\Pr{}_M(F, k) = \Pr{}_M \left(\{ s_0 s_1 \ldots s_k \ldots \in S^\omega \mid s_0 = s_{init} \ \wedge \ s_k \in F \} \right) .$$

Note that this formulation of the problem asks for the probability of reaching F after *exactly* k steps. The computation of the probability of reaching F in k or fewer steps is a variation of the problem requiring $\exists i. \ s_i \in F$ instead of $s_k \in F$.

2.2 Symbolic Markov Chains

We aim to analyze systems with large state spaces. Hence we begin with a symbolic encoding of the state space and the probabilistic transition function. Figure 1 shows an example system. The state space is described by variables $v_1, v_2, \ldots, v_n$ with specified finite domains. The transition function is defined by a list of actions, each of them describing a probability distribution over successor states. Actions have the following form:

$$(guard) \rightarrow p_1 : (update_1) + \cdots + p_m : (update_m) \;;$$

The $guard : S \rightarrow \mathbb{B}$ is a predicate on states that indicates whether the action is *enabled*. If the system has multiple actions, their guards need to partition the (reachable) state space. Thus, whenever a guard holds in a state, it is executed.

Intuitively, when executing an action, one of its *probabilistic choices*, which are separated by the symbol $+$, is selected at random. The probability distribution over the probabilistic choices is defined by the expressions $p_1, \ldots, p_m : S \rightarrow [0,1] \cap \mathbb{Q}$. Each action a and probabilistic choice p entails a *unique* successor state given by an update function $update_{a,p} : S \times S \rightarrow \mathbb{B}$ with $update_{a,p}(s,s') \wedge update_{a,p}(s,s'') \implies s' = s''$.

The description of Markov chains in terms of actions is inspired by the PRISM input language [3], as it proved to be flexible enough for a wide range of application areas, such as distributed algorithms, communication protocols, security, dependability, and biology. The PRISM input language supports additional features, like the parallel composition of multiple modules, but for the sake of simplicity, we restrict the discussion to the features described above. Our implementation presented in Section 6 does support modules, however.

2.3 The Markov Chain Entailed by a Symbolic Markov Chain

The state space of the Markov chain entailed by a symbolic Markov chain is simply the cross product of the domains of the variables $v_1, v_2, \ldots, v_n$. The initial state s_{init} is fixed by an expression in the symbolic model. To construct the transition relation, we first consider the execution of a particular action a in a state s. The probabilistic choices of a with their probabilities $p_1(s), \ldots, p_m(s)$, respectively, define a partitioning of the interval $[0,1]$ into the sub-intervals

$$I_{a,p_i}(s) = \left[\sum_{j=1}^{i-1} p_j(s), \sum_{j=1}^{i} p_j(s) \right)$$

for $i < m$ and $I_{a,p_m} = \left[\sum_{j<m} p_j, 1 \right]$. To execute a step in the model, we draw a value r from the interval $[0,1]$ uniformly at random and then proceed according to the deterministic transition relation. For a pair of states s and s' and a given random value $r \in [0,1]$, the transition relation is defined as

$$T(s,s',r) = \bigwedge_{1 \leq j \leq n} \left[guard_{a_j}(s) \implies \bigwedge_{1 \leq i \leq m} \left(r \in I_{a_j,p_i}(s) \implies update_{a_j,p_i}(s,s') \right) \right] .$$

That is, we determine which action a is enabled and then apply the update function of the probabilistic choice belonging to the sub-interval the random number r falls in.

The probabilistic transition function of the entailed Markov chain, is thus

$$P(s, s') = \int_0^1 T(s, s', r) \, dr \, , \tag{1}$$

where $T(s, s', r)$ is interpreted as 1 iff it holds true.

Other works in the area (e.g. [7, Definition 5.2]) define the probabilistic transition relation as a sum of the probabilities of all probabilistic choices that result in the specified state. Our definition untangles the possible system behaviors and the measure. This allows us to formulate a conceptually simple approximation algorithm (Section 3). Of course, both approaches to define the entailed Markov chain result in the same system behavior.

3   Incremental Symbolic Approximation

In this section, we present a method to incrementally approximate the bounded reachability probability for a given (symbolic) Markov chain. It is based on a characterization of the bounded reachability probability as an integral over a propositional formula, similar to the formulation for the one-step probabilistic transition given in Subsection 2.3.

We begin by characterizing executions of length k, that is legal combinations of execution prefixes $\bar{s} = s_0 s_1 \ldots s_k$ of sequences of random decisions $\bar{r} = r_1 r_2 \ldots r_k$; we define

$$T^k(\bar{s}, \bar{r}) = \left(s_0 = s_{init} \ \wedge \ \bigwedge_{0 \leq i \leq k-1} T(s_i, s_{i+1}, r_{i+1}) \right) . \tag{2}$$

Note that for all sequences $\bar{r}$ there is exactly one sequence of states that fulfills this condition. We are interested in all sequences of random decisions that lead to a given goal region F, i.e.,

$$T^k(\bar{r}, F) = \exists \bar{s} \in S^{k+1}. \, T^k(\bar{s}, \bar{r}) \ \wedge \ s_k \in F \, .$$

Proposition 1. *Let M be a DTMC entailed by a symbolic Markov chain. For the bounded reachability probability for a given goal region F and step number k it holds that*

$$\Pr{}_M(F, k) = \int_0^1 \ldots \int_0^1 T^k(r_0 \ldots r_k, F) \, dr_0 \ldots \, dr_k \, .$$

3.1   Identifying Cubes in the Probability Space

Proposition 1 leads to a new view on the problem. Instead of considering how the probability distributions over the state space evolve over time, we consider the probability space over all sequences of random decisions. This 'state-less' representation of the problem helps to attack the problem for models beyond the scale at which their state space can be represented explicitly or via (MT)BDDs.

We propose to exploit the additivity of the probability measure at the level of traces, i.e., to search for subsets $R_1, \ldots, R_n \subseteq [0,1]^k$ with $\forall i \forall \bar{r} \in R_i. \ T^k(\bar{r}, F)$ and then to count them separately:

$$\Pr{}_M(F, k) = \sum_i \left(\int_{R_i} 1 \ d\bar{r} \right) \ + \ \int_0^1 \ldots \int_0^1 T^k(\bar{r}, F) \wedge \bigwedge_i \bar{r} \notin R_i \ d\bar{r} \ .$$

It is important to pick sets R_i that are easy to integrate. By choosing them to be disjoint (closed and/or open) rectangles in $[0,1]^k$, we are able to obtain an arbitrarily close approximation and it is easy to determine the volume of each R_i. To see this, note that the space $T^k(\bar{r}, F)$ is the finite disjoint union of the sets $R(\bar{s}) = \{\bar{r} \in [0,1]^k \mid T^k(\bar{s}, \bar{r})\}$ with $s_k \in F$. The sets $R(\bar{s})$ are in general closed and open rectangles, as in each dimension they are defined by an upper bound and a lower bound given by the expressions $p_i(s)$ in the system description.

In practice it is of course desirable to find larger rectangles. Our proposal is essentially a greedy algorithm that searches for the next largest rectangle in a system: Check, for increasing rectangle sizes x, whether a rectangle of that size still exists; which translates to

$$\exists \bar{l}, \bar{u} \in [0,1]^k. \ x \leq \prod_{1 \leq i \leq k} u_i - l_i \ \wedge$$
$$\forall \bar{r} \in [0,1]^k. \left(\bigwedge_{1 \leq i \leq k} l_i \leq r_i \leq u_i \right) \implies T^k(\bar{r}, F) \wedge \bigwedge_i \bar{r} \notin R_i$$
$$\tag{3}$$

Whenever we find a rectangle that satisfies the conditions above, we add it to the set of rectangles R_i and repeat the process. If no rectangle exists, we reduce the size of the rectangle to search for. It is clear that we can stop the process at any time and obtain an under-approximation, i.e., $\sum_i \left(\int_{R_i} 1 \ d\bar{r} \right) \leq \Pr{}_M(F, k)$.

Note that this method has an advantage over enumerating paths through the system, if there are multiple probabilistic choices that do not change the fact that the executions reach the goal region with the same probabilistic choices in other steps—the sequence of states visited may be different though.

4 A Finite-Precision Abstraction

Our problem formulation of Section 3 is not very amenable to efficient solving with automatic methods; to achieve this goal, we employ a layer of automatic abstraction refinement, where each abstraction is obtained by bounding the precision of the analysis. In practice, we encode each of the sub-problems in the SMT theory of uninterpreted functions and bit-vectors (SMT UFBV) as this theory offers an efficient quantifier elimination/instantiation strategy [16].

To encode the problem in this purely discrete theory, we discretize the random variables according to a precision parameter h. We propose a symbolic discretization technique on the level of the formula $T^k(\bar{s}, \bar{r})$ that maintains the conciseness of the representation. That is, we do not need to consider every state or transition of the entailed DTMC, but the technique works directly on the symbolic description. This discretization preserves the probability measure up to an arbitrarily small error.

First, we replace the real valued variables $r \in [0, 1]$ by discrete variables $r \in \{0, \ldots, 2^h - 1\}$, where each of the discretization levels now corresponds to a small portion ($\frac{1}{2^h}$) of the probability mass. Second, for every action of the symbolic Markov chain with m probabilistic choices, we discretize the intervals $I_{a,i}$ introduced in Section 2.3 according to a precision parameter h:

$$\lceil I_{a_j, p_i} \rceil_h (s) = \left[\left\lfloor 2^h \cdot \sum_{j=1}^{i-1} p_j(s) \right\rfloor, \left\lceil 2^h \cdot \sum_{j=1}^{i} p_j(s) \right\rceil \right) ,$$

for $i < m$ and $\lceil I_{a_j, p_m} \rceil_h = \left[\left\lfloor 2^h \cdot \sum_{j=1}^{m} p_j(s) \right\rfloor, 2^h - 1 \right]$.

This simplifies the encoding of the one-step transition relation $T(s, s', r)$ to

$$\bigwedge_{1 \leq j \leq n} guard_{a_j}(s) \implies \bigwedge_{1 \leq i \leq m} \left(r \in \lceil I_{a_j, p_i} \rceil_h (s) \implies update_{a_j, p_i}(s, s') \right) ,$$

which is a formula over a purely discrete space. The probability of a particular probabilistic choice now approximately corresponds to the number of values for r for which the transition relation holds true and shows this choice.

Due to the overlapping intervals $\lceil I_{a_j, p_i} \rceil_h$, some of the discretization levels are assigned to multiple intervals, but otherwise this transformation maintains a clear correspondence of the values of r. Thus, the approximated transition relation now represents an *over-approximation* of the original transition relation, or, in other words, the formula $T_h^k(\bar{s}, \bar{r}) = \left(s_0 = s_{init} \wedge \bigwedge_{0 \leq i \leq k-1} T(s_i, s_{i+1}, r_{i+1}) \right)$ is a relaxation of $T^k(\bar{s}, \bar{r})$ and we define $T_h^k(\bar{r}, F)$ to be $\{\bar{r} \in [0, 1]^k \mid \exists \bar{s} \in S^{k+1}. T_h^k(\bar{s}, \lfloor \bar{r} \cdot 2^h \rfloor) \wedge s_k \in F\} \supseteq T^k(\bar{r}, F)$.

Replacing $T^k(\bar{r}, F)$ by $T_h^k(\bar{r}, F)$ in Equation 3 does not result in the desired approximation, as for the incremental symbolic search for *under-approximations of the probability*, we are interested in an under-approximation of the transition relation. We use the duality of the search for F and its complement $\bar{F}$ (that is $Pr_M(F, k) = 1 - Pr_M(\bar{F}, k)$) to derive an under-approximation: we replace $T^k(\bar{r}, F)$ by $\neg T_h^k(\bar{r}, \bar{F})$. The reason for not starting with an under-approximation right away is to avoid the additional quantifier alternation that lures in the set $T^k(\bar{r}, F)$. In this way, the discretized version of Equation 3 has only one quantifier alternation from an existential quantifier to a universal quantifier.

4.1   Precision

The total probability mass affected by this approximation within one step of the transition relation, is the probability of the union of all ambiguous discretization levels. For a given action with m probabilistic choices, there can be at most

$m - 1$ ambiguous discretization levels, hence the quality of the approximation for executing action a is $\frac{m-1}{2^h}$. The affected probability mass in one step of the system is easily obtained by considering the action with the maximal number of probabilistic choices.

When considering k steps, an obvious upper bound of the probability mass affected by the approximation is $\sum_{0 < i \leq k} \frac{m-1}{2^h} \cdot (1 - \frac{m-1}{2^h})^{i-1}$, which is smaller than $k \cdot \frac{m-1}{2^h}$. As we are free to choose the parameter h, it is feasible to keep the amount of affected probability mass arbitrarily small, because we have $\lim_{h \to \infty} \Pr_{M_h}(F, k) = \Pr_M(F, k)$.

Proposition 2. *For a given symbolic Markov chain M, its discretized variant M_h, a step number k, and a goal region F, we have*

$$\Pr_{M_h}(F, k) \leq \Pr_M(F, k) \text{ and}$$
$$\Pr_{M_h}(F, k) + k \cdot (\tfrac{m-1}{2^h}) \geq \Pr_{M_h}(F \cup \{s^*\}, k) .$$

As a consequence, the finite-precision abstraction of an under-approximation of the bounded reachability probability as discussed in Section 3 is still an under-approximation of the bounded reachability probability.

5   Implementation and Optimizations

We implemented our technique in a prototype model checker named *pZ3* to evaluate its practical efficacy. We use the Z3 theorem prover (specifically its theory for SMT UFBV [16]) as a back-end to solve the generated SMT instances. The input to the tool is a PRISM model file, a predicate F on the state space that represents the goal region, a step number $k \in \mathbb{N}$, and a target probability $p^* \in [0, 1]$. The tool then determines whether the probability to reach a state satisfying F is larger than or equal to p^*.

5.1   The Basic Encoding

By the *basic encoding*, we refer to the direct encoding of Eq. 3 in the theory of bit-vectors with quantifiers, using the finite-precision abstraction presented in Section 4. Thereby we completely rely on the SMT solver's ability to handle the quantifiers. We chose the theory of bit-vectors over a pure SAT encoding to make use of the word-level reasoning and optimizations of the SMT solver.

We omit the details of the translation of Eq. 3 into bit-vectors as it is straight-forward. However, it is interesting to note that the size of the generated SMT instance is (1) *logarithmic* in the domains of the variables, and (2) *linear* in the number of variables, the number of actions, the precision parameter h, number of updates, and the number of steps.

Note that the PRISM language supports modules that can perform actions jointly via a synchronization mechanism. We encode such synchronized actions of multiple modules compactly to avoid enumerating the exponential number of synchronizations actions.

This basic encoding, while correct, challenges the current state-of-the-art in SMT-solving as it produces large and complex quantified formulas that cannot be quickly solved. In the following, we discuss optimizations, first and foremost a custom quantifier elimination strategy, that enables checking Eq. 3 effectively for large Markov chains.

5.2   Custom Quantifier Elimination

To improve the performance of the SMT solver, we implemented a customized quantifier elimination procedure that relies on the notions of *example paths* and *close counter-examples*. The idea is to not search for a sufficiently large rectangle directly, but instead we pick a local environment by fixing an *example path* that leads to the goal region and only search for rectangles that contain this path, such that we find the largest rectangle containing at least this path. Abstractly, we pick candidate rectangles and check whether they are valid rectangles by searching for a path inside the rectangle that does *not* lead to the goal region, i.e., a counter-example. If we find such a counter-example, we remember it and generate a new candidate rectangle. This procedure is similar to what a strategy like model-based quantifier instantiation [16, 17] does with a problem like Eq. 3.

However, we may have to perform many queries to find a rectangle that does not contain a counter-example (i.e., a rectangle only containing paths that lead to the goal region). So, when searching for counter-examples, it is highly beneficial to rule out as many candidate rectangles as possible. Here, we exploit the fact that the rectangles we are looking for are convex. Therefore, counter-examples that are 'closer' to the example path rule out more candidate rectangles than those that are strictly further away. As a measure of distance between paths, we use the Hamming distance of the bit-strings that represent the random choices, as follows: Instead of encoding the random choices of each step by a bit-vector of length h, we consider these as h independent Boolean variables, such that we are able to compute the Hamming distance between two different instantiations of those variables. This entails a change of view from a k-dimensional space where each dimension has 2^h values, to a $k \cdot h$-dimensional space with Boolean values and it slightly changes the notion of rectangles: A rectangle in the bit-vector representation is not necessarily a rectangle in the Boolean representation, and vice versa. (Rectangles in the Boolean space are also called *cubes*.) Currently, we only support this restricted notion of shapes, but in general any type of convex polygonal shape can be used.

Using these definitions, we search for those counter-examples that are *closest* to the example path. We call these counter-examples *close counter-examples*. Finally, we provide a sketch of the process in Algorithm 1, which uses the following functions:

findPath: Yields a path that starts with the initial state, follows the k-step transition relation and ends up in a state satisfying F. Paths that were already counted in previous runs of the outer loop are excluded by $\neg rectangles$. This is essentially an SMT-based bounded model checking query. We utilize both

Data: Initial states I, k-step transition relation T^k, goal states F, target
      probability p^*
Result: *true* if the probability to reach F from I via T^k is at least p^*, *false*
      otherwise.
rectangles := $\emptyset$;
$p := 0$;
while $p < p^*$ **do**
    path := findPath $(I, T^k, F, \neg rectangles)$;
    if *path* $\neq \emptyset$ **then**
        closeCEs := $\emptyset$;
        rectangleFound := *false*;
        while *not rectangleFound* **do**
            rectangle := findCandidateRectangle(path, closeCEs);
            closeCE := findClosestCE(rectangle, path, I, T^k, $\neg F$);
            if *closeCE* $\neq \emptyset$ **then**
                closeCEs := closeCEs $\cup$ { closeCE };
            else
                $p := p +$ computeVolume(rectangle, rectangles);
                rectangles := rectangles $\cup$ { rectangle } ;
                rectangleFound := *true*;
            end
        end
    else
        return *false*;
    end
end
return *true*;

Algorithm 1. Quantifier elimination based on close counter-examples

under- and over-approximations as described in Section 4. Searching for paths
in the over-approximation results in example paths that share their sequence of
random decisions with a second path that does not lead to the goal region. As
an optimization, we also search for example paths in the under-approximation
of the transition relation that does not allow for overlapping intervals.

findCandidateRectangle: Finds a rectangle that contains the example path and
avoids all counter-examples. Note that this check is completely agnostic to the
transition relation.

findClosestCE: This function iteratively searches for counter-examples of increasing Hamming distance, starting with distance 0. Finding close counter-
examples seems to be a hard task for SMT solvers—in many models this is
harder than finding a path that is not related by distance to the original path.
Typically our tool spends over half of its run time in this routine.

Note that, while we require example paths to not be covered by any of the
identified rectangles, we do not require the rectangles to be intersection-free. This

is an optimization that tries to avoid the fragmentation of the remaining parts of the probability space. The function *compute Volume(...)* finally computes the volume that the new rectangle adds to the union of all rectangles.

6   Experimental Evaluation

We conducted a set of experiments to evaluate our technique and to determine its effectiveness in verifying the bounded reachability probability problem and its performance in relation to existing approaches.

Models. To obtain a benchmark set that is not biased toward our tool, we chose to consider all Markov chain models in the benchmark set that is delivered with the PRISM model checker. Out of those, we picked all models that come with a bounded reachability specification. This set comprises the *Bounded retransmission protocol* (BRP) [18], the Crowds protocol (CROWDS) [19], a contract signing protocol (EGL) [20], the self-stabilization protocol (HERMAN) [21], a model of von Neumann's NAND multiplexing [22], and a synchronous leader election protocol (LEADER) [23].[1] For each of these models, we considered multiple parameter settings, that, for example, control the number of participants in the protocol or the minimal probability that must be proven. Most of the instances considered satisfy the specified probability bounds. The full list of experiments can be found in the accompanying technical report [24].

Experimental Setup. All experiments were performed on machines with two Intel Xeon L5420 quad core processors (2.5GHz, 16GB RAM). All tools were limited to 2GB of memory and the time limit was set to 2 hours (7200s).

Comparison to the PRISM model checker. For the comparison, we configured PRISM to use its symbolic MTBDD engine and we extended the available memory of PRISM's BDD library to 2GB. Figure 2 summarizes the comparison of pZ3 to PRISM. From this plot it is evident that pZ3 solves many of the large problem instances, for which PRISM runs out of time or memory. On Markov chains that are small or of moderate size, however, PRISM has a clear advantage. Especially for the models of the leader election protocol, the bounded retransmission protocol and the self-stabilization protocol, we observe that scaling the model parameters has little effect on the run time of pZ3, whereas PRISM exceeds the time or memory limits. For the models LEADER and HERMAN, all reachable states can be reached within the considered step bound ($> 10^{12}$ states in case of the HERMAN model), suggesting that the advantage of pZ3 is due to its use of a symbolic reasoning engine rather than a variant of state space enumeration.

---

[1] These models and a detailed description for each of them can be found at http://www.prismmodelchecker.org/casestudies/

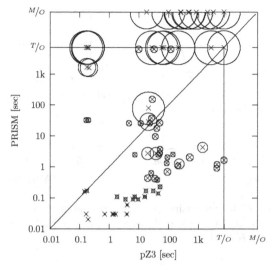

Fig. 2. The performance of PRISM vs the performance of pZ3 on all models and parameter settings. The size of the circles around datapoints indicate the logarithm of the size of the state space. For the full list of experiments, see the accompanying technical report [24].

Scalability in the target probability. Figure 3a displays the scalability in the target probability. As expected for an incremental method, the run time increases when we specify a target probability close to the actual bound. The approximation quality that can be achieved varies greatly for the different models: While for the model of the leader election protocol 99% of the probability mass is found in acceptable time, in other models only a fraction of the actual probability mass is found by pZ3.

Scalability in the precision parameter. In theory, the precision parameter h plays an important role in the quality of the results as presented in Section 4. The probabilities computed by our approach are always sound under-approximations of the bounded reachability probability; regardless of the precision parameter. For the models in the PRISM benchmark suite, small values of the precision parameters, between $h = 1$ and $h = 8$, are often enough to verify many models. Nevertheless, for some models the precision might be an issue. Figure 3b shows the sensitivity or our approach with respect to h on the leader election protocol with 4 participants and 11 probabilistic alternatives each, where it has only a moderate effect on the run time. The choice of 11 probabilistic alternatives ensures that increasing the precision parameter actually increases the maximal probability that can be detected by our approach.

7   Related Work

The first formulation of the bounded reachability problem for MDPs goes back to Bellman [25] and Shapley [26]. These are the foundation of the numerical

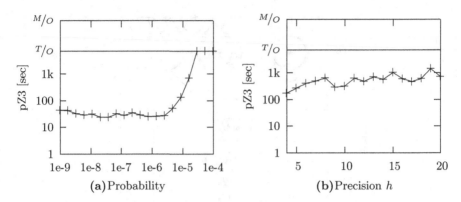

Fig. 3. Run time of pZ3 for increasing (a) target probability on the BRP model with parameters $N = 16$ and $MAX = 2$ and (b) precision on the leader election model with parameters $N = 4$ and $K = 11$

methods included in MRMC [4], IscasMC[27], and Murphi [28] and enable precise model checking, but consider states and transition probabilities explicitly and so do not scale to problem sizes where the state space is not efficiently enumerable. Simulation based techniques and statistical model checking are well suited to explore the likelihood of relatively likely events in large executable systems. However, when the events are very unlikely, simulation based techniques struggle to produce results with small margins of error.

The first symbolic approach to analysis of MDPs is based on MTBDDs [29]. Experiments with the (MTBDD-based) tool PRISM show that the approach is limited to fairly small state spaces, compared to other symbolic techniques in general automated (program) verification.

Abstract interpretation [30] and general static analysis are widely employed techniques for approximate analysis of systems, but existing frameworks based thereupon are often limited to software-specific behavior (like numerical analysis) and their precisions strongly depends on the choice of abstract domains. Esparza and Gaiser [31], basing their work upon that of Hermanns et al. [5] and Kattenbelt et al. [32], as well as Monniaux [33] give a first taste of how abstract interpretation can be employed effectively in the probabilistic setting.

Fränzle et al. [34] proposed to encode the bounded reachability problem of Markov chains and Markov decision processes into Stochastic SAT (SSAT) Stochastic SMT (SSMT), hoping to replicate the tremendous progress SAT and SMT solvers brought to other fields. The proposed algorithms for SSAT and SSMT branch over the probabilistic decisions and recursively add up the reachability probability of the individual branches. This requires the exploration of an exponential number of probabilistic branches and hence the number of steps we can explore with this approach is limited, even for small models (cf. [7, Section 6.7]). SSAT and SSMT-based analysis of Markov chains is similar to our approach, but our method does not try do develop a specialized algorithm to solve SSAT instances; it instead builds upon general purpose SMT solvers. We make use of quantified theories to search for *sets of paths* that lead to the goal region.

Counterexample generation is a closely related branch of research [35, 36]. It is concerned with generating proof objects or counter-examples to the bounded reachability problem (or alternatively reward problems) that are—in principle—human readable. Also symbolic approaches have been explored for generating counter-examples for Markov chains [12–14]. Similar to the approach discussed here, these works present methods to iteratively find evidence for the probability of a given event in probabilistic systems. These methods mostly enumerate single paths (possibly with cycles [12] or fragments of paths [14]) such that they require a large number of calls to a SAT or SMT solver. In contrast, our method detects large sets of paths with few calls to a solver and builds on a fundamentally different representation: Instead of considering paths as sequences of states of the Markov chain, we consider paths to be sequences of random decisions.

Since the presented technique is not the first symbolic approach able to solve the bounded reachability probability problem in Markov chains, a comparison of all approaches would be in order. However, to the best of our knowledge there is no tool besides PRISM that (1) symbolically analyzes Markov chains (2) is publicly available and (3) is able to parse PRISM files. A first impression of the relative performance of recent counter-example generation techniques to the technique presented here can be obtained through the data presented in this work and in [14] (see also the accompanying technical report [24]). Both works consider the same models, though the model parameters in this work are often considerably higher. For example the leader election protocol seems to be not amendable to enumerating paths, whereas the search for sets of paths performs well. For the crowds protocol, however, enumerating paths (potentially with cycles and path fragments) by many SAT or SMT calls seems considerably faster than searching for sets of paths. This raises the question on combinations or generalizations of the methods to combine the best of both worlds.

8 Conclusion

We present a novel approach to iteratively approximate the bounded reachability probability in large Markov chains, which is based on a novel problem encoding in quantified SMT theories. We employ a finite-precision abstraction to obtain a discrete problem encoding. A specialized quantifier elimination strategy is given to effectively dispatch the encoded formulas. We demonstrate the feasibility of the approach on the set of benchmark models of the PRISM model checker. Especially for large models our tool is able to outperform PRISM, suggesting that our method is a suitable, complementary approach in cases where existing methods do not scale.

Acknowledgments. This work was partly supported by the German Research Foundation (DFG) as part of the Transregional Collaborative Research Center "Automatic Verification and Analysis of Complex Systems" (SFB/TR 14 AVACS, www.avacs.org).

References

1. Kwiatkowska, M.Z., Norman, G., Parker, D.: Using probabilistic model checking in systems biology. SIGMETRICS Performance Evaluation Review 35(4) (2008)
2. Hansson, H., Jonsson, B.: A logic for reasoning about time and reliability. Formal Aspects of Computing 6(5), 512–535 (1994)
3. Kwiatkowska, M., Norman, G., Parker, D.: PRISM 4.0: Verification of probabilistic real-time systems. In: Gopalakrishnan, G., Qadeer, S. (eds.) CAV 2011. LNCS, vol. 6806, pp. 585–591. Springer, Heidelberg (2011)
4. Katoen, J.P., Zapreev, I.S., Hahn, E.M., Hermanns, H., Jansen, D.N.: The ins and outs of the probabilistic model checker MRMC. In: Proc. of QEST, pp. 167–176. IEEE Computer Society (2009), www.mrmc-tool.org
5. Hermanns, H., Wachter, B., Zhang, L.: Probabilistic CEGAR. In: Gupta, A., Malik, S. (eds.) CAV 2008. LNCS, vol. 5123, pp. 162–175. Springer, Heidelberg (2008)
6. Fränzle, M., Teige, T., Eggers, A.: Engineering constraint solvers for automatic analysis of probabilistic hybrid automata. Journal of Logic and Algebraic Programming 79(7), 436–466 (2010)
7. Teige, T.: Stochastic Satisfiability Modulo Theories: A Symbolic Technique for the Analysis of Probabilistic Hybrid Systems. Doctoral dissertation, Carl von Ossietzky Universität Oldenburg, Department of Computing Science, Germany (2012)
8. Teige, T., Fränzle, M.: Generalized Craig interpolation for stochastic Boolean satisfiability problems. In: Abdulla, P.A., Leino, K.R.M. (eds.) TACAS 2011. LNCS, vol. 6605, pp. 158–172. Springer, Heidelberg (2011)
9. Kwiatkowska, M., Norman, G., Parker, D.: Symmetry reduction for probabilistic model checking. In: Ball, T., Jones, R.B. (eds.) CAV 2006. LNCS, vol. 4144, pp. 234–248. Springer, Heidelberg (2006)
10. Katoen, J.-P., Kemna, T., Zapreev, I., Jansen, D.N.: Bisimulation minimisation mostly speeds up probabilistic model checking. In: Grumberg, O., Huth, M. (eds.) TACAS 2007. LNCS, vol. 4424, pp. 87–101. Springer, Heidelberg (2007)
11. Dehnert, C., Katoen, J.-P., Parker, D.: SMT-based bisimulation minimisation of Markov models. In: Giacobazzi, R., Berdine, J., Mastroeni, I. (eds.) VMCAI 2013. LNCS, vol. 7737, pp. 28–47. Springer, Heidelberg (2013)
12. Wimmer, R., Braitling, B., Becker, B.: Counterexample generation for discrete-time markov chains using bounded model checking. In: Jones, N.D., Müller-Olm, M. (eds.) VMCAI 2009. LNCS, vol. 5403, pp. 366–380. Springer, Heidelberg (2009)
13. Braitling, B., Wimmer, R., Becker, B., Jansen, N., Ábrahám, E.: Counterexample generation for Markov chains using SMT-based bounded model checking. In: Bruni, R., Dingel, J. (eds.) FMOODS/FORTE 2011. LNCS, vol. 6722, pp. 75–89. Springer, Heidelberg (2011)
14. Jansen, N., Ábrahám, E., Zajzon, B., Wimmer, R., Schuster, J., Katoen, J.-P., Becker, B.: Symbolic counterexample generation for discrete-time Markov chains. In: Păsăreanu, C.S., Salaün, G. (eds.) FACS 2012. LNCS, vol. 7684, pp. 134–151. Springer, Heidelberg (2013)
15. Ash, R., Doléans-Dade, C.: Probability and Measure Theory. Harcourt/Academic Press (2000)
16. Wintersteiger, C.M., Hamadi, Y., de Moura, L.M.: Efficiently solving quantified bit-vector formulas. Proc. of FMSD 42(1) (2013)
17. Ge, Y., de Moura, L.: Complete instantiation for quantified formulas in satisfiability modulo theories. In: Bouajjani, A., Maler, O. (eds.) CAV 2009. LNCS, vol. 5643, pp. 306–320. Springer, Heidelberg (2009)

18. Helmink, L., Sellink, M., Vaandrager, F.: Proof-checking a data link protocol. In: Barendregt, H., Nipkow, T. (eds.) TYPES 1993. LNCS, vol. 806, pp. 127–165. Springer, Heidelberg (1994)

19. Reiter, M., Rubin, A.: Crowds: Anonymity for web transactions. ACM Transactions on Information and System Security (TISSEC) 1(1), 66–92 (1998)

20. Even, S., Goldreich, O., Lempel, A.: A randomized protocol for signing contracts. Communications of the ACM 28(6), 637–647 (1985)

21. Herman, T.: Probabilistic self-stabilization. Information Processing Letters 35(2), 63–67 (1990)

22. von Neumann, J.: Probabilistic logics and synthesis of reliable organisms from unreliable components. In: Shannon, C., McCarthy, J. (eds.) Proc. of Automata Studies, pp. 43–98. Princeton University Press (1956)

23. Itai, A., Rodeh, M.: Symmetry breaking in distributed networks. Information and Computation 88(1) (1990)

24. Rabe, M.N., Wintersteiger, C.M., Kugler, H., Yordanov, B., Hamadi, Y.: Symbolic approximation of the bounded reachability probability in large Markov chains. Technical Report MSR-TR-2014-74, Microsoft Research (2014)

25. Bellman, R.: Dynamic programming. Princeton University Press (1957)

26. Shapley, L.S.: Stochastic games. Proceedings of the National Academy of Sciences 39(10) (1953)

27. Hahn, E.M., Li, Y., Schewe, S., Turrini, A., Zhang, L.: iscasMc: A web-based probabilistic model checker. In: Jones, C., Pihlajasaari, P., Sun, J. (eds.) FM 2014. LNCS, vol. 8442, pp. 312–317. Springer, Heidelberg (2014)

28. Della Penna, G., Intrigila, B., Melatti, I., Tronci, E., Zilli, M.V.: Bounded probabilistic model checking with the Murφ verifier. In: Hu, A.J., Martin, A.K. (eds.) FMCAD 2004. LNCS, vol. 3312, pp. 214–229. Springer, Heidelberg (2004)

29. de Alfaro, L., Kwiatkowska, M., Norman, G., Parker, D., Segala, R.: Symbolic model checking of probabilistic processes using MTBDDs and the Kronecker representation. In: Graf, S. (ed.) TACAS 2000. LNCS, vol. 1785, pp. 395–410. Springer, Heidelberg (2000)

30. Cousot, P., Cousot, R.: Abstract interpretation: A unified lattice model for static analysis of programs by construction or approximation of fixpoints. In: Proc. of POPL. ACM (1977)

31. Esparza, J., Gaiser, A.: Probabilistic abstractions with arbitrary domains. In: Yahav, E. (ed.) SAS 2011. LNCS, vol. 6887, pp. 334–350. Springer, Heidelberg (2011)

32. Kattenbelt, M., Kwiatkowska, M., Norman, G., Parker, D.: Abstraction refinement for probabilistic software. In: Jones, N.D., Müller-Olm, M. (eds.) VMCAI 2009. LNCS, vol. 5403, pp. 182–197. Springer, Heidelberg (2009)

33. Monniaux, D.: Abstract interpretation of probabilistic semantics. In: Palsberg, J. (ed.) SAS 2000. LNCS, vol. 1824, pp. 322–340. Springer, Heidelberg (2000)

34. Fränzle, M., Hermanns, H., Teige, T.: Stochastic satisfiability modulo theory: A novel technique for the analysis of probabilistic hybrid systems. In: Egerstedt, M., Mishra, B. (eds.) HSCC 2008. LNCS, vol. 4981, pp. 172–186. Springer, Heidelberg (2008)

35. Andrés, M.E., D'Argenio, P., van Rossum, P.: Significant diagnostic counterexamples in probabilistic model checking. In: Chockler, H., Hu, A.J. (eds.) HVC 2008. LNCS, vol. 5394, pp. 129–148. Springer, Heidelberg (2009)

36. Han, T., Katoen, J.P., Berteun, D.: Counterexample generation in probabilistic model checking. IEEE Transactions on Software Engineering 35(2), 241–257 (2009)

Accelerating
Parametric Probabilistic Verification[*]

Nils Jansen[1], Florian Corzilius[1], Matthias Volk[1], Ralf Wimmer[2],
Erika Ábrahám[1], Joost-Pieter Katoen[1], and Bernd Becker[2]

[1] RWTH Aachen University, Germany
{nils.jansen,corzilius,volk,abraham,katoen}@cs.rwth-aachen.de
[2] Albert-Ludwigs-University Freiburg, Germany
{wimmer,becker}@informatik.uni-freiburg.de

Abstract. We present a novel method for computing reachability probabilities of parametric discrete-time Markov chains whose transition probabilities are fractions of polynomials over a set of parameters. Our algorithm is based on two key ingredients: a graph decomposition into strongly connected subgraphs combined with a novel factorization strategy for polynomials. Experimental evaluations show that these approaches can lead to a speed-up of up to several orders of magnitude in comparison to existing approaches.

1   Introduction

Discrete-time Markov chains (*DTMCs*) are a widely used modeling formalism for systems exhibiting probabilistic behavior. Their applicability ranges from distributed computing to security and systems biology. Efficient algorithms exist to compute measures like: "What is the probability that our communication protocol terminates successfully if messages are lost with probability 0.05?". However, often actual system parameters like costs, faultiness, reliability and so on are not given explicitly. For the design of systems incorporating random behavior, this might even not be possible at an early design stage. In model-based performance analysis, the research field of *fitting* [1], where—intuitively—probability distributions are generated from experimental measurements, mirrors the difficulties in obtaining such concrete values.

This calls for treating probabilities as parameters and motivates to consider *parametric* DTMCs (PDTMCs), where transition probabilities are (rational) functions in terms of the system's parameters. Using these functions, one can, e. g., find appropriate values of the parameters such that certain properties are satisfied or analyze the sensitivity of reachability probabilities to small changes in the parameters. Computing reachability probabilities for DTMCs is typically

---

[*] This work was supported by the DFG Research Training Group 1298 (AlgoSyn), the DFG Transregional Collaborative Research Center AVACS (SFB/TR 14), the EU FP7-project MoVeS, the FP7-IRSES project MEALS and by the Excellence Initiative of the German federal and state government.

G. Norman and W. Sanders (Eds.): QEST 2014, LNCS 8657, pp. 404–420, 2014.

done by solving linear equation systems. This is not feasible for PDTMCs, since the resulting equation system is non-linear. Instead, approaches based on *state elimination* have been proposed [2,3]. The idea is to replace states and their incident transitions by direct transitions from the predecessors to the successors. Eliminating states iteratively leads to a model having only initial and absorbing states, where transitions between these states carry—as rational functions over the model parameters—the probability of reaching the absorbing states from the initial states. The efficiency of such methods strongly depends on the order in which states are eliminated and on the representation of rational functions.

Related work. The idea of constructing a regular expression representing a DTMC's behavior originates from Daws [2]. He uses state elimination to generate regular expressions describing the paths from the initial states to the absorbing states of a DTMC. Hahn *et al.* [3] apply this idea to PDTMCs to obtain rational functions for reachability and expected reward properties. They improve the efficiency of the construction by common heuristics for the generation of regular expressions [4] to guide the elimination of states. Additionally, they simplify the rational functions. These ideas have been extended to Markov decision processes [5]. The main problem is that the reachability probabilities depend on the chosen scheduler to resolve the nondeterminism. When maximizing or minimizing these probabilities, the optimal scheduler generally depends on the values of the parameters. These concepts are implemented in PARAM [6] and recently also in PRISM [7], which are—to the best of our knowledge—the only available tools for computing reachability probabilities of PDTMCs.

Several authors have considered the related problem of parameter synthesis: for which parameter instances does a given (LTL or PCTL) formula hold? For instance, Han *et al.* [8] considered this problem for timed reachability in continuous-time Markov chains, Pugelli *et al.* [9] for Markov decision processes, and Benedikt *et al.* [10] for ω-regular properties of interval Markov chains.

Contributions of this paper. In this paper we improve the computation of reachability probabilities for PDTMCs [2,3] in two important ways. First, we introduce a state elimination strategy based on a *recursive graph decomposition* of the PDTMC into strongly connected subgraphs. Each (sub-)SCC is replaced by abstract transitions that lead from its ingoing states to its outgoing states. The resulting rational functions describe exactly the probability of entering the SCC and leaving it eventually. Secondly, we give a novel method to perform arithmetic operations directly on a *factorization of polynomials*. As many benchmarks have a symmetric structure, identical polynomials occur very often; therefore a maintenance of partial factorizations often speeds up the cancelation of rational functions. Although presented in the context of parametric Markov chains, this constitutes a generic method for representing and manipulating polynomials and rational functions or is well-suited for other applications as well. The experiments show that using our techniques yields a speed-up of up to three orders of magnitude compared to [3] on many benchmarks.

An extended version of this paper including all proofs can be found in [11].

2 Preliminaries

Definition 1 (Discrete-time Markov chain). *A discrete-time Markov chain (DTMC) is a tuple $\mathcal{D} = (S, I, P)$ with a non-empty finite set S of states, an initial distribution $I : S \rightarrow [0, 1] \subseteq \mathbb{R}$ with $\sum_{s \in S} I(s) = 1$, and a transition probability matrix $P : S \times S \rightarrow [0, 1] \subseteq \mathbb{R}$ with $\sum_{s' \in S} P(s, s') = 1$ for all $s \in S$.*

The states $S_I = \{s_I \in S \mid I(s_I) > 0\}$ are called *initial states*. A *transition* leads from a state $s \in S$ to a state $s' \in S$ iff $P(s, s') > 0$. The set of *successor states* of $s \in S$ is $\mathrm{succ}(s) = \{s' \in S \mid P(s, s') > 0\}$. A *path* of $\mathcal{D}$ is a finite sequence $\pi = s_0 s_1 \ldots s_n$ of states $s_i \in S$ such that $P(s_i, s_{i+1}) > 0$ for all $0 \leq i < n$. The set $\mathrm{Paths}^{\mathcal{D}}$ contains all paths of $\mathcal{D}$, $\mathrm{Paths}^{\mathcal{D}}(s)$ those starting in $s \in S$, and $\mathrm{Paths}^{\mathcal{D}}(s, t)$ those starting in s and ending in t. We generalize this to sets $S', S'' \subseteq S$ of states by $\mathrm{Paths}^{\mathcal{D}}(S', S'') = \bigcup_{s' \in S'} \bigcup_{s'' \in S''} \mathrm{Paths}^{\mathcal{D}}(s', s'')$. A state t is *reachable* from s iff $\mathrm{Paths}^{\mathcal{D}}(s, t) \neq \emptyset$.

The *probability measure* $\mathrm{Pr}^{\mathcal{D}}$ for paths satisfies

$$\mathrm{Pr}^{\mathcal{D}}(s_0 \ldots s_n) = \prod_{i=0}^{n-1} P(s_i, s_{i+1})$$

and $\mathrm{Pr}^{\mathcal{D}}(\{\pi_1, \pi_2\}) = \mathrm{Pr}^{\mathcal{D}}(\pi_1) + \mathrm{Pr}^{\mathcal{D}}(\pi_2)$ for all $\pi_1, \pi_2 \in \mathrm{Paths}^{\mathcal{D}}$ not being the prefix of each other. In general, for $R \subseteq \mathrm{Paths}^{\mathcal{D}}$ we have $\mathrm{Pr}^{\mathcal{D}}(R) = \sum_{\pi \in R'} \mathrm{Pr}^{\mathcal{D}}(\pi)$ with $R' = \{\pi \in R \mid \forall \pi' \in R.\ \pi'$ is not a proper prefix of $\pi\}$. We often omit the superscript $\mathcal{D}$ if it is clear from the context. For more details see, e. g., [12].

For a DTMC $\mathcal{D} = (S, I, P)$ and some $K \subseteq S$ we define the set of *input states* of K by $\mathrm{Inp}(K) = \{s \in K \mid I(s) > 0 \vee \exists s' \in S \setminus K.\ P(s', s) > 0\}$, i. e., the states inside K that have an incoming transition from outside K. Analogously, we define the set of *output states* of K by $\mathrm{Out}(K) = \{s \in S \setminus K \mid \exists s' \in K.\ P(s', s) > 0\}$, i. e., the states outside K that have an incoming transition from a state inside K. The set of *inner states* of K is given by $K \setminus \mathrm{Inp}(K)$.

We call a state set $S' \subseteq S$ *absorbing* iff there is a state $s' \in S'$ from which no state outside S' is reachable in $\mathcal{D}$, i. e., iff $\mathrm{Paths}^{\mathcal{D}}(\{s'\}, S \setminus S') = \emptyset$. A state $s \in S$ is absorbing if $\{s\}$ is absorbing.

A set $S' \subseteq S$ induces a *strongly connected subgraph (SCS)* of $\mathcal{D}$ iff for all $s, t \in S'$ there is a path from s to t visiting only states from S'. A *strongly connected component (SCC)* of $\mathcal{D}$ is a maximal (w. r. t. $\subseteq$) SCS of S. An SCC S' is called *bottom* if $\mathrm{Out}(S') = \emptyset$ holds. The probability of eventually reaching a bottom SCC in a finite DTMC is always 1 [12, Chap. 10.1].

We consider *probabilistic reachability properties*, putting bounds on the probability $\sum_{s_I \in S_I} I(s_I) \cdot \mathrm{Pr}^{\mathcal{D}}(\mathrm{Paths}^{\mathcal{D}}(s_I, T))$ to eventually reach a set $T \subseteq S$ of states from the initial states. It is well-known that this suffices for checking arbitrary ω-regular properties, see [12, Chap. 10.3] for the details.

The probability of reaching a bottom SCC equals the probability of reaching one of its input states. Therefore, we can make all input states of bottom SCCs absorbing, without loss of information. Furthermore, if we are interested in the probability to reach a given state, also this state can be made absorbing without modifying the reachability probability of interest. Therefore, in the following we consider only models whose bottom SCCs are single absorbing states forming the set T of *target* states, whose reachability probabilities are of interest.

2.1 Parametric Markov Chains

To add parameters to DTMCs, we allow arbitrary rational functions in the definition of probability distributions [6].

Definition 2 (Polynomial and rational function). *Let* $V = \{x_1, \ldots, x_n\}$ *be a finite set of* variables *with domain* $\mathbb{R}$. *A polynomial* g *over* V *is a sum of* monomials, *which are products of variables in* V *and a coefficient in* $\mathbb{Z}$:

$$g = a_1 \cdot x_1^{e_{1,1}} \cdot \ldots \cdot x_n^{e_{1,n}} + \cdots + a_m \cdot x_1^{e_{m,1}} \cdot \ldots \cdot x_n^{e_{m,n}},$$

where $e_{i,j} \in \mathbb{N}_0 = \mathbb{N} \cup \{0\}$ *and* $a_i \in \mathbb{Z}$ *for all* $1 \leq i \leq m$ *and* $1 \leq j \leq n$. $\mathbb{Z}[x_1, \ldots, x_n]$ *denotes the set of polynomials over* $V = \{x_1, \ldots, x_n\}$. *A rational function over* V *is a quotient* $f = \frac{g_1}{g_2}$ *of two polynomials* g_1, g_2 *over* V *with* $g_2 \neq 0^1$. *We use* $\mathcal{F}_V = \left\{ \frac{g_1}{g_2} \mid g_1, g_2 \in \mathbb{Z}[x_1, \ldots, x_n] \wedge g_2 \neq 0 \right\}$ *to denote the set of rational functions over* V.

Definition 3 (PDTMC). *A* parametric discrete-time Markov chain *(PDTMC) is a tuple* $\mathcal{M} = (S, V, I, P)$ *with a finite set of states* S, *a finite set of parameters* $V = \{x_1, \ldots, x_n\}$ *with domain* $\mathbb{R}$, *an initial distribution* $I : S \rightarrow \mathcal{F}_V$, *and a parametric transition probability matrix* $P : S \times S \rightarrow \mathcal{F}_V$.

The *underlying graph* $\mathcal{G}_\mathcal{M} = (S, \mathcal{D}_P)$ of a (P)DTMC $\mathcal{M} = (S, V, I, P)$ is given by $\mathcal{D}_P = \left\{ (s, s') \in S \times S \mid P(s, s') \neq 0 \right\}$. As for DTMCs, we assume that all bottom SCCs of considered PDTMCs are single absorbing states.

Definition 4 (Evaluated PDTMC). *An evaluation* u *of* V *is a function* $u : V \rightarrow \mathbb{R}$. *The evaluation* $g[u]$ *of a polynomial* $g \in \mathbb{Z}[x_1, \ldots, x_n]$ *under* $u : V \rightarrow \mathbb{R}$ *substitutes each* $x \in V$ *by* $u(x)$, *using the standard semantics for* $+$ *and* $\cdot$. *For* $f = \frac{g_1}{g_2} \in \mathcal{F}_V$ *we define* $f[u] = \frac{g_1[u]}{g_2[u]} \in \mathbb{R}$ *if* $g_2[u] \neq 0$.

For a PDTMC $\mathcal{M} = (S, V, I, P)$ *and an evaluation* u, *the* evaluated PDTMC *is the DTMC* $\mathcal{D} = (S_u, I_u, P_u)$ *given by* $S_u = S$ *and for all* $s, s' \in S_u$, $I_u(s) = I(s)[u]$ *and* $P_u(s, s') = P(s, s')[u]$ *if the evaluations are defined and* 0 *otherwise.*

An evaluation u substitutes the parameters by real numbers. Well-defined probability measures are induced under the following conditions:

Definition 5 (Well-defined evaluation). *An evaluation* u *is* well-defined *for a PDTMC* $\mathcal{M} = (S, V, I, P)$ *if for the evaluated PDTMC* $\mathcal{D} = (S_u, I_u, P_u)$ *it holds that*

- $I_u : S_u \rightarrow [0, 1]$ *with* $\sum_{s \in S_u} I_u(s) = 1$, *and*
- $P_u : S_u \times S_u \rightarrow [0, 1]$ *with* $\sum_{s' \in S_u} P_u(s, s') = 1$ *for all* $s \in S_u$.

---

[1] $g_2 \neq 0$ means that g_2 cannot be simplified to 0.

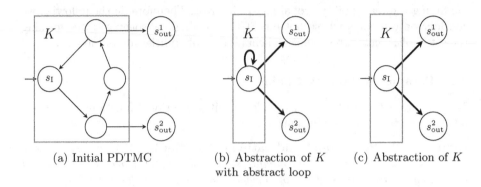

(a) Initial PDTMC   (b) Abstraction of K (c) Abstraction of K
                               with abstract loop

Fig. 1. The concept of PDTMC abstraction

An evaluation u is called graph preserving *if is well-defined and it holds that*

$$\forall s, s' \in S : P(s, s') \neq 0 \implies P(s, s')[u] > 0.$$

Note that $P(s, s')[u] > 0$ implies that no division by 0 will occur, which will be ensured during the model checking algorithm by requiring a graph preserving evaluation u, i.e., $\mathcal{G}_\mathcal{M} = \mathcal{G}_{\mathcal{M}_u}$. This is necessary, otherwise altering the graph could make reachable states unreachable, thereby changing reachability probabilities.

Definition 6. *Given a PDTMC $\mathcal{M} = (S, V, I, P)$ with absorbing states $T \subseteq S$, the* parametric probabilistic model checking problem *is to find for each initial state $s_I \in S_I$ and each $t \in T$ a rational function $f_{s_I, t} \in \mathcal{F}_V$ such that for all graph-preserving evaluations $u : V \to \mathbb{R}$ and the evaluated PDTMC $\mathcal{D} = (S_u, I_u, P_u)$ it holds that $f_{s_I, t}[u] = \mathrm{Pr}^{\mathcal{M}_u}(\mathrm{Paths}^{\mathcal{M}_u}(s_I, t)).$*

Given the functions $f_{s_I, t}$ for $s_I \in S_I$ and $t \in T$, the probability of reaching a state in T from an initial state is $\sum_{s_I \in S_I} I(s_I) \cdot \left(\sum_{t \in T} f_{s_I, t} \right).$

3   Parametric Model Checking by SCC Decomposition

In this section we present our algorithmic approach to apply model checking to PDTMCs. Let $\mathcal{M} = (S, V, I, P)$ be a PDTMC with absorbing state set $T \subseteq S$. For each initial state $s_I \in S_I$ and each target state $t \in T$ we compute a rational function $f_{s_I, t}$ over the parameters V which describes the probability of reaching t from s_I. We do this using *hierarchical graph decomposition*, inspired by a method for computing reachability probabilities in the non-parametric case [13].

3.1   PDTMC Abstraction

The basic concept of our model checking approach is to replace a non-absorbing subset $K \subseteq S$ of states and all transitions between them by transitions directly

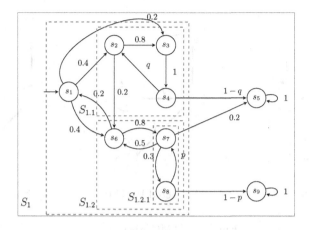

Fig. 2. Example PDTMC and its SCC decomposition

leading from the input states $\mathrm{Inp}(K)$ of K to the output states $\mathrm{Out}(K)$ of K, carrying the accumulated probabilities of all paths between the given input and output states in K. This concept is illustrated in Figure 1: In 1(a), K has one input state s_I and two output states s_out^1, s_out^2. The abstraction in 1(c) hides every state of K except for s_I; all transitions are directly leading to the output states.

As we need a probability measure for arbitrary subsets of states, we first define sub-PDTMCs induced by such subsets.

Definition 7 (Induced PDTMC). *Given a PDTMC $\mathcal{M} = (S, V, I, P)$ and a non-absorbing subset $K \subseteq S$ of states, the PDTMC induced by $\mathcal{M}$ and K is given by $\mathcal{M}^K = (S^K, V^K, I^K, P^K)$ with $S^K = K \cup \mathrm{Out}(K)$, $V^K = V$, and for all $s, s' \in S^K$, $I^K(s) \neq 0 \iff s \in \mathrm{Inp}(K)$ and*

$$
P^K(s, s') = \begin{cases} P(s, s'), & \textit{if } s \in K, s' \in S^K, \\ 1, & \textit{if } s = s' \in \mathrm{Out}(K), \\ 0, & \textit{otherwise.} \end{cases}
$$

Intuitively, all incoming and outgoing transitions are preserved for inner states of K while the output states are made absorbing. We allow an arbitrary input distribution I^K with the only constraint that $I^K(s) \neq 0$ iff s is an input state of K.

Example 1. Consider the PDTMC $\mathcal{M}$ in Figure 2 and the state set $K = \{s_7, s_8\}$ with input states $\mathrm{Inp}(K) = \{s_7\}$ and output states $\mathrm{Out}(K) = \{s_5, s_6, s_9\}$. The PDTMC $\mathcal{M}^K = (S^K, V^K, I^K, P^K)$ induced by $\mathcal{M}$ and K is shown in Figure 3(a).

Note that, since K is non-absorbing, the probability of eventually reaching one of the output states is 1. The probability of reaching an output state t from an

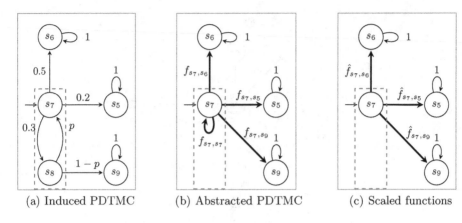

(a) Induced PDTMC   (b) Abstracted PDTMC   (c) Scaled functions

Fig. 3. PDTMC Abstraction

input state s is determined by the accumulated probability of all paths $\mathrm{Paths}(s,t)$ from s to t. Those paths are composed by a (possibly empty) prefix looping on s and a postfix leading from s to t without returning back to s. In our abstraction this is reflected by representing the prefixes by an abstract self-loop on s with probability $f_{s,s}$ and the postfixes by abstract transitions from the input states s to the output states t with probability $f_{s,t}$ (see Figure 1(b)). If all loops in K are loops on s then $f_{s,t}$ can be easily computed as the sum of the probabilities of all loop-free paths from s to t. In the final abstraction shown in Figure 1(c), we make use of the fact that all paths from s to t can be extended with the same loops on s as a prefix. Therefore we do not need to compute the probability of looping on s, but can scale the probabilities $f_{s,t}$ such that they sum up to 1.

Definition 8 (Abstract PDTMC). *Let* $\mathcal{M} = (S, V, I, P)$ *be a PDTMC with absorbing states* $T \subseteq S$. *The* abstract PDTMC $\mathcal{M}_{\mathrm{abs}} = (S_{\mathrm{abs}}, V_{\mathrm{abs}}, I_{\mathrm{abs}}, P_{\mathrm{abs}})$ *is given by* $S_{\mathrm{abs}} = \{s \in S \mid I(s) \neq 0 \vee s \in T\}$, $V_{\mathrm{abs}} = V$, *and for all* $s, s' \in S_{\mathrm{abs}}$ *we define* $I_{\mathrm{abs}}(s) = I(s)$ *and*

$$
P_{\mathrm{abs}}(s, s') = \begin{cases} \dfrac{p_{\mathrm{abs}}^{\mathcal{M}}(s, s')}{\sum_{s'' \in T} p_{\mathrm{abs}}^{\mathcal{M}}(s, s'')}, & \text{if } I(s) > 0 \wedge s' \in T, \\ 1, & \text{if } s = s' \in T, \\ 0, & \text{otherwise.} \end{cases}
$$

with

$$
p_{\mathrm{abs}}^{\mathcal{M}}(s, s') = \mathrm{Pr}^{\mathcal{M}}\big(\{\pi = s_0 \ldots s_n \in \mathrm{Paths}^{\mathcal{M}}(s, s') \mid s_i \neq s \wedge s_i \neq s', 0 < i < n\}\big).
$$

Example 2. Consider the PDTMC $\mathcal{M}' = (S', V', I', P')$ of Figure 3(a) with initial state s_7 and target states $T' = \{s_5, s_6, s_9\}$. The first abstraction step for the probabilities $p_{\mathrm{abs}}^{\mathcal{M}}(s, s')$ is depicted in Figure 3(b) with the following probabilities:

$$f_{s_7,s_5} = p_{\text{abs}}^{\mathcal{M}'}(s_7, s_5) = 0.2 \qquad f_{s_7,s_6} = p_{\text{abs}}^{\mathcal{M}'}(s_7, s_6) = 0.5$$

$$f_{s_7,s_7} = p_{\text{abs}}^{\mathcal{M}'}(s_7, s_7) = 0.3 \cdot p \qquad f_{s_7,s_9} = p_{\text{abs}}^{\mathcal{M}'}(s_7, s_9) = 0.3 \cdot (1-p)$$

The total probabilities of reaching the output states in $\mathcal{M}'_{\text{abs}}$ are given by paths which first use the loop on s_7 arbitrarily many times (including zero times) and then take a transition to an output state. For example, using the geometric series, the probability of the set of paths leading from s_7 to s_5 is given by

$$\sum_{i=0}^{\infty} (f_{s_7,s_7})^i \cdot f_{s_7,s_5} = \frac{1}{1 - f_{s_7,s_7}} \cdot f_{s_7,s_5} .$$

As the probability of finally reaching the set of absorbing states in $\mathcal{M}'$ is 1, we can directly scale the probabilities of the outgoing edges such that their sum is equal to 1: We divide each of these probabilities by the sum of all probabilities of outgoing edges, $f_{\text{out}} = 0.2 + 0.5 + 0.3 \cdot (1-p) = 1 - 0.3p$.

Thus the abstract PDTMC $\mathcal{M}'_{\text{abs}} = (S'_{\text{abs}}, V'_{\text{abs}}, I'_{\text{abs}}, P'_{\text{abs}})$ depicted in Figure 3(c) has the states $S'_{\text{abs}} = \{s_5, s_6, s_7, s_9\}$ and edges from s_7 to all other states with the following probabilities:

$$\hat{f}_{s_7,s_5} = 0.2 \,/ f_{\text{out}} \qquad\qquad \hat{f}_{s_7,s_6} = 0.5 \,/ f_{\text{out}}$$

$$\hat{f}_{s_7,s_9} = \big(0.3 \cdot (1-p)\big) \,/ f_{\text{out}}$$

Theorem 1. *Assume a PDTMC $\mathcal{M} = (S, V, I, P)$ with absorbing states $T \subseteq S$, and let $\mathcal{M}_{\text{abs}}$ be the abstraction of $\mathcal{M}$. Then for all $s_I \in S_I$ and $t \in T$ it holds that*

$$\Pr{}^{\mathcal{M}}\big(\text{Paths}^{\mathcal{M}}(s_I, t)\big) = \Pr{}^{\mathcal{M}_{\text{abs}}}\big(\text{Paths}^{\mathcal{M}_{\text{abs}}}(s_I, t)\big) .$$

It remains to define the substitution of subsets of states by their abstractions. Intuitively, a subset of states is replaced by the abstraction as in Definition 8, while incoming transitions of the initial states of the abstraction as well as outgoing transitions of the absorbing states of the abstraction remain unmodified.

Definition 9 (Substitution). *Assume a PDTMC $\mathcal{M} = (S, V, I, P)$, a non-absorbing set $K \subseteq S$ of states, the induced PDTMC $\mathcal{M}^K = (S^K, V^K, I^K, P^K)$ and the abstraction $\mathcal{M}^K_{\text{abs}} = (S^K_{\text{abs}}, V^K_{\text{abs}}, I^K_{\text{abs}}, P^K_{\text{abs}})$. The substitution of $\mathcal{M}^K$ by its abstraction $\mathcal{M}^K_{\text{abs}}$ in $\mathcal{M}$ is given by $\mathcal{M}_{K \mapsto \text{abs}} = (S_{K \mapsto \text{abs}}, V_{K \mapsto \text{abs}}, I_{K \mapsto \text{abs}}, P_{K \mapsto \text{abs}})$ with $S_{K \mapsto \text{abs}} = (S \setminus K) \cup S^K_{\text{abs}}$, $V_{K \mapsto \text{abs}} = V$ and for all $s, s' \in S_{K \mapsto \text{abs}}$, $I_{K \mapsto \text{abs}}(s) = I(s)$ and*

$$P_{K \mapsto \text{abs}}(s, s') = \begin{cases} P(s, s'), & \text{if } s \notin K, \\ P^K_{\text{abs}}(s, s'), & \text{if } s \in K \wedge s' \in \text{Out}(K), \\ 0, & \text{otherwise.} \end{cases}$$

Due to Theorem 1, it directly follows that this substitution does not change reachability properties from the initial states to the absorbing states of a PDTMC.

Algorithm 1. Model Checking PDTMCs

---

abstract(PDTMC $\mathcal{M}$)
begin

    for all non-bottom SCCs K in $\mathcal{M}^{S \setminus \mathrm{Inp}(\mathcal{M})}$ **do**   (1)

        $\mathcal{M}_{\mathrm{abs}}^{K} := \mathbf{abstract}(\mathcal{M}^{K})$ (2)

        $\mathcal{M} := \mathcal{M}_{K \mapsto \mathrm{abs}}$ (3)

    end for   (4)

    $K := \{\textit{non-absorbing states in } \mathcal{M}\}$ (5)

    $\mathcal{M} := \mathcal{M}_{K \mapsto \mathrm{abs}}$ (6)

    return $\mathcal{M}$ (7)

end

model_check(PDTMC $\mathcal{M} = (S, V, I, P)$, $T \subseteq \{t \in S \mid P(t,t) = 1\}$)
begin

    $\mathcal{M}_{\mathrm{abs}} = (S_{\mathrm{abs}}, V_{\mathrm{abs}}, I_{\mathrm{abs}}, P_{\mathrm{abs}}) := \mathbf{abstract}(\mathcal{M})$ (8)

    return $\sum\limits_{s_I \in S_I} I(s_I) \cdot \left(\sum\limits_{t \in T} P_{\mathrm{abs}}(s_I, t) \right)$ (9)

end

Corollary 1. *Given a PDTMC $\mathcal{M}$ and a non-absorbing subset $K \subseteq S$ of states, it holds for all initial states $s_I \in S_I$ and absorbing states $t \in T$ that*

$$\mathrm{Pr}^{\mathcal{M}}\left(\mathrm{Paths}^{\mathcal{M}}(s_I, t)\right) = \mathrm{Pr}^{\mathcal{M}_{K \mapsto \mathrm{abs}}}\left(\mathrm{Paths}^{\mathcal{M}_{K \mapsto \mathrm{abs}}}(s_I, t)\right).$$

3.2 Model Checking Parametric Markov Chains

In the previous section we gave the theoretical background for our model checking algorithm. Now we describe how to compute the abstractions efficiently. As a heuristic for forming the sets of states to be abstracted, we choose an SCC-based decomposition of the graph. Algorithmically, Tarjan's algorithm [14] is used to determine the SCC structure of the graph while we do not consider bottom SCCs. Sub-SCCs inside the SCCs without their input states are determined hierarchically, until no non-trivial sub-SCCs remain.

Example 3. In Figure 2, the dashed rectangles indicate the decomposition into the SCC $S_1 = \{1, 2, 3, 4, 6, 7, 8\}$ and the sub-SCSs $S_{1.1} = \{2, 3, 4\}$, $S_{1.2} = \{6, 7, 8\}$, and $S_{1.2.1} = \{7, 8\}$ with $S_{1.1} \subset S_1$ and $S_{1.2.1} \subset S_{1.2} \subset S_1$.

The general model checking algorithm is depicted in Algorithm 1. The recursive method *abstract*(PDTMC $\mathcal{M}$) computes the abstraction $\mathcal{M}_{\mathrm{abs}}$ by iterating over all SCCs of the graph without the input states of $\mathcal{M}$ (Line 1). For each SCC K, the abstraction $\mathcal{M}_{\mathrm{abs}}^{K}$ of the induced PDTMC $\mathcal{M}^{K}$ is computed by a recursive call (Line 2, Definitions 7,8). Afterwards, $\mathcal{M}^{K}$ is substituted by its abstraction in $\mathcal{M}$ (Line 3, Definition 9). Finally, the abstraction $\mathcal{M}_{\mathrm{abs}}$ is computed and returned (Line 7, Definition 8). The method *abstract* is called by *model_check* (Line 8) which yields the abstract system $\mathcal{M}_{\mathrm{abs}}$ where transitions lead only from the initial states to the absorbing states. All transitions are labeled with a rational function for the reachability probability, as in Definition 6. The total probability is computed by building the sum of these transitions (Line 9).

For the computation of the abstract probabilities $p_{abs}^{\mathcal{M}}$, we distinguish the cases where the set K has one or multiple input states.

One input state Consider a PDTMC $\mathcal{M}^K$ induced by K with one initial state s_I and the set of absorbing states $T = \{t^1, \ldots, t^n\}$, such that $K \setminus \{s_I\}$ has no non-trivial SCCs. If there is only one absorbing state, i. e., $n = 1$, we trivially have $p_{abs}^{\mathcal{M}^K}(s_I, t^1) = 1$. Otherwise we determine the probabilities $p_{abs}^{\mathcal{M}^K}(s_I, t^i)$ for all $1 \leq i \leq n$. As $K \setminus \{s_I\}$ has no non-trivial SCSs, the set of those paths from s_I to t^i that do not return to s_I consists of finitely many loop-free paths. The probability is computed recursively for all $s \in S^K$ by:

$$p_{abs}^{\mathcal{M}^K}(s, t^i) = \begin{cases} 1, & \text{if } s = t^i, \\ \displaystyle\sum_{s' \in (\text{succ}(s) \cap K) \setminus \text{Inp}(K)} P^K(s, s') \cdot p_{abs}^{\mathcal{M}^K}(s', t^i), & \text{otherwise.} \end{cases} \quad (1)$$

These probabilities can also be computed by direct or indirect methods for solving linear equation systems[2], see [15, Chapters 3,4], or state elimination as in [3].

The probabilities of the abstract PDTMC $\mathcal{M}_{abs}^K = (S_{abs}, V_{abs}, I_{abs}, P_{abs})$ as in Definition 8 can now directly be computed, while an additional constraint is added in order to avoid divisions by zero:

$$P_{abs}^{\mathcal{M}^K}(s_I, t^i) = \begin{cases} \dfrac{p_{abs}^{\mathcal{M}^K}(s_I, t^i)}{\sum_{j=1}^n p_{abs}^{\mathcal{M}^K}(s_I, t^j)}, & \text{if } \sum_{j=1}^n p_{abs}^{\mathcal{M}^K}(s_I, t^j) \neq 0, \\ 0, & \text{otherwise.} \end{cases} \quad (2)$$

Multiple input states Given a PDTMC $\mathcal{M}^K$ with initial states $S_I = \{s_I^1, \ldots, s_I^m\}$ such that $I^K(s_I^i) > 0$ for all $1 \leq i \leq m$ and absorbing states $T = \{t^1, \ldots, t^n\}$. The idea is to maintain a copy of $\mathcal{M}^K$ for each initial state and handle the other initial states as inner states in this copy. Then, the method as described in the previous paragraph can be used. However, this would be expensive in terms of both time and memory. Therefore, we first formulate the linear equation system as in Equation (1). All variables $p_{abs}^{\mathcal{M}^K}(s, t^i)$ with $s \in K \setminus \text{Inp}(K)$ are eliminated from the equation system. Then for each initial state s_I^i the equation system is solved separately by eliminating all variables $p_{abs}^{\mathcal{M}^K}(s_I^j, t^k)$, $j \neq i$.

Algorithm 1 returns the rational functions $P_{abs}^{\mathcal{M}^K}(s_I, t)$ for all $s_I \in S_I$ and $t \in T$ as in Equation (2). To allow only graph-preserving evaluations of the parameters, we perform preprocessing where conditions are collected according to Definition 5 as well as the ones from Equation (2). These constraints can be evaluated by a *SAT-modulo- theories (SMT)* solver for non-linear real arithmetic [16]. In case the solver returns an evaluation which satisfies the resulting constraint set, the reachability property is satisfied. Otherwise, the property is violated.

---

[2] Note that these equation systems are solved by keeping the parameters as constants.

4   Factorization of Polynomials

Both the SCC-based procedure as introduced in the last section as well as mere state-elimination [3] build rational functions representing reachability probabilities. These rational functions might grow rapidly in both algorithms and thereby form one of the major bottlenecks of this methodology. As already argued in [3], the best way to stem this blow-up is the cancellation of the rational functions in every computation step, which involves—apart from *addition, multiplication,* and *division* of rational functions as illustrated in Example 2—the rather expensive calculation of the *greatest common divisor* (gcd) of two polynomials.

In this section we present a new way of handling this problem: Additional maintenance and storage of (partial) polynomial factorizations can lead to remarkable speed-ups in the gcd computation, especially when dealing with symmetrically structured benchmarks where many similar polynomials occur. We present an optimized algorithm called gcd which *operates on the (partial) factorizations* of the polynomials to compute their gcd. During the calculations, the factorizations are also refined. On this account we reformulate the arithmetic operations on rational functions such that they preserve their numerator's and denominator's factorizations, if it is possible with reasonable effort.

Factorizations. In the following we assume that polynomials are *normalized*, that is they are of the form $g = a_1 \cdot x_1^{e_{1,1}} \cdot \ldots \cdot x_n^{e_{1,n}} + \cdots + a_m \cdot x_1^{e_{m,1}} \cdot \ldots \cdot x_n^{e_{m,n}}$ with $(e_{j,1}, \ldots, e_{j,n}) \neq (e_{k,1}, \ldots, e_{k,n})$ for all $j, k \in \{1, \ldots, m\}$ with $j \neq k$ and the monomials are ordered, e. g., according to the reverse lexicographical ordering.

Definition 10 (Factorization). *A factorization $\mathcal{F}_g = \{g_1^{e_1}, \ldots, g_n^{e_n}\}$ of a polynomial $g \neq 0$ is a non-empty set[3] of factors $g_i^{e_i}$, where the bases g_i are pairwise different polynomials and the exponents are $e_i \in \mathbb{N}$ such that $g = \prod_{i=1}^{n} g_i^{e_i}$. We additionally set $\mathcal{F}_0 = \emptyset$.*

For polynomials g, h and a factorization $\mathcal{F}_g = \{g_1^{e_1}, \ldots, g_n^{e_n}\}$ of g let bases$(\mathcal{F}_g) = \{g_1, \ldots, g_n\}$ and $\exp(h, \mathcal{F}_g)$ be e_i if $g_i = h$ and 0 if $h \notin$ bases$(\mathcal{F}_g)$. As the bases are not required to be irreducible, factorizations are not unique.

We assume that bases and exponents are non-zero, $\mathcal{F}_1 = \{1^1\}$, and $1^k \notin \mathcal{F}_g$ for $g \neq 1$. For $\mathcal{F}_g = \{g_1^{e_1}, \ldots, g_n^{e_n}\}$, this is expressed by the reduction $\mathcal{F}_g^{\mathrm{red}} = \{1^1\}$ if $n > 0$ and $g_i = 1$ or $e_i = 0$ for all $1 \leq i \leq n$, and $\mathcal{F}_g^{\mathrm{red}} = \mathcal{F}_g \setminus \{g_i^{e_i} \mid g_i = 1 \vee e_i = 0\}$ otherwise.

Operations on factorizations. Instead of applying arithmetic operations on two polynomials g_1 and g_2 directly, we operate on their factorizations $\mathcal{F}_{g_1}$ and $\mathcal{F}_{g_2}$. We use the following operations on factorizations: $\mathcal{F}_{g_1} \cup_{\mathcal{F}} \mathcal{F}_{g_2}$ factorizes a (not necessarily least) common multiple of g_1 and g_2, $\mathcal{F}_{g_1} \cap_{\mathcal{F}} \mathcal{F}_{g_2}$ a (not necessarily

---

[3] We represent a factorization of a polynomial as a set; however, in the implementation we use a more efficient binary search tree instead.

greatest) common divisor, whereas the binary operations $\cdot_{\mathcal{F}}$, $:_{\mathcal{F}}$ and $+_{\mathcal{F}}$ correspond to multiplication, division[4] and addition, respectively. Due to space limitations, we omit in the remaining of this paper the trivial cases involving $\mathcal{F}_0$. Therefore we define

$$\mathcal{F}_{g_1} \cup_{\mathcal{F}} \mathcal{F}_{g_2} = \{h^{\max(\exp(h,\mathcal{F}_{g_1}),\exp(h,\mathcal{F}_{g_2}))} \mid h \in \text{bases}(\mathcal{F}_{g_1}) \cup \text{bases}(\mathcal{F}_{g_2})\}^{\text{red}}$$

$$\mathcal{F}_{g_1} \cap_{\mathcal{F}} \mathcal{F}_{g_2} = \{h^{\min(\exp(h,\mathcal{F}_{g_1}),\exp(h,\mathcal{F}_{g_2}))} \mid h{=}1 \vee h{\in}\text{bases}(\mathcal{F}_{g_1}){\cap}\text{bases}(\mathcal{F}_{g_2})\}^{\text{red}}$$

$$\mathcal{F}_{g_1} \cdot_{\mathcal{F}} \mathcal{F}_{g_2} = \{h^{\exp(h,\mathcal{F}_{g_1})+\exp(h,\mathcal{F}_{g_2})} \mid h \in \text{bases}(\mathcal{F}_{g_1}) \cup \text{bases}(\mathcal{F}_{g_2})\}^{\text{red}}$$

$$\mathcal{F}_{g_1} :_{\mathcal{F}} \mathcal{F}_{g_2} = \{h^{\max(0,e-\exp(h,\mathcal{F}_{g_2}))} \mid h^e \in \mathcal{F}_{g_1}\}^{\text{red}}$$

$$\mathcal{F}_{g_1} +_{\mathcal{F}} \mathcal{F}_{g_2} = D \cdot_{\mathcal{F}} \left\{ \left(\textstyle\prod_{g_1' \in (\mathcal{F}_{g_1} :_{\mathcal{F}} D)} g_1'\right) + \left(\textstyle\prod_{g_2' \in (\mathcal{F}_{g_2} :_{\mathcal{F}} D)} g_2'\right) \right\}^{\text{red}}$$

where $D = \mathcal{F}_{g_1} \cap_{\mathcal{F}} \mathcal{F}_{g_2}$ and $\max(a,b)$ $(\min(a,b))$ equals a if $a \geq b$ $(a \leq b)$ and b otherwise. Example 4 illustrates the application of the above operations.

Operations on rational functions. We represent a rational function $\frac{g_1}{g_2}$ by separate factorizations $\mathcal{F}_{g_1}$ and $\mathcal{F}_{g_2}$ for the numerator g_1 and the denominator g_2, respectively. For multiplication $\frac{g_1}{g_2} = \frac{h_1}{h_2} \cdot \frac{q_1}{q_2}$, we compute $\mathcal{F}_{g_1} = \mathcal{F}_{h_1} \cdot_{\mathcal{F}} \mathcal{F}_{q_1}$ and $\mathcal{F}_{g_2} = \mathcal{F}_{h_2} \cdot_{\mathcal{F}} \mathcal{F}_{q_2}$. Division is reduced to multiplication according to $\frac{h_1}{h_2} : \frac{q_1}{q_2} = \frac{h_1}{h_2} \cdot \frac{q_2}{q_1}$.

For the addition $\frac{g_1}{g_2} = \frac{h_1}{h_2} + \frac{q_1}{q_2}$, we compute g_2 with $\mathcal{F}_{g_2} = \mathcal{F}_{h_2} \cup_{\mathcal{F}} \mathcal{F}_{q_2}$ as a common multiple of h_2 and q_2, such that $g_2 = h_2 \cdot h_2'$ with $\mathcal{F}_{h_2'} = \mathcal{F}_{g_2} :_{\mathcal{F}} \mathcal{F}_{h_2}$, and $g_2 = q_2 \cdot q_2'$ with $\mathcal{F}_{q_2'} = \mathcal{F}_{g_2} :_{\mathcal{F}} \mathcal{F}_{q_2}$. For the numerator g_1 we first determine a common divisor d of h_1 and q_1 by $\mathcal{F}_d = \mathcal{F}_{h_1} \cap_{\mathcal{F}} \mathcal{F}_{q_1}$, such that $h_1 = d \cdot h_1'$ with $\mathcal{F}_{h_1'} = \mathcal{F}_{h_1} :_{\mathcal{F}} \mathcal{F}_d$, and $q_1 = d \cdot q_1'$ with $\mathcal{F}_{q_1'} = \mathcal{F}_{q_1} :_{\mathcal{F}} \mathcal{F}_d$. The numerator g_1 is $d \cdot (h_1' \cdot h_2' + q_1' \cdot q_2')$ with factorization $\mathcal{F}_d \cdot_{\mathcal{F}} (\mathcal{F}_{h_1'} \cdot_{\mathcal{F}} \mathcal{F}_{h_2'} +_{\mathcal{F}} \mathcal{F}_{q_1'} \cdot_{\mathcal{F}} \mathcal{F}_{q_2'})$.

The rational function $\frac{g_1}{g_2}$ resulting from the addition is further simplified by cancellation, i.e., dividing g_1 and g_2 by their greatest common divisor (gcd) g. Given the factorizations $\mathcal{F}_{g_1}$ and $\mathcal{F}_{g_2}$, Algorithm 2 calculates the factorizations $\mathcal{F}_g$, $\mathcal{F}_{\frac{g_1}{g}}$, and $\mathcal{F}_{\frac{g_2}{g}}$.

Intuitively, the algorithm maintains the fact that $G \cdot_{\mathcal{F}} F_1 \cdot_{\mathcal{F}} F_1'$ is a factorization of g_1, where G contains common factors of g_1 and g_2, F_1 is going to be checked whether it contains further common factors, and F_1' does not contain any common factors. In the outer while-loop, an element $r_1^{e_1}$ to be checked is taken from F_1. In the inner while-loop, a factorization $G \cdot_{\mathcal{F}} F_2 \cdot_{\mathcal{F}} F_2'$ of g_2 is maintained such that F_2' does not contain any common factors with r_1, and F_2 is still to be checked.

Now we explain the algorithm in more detail. Initially, a factorization G of a common divisor of g_1 and g_2 is set to $\mathcal{F}_{g_1} \cap_{\mathcal{F}} \mathcal{F}_{g_2}$ (Line 1). The remaining factors of g_1 and g_2 are stored in F_1 resp. F_2. The sets F_1' and F_2' contain factors of g_1 resp. g_2 whose greatest common divisor is 1 (Line 2). The algorithm now iteratively adds further common divisors of g_1 and g_2 to G until it is a factorization of their gcd. For this purpose, we consider for each factor in F_1 all factors in F_2

---

[4] $\mathcal{F}_{g_1} :_{\mathcal{F}} \mathcal{F}_{g_2}$ is a factorization of g_1/g_2 only if $\mathcal{F}_{g_1}$ and $\mathcal{F}_{g_2}$ are sufficiently refined and g_2 divides g_1.

Algorithm 2. gcd computation with factorization refinement

GCD(factorization $\mathcal{F}_{g_1}$, factorization $\mathcal{F}_{g_2}$)
begin

$\quad G := (\mathcal{F}_{g_1} \cap_{\mathcal{F}} \mathcal{F}_{g_2})$ $\qquad\qquad\qquad\qquad\qquad\qquad\qquad\qquad\qquad\qquad$ (1)

$\quad F_i := \mathcal{F}_{g_i} :_{\mathcal{F}} G$ and $F_i' := \{1^1\}$ for $i = 1, 2$ $\qquad\qquad\qquad\qquad\quad$ (2)

$\quad$ **while** exists $r_1^{e_1} \in F_1$ with $r_1 \neq 1$ **do** $\qquad\qquad\qquad\qquad\qquad\quad$ (3)

$\quad\quad F_1 := F_1 \setminus \{r_1^{e_1}\}$ $\qquad\qquad\qquad\qquad\qquad\qquad\qquad\qquad\qquad$ (4)

$\quad\quad$ **while** $r_1 \neq 1$ and exists $r_2^{e_2} \in F_2$ with $r_2 \neq 1$ **do** $\qquad\qquad$ (5)

$\quad\quad\quad F_2 := F_2 \setminus \{r_2^{e_2}\}$ $\qquad\qquad\qquad\qquad\qquad\qquad\qquad\qquad$ (6)

$\quad\quad\quad$ **if** $\neg$irreducible(r_1) $\vee$ $\neg$irreducible(r_2) **then** $g := \gcd(r_1, r_2)$ (7)

$\quad\quad\quad$ **else** $g := 1$ $\qquad\qquad\qquad\qquad\qquad\qquad\qquad\qquad\qquad\quad$ (8)

$\quad\quad\quad$ **if** $g = 1$ **then** $\qquad\qquad\qquad\qquad\qquad\qquad\qquad\qquad\qquad$ (9)

$\quad\quad\quad\quad F_2' := F_2' \cdot_{\mathcal{F}} \{r_2^{e_2}\}$ $\qquad\qquad\qquad\qquad\qquad\qquad\qquad$ (10)

$\quad\quad\quad$ **else** $\qquad\qquad\qquad\qquad\qquad\qquad\qquad\qquad\qquad\qquad\quad$ (11)

$\quad\quad\quad\quad r_1 := \frac{r_1}{g}$ $\qquad\qquad\qquad\qquad\qquad\qquad\qquad\qquad\qquad\quad$ (12)

$\quad\quad\quad\quad F_i := F_i \cdot_{\mathcal{F}} \{g^{e_i - \min(e_1, e_2)}\}$ for $i = 1, 2$ $\qquad\qquad$ (13)

$\quad\quad\quad\quad F_2' := F_2' \cdot_{\mathcal{F}} \{(\frac{r_2}{g})^{e_2}\}$ $\qquad\qquad\qquad\qquad\qquad\qquad$ (14)

$\quad\quad\quad\quad G := G \cdot_{\mathcal{F}} \{g^{\min(e_1, e_2)}\}$ $\qquad\qquad\qquad\qquad\qquad\quad$ (15)

$\quad\quad\quad$ **end if** $\qquad\qquad\qquad\qquad\qquad\qquad\qquad\qquad\qquad\qquad$ (16)

$\quad\quad$ **end while** $\qquad\qquad\qquad\qquad\qquad\qquad\qquad\qquad\qquad\quad$ (17)

$\quad\quad F_1' := F_1' \cdot_{\mathcal{F}} \{r_1^{e_1}\}$ $\qquad\qquad\qquad\qquad\qquad\qquad\qquad\quad$ (18)

$\quad\quad F_2 := F_2 \cdot_{\mathcal{F}} F_2'$ $\qquad\qquad\qquad\qquad\qquad\qquad\qquad\qquad\quad$ (19)

$\quad\quad F_2' := \{1^1\}$ $\qquad\qquad\qquad\qquad\qquad\qquad\qquad\qquad\qquad\quad$ (20)

$\quad$ **end while** $\qquad\qquad\qquad\qquad\qquad\qquad\qquad\qquad\qquad\qquad$ (21)

$\quad$ **return** (F_1', F_2, G) $\qquad\qquad\qquad\qquad\qquad\qquad\qquad\qquad\quad$ (22)

end

---

and calculate the gcd of their bases using standard gcd computation for polynomials (Line 7). Note that the main concern of Algorithm 2 is to avoid the application of this expensive operation as far as possible and to apply it to preferably simple polynomials otherwise. Where the latter is entailed by the idea of using factorizations, the former can be achieved by excluding pairs of factors for which we can cheaply decide that both are *irreducible*, i. e., they have no non-trivial divisors. If factors $r_1^{e_1} \in F_1$ and $r_2^{e_2} \in F_2$ with $g := \gcd(r_1, r_2) = 1$ are found, we just shift $r_2^{e_2}$ from F_2 to F_2' (Line 10). Otherwise, we can add $g^{\min(e_1, e_2)}$, which is the gcd of $r_1^{e_1}$ and $r_2^{e_2}$, to G and extend the factors F_1 resp. F_2, which could still contain common divisors, by $g^{e_1 - \min(e_1, e_2)}$ resp. $g^{e_2 - \min(e_1, e_2)}$ (Line 12-15). Furthermore, F_2' obtains the new factor $(\frac{r_2}{g})^{e_2}$, which has certainly no common divisor with any factor in F_1'. Finally, we set the basis r_1 to $\frac{r_1}{g}$, excluding the just found common divisor. If all factors in F_2 have been considered for common divisors with r_1, we can add it to F_1' and continue with the next factor in F_1, for which we must reconsider all factors in F_2' and, therefore, shift them to F_2 (Line 18-20). The algorithm terminates, if the last factor of F_1 has been processed, returning the factorizations $\mathcal{F}_g$, $\mathcal{F}_{\frac{g_1}{g}}$ and $\mathcal{F}_{\frac{g_2}{g}}$, which we can use to refine the factorizations of g_1 and g_2 via $\mathcal{F}_{g_1} := \mathcal{F}_{\frac{g_1}{g}} \cdot_{\mathcal{F}} G$ and $\mathcal{F}_{g_2} := \mathcal{F}_{\frac{g_2}{g}} \cdot_{\mathcal{F}} G$.

Example 4. Assume we want to apply Algorithm 2 to the factorizations $\mathcal{F}_{xyz} = \{(xyz)^1\}$ and $\mathcal{F}_{xy} = \{(x)^1, (y)^1\}$. We initialize $G = F_1' = F_2' = \{(1)^1\}$, $F_1 = \mathcal{F}_{xyz}$ and $F_2 = \mathcal{F}_{xy}$. First, we choose the factors $(r_1)^{e_1} = (xyz)^1$ and $(x)^1$ and remove them from F_1 resp. F_2. The gcd of their bases is x, hence we only update r_1 to $(yz)^1$ and G to $\{(x)^1\}$. Then we remove the next and last element $(y)^1$ from F_2. Its basis and r_1 have the gcd y and we therefore update r_1 to $(z)^1$ and G to $\{(x)^1, (y)^1\}$. Finally, we add $(z)^1$ to F_1' and return the expected result $(\{(z)^1\}, \{(1)^1\}, \{(x)^1, (y)^1\})$. Using these results, we can also refine $\mathcal{F}_{xyz} = F_1' \cdot_{\mathcal{F}} G = \{(x)^1, (y)^1, (z)^1\}$ and $\mathcal{F}_{xy} = F_2 \cdot_{\mathcal{F}} G = \{(x)^1, (y)^1\}$.

Theorem 2. *Let p_1 and p_2 be two polynomials with factorizations $\mathcal{F}_{p_1}$ resp. $\mathcal{F}_{p_2}$. Applying Algorithm 2 to these factorizations results in $\gcd(\mathcal{F}_{p_1}, \mathcal{F}_{p_2}) = (\mathcal{F}_{r_1}, \mathcal{F}_{r_2}, G)$ with G being a factorization of the greatest common divisor g of p_1 and p_2, and $\mathcal{F}_{r_1}$ and $\mathcal{F}_{r_2}$ being factorizations of $\frac{p_1}{g}$ resp. $\frac{p_2}{g}$.*

5   Experiments

We developed a C++ prototype implementation of our approach using the arithmetic library GiNaC [17]. The prototype is available on the project homepage[5]. Moreover, we implemented the state-elimination approach used by PARAM [6] using our optimized factorization approach to provide a more distinct comparison. All experiments were run on an Intel Core 2 Quad CPU 2.66 GHz with 4 GB of memory. We defined a timeout (TO) of 14 hours (50400 seconds) and a memory bound (MO) of 4 GB. We report on three case studies; a more distinct description and the specific instances we used are available at our homepage.

The *bounded retransmission protocol* (BRP) [18] models the sending of files via an unreliable network, manifested in two lossy channels for sending and acknowledging the reception. This model is parametrized in the probability of reliability of those channels. The *crowds protocol* (CROWDS) [19] is designed for anonymous network communication using random routing, parametrized in how many members are "good" or "bad" and the probability if a good member delivers a message or randomly routes it to another member. *NAND multiplexing* (NAND) [20] models how reliable computations are obtained using unreliable hardware by having a certain number of copies of a NAND unit all doing the same job. Parameters are the probabilities of faultiness of the units and of erroneous inputs. The experimental setting includes our SCC-based approach as described in Section 3 using the optimized factorization of polynomials as in Section 4 (SCC MC), the state elimination as in PARAM but also using the approach of Section 4 (STATE ELIM) and the PARAM tool itself.[6] For all instances we list the number of states and transitions; for each tool we give the running time in seconds and the memory consumption in MB; the best time is **boldfaced**. Moreover, for our approaches we list the number of polynomials which are intermediately stored.

---

[5] http://goo.gl/nS378q

[6] Note that no bisimulation reduction was applied to any of the input models, which would improve the feasibility of all approaches likewise.

Model	Graph		SCC MC			STATE ELIM			PARAM	
	States	Trans.	Time	Poly	Mem	Time	Poly	Mem	Time	Mem
BRP	3528	4611	29.05	3283	48.10	**4.33**	8179	61.17	98.99	32.90
BRP	4361	5763	511.50	4247	501.71	**6.87**	9520	78.49	191.52	58.43
BRP	7048	9219	548.73	6547	281.86	**25.05**	16435	216.05	988.28	142.66
BRP	10759	13827	147.31	9231	176.89	**85.54**	26807	682.24	3511.96	304.07
BRP	21511	27651	1602.53	18443	776.48	**718.66**	53687	3134.59	34322.60	1757.12
CROWDS	198201	348349	**60.90**	13483	140.15	243.07	27340	133.91	46380.00	227.66
CROWDS	482979	728677	**35.06**	35916	478.85	247.75	65966	297.40	TO	—
CROWDS	726379	1283297	**223.24**	36649	515.61	1632.63	73704	477.10	TO	—
CROWDS	961499	1452537	**81.88**	61299	1027.78	646.76	112452	589.21	TO	—
CROWDS	1729494	2615272	**172.59**	97655	2372.35	1515.63	178885	1063.15	TO	—
CROWDS	2888763	5127151	**852.76**	110078	2345.06	12326.80	224747	2123.96	TO	—
NAND	7393	11207	8.35	15688	114.60	17.02	140057	255.13	**5.00**	10.67
NAND	14323	21567	39.71	25504	366.79	59.60	405069	926.33	**15.26**	16.89
NAND	21253	31927	100.32	35151	795.31	121.40	665584	2050.67	**29.51**	24.45
NAND	28183	42287	208.41	44799	1405.16	218.85	925324	3708.27	**50.45**	30.47
NAND	78334	121512	**639.29**	184799	3785.11	—	—	MO	1138.82	111.58

For BRP, STATE ELIM always outperforms PARAM and SCC MC by up to two orders of magnitude. On larger instances, SCC MC is faster than PARAM while on smaller ones PARAM is faster and has a smaller memory consumption.

In contrast, the crowds protocol always induces a nested SCC structure, which is very hard for PARAM since many divisions of polynomials have to be carried out. On larger benchmarks, it is therefore outperformed by more than three orders of magnitude while SCC MC performs best. This is actually measured by the timeout; using PARAM we could not retrieve results for larger instances.

To give an example where PARAM performs mostly better than our approaches, we consider NAND. Its graph is acyclic consisting mainly of single paths leading to states that have a high number of outgoing edges, i.e., many paths join at these states and diverge again. Together with a large number of different probabilities, this involves the addition of many polynomials, whose factorizations are completely stored. The SCC approach performs better here, as for acyclic graphs just the linear equation system is solved, as described in Section 3. This seems to be superior to the state elimination as implemented in our tool. We don't know about PARAM's interior for these special cases. As a solution, our implementation offers the possibility to limit the number of stored polynomials, which decreases the memory consumption at the price of losing information about the factorizations. However, an efficient strategy to manage this bounded pool of polynomials is not yet implemented. Therefore, we refrain from presenting experimental results for this scenario.

6   Conclusion and Future Work

We presented a new approach to verify parametric Markov chains together with an improved factorization of polynomials. We were able to highly improve the scalability in comparison to existing approaches. Future work will be dedicated to the actual parameter synthesis. First, we want to incorporate interval constraint

propagation [21] in order to provide reasonable intervals for the parameters where properties are satisfied or violated. Moreover, we are going to investigate the possibility of extending our approaches to models with costs.

References

1. Su, G., Rosenblum, D.S.: Asymptotic bounds for quantitative verification of perturbed probabilistic systems. In: Groves, L., Sun, J. (eds.) ICFEM 2013. LNCS, vol. 8144, pp. 297–312. Springer, Heidelberg (2013)
2. Daws, C.: Symbolic and parametric model checking of discrete-time Markov chains. In: Liu, Z., Araki, K. (eds.) ICTAC 2004. LNCS, vol. 3407, pp. 280–294. Springer, Heidelberg (2005)
3. Hahn, E.M., Hermanns, H., Zhang, L.: Probabilistic reachability for parametric Markov models. Software Tools for Technology Transfer 13(1), 3–19 (2010)
4. Gruber, H., Johannsen, J.: Optimal lower bounds on regular expression size using communication complexity. In: Amadio, R.M. (ed.) FOSSACS 2008. LNCS, vol. 4962, pp. 273–286. Springer, Heidelberg (2008)
5. Hahn, E.M., Han, T., Zhang, L.: Synthesis for PCTL in parametric Markov decision processes. In: Bobaru, M., Havelund, K., Holzmann, G.J., Joshi, R. (eds.) NFM 2011. LNCS, vol. 6617, pp. 146–161. Springer, Heidelberg (2011)
6. Hahn, E.M., Hermanns, H., Wachter, B., Zhang, L.: PARAM: A model checker for parametric Markov models. In: Touili, T., Cook, B., Jackson, P. (eds.) CAV 2010. LNCS, vol. 6174, pp. 660–664. Springer, Heidelberg (2010)
7. Kwiatkowska, M., Norman, G., Parker, D.: PRISM 4.0: Verification of probabilistic real-time systems. In: Gopalakrishnan, G., Qadeer, S. (eds.) CAV 2011. LNCS, vol. 6806, pp. 585–591. Springer, Heidelberg (2011)
8. Han, T., Katoen, J.P., Mereacre, A.: Approximate parameter synthesis for probabilistic time-bounded reachability. In: Proc. of RTSS, pp. 173–182. IEEE CS (2008)
9. Puggelli, A., Li, W., Sangiovanni-Vincentelli, A.L., Seshia, S.A.: Polynomial-time verification of PCTL properties of MDPs with convex uncertainties. In: Sharygina, N., Veith, H. (eds.) CAV 2013. LNCS, vol. 8044, pp. 527–542. Springer, Heidelberg (2013)
10. Benedikt, M., Lenhardt, R., Worrell, J.: LTL model checking of interval Markov chains. In: Piterman, N., Smolka, S.A. (eds.) TACAS 2013. LNCS, vol. 7795, pp. 32–46. Springer, Heidelberg (2013)
11. Jansen, N., Corzilius, F., Volk, M., Wimmer, R., Ábrahám, E., Katoen, J.P., Becker, B.: Accelerating parametric probabilistic verification. CoRR abs/1312.3979 (2013)
12. Baier, C., Katoen, J.P.: Principles of Model Checking. The MIT Press (2008)
13. Ábrahám, E., Jansen, N., Wimmer, R., Katoen, J.P., Becker, B.: DTMC model checking by SCC reduction. In: Proc. of QEST, pp. 37–46. IEEE CS (2010)
14. Tarjan, R.E.: Depth-first search and linear graph algorithms. SIAM Journal on Computing 1(2), 146–160 (1972)
15. Quarteroni, A., Sacco, R., Saleri, F.: Numerical Mathematics. Springer (2000)
16. Jovanović, D., de Moura, L.: Solving non-linear arithmetic. In: Gramlich, B., Miller, D., Sattler, U. (eds.) IJCAR 2012. LNCS, vol. 7364, pp. 339–354. Springer, Heidelberg (2012)
17. Bauer, C., Frink, A., Kreckel, R.: Introduction to the GiNaC framework for symbolic computation within the C++ programming language. J. Symb. Comput. 33(1), 1–12 (2002)

18. Helmink, L., Sellink, M., Vaandrager, F.: Proof-checking a data link protocol. In: Barendregt, H., Nipkow, T. (eds.) TYPES 1993. LNCS, vol. 806, pp. 127–165. Springer, Heidelberg (1994)
19. Reiter, M.K., Rubin, A.D.: Crowds: Anonymity for web transactions. ACM Trans. on Information and System Security 1(1), 66–92 (1998)
20. Han, J., Jonker, P.: A system architecture solution for unreliable nanoelectronic devices. IEEE Transactions on Nanotechnology 1, 201–208 (2002)
21. Fränzle, M., Herde, C., Teige, T., Ratschan, S., Schubert, T.: Efficient solving of large non-linear arithmetic constraint systems with complex boolean structure. Journal on Satisfiability, Boolean Modeling, and Computation 1(3-4), 209–236 (2007)

Author Index